SO-BDQ-036

THE PENGUIN DICTIONARY OF ELECTRONICS

Carol Young was educated at Aigburth Vale High School, Liverpool, and Bristol University, where she gained an honours degree in physics. She worked for six years in medical physics, first in London and then at Addenbrookes Hospital, Cambridge. She wrote the electronics entries in *The Penguin Dictionary of Physics*. Her husband, John Young, has had many years' experience of research in electronics and has acted as consultant in the compilation of this book.

CAROL YOUNG
MARKET HOUSE BOOKS LTD

THE
PENGUIN DICTIONARY
OF
ELECTRONICS
SECOND EDITION

EDITOR: VALERIE ILLINGWORTH
CONSULTANT: JOHN YOUNG

PENGUIN BOOKS

PENGUIN BOOKS

Published by the Penguin Group
Penguin Books Ltd, 27 Wrights Lane, London W8 5TZ, England
Penguin Books USA Inc., 375 Hudson Street, New York, New York 10014, USA
Penguin Books Australia Ltd, Ringwood, Victoria, Australia
Penguin Books Canada Ltd, 10 Alcorn Avenue, Toronto, Ontario, Canada M4V 3B2
Penguin Books (NZ) Ltd, 182–190 Wairau Road, Auckland 10, New Zealand

Penguin Books Ltd, Registered Offices: Harmondsworth, Middlesex, England

First edition 1979
Second edition 1988
5 7 9 10 8 6 4

Printed in England by Clays Ltd, St Ives plc
Set in Times

For Stephen, Eleanor and Timothy

PREFACE

The New Penguin Dictionary of Electronics was first published in 1979. With its emphasis on solid-state electronics, it completely replaced the former valve-based *Penguin Dictionary of Electronics*. This second edition of the solid-state dictionary has reverted to the original title, *The Penguin Dictionary of Electronics*. It is primarily concerned with the words and terms used in electronic research and industry and in solid-state theory: however, it also contains definitions of some of the words and terms in the related fields of computing, communications, and electrical engineering together with some entries of historical interest. It should therefore be of use not only to students and teachers of electronics, physics, and related subjects but also to researchers, technicians, and technologists working in electronics or an associated field or using electronic equipment in their work.

Throughout the text there are a number of long entries in which a word of major importance is defined and discussed and in which closely associated words, printed in italics, are also defined. In these, as well as in the shorter entries, an asterisk before a word indicates an entry to which the reader should refer for further information. SI units are used throughout, although some non-SI units are used where they are generally accepted as being more convenient.

I would like to acknowledge the invaluable assistance in the preparation of this dictionary given by my husband, John Young, who is a Senior Principal Research Engineer at Standard Telecommunications Laboratory, Harlow. Without his advice and constructive criticism this task would have been much more difficult. I would also like to thank Valerie Illingworth, the editor of the book, for her help and encouragement throughout.

CAROL YOUNG, 1988

NOTES

An asterisk indicates an entry to which the reader should cross-refer for further information.

Syn. is an abbreviation for 'synonym'.

Words printed in italics in a definition are closely associated with the entry under which they appear and are defined in this position. Cross-reference has been made to them from elsewhere in the dictionary.

A

ab- A prefix to a unit, indicating its use in the obsolete *CGS electromagnetic system of units.

$$1\ abampere\ =\ 10\ \text{amperes}$$
$$1\ abvolt\ =\ 10^{-8}\ \text{volt}$$
$$1\ abohm\ =\ 10^{-9}\ \text{ohm}$$

Compare stat-.

abampere *See* ab-.

A battery *See* AB pack.

ABC *Abbrev. for* automatic brightness control. *See* television receiver.

aberration A defect in the image produced by an electronic lens system.

abohm *See* ab-.

AB pack *U.S.* A package providing a complete power source for battery-operated valves consisting of both the *A battery* (*Brit.* heater battery) supplying power for the *heater and the *B battery* (*Brit.* H.T. battery) supplying power for the anode circuit.

abrupt junction A *p-n junction in which the impurity concentration changes abruptly from acceptors to donors (*see* semiconductor). In practice such a junction may be approximately realized when one side of the junction is much more highly doped than the other, i.e. a p^+-n or n^+-p junction. This is a *one-sided abrupt junction*.

absolute ampere *See* ampere.

absolute electrometer *See* attracted-disc electrometer.

absolute ohm *See* ohm.

absolute temperature *See* thermodynamic temperature.

absolute unit If a quantity *y*, such as charge or voltage, is completely defined by a simple function of the quantities $x_1, x_2, \ldots$ (where x could be time, current, etc.) so that

$$y\ =\ f(x_1, x_2, \ldots)$$

and the unit of *y*, u_y, can be obtained from the units of x_1, x_2, etc., by the equation

$$u_y\ \propto\ f(u_{x1}, u_{x2}, \ldots)$$

then if the constant of proportionality is unity, the unit is an absolute unit. All *SI units are absolute.

absolute volt *See* volt.

absolute zero The temperature at which the energy of random motion of the particles in a system at thermal equilibrium is zero. It is the lowest temperature theoretically possibly and is the zero of the *thermodynamic temperature scale.

absorption (1) Attenuation of a radiowave due to dissipation of its energy, as by the production of heat.

(2) Attenuation of a beam of light by a crystal due to localized vibrational modes in the crystal resulting from the presence of impurity atoms. This gives rise to characteristic sharp troughs in the transmission or reflection spectra and can be used to analyse the material. Absorption can also occur due to photon-induced electron transitions between different energy bands in a semiconductor and can be used to determine the energy gap.

absorption coefficient For a travelling wave in a lossy medium, the fraction of the power lost per incremental length, given by

$$\alpha = -(\mathrm{d}I/I)/\mathrm{d}z$$

For a travelling wave in a dielectric waveguide, solutions of Maxwell's equations yield the relationships between the absorption coefficient α and the refractive index of the material with the electrical conductivity and the dielectric constant of the material, i.e. between the optical and electrical characteristics of the material.

absorption loss The magnitude of the *absorption of a radiowave, usually expressed in *nepers or *decibels. *See also* unabsorbed field strength.

abvolt *See* ab-.

a.c. *Abbrev. for* alternating current.

accelerated life test A form of *life test of a circuit or device so designed that the duration of the test is appreciably less than the normal expected life of the device. This is achieved by subjecting the item to an excessive applied stress level without altering the basic modes or mechanisms of failure or their relative prevalence. Thermal stress is a commonly applied stress. *See* Arrhenius equation.

accelerating anode *See* electron gun.

accelerating electrode Any electrode that accelerates electrons in the electron beam of an *electron tube. *See also* electron gun.

acceleration voltage In general, any voltage that produces acceleration of a beam of charged particles. The term is usually reserved for those devices in which an appreciable acceleration of an electron beam is produced, as in velocity-modulated tubes.

accelerator A machine used to accelerate charged particles or ions in an electric field in order to produce high-energy beams for the study of nuclear structure and reactions. Magnetic fields are used to focus and determine the direction of the beam. The simplest form of accelerator is a *direct-voltage accelerator,* such as the *Van de Graaff accelerator. This form consists of an ion source and a target that are held at a high potential difference. The maximum energy available is severely limited by the maximum potential difference that can be maintained between source and target.

Very high energies are achieved using machines in which the beams of particles are subjected to a series of relatively small accelerating voltages. The particles travel either in a straight line, as in *linear accelerators, or in cyclic paths. In the *cyclotron the energies available are limited by the relativistic mass increase. Higher energies are achieved using the *betatron, *synchrocyclotron, and proton *synchrotron.

acceptance angle *See* phototube.

acceptor (1) *Syn. for* series resonant circuit. *See* resonant circuit; resonant frequency.

(2) *Short for* acceptor impurity. *See* semiconductor.

acceptor level *See* semiconductor.

access time *See* memory.

accumulator (1) *Syn. for* secondary cell. *See* cell. (2) *See* register.

a.c./d.c. receiver A *radio receiver that can operate with either an alternating-current supply or a direct-current supply.

acoustic delay line *See* delay line.

acoustic feedback Unwanted *feedback of the sound output of an audiofrequency loudspeaker to a preceding part of a sound-reproduction system. The sound waves can be detected and amplified by the electronic circuits in the system; above a critical level oscillations are produced that are heard as an unpleasant howling noise from the loudspeaker.

acoustic wave *Syn.* sound wave. A wave that is transmitted through a solid, liquid, or gaseous material as a result of the mechanical vibrations of the particles forming the material. The normal mode of propagation is longitudinal, i.e. the direction of motion of the particles is parallel to the direction of propagation of the wave, and the wave therefore consists of compressions and rarefactions of the material. The term 'sound wave' is sometimes confined to those waves with a frequency falling within the audible range of the human ear, i.e. from about 20 hertz to 20 kilohertz. Waves of frequency greater than about 20 kilohertz are ultrasonic waves.

In a crystalline solid an acoustic wave is transmitted as a result of the displacement of the lattice points about their mean position, and the modes of propagation are constrained by the interatomic forces active between the lattice points. The wave is transmitted as an elastic wave through the crystal lattice. The angular frequency, ω, of the wave is related to the wave vector K by the relation

$$m\omega^2 = 2\Sigma_{p>0}C_p(1 - \cos pKa)$$

where m is the mass of an atom, C_p the force constant between planes of atoms separated by p, where p is an integer, and a is the spacing between atomic planes. The range of physically realizable waves that may be transmitted is

$$\pi > Ka > -\pi$$

The limits of this range define the first Brillouin zone for the crystal lattice, and at these limits travelling waves cannot be propagated; standing waves are formed. The energy of the lattice vibrations is quantized. The quantum of energy is the *phonon, which is analogous to the photon of energy of an electromagnetic wave. The phonon energy is given by $h\nu$, where ν is the frequency and is equal to $\omega/2\pi$.

A travelling acoustic wave in a solid can be produced by applying mechanical stress to the crystal or as a result of *magnetostriction or of the *piezoelectric effect. The resulting phonons can interact with mobile charge carriers present in the material. The interaction can be considered as an electric vector, analogous to the electric vector associated with an electromagnetic wave, that extends for about a quarter wavelength distance orthogonal to the direction of propagation of the wave.

acoustic wave device A device used in a signal-processing system in which acoustic waves are transmitted on a miniature substrate in order to perform a wide range of functions. Active and passive signal-processing devices formed on a single semiconductor chip have been produced including delay lines, attenuators, phase shifters, filters, amplifiers, oscillators, mixers, and limiters.

Bulk acoustic waves are acoustic waves propagated through the bulk substrate material. The substrate material consists of a piezoelectric semiconductor, such as cadmium sulphide. The acoustic waves are generated from electrical signals as a result of the *piezoelectric effect. The electric field vector of the acoustic wave interacts with the conduction electrons of the semiconductor, which have a drift velocity due to an external applied d.c. electric field. At a sufficient value of the drift velocity the kinetic energy of the drift electrons is converted to radiofrequency energy as a result of the interaction with the acoustic field, and amplification of the original signals can result.

Surface acoustic waves are propagated along the surface of a substrate. The associated electric field extends for a short distance out of the surface and can interact with the conduction electrons of a separate semiconductor placed just above the surface. The physical separation of the acoustic substrate and the semiconductor allows the materials to be chosen so that the energy dissipation in the system is minimized. The acoustic material is a piezoelectric material that has a high electromechanical coupling coefficient and low acoustic loss. The semiconductor material is one that has high mo-

bility electrons, optimum resistivity, and low d.c. power requirement so that the optimum efficiency is obtained.

action current (or **potential**) A very small current wave (or potential
wave) associated with nerve impulses.

activated cathode *Syn. for* coated cathode. *See* thermionic cathode.

activation analysis *See* neutron activation analysis.

active Denoting any device, component, or circuit that introduces
*gain or has a directional function. In practice any item except pure
resistance, capacitance, inductance, or a combination of these three
is active. *Compare* passive.

active aerial *Syns.* primary radiator; driven aerial. *See* directive aerial.

active area *See* metal rectifier.

active component (1) *See* active. (2) *See* active current. (3) *See* active
volt-amperes. (4) *See* active voltage.

active current *Syns.* active component, energy component, power component, in-phase component of the current. The component of the
current that is in *phase with the voltage, alternating current and
voltage being regarded as vector quantities. *Compare* reactive current.

active filter *See* filter.

active interval *Syn.* trace interval. *See* sawtooth waveform.

active layer *Syn.* gain region. *See* injection laser.

active load A *load that is formed from an active device, particularly
an MOS transistor. The active device is not used for its inherent
gain, but the resistance of the device is utilized to form the load.
MOS transistors are particularly useful as active load devices since
no extra processing stops are required to form resistors, during the
manufacture of MOS integrated circuits.

active network *See* network; active.

active region (1) *Syn. for* active layer. *See* injection laser.
(2) *See* transistor.

active satellite *See* communications satellite.

active transducer *See* transducer.

active voltage *Syns.* active component, energy component, power component, in-phase component of the voltage. The component of the
voltage that is in *phase with the current, alternating current and
voltage being regarded as vector quantities. *Compare* reactive voltage.

active volt-amperes *Syns.* active component, energy component, power
component, in-phase component of the volt-amperes. The product
of the voltage and the *active current or of the current and the
*active voltage. It is equal to the power in watts. *Compare* reactive
volt-amperes.

activity The ratio of the peak value of the oscillations in a *piezoelectric crystal to the peak value of the exciting voltage.

actuacting transfer function *See* feedback control loop.

actuator A device that is used to calibrate electronic equipment, to bring such equipment into operation, or, in a control system, to convert an electrical signal into the appropriate mechanical energy. The last application is a special case of a transducer. When a device, such as a microphone, is calibrated by applying a known electrostatic force to it the actuator is described as an *electrostatic actuator*.

adaptive radio microphone equalization An *equalization system used to compensate for variations in the signal strength at the receiver for a radio microphone. Variations of signal strength can arise from interference from props or other artefacts in the studio or may be due to standing waves in the studio. The method of equalization employs four aerial receivers, instead of just one, each feeding a variable gain amplifier. The gain of each amplifier is a function of the received IF signal in order to maintain a constant signal to noise ratio. The individual outputs are combined to produce the output signal from the receiver.

Adcock direction finder *Syn.* Adcock antenna. A radio direction finder consisting of a number of spaced vertical *aerials. The errors due to the horizontally polarized components of the received waves are effectively eliminated as such components have only a minimal effect on the observed bearings.

a-d converter (or **A/D converter**) *Abbrev. for* analog/digital converter.

adder A circuit in a digital *computer that performs mathematical addition. It normally contains several identical sections each of which add the corresponding *bits of the two numbers to be added together with a carry digit from the preceding section and produce an output corresponding to the sum of the bits and a carry digit for the next section.

A *half-adder* is a circuit that adds two bits only and produces two outputs; the outputs must be suitably combined in another half-adder in order to produce the correct outputs for all possible combinations of inputs.

If two numbers each consisting of x bits are to be added a full adder circuit requires $2x$ inputs to x identical sections and $(x + 1)$ outputs in order to perform the addition.

address (1) A number that identifies a unique *memory location in computer *memory. Memories may be *word-addressable* or *byte-addressable* depending on the nature of the smallest addressable unit of store.

(2) A number that identifies a particular input/output channel through which the *central processing unit of a computer communicates with its peripheral devices.

adiabatic demagnetization *See* magnetocaloric effect.

admittance Symbol: Y; unit: siemens. The reciprocal of *impedance. It is a complex quantity given by

$$Y = G + iB$$

where G is the *conductance and B the *susceptance. Since impedance, Z, is given by

$$Z = R + iX,$$

where R and X are the resistance and reactance, respectively, then

$$Y = 1/Z = 1/(R + iX)$$
$$= (R - iX)/(R^2 + X^2)$$

admittance gap A gap in the wall of a *cavity resonator that allows it to be excited by a source of radiofrequency energy, such as a velocity-modulated electron beam, or that allows it to affect such a source.

ADP *Abbrev. for* automatic *data processing.

aeolight *U.S.* A cold-cathode glow-discharge lamp (*see* gas-discharge tube) that is filled with a mixture of permanent gases. The intensity of illumination produced varies with the applied signal and it is used as a modulating light for sound recording.

aerial *Syn.* antenna (mainly U.S.). The part of a radio system that radiates energy into space (*transmitting aerial*) or receives energy from space (*receiving aerial*). An aerial together with its *feeders and all its supports is known as an *aerial system*. There is a great variety of specially designed aerials, most of which are described by their shape, e.g. umbrella, clover leaf, H, L, T, cigar, and corner aerials. The most important types of aerial are *dipole and *directive aerials. *See also* Yagi aerial.

aerial array *Syn.* beam aerial. An arrangement of radiating or receiving elements so spaced and connected to produce directional effects. Very great directivity and consequently large *aerial gain can be produced by suitable design. An array of elements along a horizontal line is referred to either as a *broadside array* or *end-fire array* depending on whether the directivity is in the horizontal plane at right angles to or along the line of the array, respectively. Arrays are commonly designed to have both horizontal and vertical directivity. The horizontal directivity is determined by the horizontal arrangement of aerial elements while the vertical directivity is dependent on the number of elements arranged in tiers (or stacks) one above the other.

The performance of a *directive aerial or an aerial array is indicated by the *radiation pattern of the system. The direction of maximum transmission or reception is given by the major lobe of the radiation pattern (*see also* steerable aerial).

aerial current The root-mean-square value of the current measured at a specified point in an aerial, usually either at the feed point or at the current maximum.

aerial efficiency *Syn.* radiation efficiency. The ratio of the power radiated by an aerial, at a specified frequency, to the total power supplied to it.

aerial feed-point impedance The *impedance of an aerial at the point at which it is fed. The real part of this impedance is the aerial feed-point resistance and the imaginary part is the aerial feed-point reactance.

aerial gain (1) (in transmission) The ratio of the power that must be supplied to a reference aerial compared to the power supplied to the aerial under consideration in order that they produce exactly similar field strengths at the same distance and in the same specified direction (usually the direction of maximum radiation).

(2) (in reception) The ratio of the signal power produced at the receiver input by the given aerial to that produced by a reference aerial under similar receiving conditions and transmitted power.

In both cases the reference aerial must be specified.

aerial radiation resistance The power radiated by an aerial divided by the mean square value of the current at a given specified reference point on the aerial, usually the feed point or a current antinode. This resistance takes into account the energy consumed by the aerial system as a result of radiation.

aerial resistance The total power supplied to an aerial divided by the mean square value of the current at a given specified reference point on the aerial, usually the feed point or a current antinode. This resistance takes into account the energy consumed by the aerial system as a result of radiation and other losses.

aerial system *See* aerial.

AES *Abbrev. for* Auger electron spectroscopy.

a.f. *Abbrev. for* audiofrequency.

a.f.c. *Abbrev. for* automatic frequency control.

afterglow *See* persistence.

a.g.c. *Abbrev. for* automatic gain control.

ageing *Syn. for* burn-in. *See* failure rate.

agglomerate cell *See* Leclanché cell.

air-break (of a switch, *circuit-breaker, etc.) Having contacts that separate in air. *Compare* oil-break.

air bridge *Syn.* plated bridge. A method of forming interconnections or *crossovers in gallium arsenide analog devices and MMICs (monolithic microwave *integrated circuits). The bridge consists of a layer of plated metal often with no material other than air between it and the slice. Advantages include a low parasitic capacitance, immunity to edge profile problems, and the ability to carry substantial current. The major steps in the formation of an air bridge are shown in the diagram. A layer of *resist is deposited on the slice and processed to produce the required pattern of interconnections. This is then covered by a very thin layer of metal, usually using *sputtering

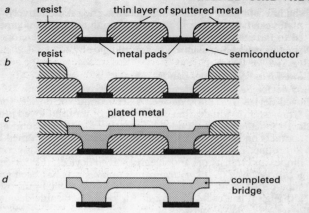

a   resist   thin layer of sputtered metal

b   resist   metal pads   semiconductor

c   plated metal

d   completed bridge

Steps in the formation of an air bridge

techniques (Fig. *a*). A second layer of resist is added, and processed to leave the thin metal layer exposed in those areas where the plating is to take place (Fig. *b*). The slice is then plated (Fig. *c*). The presence of the thin metal layer ensures that the plating current is carried to all parts of the slice. Finally the layers of photoresist and thin metal are removed to leave the required interconnections (Fig. *d*).

air capacitor A *capacitor that uses air as the main dielectric.

air gap *See* gap.

airport surveillance radar (ASR) *See* precision approach radar.

Alcomax *Tradename* A material used for permanent magnets because of its exceptionally high coercivity. It consists of an alloy of iron, nickel, aluminium, cobalt, and copper.

ALGOL–60 *See* programming language.

aliasing An effect that occurs when an analog signal f(*t*) is sampled digitally at a sampling frequency less than twice the signal frequency: a signal f(*t'*) is retrieved from the sampled information that differs from the original input signal. The retrieved signal – the *alias signal* – has a frequency that corresponds to the harmonics of the high-frequency components of f(*t*).

aligned-grid valve *Syn.* beam-power valve. A type of *thermionic valve in which the power-handling capacity of the valve is increased by arranging for only a very small fraction of the total space current to be intercepted by the screen grid. This may be achieved by placing the grids so that the conductors forming the screen grid are in the

9

shadow of those forming the control grid. Alternatively, special plates can be used to split the beam into thin pencils that are directed to fall mainly in the spaces between the grid conductors.

alignment marks *See* electron beam lithography.

alive *Syn. for* live. *See* dead.

allowed band *See* energy bands.

alloy device *See* alloyed device. *See also* alloyed junction.

alloyed device *Syn.* alloy device. A semiconductor device, such as a *transistor or *diode, that contains one or more *alloyed junctions.

alloyed junction A semiconductor junction formed by bonding metal contacts on to a wafer of semiconductor material and then heating to produce an alloy: it is a method commonly used for germanium diodes and transistors and in early silicon devices, the devices being termed *alloyed* (or *alloy*) *transistors* or *diodes*. It has been generally superceded by the *planar process for all but special-purpose items, although it is a useful technique for gallium arsenide devices.

all-pass network *See* network.

Alnico *Tradename* A material used for high-energy permanent magnets. It is an alloy of nickel, iron, aluminium, cobalt, and copper.

alpha current factor *Syn. for* common-base forward-current transfer ratio. *See* transistor.

alpha cut-off frequency The frequency at which the common-base forward-current transfer ratio, α, of a bipolar junction *transistor has fallen to $1/\sqrt{2}$ (i.e. 0.707) of the low-frequency value.

alphanumeric Ordering or ordered by both letters and numbers.

alpha particle (α-particle) The nucleus of a helium (^{4}He) atom. The positively charged particle is very stable, having two protons and two neutrons.

alpha rays A stream of *alpha particles that have energies characteristic of the emitting radioactive substance.

alternating current (a.c.) An electric current whose direction in the circuit is periodically reversed with a *frequency, f, independent of the circuit constants. In the simplest form the instantaneous current varies with time:

$$I = I_0 \sin 2\pi ft,$$

where I_0 is the *peak value of the current. A.c. is measured either by its peak value, its *root-mean-square value, or more rarely by the *current average* – the algebraic average of the current during one positive half cycle.

alternating-current bias *Syn. for* magnetic bias. *See* magnetic recording.

alternating-current generator A *generator for producing alternating currents or voltages. *See* induction generator; synchronous alternating-current generator.

10

alternating-current Josephson effect *See* Josephson effect.

alternating-current motor A *motor that requires alternating current for its operation.

alternating-current resistance *See* effective resistance.

alternating-current transmission A method of transmission used in television in which the direct-current component of the luminance signal (*see* colour television) is not transmitted. *A direct-current restorer must be used in this form of transmission. *Compare* direct-current transmission.

alternator *See* synchronous alternating-current generator.

alumina Aluminium oxide, symbol: Al_2O_3. In solid-state electronics it is used as a dielectric, as in thin-film capacitors, or as the gate dielectric in *MIS transistors. In valve electronics it was used as an insulator due to its excellent electrical and thermal resistance.

aluminium Symbol: Al. A metal, atomic number 13, that is extensively used in electronic equipment and devices. It is a good conductor and is ductile, malleable, resistant to corrosion, lightweight, easily evaporated on to surfaces, and abundant, making it the element of choice in both micro- and macroelectronic applications.

aluminium antimonide Symbol: AlSb. A semiconductor with useful properties up to operating temperatures of 500°C.

aluminium gate circuit *See* MOS integrated circuit.

aluminized screen *See* cathode-ray tube.

a.m. (or **AM**) *Abbrev. for* amplitude modulation.

American wire gauge *Syn.* Brown and Sharpe wire gauge. *See* wire gauge.

AM/FM receiver A *radio receiver that detects both amplitude-modulated and frequency-modulated signals.

ammeter An indicating instrument for measuring current. The most common types are *moving-coil and *moving-iron instruments and the *thermoammeter. Most ammeters are shunted *galvanometers.

amp *Short for* ampere.

ampere Symbol: A. The *SI unit of electric *current defined as the constant current that, if maintained in two straight parallel conductors of infinite length and negligible cross section and placed one metre apart in a vacuum, would produce between these conductors a force equal to 2×10^{-7} newton per metre of length.

    This unit was once termed the *absolute ampere* and replaced the *international ampere as the standard of electric current in 1948.

ampere balance *See* Kelvin balance.

Ampere–Laplace law *See* Ampere's law.

ampere per metre Symbol: A/m. The *SI unit of *magnetic field strength. It is the magnetic field strength in the interior of an elongated uniformly wound solenoid that is excited with a linear current density in its winding of one ampere per metre of axial distance.

Ampere's circuital theorem *See* Ampere's law.

Ampere's law (1) *Syn.* Ampere–Laplace law. The force between two parallel current-carrying conductors in free space is given by

$$dF = \mu_0 I_1 ds_1 I_2 ds_2 \sin\theta / 4\pi r^2$$

where I_1 and I_2 are the currents, ds_1 and ds_2 the incremental lengths, r the distance between the incremental lengths and θ the angle (*see* diagram); μ_0 is the *permeability of free space. Ampere's law thus relates electrical and mechanical phenomena. *See also* Coulomb's law.

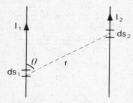

Ampere's law

(2) *Syn.* Ampere's circuital theorem. The work done in traversing a closed circuit that encloses a current I is given by

$$\oint B.ds = \mu_0 I$$

where μ_0 is the *permeability of free space, B is the magnetic flux density, and ds an incremental length.

The total current flowing is given by the integral of the current density flowing in the area bounded by the loop. In a medium of *magnetization M, the total current density, j_T, is given by the sum of the real current density, j, and the equivalent magnetization current density, j_M, where

$$j_M = \text{curl } M$$

Since

$$\oint B.ds = \int \text{curl } B.dA$$

where dA is an increment of area, then

$$\int \text{curl } B.dA = \mu_0 \int j_T.dA$$

and thus

$$\text{curl}(B - \mu_0 M) = \mu_0 j$$

Since H, the magnetic field strength, is defined as

$$\mu_0 H = B - \mu_0 M$$

12

Ampere's law, in differential form, may be written as

$$\text{curl } \boldsymbol{H} = \boldsymbol{j}$$

ampere-turn Symbol: At. A unit of *magnetomotive force equal to the product, NI, of the total number of turns, N, in a coil through which a current, I, is flowing.

amplification The reproduction of an electrical signal, usually at an increased intensity, by an electronic device.

amplification factor Symbol: μ. In an active electronic device used to deliver a constant current to a load, the ratio of the incremental change in output voltage required to maintain the constancy of the current to a corresponding incremental change in the input voltage. μ is equal to $-A_v$, where A_v is the voltage *gain of the device.

amplifier A device that produces an electrical output that is a function of the corresponding electrical input parameter and increases the magnitude of the input by means of energy drawn from an external source, i.e. it introduces *gain. A *linear amplifier* is one in which the instantaneous output signal is a linear function of the corresponding input signal; otherwise the amplifier is described as *nonlinear.*

Most practical amplifiers are alternating-current amplifiers and consist of several small-gain *amplifier stages coupled together to produce a substantial overall gain. Negative *feedback is commonly used to provide stability and to prevent the amplifier behaving as an *oscillator.

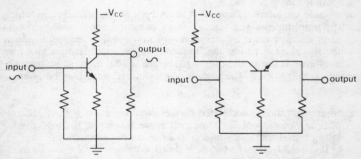

a Common-emitter n-p-n transistor amplifier   *b* Common-base n-p-n transistor

The nomenclature of amplifiers depends on their application, construction, and method of operation according to context. The simple amplifier shown in Fig. *a* can be described as a common-emitter, or linear, or *class A, or audiofrequency amplifier. An amplifier that produces an increased e.m.f. operating into a high impedance is a *voltage amplifier.* A *transistor in *common-base

13

connection (Fig. *b*) acts as a simple voltage amplifier operating with low input impedance and high output impedance. One that produces an appreciable current flow into a relatively low impedance or a large increase in output power is a *power amplifier*.

Alternating-current amplifiers are described either by the range of frequencies amplified, i.e. they are either *wideband or *tuned amplifiers, or by the region of the electromagnetic spectrum in which·they operate, as with audiofrequency or radiofrequency amplifiers.

Direct-current may be amplified directly using a *direct-coupled (d.c.) amplifier or indirectly using a *chopper amplifier. *See also* class A, class AB, class B, class C, and class D amplifiers.

amplifier stage A relatively small-gain amplifier that is coupled to other similar devices in cascade, i.e. the output of one stage forms the input of the next, in order to provide a large overall gain. The *stage efficiency* is defined as the ratio of the useful power output to the load of the amplifier stage, to the power supplied at the input.

amplitude (1) Strictly, the *peak value of an alternating current or wave in the positive or negative direction. The difference between extreme values in a complete cycle is the *peak-to-peak amplitude*.

(2)The value of an alternating current or wave in the positive or negative direction at a particular moment.

(3) *See* wave.

amplitude distortion *See* distortion.

amplitude fading *See* fading.

amplitude modulation (a.m. or AM) A type of *modulation in which the amplitude of the *carrier wave is varied above and below its unmodulated value by an amount proportional to the amplitude of the signal wave and at the frequency of the modulating signal (*see* diagram). If the modulating signal is sinusoidal then the instantaneous amplitude *e*, of the amplitude-modulated wave may be given as

$$e = (A + B \sin pt) \sin \omega t$$

where A is the unmodulated carrier amplitude and B the peak amplitude variation of the composite wave, p is equal to $2\pi \times$ modulating signal frequency and ω to $2\pi \times$ carrier frequency.

If the *modulation factor, m,* is defined by $m = B/A$ then the modulated wave may be given as

$$e = (1 + m \sin pt)A \sin \omega t$$

The modulation factor is sometimes quoted as a percentage of the carrier signal amplitude and is then termed the *percentage modulation*.

Amplitude modulation may be achieved using a radiofrequency *class C amplifier or *oscillator. The carrier wave is produced in the amplifier or oscillator; the modulating signal is superimposed on it

14

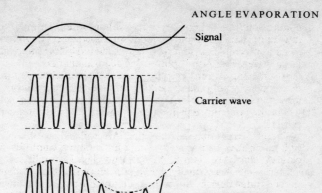

Amplitude modulation

by varying either the anode voltage (*anode modulation*) or the grid bias (*grid modulation*) in proportion to the modulating signal. The output of the device is then an amplitude-modulated radiofrequency wave. *Compare* frequency modulation.

AM receiver A *radio receiver that detects amplitude-modulated signals.

analog circuit *See* linear circuit.

analog computer *See* computer.

analog delay line *See* delay line.

analog/digital converter *Syn.* digitizer. A device for converting a continuously varying signal, such as voltage or frequency, into a series of numbers suitable for use by digital equipment, such as a *computer.

analog gate *See* gate.

anaphoresis *See* electrophoresis.

AND circuit (or **gate**) *See* logic circuit.

Anderson bridge A *bridge, modified from the *Maxwell bridge, in which an inductance, L, is compared with a capacitance, C. Using a *null-point detector instrument I (*see* diagram), R_4 and X are adjusted until, when the bridge is balanced,

$$R_2R_3 = R_1R_4$$

$$L = C[R_2R_3 + (R_3 + R_4)X]$$

angle evaporation A method sometimes used to form the gate electrodes of FETs in integrated circuits. The metal is evaporated onto

15

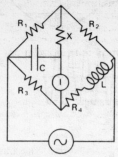

Anderson bridge

the slice at an angle (*see* diagram), resulting in a gate with a smaller length than at normal incidence. This technique is only suitable for certain applications.

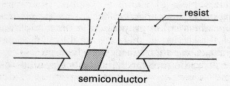

resist

semiconductor

Gate produced using angle evaporation

angle modulation *See* modulation.

angle of flow The portion of the cycle of an alternating voltage, expressed as an angle, during which current flows.

angstrom Symbol: Å. A unit of length equal to 10^{-10} metre.

angular frequency Symbol: ω. The frequency of a periodic phenomenon expressed in radians per second. It is equal to the frequency in hertz times 2π.

anion An ion that carries a negative charge and, in electrolysis, moves towards the anode, i.e. travels against the direction of conventional current. *Compare* cation.

anisotropic Denoting crystalline material whose properties, including conductivity, permittivity, and permeability, vary with direction relative to the crystal axes as a result of the crystal structure.

anode The positive *electrode of an electrolytic cell, discharge tube, valve, or solid-state rectifier; the electrode by which electrons leave (and conventional current enters) a system. *Compare* cathode.

16

anode current The current flowing from the *anode to the *cathode of a device such as a solid-state rectifier.

anode dark space *See diagram at* gas-discharge tube.

anode glow *See diagram at* gas-discharge tube.

anode modulation *See* amplitude modulation.

anode rays *See* gas-discharge tube.

anode shield *See* mercury-arc rectifier.

anode stopper *See* parasitic oscillations.

anode-voltage-stabilized camera tube *Syn. for* high-electron-velocity camera tube. *See* camera tube. *See also* iconoscope.

anodic etching *See* etching.

antenna *See* aerial.

anticapacitance switch A *switch that is designed to present the minimum possible series capacitance to a circuit when in the open position.

anticathode *Syn. for* target electrode. *See* X-ray tube.

anticoincidence circuit A circuit with two or more input terminals that produces an output signal when a signal is received by only one input; there is no output when a signal is received by each input either simultaneously or within a specified time interval, Δt. An electronic counter incorporating such a circuit will record single events and is termed an *anticoincidence counter*. *See also* coincidence circuit.

antiferromagnetism An effect observed in certain solids whose magnetic properties change at a certain characteristic temperature, T_N, known as the *Néel temperature*. Antiferromagnetism occurs in materials that have a permanent molecular *magnetic moment associated with unpaired electron spins. At temperatures above the Néel temperature thermal agitation causes the spins to be randomly orientated throughout the material, which becomes paramagnetic: it obeys the Curie-Weiss law approximately but is characterized by a negative Weiss constant, θ, i.e.

$$\chi = C/(T + \theta),$$

where χ is the *susceptibility and C a constant (*see* paramagnetism).

At temperatures below the Néel temperature interatomic exchange forces (*see* ferromagnetism) cause the spins to tend to line up in an antiparallel array. The susceptibility of the material depends on the crystalline structure, the temperature, and, in a single crystal, the direction of application of the external magnetic flux. In a single crystal of antiferromagnetic material, the antiparallel arrangement of spins can be considered as two interlocking sublattices, each spontaneously and equally magnetized in opposite directions (Fig. *a*). The sublattices may be represented by magnetic vectors M_A and M_B, each of which produces a flux experienced by the other. Thus the flux experienced by sublattice A may be represented as

$$B_A = -\lambda M_B$$

where λ is a constant.

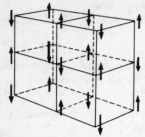

The equivalent magnetization vectors

a **Antiparallel arrangement of magnetic moments in simple cubic lattice**

If an external magnetic flux density, **B**, is applied in a direction parallel to the direction of the spins, at temperatures near the Néel temperature the efficiency of the interaction is reduced by thermal effects: the spins are not all completely antiparallel so the material has a small positive magnetic susceptibility, $\chi_\parallel$, falling from a maximum at the Néel temperature as the temperature is reduced. At temperatures near absolute zero the antiparallel array becomes saturated due to the greater efficiency of the interaction and the susceptibility falls to zero (Fig. *b*).

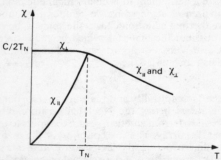

b **Variation of susceptibility with temperature in a single antiferromagnetic crystal**

If the magnetic flux is applied in a direction perpendicular to the spins, each magnetic vector turns towards B (Fig. *c*). The angle of rotation, α, is determined by the components of the magnetization vectors M_x and M_z and the components of flux B_x and B_z and is

18

independent of the temperature. The overall magnetization of the specimen is $2M_x$, where M_x is the component of each sublattice in the direction of B; it can be shown that

$$2M_x = B/\lambda$$

and that

$$\lambda = 2\mu_0 T_N/C$$

where μ_0 is the *permeability of free space and C the Curie constant from the Curie-Weiss law for the material. The susceptibility $\chi_\perp$ is thus equal to $C/2T_N$, i.e. it is not dependent on temperature below the Néel temperature (Fig. b).

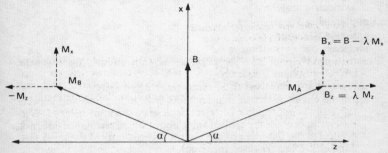

c Effect of external flux, B, applied perpendicular to magnetic moments

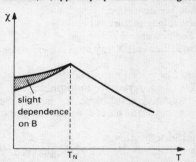

d Variation of susceptibility with temperature
of polycrystalline specimen

The behaviour as described above assumes two equal sublattices. Some crystal structures, e.g. face-centred cubic, have more than two sublattices and the stable saturation array therefore may not be completely antiparallel; this leads to more complicated behaviour.

19

In a polycrystalline sample the susceptibility is a compromise between the two extreme conditions described above, with a small dependence on magnetic field strength at low temperatures, and antiferromagnetic materials are characterized by a maximum susceptibility at the Néel temperature (Fig. *d*).

antihunting circuit A circuit designed to prevent oscillation in a *feedback control loop, thus stabilizing it.

anti-interference aerial system A receiving *aerial system designed so that only the aerial itself abstracts energy from the incident radiation and no energy is collected by any feeders or supports of the system. The effects of interference can be greatly reduced by locating the aerial remotely from sources of man-made interference, such as electric motors.

antijamming *See* jamming.

antinode *See* node.

antiphase *See* opposition.

antiresonant circuit *Syn. for* parallel resonant circuit. *See* resonant circuit; resonant frequency.

anti-transmit-receive switch *See* transmit-receive switch.

aperiodic circuit A circuit that contains both inductance and capacitance but is not capable of *resonance since the total energy loss in the circuit exceeds the critical value above which resonance does not occur, i.e. the damping of the circuit exceeds critical damping (*see* damped). *Compare* resonant circuit.

aperiodic damping *Syn. for* overdamping. *See* damped.

aperture distortion *See* distortion.

aperture grille *See* colour picture tube.

apparent power *See* power.

appearance potential *See* soft X-ray appearance potential spectroscopy.

Applegate diagram *See* velocity modulation.

Appleton layer *Syn. for* F-layer. *See* ionosphere.

apple tube A *colour picture tube in which the three colour *phosphors are arranged as narrow vertical stripes on the television screen.

applicator *See* dielectric heating.

arbitrary unit A unit, such as the *kilogram, that is defined in terms of a prototype.

arc A luminous electric discharge in a gas with a characteristically high current density and low potential gradient. An arc occurs (is struck) when the gas along its path becomes a *plasma. The arc is drawn from a localized spot on the cathode (the *arc spot* or *cathode spot*) resulting in heating of the cathode. Thermionic emission occurs with a resultant drop in potential across the tube. Arc-discharge tubes can therefore carry very large currents at voltage drops of tens of volts only. *See also* gas-discharge tube. *Compare* spark.

arc-back An *arc struck from the anode to the cathode in a gaseous rectifier valve (i.e. in the reverse direction). Arc-backs most commonly occur under fault conditions.

arc baffle *See* mercury-arc rectifier.

arc converter A device by means of which alternating current is generated from an electric arc.

arc discharge *See* gas-discharge tube.

arc-discharge tube *See* gas-discharge tube.

arcing contacts Auxiliary contacts in any type of *circuit-breaker switch that operate in conjunction with the main contacts, closing before and opening after them, so as to protect them from arc damage.

arcing horn *Syn.* protective horn. A horn-shaped conductor fitted to an insulator in order to prevent damage to the latter in the event of a power *arc arising from fault conditions. It is also an element of a *horn gap.

arc lamp A lamp that utilizes the brilliant light accompanying an *arc as a source of illumination. The colour of the light produced will depend on the gas inside the discharge tube.

arcover *See* flashover.

arc spot *See* arc.

arc welding A form of welding in which an arc discharge is developed between the metal parts to be joined together and a filler rod of the same material. The heat generated by the arc causes the metals to melt at the site to be joined, and they then fuse together.

argon Symbol: Ar. An inert gas, atomic number 18, that is extensively used as the gas in gas-filled tubes.

arithmetic unit *See* central processing unit.

arm *See* bridge.

armature (1) *Syn.* rotor. The rotating part of an electric *generator or *motor.

(2) Any moving part in electrical equipment that closes a magnetic circuit or that has a voltage induced in it by a magnetic field. An example is the moving contact in an electromagnetic *relay.

(3) *See* keeper.

armature relay *See* relay.

Arrhenius equation A physical model that expresses the time rate of degradation, $R(T)$, of some device parameter as a function of operating temperature, T:

$$R(T) = A \exp(-eA_E/kT)$$

where e is electronic charge and k is the Boltzmann constant. The constant A_E is a measure of the average behaviour of a population of

21

items and is the 'empirical' activation energy of the physical process causing component degradation. A is a constant and is the intercept of the plot of logarithm of degradation rate versus reciprocal temperature.

The Arrhenius equation is used as a basis for high-stress *accelerated life tests. It is a more commonly used model than the *Eyring reaction rate equation*, which relates the time rate of degradation $R(T, S)$ to both the thermal stress, T, and a nonthermal stress, S:

$$R(T, S) = A \exp(-[eA_E - DS]/kT)$$

where D is a constant.

arsenic Symbol: As. An element, atomic number 33, that belongs to group 5A of the periodic table and is thus used as a donor impurity in *semiconductors and in 3–5 compound semiconductors.

artificial aerial *U.S. syn.* dummy antenna. A device that simulates all the electrical characteristics of an actual aerial except that the energy supplied to it is not radiated. (It is usually dissipated as heat in a resistor.) It allows adjustments to be made to a *transmitter or *receiver before connecting up to the actual aerial.

artificial line An electrical network consisting of resistance, inductance, and capacitance that simulates the characteristics of a *transmission line at any particular frequency.

artificial satellite *See* satellite.

artwork The required patterns for integrated-circuit manufacture produced in a form suitable for reduction to masks. The pattern is formed of alternate clear and opaque sections, usually in a form of Mylar film.

ashering *See* scumming.

ASIC *Abbrev. for* application specific integrated circuit. An integrated circuit designed for a specific application, rather than a generalized mass-produced circuit.

aspect ratio (1) The ratio of the width of a television picture to the height. Most countries, including the U.K. and U.S., have adopted an aspect ratio of 4:3.

(2) The ratio of the width to the length of the channel in *MOS *field-effect transistors.

ASR *Abbrev. for* airport surveillance radar. *See* precision approach radar.

assembler *See* programming language.

assembly A number of electronic components, parts, devices, or subassemblies (or a combination thereof) that are electrically and mechanically connected and that perform a specific function. When used, together with other assemblies, as part of a piece of electrical equipment the assembly may be removed or replaced as a whole, allowing easy maintenance and repair of the electrical equipment.

assessed failure rate *See* failure rate.

assessed mean life *See* mean life.
assisted lift-off *See* lift-off.
astable multivibrator *See* multivibrator.

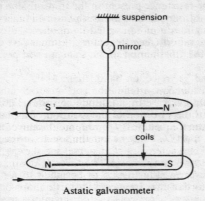

suspension

mirror

S'   N'

coils

N   S

Astatic galvanometer

astatic galvanometer *Syn.* Broca galvanometer. A very sensitive type of moving-magnet *galvanometer that has two magnetic needles, NS and N'S', made as nearly equal as possible and suspended on the same axis in opposition. When the magnets are deflected from their stable point the restoring couples on them due to the earth's magnetic field act in opposite directions with a very small resultant couple. The coils carrying the current to be measured are wound round the magnets in opposite senses (*see* diagram) so that the couples, due to magnetic fields produced by the coils, add. Very small currents cause a noticeable deflection of the magnets that is detected by means of a light spot reflected along a scale from a small mirror attached to the suspending fibre.

Aston dark space *See diagram at* gas-discharge tube.

asymmetrical Denoting a periodic quantity that has a direct component.

asynchronous logic *See* synchronous logic.

atmospheric noise *See* radio noise.

atomic number Symbol: Z. The number of electrons moving around the nucleus of an atom, which is equal to the number of protons in the nucleus. The position of an element in the *periodic table, and hence its chemical properties, is determined by atomic number.

ATR switch *Abbrev. for* anti-transmit-receive switch. *See* transmit-receive switch.

attenuation The reduction in magnitude of any electrical parameter of a signal, particularly electromagnetic radiation, on passing along any transmission path. The amount of attenuation is given by the ratio of the value of the parameter at the output to the corresponding value at the input under specified conditions. Attenuation results from the resistance present in the transmission path. It can be deliberately introduced into a transmission channel in order to reduce the magnitude of unwanted components of the parameter under consideration (*see* attenuator). Attenuation can also result from unwanted *dissipation in the transmission path. *See also* absorption.

attenuation band *See* rejection band; filter.

attenuation constant *Syn.* attenuation coefficient. Symbol: α. The rate of exponential decrease in amplitude of voltage, current, or field-component in the direction of propagation of a plane progressive wave, at a given frequency. If I_2 and I_1 are the currents at two points a distance d apart (I_1 being nearer the source of the wave), then

$$I_2 = I_1 \exp(-\alpha d)$$

α is usually expressed in *nepers or *decibels.

attenuation distortion *Syn.* frequency distortion. *See* distortion.

attenuation equalizer A *network that throughout a specified frequency band provides compensation for attenuation *distortion.

attenuator A *network or *transducer designed to produce distortionless. attenuation of an electrical signal. It may be variable or fixed. (The latter is also called a *pad*.) Attenuators are usually calibrated in *decibels.

atto- Symbol: a. A prefix to a unit, denoting a submultiple of 10^{-18} of that unit.

attracted-disc electrometer *Syn.* absolute electrometer. An electrometer that measures potential difference in terms of fundamental mechanical quantities. The potential difference to be measured is applied across two parallel metal discs and the force of attraction between them is measured.

audio device *Short for* audiofrequency device. *See* audiofrequency.

audio effect *Short for* audiofrequency effect. *See* audiofrequency.

audiofrequency (a.f.) Any frequency to which a normal human ear responds. The audible range in practice extends from about 20 to 20 000 Hz. Intelligible speech can be obtained in a communication system if a frequency range from about 300 to 3400 Hz is reproduced; any frequency in this range is termed a *voice frequency*. Any electronic device, such as an *amplifier, *choke, or *transformer, that operates in this range is known as an *audiofrequency device* or *audio device;* similarly any effect such as *distortion, that involves audiofrequencies is termed an *audiofrequency effect* or *audio effect*.

audiometer An instrument used to measure both hearing loss due to deafness and the masking produced by noise. Many forms exist, the most common being a system involving the production of a sound of known frequency and intensity in a telephone earpiece. Frequency and intensity are both variable, and the instrument may be calibrated to read hearing loss directly in operation.

audio signal *See* television.

Auger electron spectroscopy (AES) A method of spectroscopy that detects the electrons produced by the *Auger process. The Auger electrons have a mean free path of only 1–3 nanometres, and their energy is typical of the material producing them. The Auger process dominates in low atomic number materials. Auger electron spectroscopy therefore is useful in the detection of low atomic number atoms in the surface of a semiconductor. Excess electrons are produced in the material by exciting the atoms with an electron beam, and the resulting Auger electrons are detected. The energy spectrum of the detected electrons can be correlated to the atomic species in the material. A *scanning Auger microprobe* (SAM) uses a focused electron beam to excite the material, which is then scanned across the surface. Lateral resolution using a SAM probe is limited by the diameter of the electron beam used, and can be as good as 50 nanometres. The composition of a material as a function of depth below the surface can be determined by using an ion beam to sputter material from the surface (sputter *etching) and a SAM to continuously detect the Auger electron spectra produced.

AES is a flexible analytical tool. It is used to detect physical defects due to unwanted particles in a material, to detect contaminants causing high contact resistance, to examine bonds between an IC and external leads, to study the composition of thin films, and to provide information about the chemical state of materials (for example, elemental and oxidized silicon produce different spectra).

Auger process *See* recombination processes.

autocatalytic plating *See* electroless plating.

autodyne mixer *See* superheterodyne reception.

autodyne oscillator *See* beat reception.

autoemission *Syn.* autoelectronic emission. *See* field emission.

automatic brightness control *See* television receiver.

automatic contrast control *See* television receiver.

automatic control Any system or device that carries out operations automatically in response to the output of other electronic devices, such as sensing devices, computers, or discriminator circuits. *Feedback of the output signal is frequently employed to provide the controlling signal.

An *automatic controller* measures any variable quantity or condition and produces an output designed to correct any deviation from the desired value. The *threshold signal* is the minimum input signal to which the control system responds by producing a corrective signal. A system that is actuated by electrical signals is known as an *electric controller*. If the electrical signals are used to excite electromagnets that determine all the basic functions of the device, it is termed a *magnetic controller*. In electrical switching, a desired performance is maintained by automatically opening and closing *switches in a given sequence.

automatic data processing (ADP) *See* data processing.

automatic direction finding *See* direction finding.

automatic frequency control (a.f.c.) A device to maintain automatically the frequency of any source of alternating voltage within specified limits. The device is 'error-operated' and usually consists of two parts: one part is a frequency *discriminator that compares the actual and desired frequencies and produces a d.c. output voltage proportional to the difference between them, with sign determined by the direction of drift; the other part is a *reactor that forms part of the *oscillator tuned circuit and is controlled by the discriminator output in such a way as to correct the frequency error.

automatic gain control (a.g.c.) *Syn.* automatic volume control. A device for holding the output volume of a radio receiver substantially constant despite variations in the input signal. The term also covers the process involved. A variable gain element in the receiver is controlled by a voltage derived from the input signal. Variations in the size of the input signal cause compensatory changes in gain in the receiver. *Biased automatic gain control* is a process that comes into operation only for signals above a predetermined level. *Quiet automatic gain control* is a combination of biased a.g.c. and *muting, usually designed to inhibit any output when input signals are received that are too weak to actuate the gain control device.

automatic grid bias *See* grid bias.

automatic noise limiter A device in a radio receiver designed to limit the effect of impulse *noise.

automatic tracking A method of holding a radar beam locked on target while the target range is being determined.

automatic tuning control A type of *automatic frequency control in a radio receiver that adjusts the tuning to the correct setting for a given received signal when the manually operated adjustment has been set only approximately for that signal. It holds the tuning at the correct setting despite any small drift in the components of the receiver.

automatic volume compressor *See* volume compressor.

automatic volume control (a.v.c.) *See* automatic gain control.

automatic volume expander *See* volume compressor.

autotransductor A *transductor in which the main current and control current are carried in the same windings.

autotransformer A *transformer that has a single winding, tapped at intervals, rather than two or more independent windings. Part of the winding is thus common to both primary and secondary circuits (*see* diagram). The voltage V_2 across a tapped section is related to the total applied voltage V_1 by

$$V_2/V_1 \propto n_2/n_1$$

where n_2 is the number of turns in the tapped section and n_1 the total number in the winding.

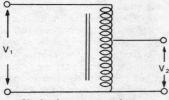

Single-phase autotransformer

available power *See* maximum power theorem.

avalanche *Syn.* Townsend avalanche. A cumulative ionization process in which a single particle or photon produces several ions, each of which in turn gains sufficient energy from an accelerating field to produce more ions, and so on. A large number of charged particles is thus produced from the initial event. The phenomenon is utilized in the *Geiger counter, the avalanche *photodiode, and *IMPATT diodes.

avalanche breakdown A type of *breakdown that occurs in a *semiconductor. It is caused by the cumulative multiplication of free charge carriers under the action of an applied electric field. Some of the carriers gain enough energy to liberate new electron-hole pairs by *impact ionization, which in turn can generate further pairs, i.e. an *avalanche takes place.

The avalanche process occurs when a *critical field* is reached across the semiconductor. The critical field is the voltage gradient that just causes avalanche in a given specimen of semiconductor and is typical for the material used. For a given semiconductor the value of the applied voltage required to produce the critical field – the breakdown voltage – is a function of the *doping concentration and the thickness of the specimen.

Avalanche breakdown is the breakdown that occurs in a reverse-biased *p-n junction. For a given junction the breakdown voltage is a function of the doping concentrations on each side of the junction and on the width of the *depletion layers associated with the junction. A more highly doped junction has a lower value of breakdown voltage than a junction with lighter doping. The avalanche process is initiated by carriers contributing to the small reverse saturation current. *Compare* Zener breakdown.

avalanche photodiode *See* photodiode.

a.v.c. *Abbrev. for* automatic volume control. *See* automatic gain control.

avionics Electronics applied to aeronautics and astronautics.

Ayrton shunt *See* universal shunt.

B

back electromotive force (back e.m.f.) An e.m.f. that opposes the normal flow of current in a circuit.

backgating *Syn.* backside gating. A phenomenon that arises in closely packed monolithic *integrated circuits where a negative bias on an *ohmic contact can affect nearby *FET devices. The substrate can be biased and acts as a *gate on the back of the FET, affecting the source-drain current.

background counts Counts registered by a radiation *counter in the absence of the radiation source to be measured. These counts may arise from other radiation sources, naturally occurring background radiation, contamination of the counter itself, or spurious signals in the electronic circuitry of the counter or from a combination of these.

background noise *Syn. for* random noise. *See* noise.

back heating *See* magnetron.

backlash The incomplete rectification of an alternating current in a thermionic valve due to the presence of positive ions in the residual gas in the valve.

back-layer photovoltaic cell *Syn. for* back-wall photovoltaic cell. *See* photovoltaic cell.

back lobe *See* radiation pattern.

back plate *Syn. for* signal electrode. *See* vidicon; iconoscope.

back porch *See* television.

back-wall photovoltaic cell *See* photovoltaic cell.

backward diode *Syn.* unitunnel diode *See* tunnel diode.

backward-wave oscillator *See* travelling-wave tube.

balance controls (1) Variable components used in electrical instruments and bridges in order to obtain conditions of electrical equilibrium in the circuit.

(2) Variable components used to equalize the outputs of two or more similar circuits, such as those in a stereophonic sound reproduction system.

balanced amplifier *Syn. for* push-pull amplifier. *See* push-pull operation.

balanced line *See* transmission line.

balanced quadripole *See* quadripole.

balanced-wire circuit A circuit having two sides that are symmetrical with respect to earth and other conductors and are electrically alike.

balance method *See* null method.

ballast lamp *See* ballast resistor.

ballast resistor A resistor manufactured from material having a large positive *temperature coefficient of resistance. It is constructed so as to have a substantially constant current over a range of voltages, and may be used as a current *regulator; it is connected in series with a circuit to absorb small changes in the applied voltage and stabilize the current in the circuit. One consisting of a resistor seated in an evacuated or gas-filled envelope of glass or metal is called a *ballast lamp. Compare* barretter; thermistor.

ballistic galvanometer A *galvanometer that measures the quantity of electricity Q, flowing during the passage of a transient current I, where

$$Q = \int_0^\infty I \mathrm{d}t$$

The value of Q is deduced from the deflection, θ, of the moving part of the instrument. For a moving-coil instrument, $Q \propto \theta$ and for a suspended magnet type $Q \propto \sin\frac{1}{2}\theta$.

balun *Syn.* balancing transformer. Acronym from *bal*anced *un*balanced. A device that is used to couple a balanced impedance, such as an aerial, to an unbalanced transmission line, such as coaxial cable. A balun is required in order to prevent asymmetrical loading of the balanced impedance and the induction of currents on the exterior of the unbalanced transmission line.

A typical example is a quarter-wavelength balun, which consists of a conducting cylinder open-circuited at one end (corresponding to the end of the line) and electrically connected to the outer conductor of the coaxial cable at the other. The cylinder is used either as an outer sleeve completely surrounding the cable or as an open stub parallel to it. It has an electrical length of one quarter wavelength. It places a very high decoupling impedance between the conductors of the cable at the open end, thus preventing induction of currents on the outer conductor, and prevents a symmetrical load to the impedance.

banana jack-and-plug A single conductor jack-and-plug system in which the plug, with a sprung metal tip, somewhat resembles a banana in shape.

banana tube A special type of *colour picture tube having the three colour *phosphors in the form of thin stripes on the screen. The line scan is produced inside the tube and the field scan is produced by cylindrical lenses mounted around the tube.

band (1) A specific range of frequencies used in communications for a definite purpose, such as the long-wave band in radio; certain frequencies within the band are assigned to different transmitting stations and the receiver may be tuned to any desired frequency within the band. *See also* frequency band.

(2) A closely spaced group of atomic energy levels. *See* energy bands.

(3) A closely spaced group of molecular energy levels that appear as fluted bands separated by dark spaces in the spectra of compounds. Under higher resolution the bands resolve into fine spectral lines.

band-edge energy *See* semiconductor.

bandgap *Syn. for* forbidden band. *See* energy levels; direct gap semiconductor.

band-pass filter *See* filter.

band-stop filter *Syn.* band-rejection filter. *See* filter.

band switch *See* turret tuner.

band-to-band recombination *See* recombination processes.

bandwidth (1) The band of frequencies occupied by a transmitted modulated signal (*see* modulation) and lying to each side of the *carrier-wave frequency.

(2) The amount of deviation of frequency that an *aerial array is capable of handling without a mismatch.

(3) The band of frequencies over which the power amplification in an amplifier falls within a specified fraction (usually one half) of the maximum value.

(4) *See* receiver.

bank (1) A number of devices of the same kind, connected so as to act together.

(2) An assembly of fixed contacts that is used in automatic switching in telephony to form a rigid unit in a selector or similar device with which wipers engage.

Barkhausen effect An effect observed in ferromagnetic materials whereby the magnetization of the specimen proceeds as a series of finite jumps when the magnetizing flux is increased steadily. The effect supports the domain theory of *ferromagnetism: the spin magnetic moments present in the material can only have certain allowed orientations; the minute jumps correspond to the spins changing from one allowed orientation to the next. If all possible directions were allowed the magnetization would proceed smoothly.

The effect can be demonstrated by winding the specimen with two coils. When the current in the primary coil is increased steadily to produce a smoothly increasing magnetic flux density, the fluctuations in the magnetization can be shown by connecting the secondary coil to a sensitive cathode-ray oscillograph.

barrage reception A method of reception used in telecommunications in which the receiving aerial consists of an array of several *directive aerials of different orientations. The received signal is input from selected aerials in the array that are chosen so as to minimize interference from a particular direction.

barrel distortion *See* distortion.

barretter A device consisting of a sensitive metallic resistor, usually enclosed in a glass bulb, with a positive *temperature coefficient of resistance. It is constructed to give a constant voltage over a range of currents and may be used as a voltage *regulator; it is connected in series with a circuit and stabilizes the voltage by absorbing small changes in the applied current. *Compare* ballast resistor.

barrier-grid storage tube *See* storage tube.

barrier height *Short for* Schottky barrier height. *See* Schottky diode, Schottky effect.

barrier-layer photocell *Syns.* rectifier photocell; blocking-layer photocell; photronic cell. *See* photovoltaic cell.

base (1) *Short for* base region. The region in a bipolar junction transistor between the *emitter and *collector into which minority carriers are injected. The electrode attached to the base is the *base electrode*. *See also* transistor; semiconductor.

(2) *Short for* base electrode.

baseband *See* carrier wave.

base electrode *See* base.

base level (of a pulse) *See* pulse.

base limiter *Syn.* inverse limiter. *See* limiter.

base region *See* base.

base spreading resistance *See* spreading resistance; transistor parameters.

base stopper *See* parasitic oscillations.

base units *See* SI units.

BASIC *See* programming language.

basic frequency In any oscillatory signal composed of several different sinusoidal components, the frequency of the component considered to be the most important.

basic noise *See* noise.

bass boost *Syn.* bass compensation. *See* radio receiver; equalization.

bass response *See* radio receiver.

batch processing A method of operation used with a large computer system in which a number of previously prepared programs are collected together and input to the computer as a unit. The computer then performs the operations as time becomes available in the system. The programs forming the batch are input from a central input/output device or from a small number of *remote job entry* locations in which the input/output devices are situated at a remote location but are noninteractive. *Compare* interactive.

battery A source of direct current or voltage that consists of two or more electrolytic *cells connected together and used as a single unit.

A *floating battery* is formed from secondary cells and is connected simultaneously to a discharging circuit and a charging circuit. The current in the charging circuit is adjusted so as to balance the loss of charge from the battery to the discharging circuit driven by it. A constant level of charge is therefore maintained in the battery. A floating battery is often used to provide a constant e.m.f. in the discharge circuit, despite fluctuations in the electrical mains supply.

A *dry battery* is a relatively small portable battery made up from dry cells.

baud A unit of telegraph signalling speed equal to one unit element per second. Thus if the duration of the unit element is $1/n$ seconds then the speed of transmission of successive signals is n bauds.

bayonet fitting A type of pin-and-socket fitting in which the base of a lamp or tube has two pins, diametrically opposite each other, that can be inserted into the socket and rotated in slots in the socket so that the device is held in position.

B battery *See* AB pack.

BBD *Abbrev. for* bucket-brigade device.

BCCD *Abbrev. for* bulk-channel charge-coupled device. *See* charge-coupled device.

BCS theory *See* superconductivity.

b.d.v. *Abbrev. for* breakdown voltage.

beacon A signal station or the signal transmitted by such a station, which acts as a reference point. A beacon that transmits an identifiable signal is a *code beacon.* A *homing beacon* is one that guides an object, such as an aircraft, to a target, such as an airport. At the airport, signals from the *localizer beacon* associated with the instrument landing system are picked up enabling the aircraft to be guided to land. A beacon that employs radar signals is a *radar beacon* and one employing radiofrequency waves is a *radio beacon.* The receiver used for detecting the signals from a beacon is a *beacon receiver.*

bead thermistor *See* thermistor.

beam A narrow stream of essentially unidirectional electromagnetic radiation (as in a radiowave) or charged particles (as in an electron beam).

beam aerial *See* aerial array.

beam angle *See* cathode-ray tube.

beam bending An unwanted effect that occurs in television *camera tubes. The electron beam used to scan the target area can be deflected from its intended position by the electrostatic charges stored on the target. This can result in misalignment of the picture image in the receiver with respect to the original optical image.

beam coupling The production in a circuit of an alternating current between two electrodes upon passage of an intensity-modulated electron beam. The *beam-coupling coefficient* is the ratio of the alternating current produced to the beam current.

beam lead A connecting lead on a silicon *chip that is formed chemically and cantilevered across a void or space on the chip. The lead may be cantilevered either from the chip to the interconnection pattern or vice versa.

beam maser *See* maser.

beam power valve *See* aligned-grid valve.

beam switching *See* lobe switching.

beat frequency *See* beats.

beat-frequency oscillator (bfo) *See* beats.

beating *See* beats.

beat note *See* beats.

beat oscillator *See* beat reception.

beat reception *Syn.* heterodyne reception. A method of radio reception that employs beating. The received radiofrequency (r.f.) oscillations are combined with r.f. oscillations generated separately in the receiver by a *beat oscillator* to produce *beats (usually audiofrequency). The combined oscillations are then detected, amplified, and rendered audible. An *autodyne oscillator* is one in which the beat oscillator also performs the function of detector and amplifier. *Compare* superheterodyne reception.

beats The periodic signal produced by interference when two signals of slightly different frequencies are combined. The amplitude is equal to the sum of the original amplitudes; the frequency (the *beat frequency*) is equal to the difference between the original frequencies. The production of beats is termed *beating* and is achieved by using a *beat-frequency oscillator* (bfo). This device incorporates two radiofrequency oscillators, one producing a fixed frequency wave and the other a variable frequency; the output is produced by beating together the two frequencies.

bel Symbol: B. *See* decibel.

beryllium oxide *Syn.* beryllia. Symbol: BeO. An insulating material that has a high thermal conductivity (about half that of copper) and is used in heat sinks.

beta circuit *See* feedback.

beta-current gain factor *Syn.* common-emitter forward-current transfer ratio. Symbol: β or h_{fe}. The short-circuit current-amplification factor in a bipolar transistor with common-emitter connection:

$$\beta = (\partial I_\text{C}/\partial I_\text{B}) \qquad V_{\text{CE}} \text{ constant}$$

where I_C is the collector current and I_B is the base current; the collector voltage, V_{CE}, is constant. β is always greater than unity and practical values up to 500 are used.

beta cut-off frequency The frequency at which the *beta-current gain factor has fallen to $1/\sqrt{2}$ of its low frequency value. The beta cut-off frequency is considerably lower than the *alpha cut-off frequency.

beta decay The spontaneous transformation of a nuclide into one of its isobars with the emission of an electron plus antineutrino or a positron plus neutrino. The mass of the nucleus is unchanged but the atomic number is changed by $+1$ or -1.

34

beta particle (β-particle) An electron or positron emitted by a radioactive nucleus during *beta decay. A stream of beta particles of varying energies is a *beta ray* (β-ray).

beta ray *See* beta particle.

betatron A cyclic *accelerator that employs magnetic induction to produce high-energy electrons. A magnetic field is used to deflect the electrons into a circular orbit. The magnetic orbital flux increases with time, giving rise to an induced circumferential electric field that accelerates the electrons.

bfo *Abbrev. for* beat-frequency oscillator. *See* beats.

bias *Short for* bias voltage. A voltage applied to an electronic device to ensure that it operates on a particular portion of its characteristic curve.

biased automatic gain control *Syn.* delayed automatic gain control. *See* automatic gain control.

bias voltage *See* bias.

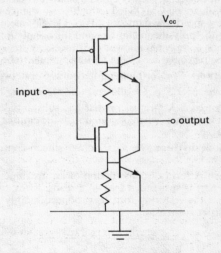

Bi–CMOS output buffer

bi-CMOS *Syn.* merged CMOS/bipolar. Integrated circuits that contain both bipolar junction *transistors and *complementary MOS transistors (CMOS). The combination of both types of device on the same chip has a wide variety of functions and allows the advantages of both processes to be exploited. Bipolar circuits are inherently

35

faster than CMOS circuits and have much better analog performance. Bipolar transistors are preferred for *operational amplifiers, *comparators, *multipliers, and high-speed logic circuits, such as *emitter-coupled logic (ECL). CMOS circuits are preferred where low power dissipation and high *packing densities are required, as with memory counters, registers, and random logic. Merging the two types of circuit allows combinations of different circuits to be formed on the same chip. It also allows the production of circuits where the characteristics of both types of device are needed, for example mixed analog-digital circuits or logic circuits where part of the circuit demands the high speed of bipolar logic, such as fast clock circuits or input/output buffers. A merged logic output buffer is shown in the diagram. Bi-CMOS circuits provide improved system performance and a reduction in the number of components, hence reduced chip sizes and lower costs.

One difficulty of producing bi-CMOS circuits arose from the complexity of the processing steps required to produce the two different types of device, and early circuits based on standard bipolar technology with added CMOS led to high-performance bipolar elements but low-speed low-density CMOS. Development of bipolar transistors with *polysilicon emitters and the use of *ion implantation rather than epitaxy has allowed the processing steps to be minimized and the performance of the CMOS circuits to be optimized.

bidirectional network *Syn. for* bilateral network. *See* network.

bidirectional transducer *Syn. for* bilateral transducer. *See* transducer.

bidirectional transistor A *transistor that has substantially the same electrical characteristics when operated with the emitter and collector interchanged.

bidirectional triode thyristor *Syn. for* triac. *See* silicon-controlled rectifier.

bifilar suspension A form of construction of an instrument in which the movable part is suspended by two threads, wires, or strips, arranged so that the restoring force is produced mainly by gravity. *Compare* unifilar suspension.

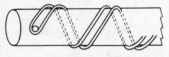

Bifilar winding

bifilar winding A method of winding consisting of two contiguous insulated conductors connected so that they carry the same current in opposite directions. This results in a negligible magnetic field being produced. The technique is commonly used to wind noninductive resistors (*see* diagram).

bilateral network *Syn.* bidirectional network. *See* network.

bilateral transducer *Syn.* bidirectional transducer. *See* transducer.

bilevel resist *See* multilevel resist.

bimetallic strip A device consisting of two metals with different coefficients of expansion riveted together; an increase in temperature causes the strip to bend. One end of the strip is held rigidly and the movement of the other end may be used to open and close contacts, particularly in thermostats, or to move the pointer of a pointer-type thermometer. A thermal instrument that utilizes the deformation of a bimetallic element, heated directly or indirectly by an electric current, is a *bimetallic instrument.*

binary code *See* binary notation.

binary logic circuit *See* logic circuit.

binary notation A method of numerical representation with two as the base and thus having only two digits (0 and 1). These symbols may be easily represented electronically by two voltage levels in a circuit (*see* logic circuit) and binary notation is therefore used in *computers. A *binary code* has each element represented by one or other of two distinct states, values, or numbers. The radix point in a binary system is the *binary point* and is analogous to the decimal point in decimal notation.

binary point *See* binary notation.

binary scaler *See* scaler.

binding energy (1) Symbol: E_B. The total energy released when protons and neutrons bind together to form a nucleus.

(2) The energy required to release an extranuclear electron from a nucleus. The energy required to strip all the extranuclear electrons from a nucleus is the *total electron binding energy. See* ionization potential.

biot Symbol: Bi. A unit of current in the obsolete CGS electromagnetic system of units. It is equal to 10 amperes.

Biot–Savart law The magnetic flux density, *B,* at a point distance *r*

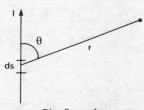

Biot-Savart law

from a current-carrying conductor in free space when a current I flows in an element of length ds is given by

$$d\boldsymbol{B} = \mu_0 I (\mathrm{d}s \times \boldsymbol{r})/4\pi r^3$$
$$= \mu_0 I \mathrm{d}s \, \sin\theta/4\pi r^2$$

where μ_0 is the permeability of free space (*see* diagram). Integration of the Biot–Savart law gives Ampere's circuital theorem (*see* Ampere's law).

bipolar integrated circuit A type of monolithic *integrated circuit based on *bipolar transistors. A section of a typical circuit is shown in the diagram. The substrate (E) is formed from a wafer of *semiconductor (p-type is shown) and has buried n⁺ (highly doped) regions selectively diffused into it. These regions serve to reduce the collector series resistance of the completed transistors. The n-type expitaxial layer (D) is then grown on the substrate to a typical thickness of 7 μm, and this in turn has an insulating oxide layer (B) grown on to it. *Photolithography is used to etch the oxide layer in the desired positions and isolating diffusions (C), of the same semiconductor type as the substrate, are made to isolate the individual components from each other. The individual components are formed by oxide growth, photoetching, and diffusion of the appropriate type of impurity, in turn, into the epitaxial layer. A final passivating oxide layer is grown. This has windows etched into it to enable contacts to be made to the semiconductor; the desired interconnec-

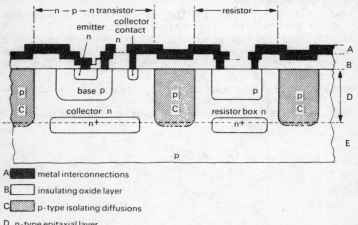

A ■ metal interconnections

B ☐ insulating oxide layer

C ▨ p-type isolating diffusions

D n-type epitaxial layer

E p-type substrate

Cross section of typical bipolar integrated circuit

tion pattern is made by evaporating a metal layer (A), usually aluminium, on to the oxide and etching.

The substrate is held at the most negative potential possible, thus causing the collector-substrate junctions to be reverse biased and preventing current flow across them. This ensures isolation of the individual components. The n-type region surrounding the resistor – the resistor box – is held at the most positive potential possible, thus preventing current flow across the resistor-resistor box junction by ensuring that it too is reverse biased.

bipolar junction transistor *See* transistor.

bipolar transistor A *transistor such as a junction transistor in which both electrons and holes play an essential role in its operation. In common usage a bipolar junction transistor is usually referred to simply as a *transistor*.

biscuit *See* recording of sound.

bistable *Short for* bistable multivibrator. A circuit having two stable states. *See* flip-flop.

bit Acronym from *bi*nary digi*t*. One of the digits 0 or 1 used in *binary notation. It is the basic unit of information in a computer or data processing system.

bit line *See* random-access memory.

Bitter patterns Experimental patterns demonstrating the presence of magnetic domains in ferromagnetic and ferrimagnetic materials. A magnetic fluid is applied to the surface of the sample. The magnetic particles in the fluid concentrate at the boundaries between the domains where discontinuities in the magnetic field are present (*see* diagram). The spiky 'fir-tree' domains shown in the diagram are closure domains arising because the crystal surface is not exactly a crystal plane. The direction of magnetization of a domain can be found using a Bitter pattern by making small scratches in the surface of the domain.

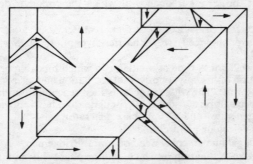

Schematic abstraction from typical Bitter pattern

When observed under a microscope the pattern is seen to change when an external magnetic flux is applied to the sample.

black-and-white television *Syn. for* monochrome television. *See* television.

black box Any self-contained unit or part of an electronic device that may be treated as a single package. Such a circuit may be approached mathematically in terms of its input and output characteristics, irrespective of its internal elements.

black level *See* television.

black-out A temporary loss of sensitivity of any electronic device following the passage of an intense transient signal.

black-out point *See* cut-off.

black recording *See* recording of sound.

blanking The rendering of a device or channel ineffective or inoperative for a desired time. For example, blanking eliminates the return trace from the screen of a cathode-ray tube. There are usually two blanking components to eliminate the horizontal and vertical components of the return trace.

blanking level *See* television.

blasting A form of *distortion in audio and radio receivers due to extreme overloading of the amplifying circuits.

blemish *See* storage tube.

Bloch walls *See* ferromagnetism.

block access *See* random-access memory.

blocked impedance The *impedance measured when the mechanical motion of an electromechanical or acoustic *transducer is blocked. *Compare* motional impedance.

blocking capacitor A *capacitor included in a circuit in order to limit the flow of direct current and low-frequency alternating current while allowing the passage of higher-frequency alternating current. The capacitance is chosen so that there is a relatively small *reactance at the lowest frequency at which the circuit is designed to operate.

blocking current *See* silicon-controlled rectifier.

blocking-layer photocell *Syn. for* barrier-layer photocell. *See* photovoltaic cell.

blocking oscillator A type of oscillator in which blocking occurs after completion of (usually) one cycle of oscillation and lasts for a predetermined time. The whole process is then repeated. Fundamentally it is a special type of *squegging oscillator and has applications as a *pulse generator and a *time-base generator.

blue gun *See* colour picture tube.

boat grown gallium arsenide *See* horizontal Bridgeman.

Bode diagram A diagram in which gain or phase shift in a feedback-control system is plotted against frequency. It is so called because of

Bode's theorem, which shows the interdependence of phase angle and rate of change of gain of a network at a desired frequency.

Bode equalizer An *attenuation equalizer designed so that one simple control varies the amount of equalization in the same proportion for all frequencies within the range.

Bode's theorem *See* Bode diagram.

body capacitance A *capacitance caused by the proximity of a human body to a circuit.

Boella effect A reduction in the *effective resistance of fixed composition *resistors when operated at VHF or higher frequencies, because of dielectric losses.

bolometer A small resistive element capable of absorbing electromagnetic power. The resulting temperature rise is used to measure the power absorbed.

Boltzmann constant Symbol: k. A fundamental constant having the value $1.380\,41 \times 10^{-23}$ joule per kelvin.

In a system in which n_1 particles have an energy E_1 and n_2 particles have an energy E_2 then

$$n_1 = n_2 \exp[-(E_1 - E_2)/kT]$$

where T is the thermodynamic temperature. This is the *Boltzmann distribution law.*

Boltzmann distribution law *See* Boltzmann constant.

bonded Denoting metal parts in a circuit that are connected together electrically so that they are at a common voltage.

bonded silvered mica capacitor *See* mica capacitor.

bonding pads Metal pads arranged on a semiconductor *chip (usually around the edge) to which wires may be bonded so that electrical connection can be made to the component(s) or circuit(s) on the chip. Bonding is usually effected by thermocompression or ultrasonic bonding. *See also* tape automated bonding; wire bonding.

booster (1) A generator or transformer inserted in a circuit in order to increase (*positive booster*) or decrease (*negative booster*) the magnitude or to change the phase of the voltage acting in the circuit.

(2) A repeater station that amplifies and retransmits a broadcast signal received from a main station, with or without a change of frequency.

bootstrapping A technique used in a variety of applications in which a capacitor – the *bootstrap capacitor* – is used to provide 100% positive *feedback for alternating currents across an amplifier stage of unity gain or less. Bootstrapping is used for control of the output signals by using the positive feedback to control the conditions in the input circuit in a desired manner.

Bootstrapping is commonly used in circuits that generate a linear time base, particularly in a sawtooth generator. A simple sawtooth generator consists of a capacitor that is charged by means of an

41

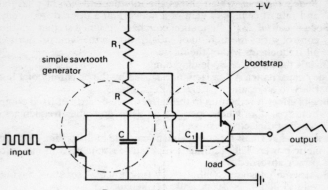

a Bootstrapped sawtooth generator

input load resistor and discharged by a periodic step voltage. As the capacitor is charged, the voltage increases exponentially and as the voltage increases, that across the input load (and hence the charging current) drops correspondingly. The output is approximately linear provided that only a small portion of the charging characteristic is used. The linearity may be improved by using a bootstrap circuit to maintain a constant charging current. A typical circuit is shown in Fig. *a.* The output is taken from an *emitter follower, capacitively coupled via the bootstrap capacitor C_1 to the input load resistor R. As the output voltage rises, the voltage at the node between R and R_1 also rises; the voltage across R and hence the charging current is therefore maintained substantially constant.

Bootstrapping may also be used in *MOS logic circuits in order to optimize the voltage swing between the high and low logic levels. A typical bootstrapped circuit is shown in Fig. *b.* The output voltage V_o at point X is high when a low logic level is applied to the gate of the transistor T_s and is determined by the value of the gate and threshold voltages of the transistor T_1. In the absence of the bootstrap capacitor, C, and the load transistor, T_L, the gate voltage V_{G1} of T_1 is equal to V_{DD} and

$$V_o = V_{DD} - V_T$$

where V_T is the threshold voltage. If a load transistor T_L is present then

$$V_o = V_{DD} - 2V_T$$

If V_{G1} can be increased to a value greater than V_{DD}, then V_o can also rise to a maximum possible value equal to V_{DD}. Bootstrapping is

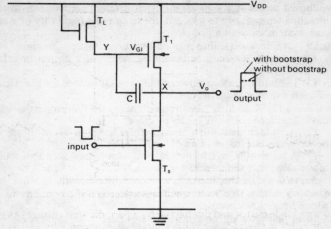

b Bootstrapping in MOS circuits

used to achieve this effect. A bootstrap capacitor, C, is connected between points X and Y (Fig. *b*). As V_o rises from the low logic level, V_{G1} also rises because of the positive feedback provided by C. This causes T_L to switch off and point B is thus isolated from the supply bus. V_{G1} can therefore rise to a value greater than V_{DD} and as V_{G1} increases, V_o also increases until it reaches V_{DD}. The total voltage swing at X is therefore optimized.

Bootstrapping is also used in high input impedance amplifiers, such as an emitter follower or a field-effect transistor stage, in order to maximize the a.c. input impedance. The a.c. input signal to the base or gate electrode can tend to flow through the bias resistors used to provide the d.c. bias for the base or gate unless bootstrapping is provided to prevent this effect.

borocarbon resistor *See* boron resistor.

boron resistor *Syn.* borocaron resistor. A resistive *film resistor that has a small percentage of boron introduced into the carbon film to add stability.

bottoming The operation of an electronic device in such a way that, when in the conducting state, the lowest instantaneous output voltage is determined by the device characteristics rather than the input voltage.

branch *See* network.

branch point *Syn.* node. *See* network.

43

breadboard model A rough assembly of discrete components, often attached temporarily to a board, for testing the feasibility of a circuit, system, or design principle.

break (1) An accidental interruption of a broadcast programme.

(2) To open a circuit suddenly by means of a switch or other device.

(3) The minimum gap between the contacts needed for the circuit of such a device to be open.

breakdown (1) A sudden catastrophic change in the properties of a device rendering it unfit for its purpose.

(2) A sudden disruptive electrical discharge in an insulator or between the electrodes of an *electron tube.

(3) A sudden change from a high dynamic resistance to a much lower value in a semiconductor device.

See also avalanche, thermal, and Zener breakdown.

breakdown voltage (b.d.v.) The voltage under specified conditions at which breakdown occurs.

breakover point *See* silicon-controlled rectifier.

breakover voltage *See* pnpn device.

bremsstrahlung Electromagnetic radiation produced when an electron is suddenly decelerated by a nuclear field. *See* X-rays.

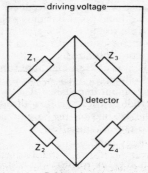

Bridge circuit

bridge An assembly of at least four circuit elements, such as resistors, capacitors, etc., together with a current source and a null point detecting device (*see* diagram). Each of the circuit elements is arranged in one *arm* of the bridge. When the bridge is balanced, i.e. zero response is obtained from the null detector, there is a calculable relationship between the values of elements in the arms given by

$$(Z_1/Z_2) = (Z_3/Z_4)$$

44

An unknown element may therefore be very precisely measured by comparison with known standards. The current source may produce either direct or alternating current. Bridge networks form the basis of many measuring instruments.

For resistance measurements *see* Wheatstone bridge, post-office box, Kelvin double bridge, Carey–Foster bridge. For capacitance measurements *see* Wien bridge, de Sauty bridge, Schering bridge. For inductance measurements *see* Anderson bridge, Hay bridge, Maxwell bridge, Owen bridge. For mutual inductance measurements *see* Campbell bridge, Felici balance, Hartshorn bridge.

See also resonance bridge; Wagner earth connection.

bridged-H network *See* quadripole.

bridged-T network *See* quadripole.

Bridgeman method *See* horizontal Bridgeman.

bridge network *See* network.

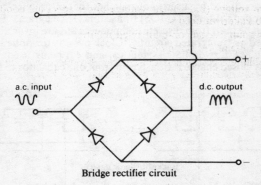

Bridge rectifier circuit

bridge rectifier A *full-wave rectifier circuit in the form of a bridge, with a rectifier in each arm (*see* diagram).

brightness control *See* television receiver.

Brillouin zone *See* energy bands.

British Standards Institute (BSI) *See* standardization.

British Standard wire gauge (NBS) *See* wire gauge.

broadband *See* wideband.

broadcasting Radio or television transmission to the public. Specific *frequency bands are available for public broadcasts and are assigned in accordance with international agreements.

broadside array *See* aerial array.

Broca galvanometer *See* astatic galvanometer.

Brown and Sharpe wire gauge (B & S) *Syn. for* American wire gauge. *See* wire gauge.

Brownian-motion noise *Syn. for* thermal noise. *See* noise.

brush A conductor made from specially prepared carbon that provides electrical contact with a moving conductor, usually between the rotating and stationary parts of electrical machines. The contact resistance between the brush and the moving surface is the *brush contact resistance*. In electrical machines this resistance usually decreases with increasing current density resulting in a roughly constant voltage drop at the contact surface.

brush contact resistance *See* brush.

brush discharge A luminous electrical discharge that appears as a number of branching threads surrounding a conductor. Brush discharge occurs when the electric field around the conductor is not sufficiently large to produce a *spark.

BSI *Abbrev. for* British Standards Institute. *See* standardization.

bubble memory *See* magnetic bubble memory.

bucket-brigade device (BBD) A device that consists of a number of capacitors linked by a series of switches that in practice consist of *bipolar or *field-effect (MOS) transistors (*see* diagram). These circuits can be built from discrete components but are invariably manufactured as *integrated circuits. *Clock pulses are applied to close the switches, a two-phase system (ϕ_1 and ϕ_2) being used. As each switch is closed charge is transferred from one capacitor to the next.

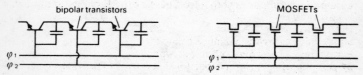

Bucket-brigade devices

Bucket-brigade devices are frequently used as *delay lines in both digital and analog systems since the amount of charge stored may vary continuously from zero to the limit set by the magnitude of the capacitance and the operating voltage. The capacitors are provided in practice by using the collector-base capacitance or drain-gate capacitance of the transistors used in the circuit.

buffer An isolating circuit interposed between two circuits to minimize reaction from the output to the input. Usually it has a high input impedance and low output impedance. It may be used to handle a large *fan-out or to convert input and output voltage levels.

buffer memory *See* computer; memory.

bug An error or fault in a computer *program or in computer equipment. To *debug* the program or system is to find and correct any errors.

build-up time The time required for a current to rise to its maximum value in an electronic circuit or device.

46

built-in field *Syn.* diffusion potential. *See* p-n junction.

bulk acoustic wave *See* acoustic wave device.

bulk-channel charge-coupled device *Syn.* buried-channel charge-coupled device. *See* charge-coupled device.

bulk lifetime *Syn.* volume lifetime. *See* semiconductor.

buncher *See* velocity modulation; klystron.

bunching *See* velocity modulation.

buried-channel charge-coupled device *Syn. for* bulk-channel charge-coupled device. *See* charge-coupled device.

buried layer A layer of high-conductivity *semiconductor material diffused into the substrate layer during the manufacture of *bipolar integrated circuits and *transistors. The buried layer is located below the *collector and serves to reduce collector resistance.

burn-in *Syn.* ageing. *See* failure rate.

burning-on *See* metallizing.

burst signal *Syn. for* colour burst. *See* colour television.

bus *Syns.* busbar; busline. A conductor having low impedance or high current-carrying capacity to which two or more circuits can be separately connected or which can connect several like points in a system, as with an *earth bus*. Buses frequently feed power to various points. The name was first applied to metal bars of rectangular cross section supported on porcelain insulators.

bushing An insulator used to form a passage for a conductor through a partition.

Butterworth filter *See* filter.

button mica capacitor *See* mica capacitor.

button microphone *See* carbon microphone.

by-pass capacitor A shunt capacitor that is used to provide a comparatively low impedance path for alternating current. The magnitude of the capacitance determines the frequency that it passes. Such a capacitor is used to prevent a.c. signals reaching a particular point in a circuit, or to separate out a desired a.c. component.

byte A group of *bits taken together and treated as a unit in a digital *computer. A byte is usually shorter than a *word.

byte-addressable *See* address.

C

cable An assembly of conductors that has some degree of flexibility; the conductors are insulated from each other and enclosed in a common binding or sheath. The most common types of cable are *paired cable and *coaxial cable although other configurations may be used. A *uniform cable* is one that contains the same electrical characteristics at all points along its length. A *composite cable* is one in which the gauge of the conductors and/or the type of construction is not the same throughout its cross section.

cable track locator A device used to locate buried telephone cables. An audiofrequency oscillator, known as a bleeper, is used to send a signal along the cable in question and the bleeps are detected using a portable receiver and headset. The position of the buried cable is along the track corresponding to the maximum detected signal intensity.

cadmium cell *See* Weston standard cell.

calculator A device that carries out logical and arithmetical operations of all kinds. The term is usually reserved for small machines capable of carrying out comparatively simple arithmetical tasks. These consist of integrated logic circuits that perform the arithmetic, a keyboard for data input, and an illuminated display such as a *liquid-crystal display or *light-emitting diodes. The earlier electromechanical calculators have now been almost entirely superseded.

calibration The determination of the relation between the indicated value on a measuring instrument and the true value of the quantity to be measured. The true value is the value that would be obtained if all sources of error were eliminated.

camera tube The device contained in a *television camera that acts as an optical-electrical *transducer and converts the *optical image* of the scene to be transmitted into electrical *video signals*. Most camera tubes are *electron tubes: the two basic types of tube are the *image orthicon and the *vidicon tubes from which many other tubes have been developed.

A wide variety of camera tubes is available. The main differences between them are in the target material and the velocity of the electrons used to produce the video signal. *Photoemissive camera tubes* have a target coated with photoemissive material (*see* photoemission). *Photoconductive camera tubes* are coated with material that exhibits *photoconductivity. *Low-electron-velocity tubes* are most often used but *high-electron-velocity tubes,* such as the iconoscope, are also produced.

The performance of camera tubes depends very greatly on the scanning system employed. Beam alignment is achieved by using small coils to ensure that the electron beam emerging from the electron gun is central. Deflection of the beam is provided by deflection coils controlling the horizontal and vertical directions. These coils are supplied with a *sawtooth voltage causing a linear scan with very rapid return to the start of the scanning position. Focusing coils are also provided to ensure a small cross section when the beam reaches the target; an extra electrode ensures that the beam is essentially perpendicular to the target surface. In low-electron-velocity tubes this electrode decelerates the beam so that it is essentially stationary at the target.

A *solid-state camera has also been developed in which the transduction element is an array of *charge-coupled devices. This type of camera is very much smaller and lighter than those containing electron tubes and is typically the size of an ordinary hand-held photographic camera.

See also image dissector.

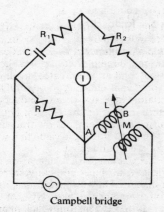

Campbell bridge

Campbell bridge An a.c. *bridge that is used to measure a mutual inductance, M, by comparison with a standard capacitor, C. The resistances R_1 and R_2 are varied until null deflection is obtained on the indicating instrument I (*see* diagram). At balance

$$L/M = (R + R_1)/R$$
$$M/C = RR_2$$

where L is the self-inductance of the coil AB.

49

canned Denoting a device that is completely enclosed and sealed by a metal sheath.

capacitance Symbol: C; unit: farad. The property of an isolated *conductor or a set of conductors and *insulators whereby it stores electric charge. A charge of Q coulombs will increase the voltage of an isolated conductor by V volts. The capacitance is defined as the ratio Q/V and is determined by the size and shape of the conductor. It is constant for a given isolated conductor.

If the isolated conductor is placed near a second conductor or a semiconductor but is separated from it by air or some other insulator, the system forms a *capacitor. An electric field is produced across the system and the potential difference between the conductors is determined by this field. The capacitance, C, is defined as the ratio of the charge on either conductor to the potential difference between them. *See also* mutual capacitance; impedance; series; parallel.

capacitance integrator *See* integrator.

capacitance strain gauge *Syn. for* variable capacitance gauge. *See* strain gauge.

capacitive coupling *See* coupling.

capacitive feedback *See* feedback.

capacitive load *See* leading load.

capacitive reactance *See* reactance.

capacitive tuning *See* tuned circuit.

Electrolytic   Fixed   Variable

Capacitor symbols

capacitor A component that has an appreciable *capacitance. It consists of an arrangement of at least two conductors or semiconductors separated by a dielectric (an insulator). The conductors or semiconductors are known as *electrodes* or *plates*. The value of the capacitance of a given device depends on the size and shape of the electrodes, the separation between them, and the relative *permittivity of the dielectric. Most types of capacitor have a value, which may be variable, determined by the geometry of the device. Some capacitors however have a value that is also a function of the voltage across the device or of the operating frequency. The dielectric material may be solid, liquid, or gaseous. *See* blocking, ceramic, electrolytic, foil, mica, MOS, paper and plastic-film capacitors; varactor.

capacitor microphone *Syn.* electrostatic microphone. A type of microphone in which a diaphragm forms one plate of a capacitor. Variations of sound pressure cause movements of the diaphragm and these alter the capacitance. Corresponding changes in potential difference across the capacitor are thus produced.

capacitor pick-up *See* pick-up.

capacity *Obsolete syn. for* capacitance.

carbon Symbol: C. A nonmetal, atomic number 6, that exists in two allotropic crystalline forms: diamond and graphite. In diamond form the *resistivity (5×10^{14} ohm cm) falls within the range of insulators. In graphite form it is a poor conductor, resistivity about 1.4×10^{-3} ohm cm, and in granular form exhibits a variation of resistance with pressure.

In graphite form it is used to form resistors, in microphones, and as a filament of electric lights.

carbon-composition resistor *Syn. for* carbon resistor. *See* resistor.

carbon-film resistor *See* film resistor.

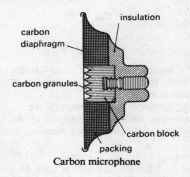

Carbon microphone

carbon microphone A microphone that utilizes the variation of contact resistance of granular carbon with applied pressure. There are several types of carbon microphone, including the telephone transmitter (*see* diagram), in which sound-pressure variations are transmitted to the carbon granules through a diaphragm; the corresponding changes of resistance are detected by fluctuations in a current passed through the granules. The *button microphone* is much smaller than the telephone transmitter and has a small mica diaphagm. It is designed to screw on to a much larger diaphragm. The *Reicz microphone* has the granules contained in a cavity behind a thin mica or oiled silk diaphragm of large area.

carbon resistor *Syn.* carbon-composition resistor. *See* resistor.

carcinotron A crossed-field type of backward-wave oscillator. *See* travelling-wave tube.

Cardew voltmeter *See* thermoammeter.

card punch *See* punched card.

card reader *See* punched card.

Carey–Foster bridge A modification of the *Wheatstone bridge that measures the difference in resistance between two nearly equal resistances. The resistances are placed in the ratio arms of the bridge and the balance point found on a slide-wire. The resistances are then switched and a new balance point found. The resistance difference is proportional to the distance between the balance points.

carrier (1) *Short for* charge carrier. A mobile electron or *hole that transports charge through a metal or a *semiconductor and is responsible for its conductivity.

(2) *Short for* carrier wave.

carrier concentration The number of charge *carriers in a *semiconductor per unit volume, in practice usually quoted as numbers per cubic centimetre. In an intrinsic (i-type) semiconductor the number of holes, p, and of electrons, n, is equal:

$$n = p = n_i$$

where

$$n_i^2 = N_c N_v \exp(-E_g/kT)$$

where N_c and N_v are the densities of energy states in the conduction and valence bands respectively, E_g is the difference in energy between the conduction and valence bands, k is the Boltzmann constant, and T the thermodynamic temperature.

In an extrinsic semiconductor the electrical neutrality of the sample is preserved and in a sample that contains impurities,

$$N_A^- + n = N_D^+ + p$$

where N_A^- and N_D^+ are the numbers of ionized acceptor and donor impurities, respectively. At relatively high temperatures most of the impurity atoms are ionized and so

$$n + N_A \simeq p + N_D$$

where N_A and N_D are the remaining numbers of nonionized acceptors and donors. The product $np = n_i^2$ is independent of added impurities.

In an n-type semiconductor the number of electrons, n_n, at thermal equilibrium is given by

$$n_n \simeq N_D \text{ if}$$
$$|N_D - N_A| >> n_i \text{ and}$$
$$N_D >> N_A$$

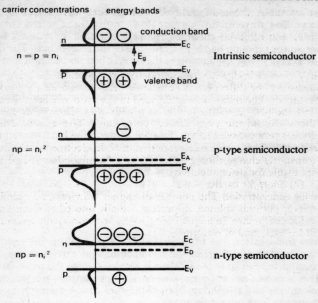

The number of holes, p_n, in an n-type semiconductor is given by

$$p_n = n_i^2/n_n$$

In a p-type semiconductor the situation is reversed:

$$p_p \simeq N_A \text{ if}$$
$$|N_A - N_D| >> n_i \text{ and}$$
$$N_A >> N_D;$$
$$n_p = n_i^2/p_p$$

The carrier concentrations of i-, n-, and p-type semiconductors are shown schematically in the diagram.

carrier frequency The frequency of a *carrier wave.

carrier mobility *See* drift mobility.

carrier noise *See* noise.

carrier storage *Syns.* charge storage; storage effect. An effect occurring in a *p-n junction under forward bias. In a semiconductor with a relatively long bulk lifetime, excess minority carriers injected across the junction remain near the junction as a net concentration of charge. When *reverse bias is applied to the junction the effect of the carriers stored near the junction is to produce a reverse current across the junction substantially greater than the normal reverse saturation current; this current flows until all the stored charges have been removed either by recombination or by crossing back across the junction under the influence of the reverse bias. The time interval between application of the reverse bias and cessation of the reverse current surge is the *storage time* of the junction.

When several diodes are used together in a rectifier circuit that also contains inductive elements, the carrier storage can result in undesirable transients in the circuit and limits the frequency at which such circuits can be operated. The storage effect is however utilized in the *charge-storage diode.

carrier suppression *See* carrier transmission.

carrier system A transmission system used in telephony and telegraphy in which modulated *carrier waves are transmitted along lines or cables. The multichannel carrier system in telephony allows many simultaneous independent signals to be transmitted on the same circuit.

carrier telegraphy *See* telegraphy.

carrier telephony *See* telephony.

carrier transmission The transmission of a signal that is the result of *modulation of a *carrier wave. Sometimes the carrier wave is not transmitted but only the sidebands resulting from the modulation. This is known as *carrier suppression.* Carrier transmission is used in telegraphy and telephony. These forms are carrier *telegraphy and carrier *telephony, respectively.

carrier wave *Syn.* carrier. The wave that is intended to be modulated in *modulation, or, in a modulated wave, the carrier-frequency spectral component. The process of modulation produces spectral components falling into frequency bands at either the upper or lower side of the carrier frequency. These are *sidebands,* denoted *upper* or *lower sideband* according to whether the frequency range is above or below the carrier frequency. A sideband in which some of the spectral components are greatly attenuated is a *vestigial sideband.* In general these components correspond to the highest frequency in the modulating signals. A single frequency in a sideband is a *side frequency.* The *baseband* is the frequency band occupied by all the transmitted modulating signals.

cascade A method of connecting in series a chain of electronic circuits or devices so that the output of one is the input of the next.

cascade control An automatic control system in computer technology in which each control unit controls the succeeding unit and is controlled by the preceding one in the chain.

cascoded circuit *See* emitter-coupled logic.

cataphoresis *See* electrophoresis.

catastrophic failure *See* failure.

catcher *See* klystron.

Catching diodes for voltage limitation

catching diode *Syn.* clamping diode. A *diode used to limit the voltage at a point in a circuit. Two diodes can be used to keep the voltage within specific limits (*see* diagram), i.e. within $\pm V_d$, where V_d is the *diode forward voltage.

cathode The negative *electrode of an electrolytic cell, discharge tube, valve, or solid-state rectifier. The electrode by which electrons enter (and conventional current leaves) a system. *Compare* anode.

cathode dark space *Syn. for* Crookes dark space. *See* gas-discharge tube.

cathode disintegration *See* cathode sputtering.

cathode follower *See* emitter follower.

cathode glow *See diagram at* gas-discharge tube.

cathode-ray direction finding *See* direction finding.

cathode-ray oscilloscope (CRO) An instrument in which a variety of electrical signals are presented on the screen of a *cathode-ray tube for examination. The signal under examination is used to deflect the electron beam of the CRT in one direction (usually the vertical) while another known signal is used in the other direction. The composite signal is shown on the screen. Visualization of the input signal is achieved using the output from a sweep generator, usually called a time-base generator (*see* time base) and selecting the appropriate sweep speed. The sweep may be generated as a *sawtooth waveform or initiated by an external trigger pulse.

More complicated cathode-ray oscilloscopes often include a delayed trigger, access to the X-deflection plates, and beam-intensity modulation facilities.

cathode rays *See* gas-discharge tube.

cathode-ray tube (CRT) A funnel-shaped electron tube that converts electrical signals into a visible form. All CRTs have an *electron gun to produce an electron beam, a grid that varies the electron beam intensity and hence the brightness, and a luminescent screen to produce the display. The electron beam is moved across the screen either by deflection plates or magnets. The *deflection sensitivity* of the tube is the distance moved by the spot on the screen per unit change in the deflecting field. Focusing of the beam may also be done electrostatically or electromagnetically or by a combination of methods (*see* diagrams). A greater degree of focusing is required

55

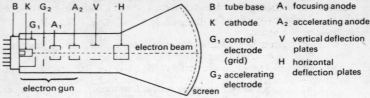

B	tube base	A₁	focusing anode
K	cathode	A₂	accelerating anode
G₁	control electrode (grid)	V	vertical deflection plates
G₂	accelerating electrode	H	horizontal deflection plates

a Electrostatic focusing and deflection

when the electron beam is deflected towards the edges of the screen. The point at which the electron beam comes to a focus is the *crossover area* and the solid angle of the cone of electrons emerging from this area is the *beam angle*. For convenience the deflection and focusing coils are often mounted around the narrow neck of the tube as a single unit, termed a *scanning yoke*. Such an arrangement reduces the overall physical dimensions of the assembly and is particularly important when the tube contains more than one electron beam, as in the *double-beam CRT* or some forms of colour picture tube, and therefore requires more than one set of coils.

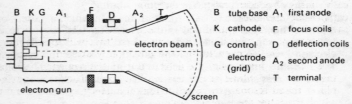

B	tube base	A₁	first anode
K	cathode	F	focus coils
G	control electrode (grid)	D	deflection coils
		A₂	second anode
		T	terminal

b Electromagnetic focusing and deflection

The screen of the CRT may be coated with aluminium on the inside and this coating held at anode potential. Such an *aluminized screen* prevents the accumulation of charge on the phosphor and improves its performance by increasing the visible output and reducing the effects of ion bombardment. In the case of a CRT in which the screen is not aluminized, the maximum potential difference that can be applied between the anode and cathode is limited to the value at which the secondary emission ratio of the screen rises to unity and is known as the *sticking potential*.

High-frequency applications, as in the *cathode-ray oscilloscope, usually employ electrostatic deflection; electromagnetic deflection is used when high-velocity electron beams are required, as in *television or *radar receivers, which need a bright display.

cathode-ray tuning indicator *See* magic eye.

cathode-ray voltmeter A voltmeter in the form of a cathode-ray tube of known sensitivity of deflection.

cathode spot *See* arc.

cathode sputtering *Syn.* cathode disintegration. The slow disintegration of the cathode in a *gas-discharge tube due to bombardment by positive ions. The phenomenon can be used for the deposition of thin metallic films on a surface. The object to be coated is placed near the cathode of a gas-discharge tube, the cathode being made from the desired metal. The whole system is formed as a demountable assembly that is evacuated to the correct pressure to form a gas tube. Typical operating conditions are at gas pressures between 1·0 and 0·01 mmHg and anode voltages between one and 20 kilovolts.

cathode-voltage-stabilized camera tube *Syn. for* low-electron-velocity camera tube. *See* camera tube. *See also* image orthicon; vidicon.

cathodic etching *See* etching.

cathodoluminescence The emission of light when substances are bombarded by cathode rays (electrons). The frequency of light emitted is characteristic of the bombarded substance.

cation An ion that carries a positive charge and, in electrolysis, moves towards the cathode, i.e. travels in the direction of conventional current. *Compare* anion.

cat's whisker *See* point contact.

cavity maser *See* maser.

cavity resonator *Syns.* rhumbatron; resonant cavity. The space within a closed or substantially closed conductor that will maintain an oscillating electromagnetic field when suitably excited externally. The resonant frequencies are determined by the size and shape of the cavity. The whole device has marked resonant effects and replaces tuned resonant circuits for high-frequency applications for which the latter are impracticable.

CCD *Abbrev. for* charge-coupled device.

CCD filter A circuit in which the ability of a *charge-coupled device to provide a precise predetermined delay time to an analog signal is utilized in order to produce a desired signal-processing function.

One class of CCD filters is formed by inserting a charge-coupled device of known delay into a circuit so that it provides a feedback (Fig. *a*) or feedforward (Fig. *b*) path. The simple circuits shown in Figs. *a* and *b* act as band-pass and band-stop (notch) filters, respectively. In both circuits, the signals that pass through the CCD array are subject to a constant delay time. The signals in which the delay time, MT_c, corresponds to a known multiple of the input signal period will be in phase with the input signal; M is the number of CCD cells and T_c the time for a single transfer. In the circuit in Fig. *a* the signals add coherently, when MT_c is an odd integral multiple of half the input period, to produce a band-pass filter. In the circuit in Fig. *b* the signals are in phase if MT_c is an integral multiple of the

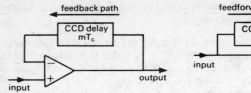

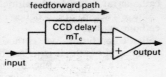

a Feedback delay:
simple recursive band-pass CCD filter

b Feedforward delay:
simple transversal band-stop filter

input period. The in-phase signals produce a minimum output from the differential amplifier resulting in a band-stop filter. Very much more complicated recursive filters can be made in order to produce desired filtering functions. *CCD multiplexing can be used so that several inputs may be processed simultaneously by one filter.

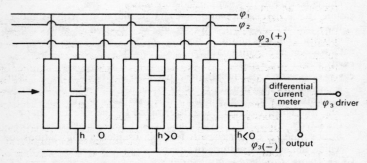

c Schematic diagram of 3-phase split-electrode transversal filter

The second main class of CCD filters utilizes the *split-electrode technique* in order to produce a desired output. In this type of transversal filter some of the gate electrodes of the CCD array are split into two parts (Fig. *c*). As a charge packet is transferred through the array, the amount of charge that is stored under the parts of the split electrode is in the ratio of the areas and, provided that the thickness is uniform, in the ratio of the lengths. The signal charge under the electrode can be sampled nondestructively using the clock line to that electrode. The stored charges cause equal and opposite charges to flow from the clock line on to the electrode. The difference in charge between the two parts of the electrode is equal to the original charge packet multiplied by a weighting factor, *h*, dependent on the areas of the two parts of the split electrode. In the case of a filter consisting of several split-electrode stages, all connected to the same clock lines, the output on the clock lines is the sum of the signals to each split electrode, i.e.

$$V_{out}(nT_c) = \Sigma_{m=1}^{N} h_m V_{in}(nT_c - mT_c)$$

where n, m are integers, T_c is the clock voltage period, and h_m the weighting factor of the mth electrode in an array of N electrodes; V_{in} is the sampled input signal at time $t = nT_c$.

A wide variety of filtering functions, particularly spectral analysis and Fourier analysis functions, can be achieved using split electrodes by choosing the appropriate weighting factors for the electrodes and the appropriate clock frequency.

CCD memory A *solid-state memory that consists of one or more *shift registers formed from *charge-coupled devices and that is used to store digital information. The CCD memory is a dynamic *serial memory and is inherently slower than *random-access memories (RAMs). The device is cheaper and more compact and is suitable for applications that are either serial in nature or that do not require the fast operating speeds of RAMs.

Several different types of CCD memory have been produced that differ in the methods of transferring the data between the storage locations. In general they fall into two categories: memories with relatively long shift registers that have long access times to information and are relatively low cost; memories consisting of a number of short shift registers in which the data is circulated synchronously in parallel and that have much faster access times. The latter are more complex and more expensive than the former. The fastest version has access times approaching those of RAMs.

The largest memory chips are manufactured using serial-parallel-serial transfer of information. Storage capacities of 64 kilobits arranged in four 16 K blocks can be produced in a single chip of silicon, together with all the necessary support circuits. In this method of operation a rectangular array of storage cells is used; a number of bits of information is input serially to the first row in the array at high speed. When this row is filled the bits are transferred in parallel at a slower rate, through the columns of the array, and output serially at high speed. Each column is provided with a sense/regeneration amplifier that automatically regenerates the information on completion of a cycle. In an array of M columns and N rows, the total cycle time is the same as that for a single shift register of MN storage locations operated at the higher clocking frequency but the number of transfers is reduced to $(M + N)$ transfers, which reduces the overall loss of signal due to the transfer inefficiency. The slower rate of clocking through the parallel sections also reduces the power required for the operation.

The number of bits of information that can be stored in the array can be increased using *electrode per bit* (E/B) operation. The basic memory cell of a CCD device normally requires two pairs of over-lapping electrodes with two-phase clocking. In E/B operation $(n - 1)$ bits can be stored in an array of n electrodes, by maintaining one empty electrode and transferring the information sequentially into the empty potential well using overlapping clock voltages. The empty potential well moves through the array as the information is transferred. This method however requires separate clocks for each electrode to maintain the integrity of each bit.

High operating speeds are achieved using a number of short CCD shift registers that are clocked in parallel, each of which is connect-ed to its own sense/refresh circuits. A particular register may be selected using suitable decoding circuits.

The *line-addressable random-access memory* (LARAM) is a block-access device that consists of a number of short CCD shift registers that share common input, output, and regeneration circuits. The data is not circulated continuously in this memory: the desired reg-ister is selected by a decoder that activates the clock to that register, the other registers remaining quiescent. The cycle time of the shift registers is comparable to the access time of RAMs. The LARAM has the disadvantage that an appreciable time is spent in refreshing the information since a single circuit serves the entire block; in addition the dark current that appears in the quiescent storage sites is nonuniform, unlike other CCD memories in which the charges are circulating continuously and in which an average value of dark cur-rent results.

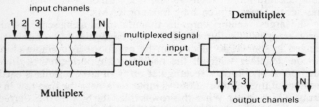

a Sample by sample CCD multiplexing

CCD multiplexer A *charge-coupled device that is used for *time-division multiplexing of a number of input signals. One method of organization is shown in Fig. *a.* Signals from the various inputs are clocked in parallel into a CCD array and then transferred sequen-tially along the array. A similar array operating synchronously is used to separate the signals at the output. *Crosstalk between the signals can be minimized by using inert channels between each sig-nal.

An alternative method is shown in Fig. *b.* Each input signal is

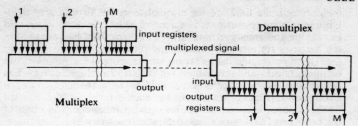

b Signal segment CCD multiplexing

used to charge a number of CCD electrodes forming an input register. When all the electrodes are full the samples in all the registers
are clocked in parallel into the main CCD array. They are then
transferred sequentially to the output. The output signal consists of
groups of several consecutive samples of each input signal. This
latter arrangement has the advantage that it may be used for continuous sampling of the input signals, since the input registers are
recharged during the output sequence of the main CCD array. The
signals are separated at the demultiplexer using a similar series of
output registers operated synchronously.

CCIR *Abbrev. for* International Radio Consultative Committee. *See*
standardization.

CCITT *Abbrev. for* International Telegraph and Telephone Consultative Committee. *See* standardization.

CD *Abbrev. for* compact disc. *See* compact disc system.

CEEFAX *Tradename. See* teletext.

cell (1) *Syns.* electrolytic cell; voltaic cell. A device that produces electricity by chemical means, consisting of a pair of plate electrodes in
an electrolyte. In a *primary cell* the chemical action is not normally
reversible, the current being produced as a result of the dissolution
of one of the plates. A *secondary cell* has reversible chemical action
and is charged by passing a current through it. The rate and direction of the chemical action is determined by the value of the external
voltage. The *volt efficiency* of a secondary cell is the ratio of the
voltage developed by it during the discharge to the average voltage
supplied to it during the recharging cycle.

The *cell internal resistance* is the resistance offered to the passage
of current inside the cell. If the open-circuit e.m.f. is E, then the
potential difference, U, across the cell when current flows is given
by

$$U = ER/(R + r)$$

where R is the external resistance and r the internal resistance.

A portable cell that has the electrolyte in the form of a nonspillable jelly or paste is known as a *dry cell.* Most cells contain a liquid electrolyte and are sometimes termed *wet cells.* Dry cells are used in the batteries for torches, portable radios, etc.

(2) Any device that can generate a direct electromotive force from a nonelectrical source of energy, particularly from light energy. Examples include *solar cells and *photovoltaic cells.

cell constant The area of the electrodes in an electrolytic *cell divided by the distance between them. This equals the resistance in ohms of the cell when filled with a liquid of unit resistivity.

cell internal resistance *See* cell.

Cellnet *Short for* cellular network. A system of *telephony that uses a radio link to connect the user to the main telephone network, and allows the use of portable battery-operated telephones. The service is run jointly by British Telecom and Securicor. The system consists of a number of transmit/receive stations (cells) each connected to a central switching station, which in turn is connected to the telephone grid. The area served by each cell is determined by local landscape. Features such as hills or clusters of tall buildings affect the area that can be covered. Signals from a user are detected by the appropriate cell and transmitted to the central switching station. As the user moves from one cell area to an adjacent one (perhaps travelling in a motor vehicle), the central switching station detects this movement and automatically switches to receive signals from the new area.

centi- Symbol: c. A prefix to a unit, denoting a submultiple of 10^{-2} of that unit: one centimetre equals 10^{-2} metre.

central processing unit (CPU) The portion of a *computer that controls the operation of the entire computer system and executes the arithmetical and logical functions contained in a particular *program.

A CPU usually consists of two units: the *control unit* organizes the data and program storage in the memory and transfers data and other information between the various parts of the system; the *arithmetic unit* executes the arithmetical and logical operations, such as addition, comparison, and multiplication.

centre tap A connection made to the electrical centre of an electronic device such as a resistor or transformer.

ceramic capacitor A capacitor utilizing a ceramic as the dielectric material. The behaviour of the capacitor is determined by the electrical properties of the ceramic used; these vary widely but most have high permittivity allowing the capacitors to be smaller than most other types.

ceramic pick-up *See* pick-up.

cermet Acronym from *cer*amic and *met*al. A mixture of an insulating material and a highly conducting material. Ceramics such as oxides and glasses are used as the insulator and metals such as chromium and silver are the conductors. Cermets are used for the production of thick films and thin films, particularly for use in *film resistors.

CGS system A system of *units, now obsolete, in which the fundamental units of mass, length, and time were the centimetre, gram, and second. In order to include and completely define electric and magnetic quantities a fourth fundamental quantity is required. As a result two mutually exclusive systems of CGS units arose: the *CGS electromagnetic units* (emu) and the *CGS electrostatic units* (esu).

In the emu system the fourth quantity was the *permeability of free space, μ_0, which was chosen to be unity. In the esu system the fourth quantity was the *permittivity of free space, ε_0, also chosen to be unity. The choice of the four fundamental quantities completely defined each system but since, by Maxwell's equations, it can be shown that

$$\mu_0\varepsilon_0 = 1/c^2$$

where c is the velocity of light, the systems were mutually exclusive. All mechanical units however, such as force and work, have the same units in either system.

To differentiate between the two systems the prefix *ab- was used to denote units in the emu system and the prefix *stat- for units in the esu system. The ratio of ab- unit to equivalent stat- unit for the primary CGS units, such as current, is equal to c or $1/c$, where c is the velocity of light in cm/s. The ratio for secondary units is some power of c.

The esu units of charge and current and the emu unit of potential were inconveniently small for practical use and a set of *practical units* was devised. These units were multiples or submultiples of the ab- and stat- units. The practical system was not however a CGS system: mechanical quantities, such as power, which is equal to current times voltage (in practical units), were not in CGS units. The practical units were defined in an arbitrary way, like the metre, and were used in the *international system of units. They were superseded by the *MKS system of *absolute units, from which the system of *SI units was developed.

channel (1) A specified frequency band or a particular path used in communications for the reception or transmission of electrical signals.

(2) A route along which information may travel or be stored in a data-processing system or a computer.

(3) (in a *field-effect transistor) The region connecting *source and *drain, the conductivity of which is modulated by the *gate voltage.

(4) (in a p-n-p junction *transistor) The spurious extension of n-type *base across the surface of the *collector to the edge of the

chip. This results in excessive leakage currents and may be overcome using a *channel stopper. The inevitable presence of positive charges at the interface of the high-resistivity p-type collector with the passivating oxide layer causes the formation of the n-type channel on the surface of the collector.

channel code *See* compact disc system.

channel stopper (1) (in a p-n-p junction *transistor) A means for limiting *channel formation by surrounding the n-type *base entirely with a ring of highly doped low resistivity p-type material.

(2) (in an *MOS integrated circuit) A region of highly doped material of the same type as the lightly doped substrate. This increases the field *threshold voltage and inhibits the formation of spurious *field-effect transistors caused when interconnections pass between adjacent *drain regions.

characteristic A relationship between two values that characterizes the behaviour of a device, circuit, or apparatus. The relations are most commonly produced for unipolar and bipolar *transistors, *thermionic valves, and *quadripoles. They are usually plotted in the form of families of graphs (*characteristic curves*) relating the currents obtained to the voltages applied for a range of operating conditions.

The *electrode characteristic* shows the relationship between current and voltage at an electrode of the device, for example, drain current versus drain voltage in a *field-effect transistor. The *transfer characteristic* shows the relationship between the current (or voltage) at one electrode and the voltage (or current) at a different electrode, for example, drain current versus gate voltage. The *static characteristic* shows, for example, collector current versus base voltage, all other applied voltages being held constant, i.e. under static conditions. The *dynamic characteristic* relates the current from one electrode and the voltage on another, under dynamic conditions. *See also* transfer parameters.

characteristic curve *See* characteristic.

characteristic impedance *See* iterative impedance; transmission line.

characteristic X-rays *See* X-rays.

charge Symbol: Q; unit: coulomb. A property possessed by some elementary particles causing them to exert forces of attraction or repulsion on each other. The types of force exerted by charged particles are differentiated by the terms *negative* and *positive,* the natural unit of negative charge being possessed by the *electron. The *proton has an equal amount of positive charge. A charged body contains an excess or lack of electrons with respect to its proton content.

Two electric charges, when close to each other, exert a force on each other. This force is given by *Coulomb's law and can be explained in terms of the *electric field surrounding the charges. If a charge, Q, is placed in an electric field E it experiences a force QE.

charge carrier *See* carrier; semiconductor.

charge-coupled device (CCD) A *charge-transfer device that consists of an array of *MOS capacitors suitably designed so that they are coupled and therefore charges can be moved through the semiconductor substrate in a controlled manner. The CCD can perform a wide variety of electronic functions. The device is essentially an analog *shift register and can be used for signal processing: it can be used to form analog or digital *serial memories (*see* CCD memory) and can also function as a dynamic filter (*see* CCD filter) The device may also be used for imaging, as in the *solid-state camera.

In any given device the amount of charge that can be stored at any given capacitor is determined by the surface potential, ϕ_S (*see* MOS capacitor), up to a maximum value, which is a function of the device geometry. In the absence of any signal charges, ϕ_S is a function of the gate voltage, the thickness of the oxide layer, and the doping level of the substrate. Injected signal charges, in the form of minority carriers, are stored at the surface of the substrate in the regions of maximum surface potential and the effective value of ϕ_S is reduced by the presence of the signal charges. If two adjacent capacitors have different values of ϕ_S the charges move to the location of maximum ϕ_S provided that the potential profiles overlap. It is convenient to represent graphically the surface potential, ϕ_S, of the empty capacitor, i.e. no signal charges present, as a potential well and to consider the charges as analogous to a fluid that fills the potential well rather as water fills a bucket.

Thermally generated minority carriers also tend to collect in the potential wells but provided that the signal charges are transferred at a rate that is very much faster than the relaxation time of the capacitor (*see* MOS capacitor) the effect of such carriers is small.

The *transfer inefficiency* is defined as the amount of the stored charge that is lost during the transfer operation and is often quoted as a percentage. The transfer efficiency is the amount of charge transferred, i.e.

$$(100 - \text{transfer inefficiency})\%$$

Since it is always less than 100% the signal charges must be regenerated after a number of transfers in order to restore the original signal level.

The devices are described as *n-channel CCDs* or *p-channel CCDs* depending on whether the minority carriers transferred are electrons or holes, respectively. N-channel devices operate with a p-type substrate and positive gate voltages; p-channel devices employ an n-type substrate and negative gate voltages. The same principles of operation apply to both types of device.

The original CCD consisted of a simple array of closely spaced metal gate electrodes (Fig. *a*). Charges are transferred between electrodes by adjusting the gate voltages so that the surface potentials change in a controlled manner. The charges move to the electrode of greatest ϕ_S. *Three-phase* clocking is required, i.e. every third electrode is clocked synchronously in order to maintain separation of individual charge packets and to prevent back transfer of charge (Fig. *b*). The simple device has the advantage of very few processing steps but the spacing of the gates is required to be very small, requiring extremely precise *photolithography, and extra circuits are required for the three-phase clocking.

Two-phase CCDs can operate with only two clock voltages. The shape of the surface potential profile under each electrode is made asymmetrical in order to provide built-in directionality of transfer. This may be achieved either by varying the thickness of the oxide layer under the electrode or by varying the doping level at the surface of the substrate by selective *ion implantation of impurities (Fig. *c*).

Overlapping-gate CCDs consist of an array of MOS capacitors with extra electrodes, known as *transfer gates,* formed between each capacitor and overlapping the storage gates (Fig. *d*). This structure, which is sometimes termed a *surface-charge transistor,* does not rely on the physical proximity of the storage electrodes in order to effect efficient charge transfer. Two-phase clocking may be used by connecting the storage gates and transfer gates in pairs, which effectively provides single capacitors with asymmetric potential wells. *Four-phase* clocking may be used, in which case the transfer gates effectively act as extra MOS capacitors, clocked independently, and the overlapping structure ensures coupling between adjacent capacitors. The storage electrodes are formed from highly doped *polysilicon and the transfer gates may be either metal or polysilicon. Three overlapping levels of polysilicon have also been used. Most CCDs are now formed as overlapping polysilicon gate two-phase n-channel devices (Fig. *d*) and have transfer efficiencies of about 99·99%. The disadvantages of the more complex processing are offset by the markedly enhanced performance and simpler clocking.

Transfer losses between storage sites in CCDs are caused by diffusion due to the mutual repulsion between charge carriers and by the existence of surface trapping states. The effect of diffusion diminishes relatively quickly as the amount of signal charge is reduced (by the diffusion). The effect of the surface states is substantially constant along the array. It can be reduced by using *bulk-channel CCDs* (BCCDs). A thin layer of opposite conductivity type is formed on the surface of the substrate either by epitaxy or ion implantation. The effect of this layer is to move the position of maximum potential below the surface (Fig. *e*) and thus prevent loss to surface traps. Trapping levels in the bulk material, however, cause some loss of charge but less than that due to the surface states.

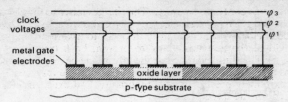

a Schematic cross section of simple n-channel CCD

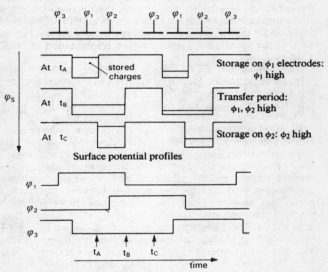

b 3-phase charge transfer showing transfer from ϕ_1 to ϕ_2 and relative time variation of clock voltages

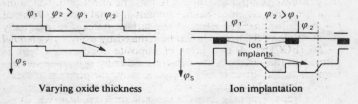

c Asymmetric potential wells

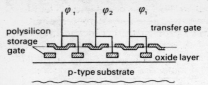

d Overlapping gate structure

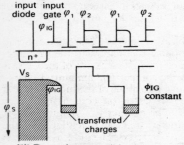

f(i) Dynamic current injection: charge transferred during ϕ_1 high

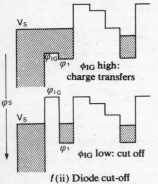

ϕ_{IG} high: charge transfers

ϕ_{IG} low: cut off

f(ii) Diode cut-off

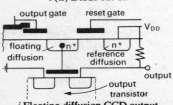

i Floating-diffusion CCD output

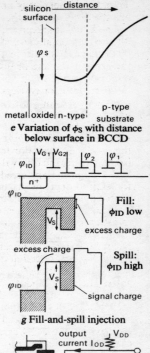

e Variation of ϕ_S with distance below surface in BCCD

g Fill-and-spill injection

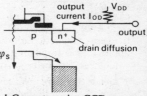

h Current-sensing CCD output

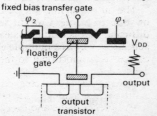

j Floating-gate amplifier CCD output

Transfer efficiencies of about 99·999% can be achieved. The BCCD has the disadvantage that an extra processing step is required.

The signal charge in a CCD is produced either by injection from a diffused region at one end of the array or, for imaging applications, by exposure to electromagnetic radiation. Charges are injected into the array using one of several methods. In each case the source of minority carriers is a source diffusion that acts essentially as an infinite source of carriers. The source-substrate junction is always reverse biased and forms an input diode. The three main methods of injection are dynamic current injection, diode cut-off, and fill-and-spill injection.

In *dynamic current injection* the input signal V_s is applied either to the source electrode or to an input gate adjacent to the gate electrode. During the interval in which the first electrode of the CCD is 'on', charge is transferred from the source. The amount of charge transferred depends on the relative potentials of the input gate and source and on the sampling interval, i.e. ϕ_1 'on' (Fig. f(i)).

Diode cut-off operates in a similar manner to dynamic current injection but the input gate is clocked to a low potential while the first CCD gate electrode is still on (Fig. f(ii)). The diode is effectively cut off before the injected charge is transferred.

Fill-and-spill injection requires two input gate electrodes. The signal voltage is applied as a potential difference between these two gates. The input diode is clocked with a voltage ϕ_{ID} so that when ϕ_{ID} is low charge is transferred to the second input gate up to ϕ_{ID}; ϕ_{ID} is then clocked high so that the excess charge returns to the source until a charge level, determined by V_s, remains (Fig. g). The first electrode of the CCD array is then clocked high so that the signal charge is transferred into the array.

The device is used for imaging by exposing an entire CCD array to electromagnetic radiation of either visible or near ultraviolet frequencies. Photo-generation of electron-hole pairs occurs at the electrodes; the majority carriers move into the bulk material and the minority carriers remain at the electrodes. The amount of charge is a function of the light intensity and the charges can then be transferred out sequentially to provide a video signal.

The output signal of a CCD is obtained either by *current sensing* or *voltage sensing*. Current sensing of the output employs a drain diffusion that is held at a steady potential (Fig. h). The CCD array acts effectively as a multiple gate MOS transistor in which the drain current is proportional to the charges transferred to the final gate electrode.

Voltage sensing is achieved by using a *floating-diffusion amplifier*. A diffusion is made in which the potential is allowed to float relative to a reference value and is determined by the charge signal reaching the final gate electrode (Fig. i). The floating diffusion is connected to the gate of an MOS transistor and the potential variations produce an output signal proportional to the signal variations. The floating diffusion is reset to the reference level using a reset gate between each signal pulse.

Nondestructive voltage sensing can be performed using a *floating-gate amplifier*. A gate electrode is allowed to float; the charges transferred produce image charges in this gate. The floating gate is connected to the gate of an MOS transistor to produce the output signal (Fig. *j*). *Distributed floating-gate amplifiers* employ several floating gate amplifiers along the array with the outputs connected in phase to provide an enhanced signal-to-noise ratio.

charge density (1) *Syn.* volume charge density. Symbol: ρ; unit: coulomb per metre cubed. The electric charge per unit volume of a medium or body.

(2) *Syn.* surface charge density. Symbol: σ; unit: coulomb per square metre. The electric charge per unit surface area of a medium or body.

charge storage *See* carrier storage.

charge-storage diode *Syns.* snap-off diode; snapback diode; step-recovery diode. A p-n junction diode in which *carrier storage is the major factor contributing to the operation of the device. The diode is designed so that most of the injected minority carriers under forward bias are stored near the junction and are immediately available for conduction when reverse bias is applied. When the diode is switched from forward to reverse bias the diode conducts in the reverse direction for a short time interval then the current is abruptly cut off when all the stored charges have been dispelled. The diode therefore remains in a low-impedance state until the cut-off occurs. The reverse voltage then builds up rapidly at a rate determined by the reverse junction capacitance and the external circuit. The cut-off occurs in the range of picoseconds and results in a fast-rising voltage wavefront that is rich in harmonics. The diode is therefore used as a harmonic generator or as a pulse former. Most charge-storage diodes are fabricated in silicon with relatively long minority-carrier lifetimes ranging from 0·5 to 5 microseconds. *Compare* fast-recovery diode.

charge-storage tube *See* storage tube.

charge-transfer device (CTD) A semiconductor device in which discrete packets of charge are transferred from one location to the next. Such devices can be used for the short-term storage of charge in a particular location provided that the storage time is short compared with the recombination time in the material. Several different types of charge-transfer device exist, the main classifications being *charge-coupled devices and *bucket-brigade devices. Applications of charge-transfer devices include short-term memory systems, *shift registers, and imaging systems. Information is usually only available for *serial access.

chart recorder *See* graphic instrument.

Chebishev filter *See* filter.

chemiluminescence *See* luminescence.

Child's law *See* thermionic valve.

chip *Syn.* die. A small piece of a single crystal of *semiconductor material containing either a single component or device or an *integrated circuit. Chips are commonly sliced from a much larger *wafer of substrate, in which many of the components or circuits have been produced in chequerboard pattern. Chips are not normally ready for use until suitably packaged.

chlorobenzene method *See* lift-off.

choke (1) *Syn.* choking coil. An *inductor used to present a relatively high impedance to alternating current. Audiofrequency and radiofrequency chokes are used in audio- and radiofrequency circuits, respectively. *Smoothing chokes* are used to reduce the amount of fluctuation in the outputs of rectifying circuits. Smoothing chokes whose impedance varies with the current passing through them are called *swinging chokes*.

(2) A groove cut to a depth of a quarter wavelength in the metal surface of a *waveguide to prevent the escape of microwave energy.

choke coupling *See* coupling.

choking coil *See* choke.

chopped impulse voltage *See* impulse voltage.

chopper *See* vibrator.

chopper amplifier An *amplifier that amplifies direct-current signals by first converting them into alternating current and then using normal a.c. amplifying techniques. The conversion is achieved using a system of *relays or a suitable *vibrator.

chrominance signal *See* colour television.

chrominance subcarrier *See* colour television.

chroming A method of producing a special effect on a video tape recording. A video tape is recorded normally and then rerecorded onto another tape. The recorded signals from the original tape have neighbouring pixels electronically aggregated to artificially reduce the resolution of the picture. Chosen sections of the picture may also have the colour electronically altered to give the desired effect.

chronotron An electronic device that measures the time interval between events. A pulse is initiated by each event and the time interval is determined by the position of the pulses along a transmission line.

chucking *See* compact disc system.

CIRC *Abbrev. for* cross interleaved Reed-Solomon code. *See* compact disc system.

circuit The combination of a number of electrical devices and conductors that, when connected together to form a conducting path, fulfil a desired function such as amplification, filtering, or oscillation. A circuit may consist of discrete components or may be an *integrated circuit. Some circuits, such as *charge-coupled devices, can only be produced in integrated form.

A circuit is *closed* when it forms a continuous path for current. When the path is broken, as by a switch, the circuit is *open*. Any constituent part of the circuit other than the interconnections is a *circuit element*.

circuit-breaker A device such as a contactor, switch, or tripping device that is used to *make or *break a circuit under normal or fault conditions. An unwanted arc is often produced as the circuit-breaker operates and this can be minimized using a *magnetic blow-out* device. This device is fitted to the circuit-breaker and produces a magnetic field in the neighbourhood of the arc, thus causing the pathlength of the arc to increase and thereby extinguishing it rapidly. When used for fault conditions an automatic break and manual make system is commonly used.

circuit diagram A diagram that represents the function and interconnections of a circuit. Each circuit element is represented by its appropriate graphical symbol. *See* Table 1, backmatter.

circuit element *See* circuit.

circuit-gap admittance *See* gap admittance.

circuit parameters *See* parameters.

circular scanning *See* scanning.

clamping diode *See* catching diode.

clamp-type mica capacitor *See* mica capacitor.

Clark cell A mercury-zinc standard cell with e.m.f. defined as 1·4345 volts at 15°C. It has been superseded by the *Weston standard cell.

class A amplifier A linear *amplifier in which output current flows over the whole of the input current cycle, i.e. the *angle of flow equals 2π. These amplifiers have low distortion but low efficiency. Distortion can occur with large-signal operation due to the device transfer characteristics becoming nonlinear (*see* diagram).

class AB amplifier A linear *amplifier in which the output current flows for more than half but less than the whole of the input cycle, i.e. the *angle of flow is between π and 2π. At low input-signal levels class AB amplifiers tend to operate as *class A and at high input-signal levels as *class B amplifiers.

class B amplifier A linear *amplifier operated so that the output current is cut off at zero input signal, i.e. the *angle of flow equals π, and a half-wave rectified output is produced. Two transistors are required in order to duplicate the input waveform successfully, each one conducting for half of the input cycle (*see* push-pull operation).

72

Class B amplifiers are highly efficient but suffer from crossover *distortion.

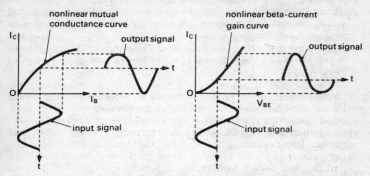

Distortion in class A amplifier with large-signal operation

class C amplifier A nonlinear *amplifier in which the output current flows for less than half the input cycle, i.e. the *angle of flow is less than π. Although more efficient than other types of amplifier, class C amplifiers introduce more *distortion.

class D amplifier An *amplifier operating by means of pulse-width modulation (*see* pulse modulation). The input signal produces a square wave modulated with respect to its *mark space ratio. *Push-pull switches are then operated by the modulated square wave so that one switch operates with a high input level and the other with a low input level. The resultant output current is proportional to the mark space ratio and hence to the input current.

In theory class D amplifiers are very highly efficient but they require an impractically high speed of switching to avoid distortion.

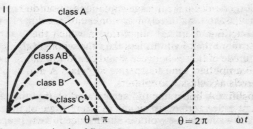

Angle of flow of output current for
different amplifier classes

clear *Syn.* reset. To restore a memory or storage device to a standard state, usually the zero state.

clipper *Syns.* voltage selector; peak limiter. *See* limiter.

clock An electronic device that generates periodic signals that are used to synchronize operations in a *computer or to monitor and measure properties of the circuits involved. The master frequency generated by a clock is the *clock frequency*. The regular pulses applied to the elements of a *logic circuit to effect logical operations are called *clock pulses*.

The use of clock pulses in order to drive any particular electronic circuit, device, or apparatus is known as *clocking* and the driven circuit, etc., is described as *clocked* or *synchronous*.

clocked circuit *See* clock.

clocked flip-flop *See* flip-flop.

clock frequency *See* clock.

clocking *See* clock.

clock pulse *See* clock.

closed circuit *See* circuit.

closed-circuit television A *television system, other than broadcast television, that forms a closed circuit between television camera and receiver. Closed-circuit television has many industrial and educational applications, for example in security systems.

cloud pulse *See* storage tube.

clutter Unwanted signals that are caused by noise, unwanted images, or echos and appear on the display of a *radar system.

CML *Abbrev. for* current mode logic. *See* emitter-coupled logic.

C/MOS (or **CMOS**) *See* complementary transistors.

C/MOS logic circuit *See* MOS logic circuit.

coarse scanning *See* radar; scanning.

coated cathode *Syn.* Wehnelt cathode. *See* thermionic cathode.

coax *Short for* coaxial cable.

coaxial cable A cable formed from two or more coaxial cylindrical conductors insulated from each other. The outermost conductor is often earthed. Coaxial cables are frequently used to transmit high-frequency signals, as in television or radio, since they produce no external fields and are not affected by them. A terminal to which a coaxial cable may be connected is a *coaxial terminal*.

coaxial pair *See* pair.

coaxial resonator *See* waveguide.

coaxial terminal *See* coaxial cable.

Cockcroft–Walton generator (or **accelerator**) A high-voltage direct-current *accelerator consisting of cascaded rectifier circuits and capacitors, to which a low-voltage alternating current is applied.

code beacon *See* beacon.

coder *See* pulse modulation.

coercive force *See* magnetic hysteresis.

coherent detection *Syn. for* synchronous detection. *See* suppressed-carrier transmission.

coherent oscillator *See* coherent radiation.

coherent radiation Radiation in which the waves are in *phase both spatially and temporally. A *coherent oscillator* is one that produces very pure well-defined oscillations, as in a *laser.

coil A conductor or conductors wound in a series of turns. Coils are used to form inductors or the windings of transformers and motors.

coil aerial *See* loop aerial.

coiled coil *See* incandescent lamp.

coil loading *See* transmission line.

coincidence circuit A circuit with two or more input terminals that produces an output signal only when an input signal is received by each input either simultaneously or within a specified time interval, Δt. An electronic counter incorporating such a circuit records events occurring together and is termed a *coincidence counter*. Such a device, if fed by the outputs of two radiation detectors (*see* diagram), may be used to determine the direction of radiation or to detect cosmic-ray showers. *See also* anticoincidence circuit.

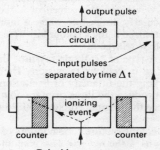

Coincidence counter

cold cathode A cathode from which *field emission of electrons occurs at ambient temperatures due to the application of a sufficiently high voltage gradient at its surface. An electron tube containing a cold cathode is a *cold-cathode tube.*

cold emission *See* field emission.

collector (1) *Short for* collector region. The region in a bipolar junction transistor into which *carriers flow from the *base through the collector junction. The electrode attached to this region is the *collector electrode. See also* transistor; semiconductor.

(2) *Short for* collector electrode.

collector-current multiplication factor The ratio of enhanced current flow in the collector of a transistor as a result of minority carriers entering from the base with sufficient energy to create extra electron-hole pairs, to the current carried by minority carriers at the collector voltage. This ratio is usually unity but under high field conditions increases rapidly as the *avalanche breakdown voltage is reached.

collector efficiency *See* common-base forward-current transfer ratio; transistor.

collector electrode *See* collector.

collector region *See* collector.

colour breakup *See* tearing.

colour burst *Syn.* burst signal. *See* colour television.

colour cell *See* colour picture tube.

colour code A method of marking electronic parts, such as resistors, with information for the user. The value, tolerance, voltage rating, and any special characteristic of the component may be indicated using coloured bands or dots painted on it. *See* Table 2, backmatter.

colour coder *See* colour television.

colour decoder *See* colour television.

colour flicker *See* flicker.

colour fringing The presence of unwanted coloured fringes round the edges of objects viewed on the screen of a *colour picture tube. Fringing is particularly noticeable when a monochrome signal is received. It is minimized using a colour killer (*see* colour television).

colour killer *See* colour television.

colour picture tube A type of *cathode-ray tube designed to produce the coloured image in *colour television. The coloured image is produced by varying the intensity of excitation of three different *phosphors that produce the three primary colours red, green, and blue and reproduce the original colours of the image by an additive colour process.

The three-gun colour picture tube consists of a configuration of three *electron guns – the *red gun, blue gun,* and *green gun* – that are tilted slightly so that the electron beams intersect just in front of the screen. Each electron beam has an individual electron lens system of focusing and is directed towards one of the three sets of colour phosphors. There are several different types of colour picture tube, the main differences being in the configuration of electron guns and arrangement of the phosphors on the screen.

One main type is the *dot matrix tube,* an example of which is the *colourtron.* It has a triangular arrangement of electron guns and has the phosphors arranged as triangular sets of coloured dots (Fig. *a*). A metal *shadow mask* is placed directly behind the screen, in the plane of intersection of the electron beams, to ensure that each beam hits the correct phosphor (Fig. *b*). The mask acts as a physical

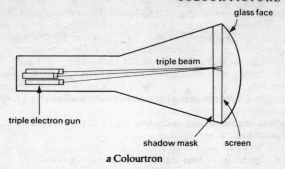

a Colourtron

barrier to the beams as they progress from one location to the next and minimizes the generation of spurious colours by excitation of the wrong phosphor.

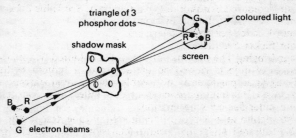

b Light production in colourtron

The other main type of three-gun colour picture tube is the *slot matrix tube,* which has the electron guns arranged in a horizontal line. The phosphors are arranged as vertical stripes on the screen (Fig. *c*) and the shadow mask is replaced by an *aperture grille* of vertical wires. This type has advantages in focusing the beams but has a smaller field of view than the triangular arrangement of electron guns.

The *Trinitron* is a type of colour picture tube that has certain advantages over three-gun tubes. It has a single electron gun with three cathodes aligned horizontally, an aperture grille, and vertically striped phosphors. The cathodes are tilted towards the centre so that the electron beams intersect twice, once within the electron lens focusing system and once at the aperture grille (Fig. *d*). This allows a single electron lens system to be used for all three beams, thus needing fewer components. The system is therefore much lighter and cheaper than the three-gun tubes. The effective diameter of the elec-

77

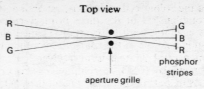

c Horizontal arrangement of electron guns

tron lens is greater and sharper focusing of the three beams is therefore possible.

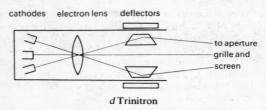

d Trinitron

Misconvergence of the electron beams as they traverse the screen increases with the distance from the centre of the screen. In the horizontal arrangement of electron guns misconvergence only occurs along the line direction rather than in both line and field directions (*see* television) as occurs in the triangular configuration. The three-cathode arrangement of the Trinitron however allows a greater lens aperture than in the three-gun arrangement. The diameter of the electron tube for a given screen size is also reduced in the Trinitron.

Cathode-ray tubes of the Trinitron type can be used for applications where multiple electron beams are required; the angle of the tube may be increased to give a relatively wide-angle colour picture tube and hence a relative reduction in the overall size of colour television receivers.

The colour quality and definition of the picture on the screen of a picture tube depends greatly on the dynamic convergence of the beams and the size of the *colour cell*. Scanning of the three electron beams across the screen is effected by a system of deflection coils to which *sawtooth waveforms are applied in synchronism with the transmitter, the flyback signal being blanked. Extra *convergence coils* are frequently used to ensure the correct convergence of the beams at the shadow mask or aperture grille. A system of *dynamic focusing* is also used in which the voltage applied to the convergence coils is varied automatically according to the relative position of the spots on the screen; this minimizes misconvergence.

78

The size of the colour cell is the smallest area on the screen that includes a complete set of the three primary colours. A smaller colour cell is available with horizontally aligned electron guns or cathodes than is possible in the triangular configuration.

colour saturation control *See* television receiver.

colour separation overlay (CSO) A technique used in colour television for superimposing part of one scene on another. When a particular colour, such as blue, occurs in one scene viewed by a camera, the output of another camera filming a different scene is automatically switched in to replace the areas of the chosen colour in the original picture. All other colours are transmitted normally from the first camera. The technique is widely used for achieving special effects.

colour television A television system that produces a coloured image on the screen of a *colour picture tube. An additive colour reproduction process is used on the screen whereby three primary colours – red, green, and blue – are combined by eye to produce a wide variety of colours. The apparent colour of the image depends on the relative intensities of the three primary colours and a properly adjusted colour *television receiver approximates the original colours of the transmitted scene.

Three separate video signals are produced by a colour *television camera. These signals are used to produce a composite signal that is broadcast and is received by a colour receiver. The receiver extracts the original video information from the composite signal and modulates the intensities of the three electron beams of the colour picture tube in order to excite the appropriate red, green, or blue phosphors on the screen.

The composite signal transmitted in colour television needs to be compatible with the large number of black and white (monochrome) receivers in use. It is therefore composed of two parts: the *luminance signal* and the *chrominance signal.* The luminance signal contains brightness information. It is obtained by combining the outputs of the three colour channels and is used for amplitude modulation of the main picture carrier frequency. This produces the black and white image. The colour information is contained in the chrominance signal, which is transmitted using a subcarrier wave at a frequency chosen to cause the least interference on a monochrome set. The chrominance signal is obtained by combining, in a *colour coder* circuit, fixed specified fractions of the separate video signals into sum and difference signals. Two *quadrature components of the chrominance signal are produced and used for amplitude modulation of the *chrominance subcarriers.* The subcarriers are suppressed at transmission. The original information is extracted from the chrominance signal in the *colour decoder* in the receiver. The *frequency overlap* is the range of transmitted frequencies that are common to both the luminance and chrominance channels.

The *composite colour signal* contains the luminance and chrominance signals; it also contains synchronizing pulses for line and field scans as well as a *colour burst* signal. The colour burst establishes a phase and amplitude reference signal that is used to demodulate the chrominance signal. In colour receivers the chrominance circuits are disabled by the *colour killer* when a black and white signal is being received. This ensures that only luminance information reaches the tube and prevents colour fringing on the image.

See also PAL; SECAM; television; colour picture tube; television camera; camera tube.

colourtron *See* colour picture tube.

Colpitt's oscillator *See* oscillator.

coma *See* distortion.

combination-tone distortion *Syn. for* intermodulation distortion. *See* distortion.

common-anode connection *See* common-collector connection.

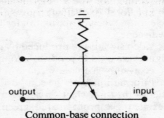

Common-base connection

common-base connection *Syn.* grounded-base connection. A method of operating a *transistor in which the *base is common to both the input and output circuits and is usually earthed (*see* diagram). The *emitter is used as the input terminal and the *collector as the output terminal. This type of connection is commonly used as a voltage amplifier stage.

The equivalent connection for a *thermionic valve is *common-grid connection* and for a field-effect transistor is common-gate connection.

common-base current gain *See* common-base forward-current transfer ratio; transistor.

common-base forward-current transfer ratio *Syns.* collector efficiency; alpha current factor; common-base current gain. Symbol: α or h_{fb}. The ratio of the collector current I_c to the emitter current I_e of a bipolar junction transistor operated in common-base connection. *See also* transistor.

common branch *Syn. for* mutual branch. *See* network.

common-cathode connection *See* common-emitter connection.

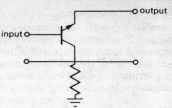

Common-collector connection

common-collector connection *Syn.* grounded-collector connection. A method of operating a *transistor in which the *collector is common to both the input and output circuits and is usually earthed (*see* diagram). The *base is used as the input terminal and the *emitter as the output terminal. This type of connection is used for the *emitter follower.

The equivalent connection for a *thermionic valve is *common-anode connection* and for a field-effect transistor is common-drain connection.

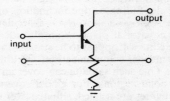

Common-emitter connection

common-emitter connection *Syn.* grounded-emitter connection. A method of operating a *transistor in which the *emitter is common to both the input and output circuits and is usually earthed (*see* diagram). The *base is used as the input terminal and the *collector as the output terminal. This type of connection is used for power amplification with a nonsaturated transistor and for switching with the transistor in saturation.

The equivalent connection for a *thermionic valve is *common-cathode connection* and for a field-effect transistor is common-source connection.

common-emitter forward-current transfer ratio *See* beta-current gain factor.

common-grid connection *See* common-base connection.

common-impedance coupling *See* coupling.

common-mode rejection ratio *See* differential amplifier.

communication *See* telecommunication.

communications satellite (comsat) An artificial unmanned satellite in earth orbit that provides high-capacity communication links between widely separated locations on earth. International telephone services and the exchange of live television programmes and news are achieved by transmitting microwave signals, suitably modulated, from an earth station to an orbiting satellite and back to another earth location.

The first satellites, including the large metallized balloons *Echo 1* and *Echo 2,* were *passive* systems; they simply reflected or scattered the microwave beam back to another earth station. Present-day systems use *active satellites,* in which the signal is amplified and its frequency changed by a *transponder before it is retransmitted to earth. The first successful active satellites included the *Telstar* and *Relay* satellites (1962–64). They were in relatively low orbits and were only in line of sight of any two earth stations for a short period each day and then had to be tracked as they moved across the sky.

A satellite in a *geosynchronous orbit* revolves from west to east in an orbital period equal to that of the earth's rotation (about 23 hours 56 minutes). It traces a figure-of-eight pattern in the sky with a latitude range depending on the inclination of the orbit to the equatorial plane. A *geostationary orbit* is a circular synchronous orbit lying in the equatorial plane at an altitude of 35 790 km. A geostationary satellite will appear to be stationary to an observer on earth. Most commercial communications satellites now lie in geostationary orbits. One such satellite can cover an extensive surface area, excluding polar regions. For global coverage at least three geostationary satellites are needed, situated over the equatorial Atlantic, Indian, and Pacific Oceans. For long east-west links at high northerly latitudes the Russians have used satellites in highly elliptical 12 hour orbits.

The International Telecommunications Satellite Consortium (*Intelsat*), established in 1964, is responsible for all non-Soviet international nonmilitary satellite communications. *Early Bird,* later renamed *Intelsat I* was launched in 1965 and had 240 voice circuits. The much heavier *Intelsat IVa,* first launched 1975, have up to 6000 circuits and a much longer life.

The *radio window covers the frequency range from 15 megahertz to 50 gigahertz, with the optimum transmission in the range 1 to 10 gigahertz. The frequency bands used by Intelsat systems are from 5·925–6·425 GHz for the earth-to-satellite path and from 3·7–4·2 GHz for the satellite-to-earth path.

*Solar cells form the primary power supply for satellites with a back-up of batteries for use during the brief periods of solar eclipses. The operational lifetime of a modern satellite should be at least five years. Redundancy of all the essential subsystems is required although it is not necessary for all the components when there is a large number present, as with the transponders.

A geostationary satellite can be maintained in a stable attitude by spinning it about an axis parallel to the earth's axis. In Intelsat IV the high-gain aerials are mounted on a platform that rotates about the spin axis but in the opposite direction. The aerials then appear stationary with respect to the earth, at their desired orientation. Parabolic reflectors allow spot-beam transmission to regions of limited size, such as W Europe, which have high communication traffic densities. Transmission and reception over a hemisphere is effected using conical horn aerials.

The earth stations must be situated some distance from terrestrial microwave relay systems to avoid radio interference. The aerials of the stations in the Intelsat system, such as that at Goonhilly in Cornwall, have apertures of 25 to 30 metres. The aerials should be steerable to compensate for perturbations of the orbits caused by gravitational effects of moon and sun. Present-day systems provide simultaneous *multiple access* to one satellite from a large number of earth stations within one coverage zone. This is achieved by *time-division multiplexing or *frequency-division multiplexing. In the former case a station does not share the transponder power with other stations and can operate at close to saturation where the transponder is most efficient.

commutation The transfer of current from one path in a circuit to a different one. The transfer is often done in a periodic and automatic manner.

commutation switch A *switch that automatically switches from open to closed and vice versa repetitively. It is most often used for *pulse modulation.

commutator *See* motor.

compact disc system (CD system) A sound reproduction system that uses light to detect audio signals made by *digital recording* on a disc. It combines robustness and the ability to be mass-produced relatively cheaply with very sophisticated control and detection systems. It differs from other sound reproduction systems in that there is no physical contact between the pick-up and the recording, thus minimizing wear; the information layer is buried below the surface of the disc, which minimizes errors in the sound reproduction due to dust or other marks on the surface, and the discs are much smaller than conventional vinyl discs – hence the term compact. Approximately 60 CD tracks would fit into one groove of a vinyl record. The information is retrieved from only one side of the disc – the 'back' or nonlabelled side – and the track spirals outwards from the centre, rather than inwards as with conventional systems. The compact disc uses a *constant linear velocity* of track relative to pick-up so the rotational speed is a function of the radius of the track, and varies as the pick-up moves across the disc. When a disc is loaded into the player it must be secured to the spindle. This process is termed *chucking*. Some types of player use a motorized drawer, others operate the clamp when the lid is lowered.

The essential component of the pick-up is a small low-power *injection laser continuously emitting coherent light, which is focused as a small spot onto the reflecting surface of the disc. The laser is required to be disabled whenever the mechanism is accessed. Small bumps in the mirrored surface, of height one quarter of the wavelength of the laser light, result in destructive interference between light reflected from the raised portion and that reflected from the normal surface, with a consequent attenuation of the reflected light. Very small bumps can be detected by monitoring the reflected light output; the bumps can be mass-produced by pressing in a similar method to producing the grooves in an ordinary vinyl recording. The audio information is sampled at a rate of 44.1 kHz, and quantized into 16-bit *words, using an *analog/digital converter. The coded digital data is recorded on the disc as a series of small bumps.

Reflected light from the disc is detected on playback by a *phototransistor to produce an electric signal corresponding to the recorded information. These signals are then converted back into audio signals. The track size is limited only by the behaviour of light, and the disc can therefore be much smaller than a conventional LP containing the same amount of audio information. A standard size of a maximum of 75 minutes playing time has been agreed by the manufacturers.

The essential parts of the pick-up system are shown in Fig. *a*. They consist of a laser and its power control system, a method of focusing the laser light onto the disc, a means of separating the reflected light from the laser light, and a phototransistor to detect the output and convert it to an electric signal. In order to maintain high-quality sound output very sophisticated error control systems are required to ensure that focusing and tracking integrity are maintained, and that the disc is rotated at the correct speed.

The resolution of the pick-up depends critically on the spot size: disc warp and irregularities in the thickness will cause an out of focus condition with resulting loss of sound quality and *crosstalk from adjacent tracks. A *focus servo system* is used to move the lens along the optical axis to keep the spot in focus. The *focus error signal* is generated using one of two main methods. The cylindrical lens method (Fig. *b*.) has a cylindrical lens placed between the beam splitter and the photodetector. The image reaching the sensor will be circular only when the focus is correct, otherwise it becomes an ellipse whose aspect ratio changes as a function of the state of focus. The sensor is split into four quadrants, connected as shown. The focus error signal is generated from the difference between the out-

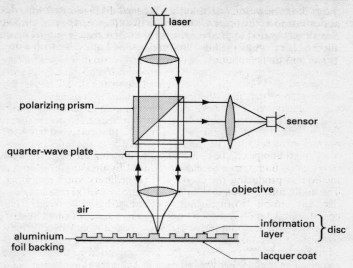

a Essential components of the optical pick-up of a CD system

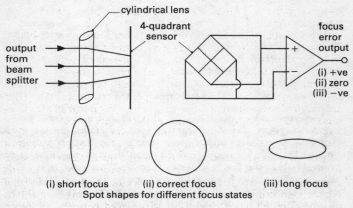

b Cylindrical lens method

puts, and the data signal is the sum of all four outputs. The knife edge and dual prism methods are the second means of generating the focus error signal. They also require split sensors, mounted be-

85

yond the focal point. In the knife edge method (Fig. *c*) a knife edge is positioned at the point of correct focus and the outputs of the two sensors compared to produce the focus error signal. The dual prism method is essentially similar but replaces the knife edge with a dual prism and three sensors.

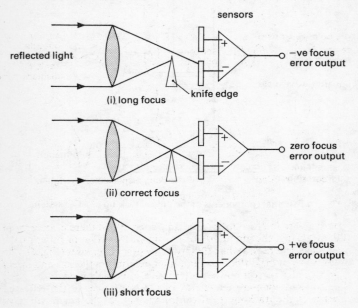

c Knife edge method

Accurate tracking of the beam is also required, and a *track-following servo system* is used to keep the spot centralized on the track. Tracking errors arise from various sources: the track separation is smaller than the accuracy to which either the player spindle or the central hole in the disc can be manufactured; a warped disc will be tilted relative to the beam at the surface and the apparent position of the track relative to the pick-up will constantly change as the disc rotates; external forces outside the CD player can induce vibrations that tend to disturb the tracking.

One method of deriving a *tracking error signal* is the three spot method in which two additional light beams are focused onto the disc track, offset one to each side of the centre line. Two extra photosensors detect the reflected light from the side spots, and the

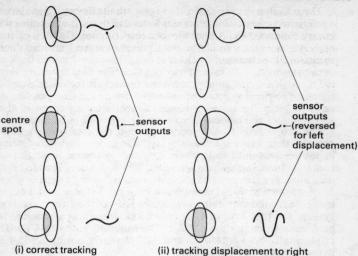

(i) correct tracking (ii) tracking displacement to right

d Three spot method

outputs are fed to a differential amplifier. When there is an error in the tracking, the amplitude of the signal from one or other of the side spots increases while that from the other decreases (Fig. *d*). The output from the differential amplifier determines the size and direction of the correction. An alternative approach is to use a split sensor and compare the energy of the two halves of the reflected beam. If the tracking is off-centre one half of the beam will have more energy than the other.

During the recording process, the audio signal is sampled and a form of *pulse code modulation is used to convert the samples into a coded form which modulates a high-frequency clock pulse. The clock operates at 4.3218 MHz. The majority of compact disc systems use *eight to fourteen modulation* (*EFM*) in which any combination of eight real data bits is uniquely described by a pattern of 14 *channel bits*. A further three *packing bits* are interposed between each pattern to separate them. The digital modulation code produced is known as the *channel code*. The transitions between ones and zeros of the channel code are used to produce bump edges in the recorded data. The bump edges are detected by the optical system to produce corresponding transitions in the replay signal. The replay signal must then be accurately decoded to produce the audio output.

87

The first step in the process is to compare the detected signal with a reference voltage. This process is termed *slicing*. This recreates the binary channel code. A phase-locked loop running at the clock frequency counts the number of clock pulses between transitions and recreates the patterns of 14 channel bits. These are decoded back to data bytes using a ROM or array of gates. The data bytes are then fed through a digital/analog converter to recreate the original audio signals. The packing bits are used to determine the start position of each 14 bit pattern, and a regular synchronizing pattern is also added to lock the readout circuits to the symbol boundaries.

The actual layout of the coded data on the disc is much more complex than a simple sequential layout in order to allow for errors in the readout data to be corrected, and to reduce noise due to contamination and surface scratches. The audio data is coded into data blocks or frames of 33 patterns, each following a synchronization pattern. The audio samples use 24 of the 33 bytes, and 8 of the bytes are redundancy bytes forming the basis of the error correction system; the first byte of each data block is used as a subcode to produce a running time display. The sampling rate of 44.1 kHz, producing 16-bit words in left and right channels, results in 176.4 K bytes of audio data per second.

The error correction system has to deal with errors due to scratches, which can affect many bits of code (*burst errors*), and random errors caused by imperfect pressing of a bump edge. This latter can result in the conversion of one pattern into another and therefore up to 8 bits of data can be in error. Extra bits are added to the coded information (*redundancy*) and can be used to correct damaged data bits when playing back. A *code word* consists of the total of data and redundant bits. The value of the redundant bits is calculated from the data itself according to a code known as the *Reed-Solomon code*. The resulting code words are then interleaved within each data frame to reduce the effect of burst errors; a large error causes slight damage to many code words rather than severe damage to one. This system is known as *cross interleaved Reed-Solomon code* (CIRC). The Reed-Solomon codes used to correct errors are based exclusively on the principle of finite elemental fields developed by Galois, and are a very powerful tool. Error correction is necessary because the effect of digital errors results in a sound rather like vehicle ignition interference in radio reception and is unacceptable to a listener. The development of compact disc systems has always taken into account that the finished product should not require any particular handling problems. Where data corruption is so bad that the error correction cannot cope, the system contains muting circuits that operate to reduce the gain of the player.

The CD player has various memory requirements for the decoding of the data: these are timebase correction, de-interleaving, temporary storage while correctors are computed, and mute delay. These are normally combined in a single *RAM. The sophisticated error correction systems have been largely responsible for the success of the CD players in the domestic environment since their introduction in 1983.

compander *See* volume compressor.

comparator A circuit, such as a *differential amplifier, that compares two inputs and produces an output that is a function of the result of the comparison.

compensated intrinsic semiconductor *See* semiconductor.

compensating leads An extra pair of leads that are used in conjunction with a sensitive measuring instrument connected to a *bridge circuit. The leads are identical to those that connect the measuring device, such as a *resistance thermometer, to the bridge and run alongside them. They are shorted at the instrument end and connected in series with the balancing element of the bridge circuit. Any resistance in the working leads is thus balanced by the compensating leads and resistance changes that occur in the working leads due to ambient conditions are also produced in the compensating leads.

compiler *See* programming language.

complementary transistors A pair of transistors of opposite type used together. A pair of n-p-n and p-n-p bipolar *transistors can be used to achieve *push-pull operation, frequently employed in *class B amplification.

Complementary MOS *field-effect transistors (*abbrev.* COS-MOS, CMOS, or C/MOS) are used for low-dissipation logic circuits. Such devices are used in conjunction with other similar stages all having essentially capacitive input impedance and therefore zero d.c. current flow. In the basic inverter (*see* diagram) the p-channel device conducts when the input is low and the n-channel device conducts when the input is high, thus giving an output that is inverted.

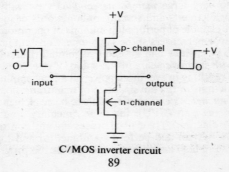

C/MOS inverter circuit

complex impedance *See* impedance.

component A discrete packaged electronic element, such as a resistor, that performs one electrical function. Primarily active electronic elements are usually termed *devices and passive ones are *component parts.

component part (1) A discrete electronic element, usually passive.

(2) An independent physical body that contains the realization of an electronic property or properties and cannot be practicably reduced or subdivided without destroying its function.

composite cable *See* cable.

composite colour signal *See* colour television.

composite conductor A conductor formed from strands of different metals, such as copper and steel.

composition resistor Usually a carbon-composition resistor. *See* resistor.

compound magnet *See* permanent magnet.

compound modulation *Syn. for* multiple modulation. *See* modulation.

compressor *Short for* volume compressor.

computer Any automatic device for the processing of information received in a prescribed and acceptable form according to a set of instructions. The instructions and information are stored in a *memory. The most widely used and most versatile of these devices is the *digital computer,* which can manipulate large amounts of information at high speed. Its input must be discrete rather than continuous and may consist of combinations of numbers, characters, and special symbols, the instructions – the *program – being written in an appropriate *programming language. The information is represented internally in *binary notation.

The development of microelectronics has allowed the corresponding development of a wide range of computers varying in size and complexity according to the required applications. Modern computers range from the *microcomputer* that contains typically a few thousand *logic circuits and a few hundred thousand *words of memory to very large *mainframe computers* typically containing 100 000 logic circuits and several million words of memory. Such devices are capable of performing millions of operations per second and can serve many users at the same time. Constant improvements in *packing densities and subsequent miniaturization of circuits, coupled with improvements in the speed of operation of the logic circuits, is resulting in ever more powerful microcomputers and dramatic reductions in the physical size of mainframe computers.

Most *computer systems* consist of three basic elements: the *central processing unit (CPU), the main *memory, and *peripheral devices (*see* diagram). The central processing unit controls the operation of the system and performs arithmetical and logical operations on the data. The main memory stores the program and the

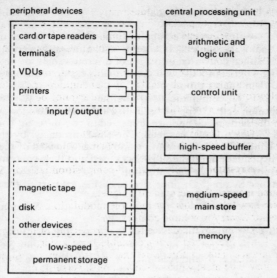

peripheral devices   central processing unit

card or tape readers   □   arithmetic and
terminals   □   logic unit
VDUs   □
printers   □   control unit

                 input / output

                                   high-speed buffer

magnetic tape   □   medium-speed
disk   □   main store
other devices   □

              low-speed   memory
          permanent storage

Elements of a computer system

data in units of *bytes or words, each of which has a unique *ad-dress, so that they may be retrieved quickly by the CPU. A *buffer memory* is employed by larger systems: it interacts directly with the CPU and transfers information at extremely high speed. The infor-mation currently in active use is held in the buffer. The peripheral devices perform *input/output and permanent storage of informa-tion. A complete computer system consists of the *hardware – the electronic and other devices – and complementary *software – the set of programs and data.

The *analog computer* is a device that accepts data as a continuous-ly varying quantity rather than as a set of discrete items required by the digital computer. It is used in scientific experiments, simulation processes, and in the control of industrial processes where a con-stantly varying quantity can be monitored. A problem is solved by physical analogy, usually electrical. The magnitudes of the variables in an equation are represented by voltages fed to circuit elements connected in such a way (*see* patching) that the input voltages inter-act according to the same equation as the original variables. The output voltage is then proportional to the numerical solution of the problem. It can solve or analyse many types of differential equations.

The *hybrid computer* combines some of the properties of digital and analog computers. It accepts a continuously varying input, which is then converted into a set of discrete values for digital processing. Digital processing is considerably faster and more sophisticated than analog processing.

comsat *Short for* communications satellite.

concentration cell A type of *cell in which the electrodes are made from the same metal and are immersed in two different concentrations of one salt of the metal. The e.m.f. produced by the metal dissolving in the weaker solution and being deposited by the stronger is dependent upon the substances used and their concentrations.

condenser *Obsolete syn. for* capacitor.

conductance Symbol: G; unit: siemens. The real part of the *admittance, *Y,* which is given by

$$Y = G + iB$$

where *B* is the *susceptance. For a direct-current circuit the conductance is the reciprocal of the resistance. For a circuit containing both resistance, *R,* and reactance, *X,* the conductance is given by

$$G = R/(R^2 + X^2)$$

conductimeter *Syn. for* conductivity meter. *See* conductivity.

conduction The transmission of electric (or heat) energy through a substance that does not itself move. In electrical conductors, such as metals, it entails the migration of *electrons; in gases and solids it results from the migration of *ions. *See also* hole conduction.

conduction band The band of energies of electrons in a solid in which the electrons can move freely under the influence of an electric field. In metals the conduction band is the highest occupied band. In *semiconductors it is a vacant band of higher energy than the *valence band. *See* energy bands.

conduction current *See* current.

conduction electrons Electrons in the *conduction band of a solid. *See* energy bands.

conductivity Symbol: κ or σ; unit: siemens per metre. The reciprocal of *resistivity. It is given by the ratio of current density, *j,* to electric field strength, *E*:

$$\kappa = j/E$$

This is an expression of *Ohms law. A *conductivity meter* is an instrument used for measuring conductivity.

conductivity meter *Syn.* conductimeter. *See* conductivity.

conductivity modulation *See* semiconductor.

conductor A material that offers a low resistance to the passage of electric current: when a potential difference is applied across it a relatively large *current flows.

cone loudspeaker *See* loudspeaker.

confusion reflector A device used to produce false signals with *radar. Strips of paper or metal foil may be used: long strips of metal foil are called *rope* or *window*.

conical scanning *See* scanning.

conjugate branches *See* network.

conjugate impedances Two *impedances given by

$$Z_1 = R + iX$$
$$Z_2 = R - iX$$

in which the resistance components, R, are equal and the reactance components, X, are equal in magnitude but opposite in sign.

constantan An alloy of copper (50 to 60 per cent) and nickel that has a very low *temperature coefficient of resistance and a comparatively high resistance. It is used with copper, silver, etc., to form *thermo-couples and also for precision wire-wound *resistors.

constant-current source A circuit that ideally has an infinitely high output *impedance so that the output current is independent of voltage. In practice sufficiently high output impedances are only achieved for a limited range of output voltages.

constant failure-rate period *See* failure rate.

constant-R network *See* quadripole.

constant-voltage source A source of voltage that produces a substantially constant value of voltage independently of the current supplied by it. An ideal voltage source has an internal impedance of zero.

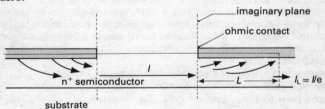

Contact resistance and transfer length

contact The bringing together of two conductors so that current may flow. The resistance at the surface of contact is the *contact resistance*. In the case of an *ohmic contact between a metal and a semiconductor the *specific contact resistance* is the contact resistance of a unit area for current flow perpendicular to the contact. In

93

the case of a planar configuration where the final direction of current flow is parallel to the plane of the metal (*see* diagram), the contact resistance is defined to be the resistance between the metal and an imaginary plane at the edge of and perpendicular to the metal, and the *transfer length*, *L*, is the distance from the edge of the metal at which the current in the semiconductor has fallen to $1/e$ of its original value.

If the conductors are made from two different materials a difference of potential will arise when they are placed in contact. This *contact potential* results from a difference in the *work functions of the two materials and is usually of the order of a few tenths of a volt. If the contact is made between two semiconductors of different polarities or between a metal and a semiconductor, a built-in field will be produced with an associated contact resistance to current flow. *See also* p-n junction; Schottky barrier.

contact lithography *See* photolithography.

contactor A type of switch used for the automatic making and breaking of a circuit and designed for frequent use.

contact potential *See* contact.

contact resistance *See* contact.

continuous duty *See* duty.

continuous loading *See* transmission line.

continuous-wave radar *See* radar.

continuous X-rays *See* X-rays.

contrast control *See* television receiver.

control characteristic *See* thyratron.

control circuit *See* transductor.

control electrode An electrode to which a signal is applied in order to produce changes in the currents of one or more of the other electrodes. In a bipolar transistor with *common-emitter connection the base electrode is the control electrode; the gate electrode is the control electrode of a *field-effect transistor; in a *thermionic valve it is the *control grid;* in a *cathode-ray tube it is the *modulator electrode.

control grid *See* control electrode; thermionic valve.

controlled-carrier modulation *See* floating-carrier modulation.

control ratio *Syn.* grid control ratio. *See* thyratron.

control unit *See* central processing unit.

convergence In a multibeam electron tube, such as a *colour picture tube, the intersection of the beams at a specified point. Convergence may be achieved electrostatically using a *convergence electrode* or electromagnetically using a *convergence magnet.* When scanning of the beams across the screen of the tube is carried out, the surface generated by the point of intersection of two or more of the electron beams is the *convergence surface.*

convergence coils *See* colour picture tube.

conversion conductance *See* mixer.

conversion gain ratio *See* frequency changer.

conversion transducer *See* frequency changer.

conversion voltage gain *See* frequency changer.

converter (1) A device for converting alternating current to direct current or vice versa.

(2) A device that changes the frequency of a signal; a *frequency changer.

(3) A device, such as an impedance converter, that has different electrical properties at its input and output and may be used to couple dissimilar circuits.

(4) A device, such as a compiler, that changes an information code.

(5) A *transducer that converts energy of one type, such as sound waves or electromagnetic radiation, into electrical energy.

Coolidge tube An early form of evacuated *X-ray tube.

Cooper pair *See* superconductivity.

coplanar process *Tradenames:* Planox; Isoplanar process; Locos. A technique used during the manufacture of LSI *MOS *integrated circuits and some forms of LSI *bipolar integrated circuits, such as I²L. Regions of relatively thick silicon dioxide are used in order to isolate device areas and to prevent *spurious MOST formation. The coplanar process was developed in order to minimize the vertical projection of the oxide layer. A layer of silicon nitride is deposited on the surface of the silicon wafer and is etched to expose the regions of the surface where thick oxide is required. As oxidation takes place, the effective silicon surface moves downwards and is replaced by a thicker layer of silicon dioxide, so that approximately one third of the oxide is below the original exposed surface level (Fig. *a*). Oxidation may be preceded by etching of the exposed silicon surface so that the final oxide surface is at the same level as the original substrate (Fig. *b*). The silicon nitride is then removed from the rest of the surface and the integrated circuits are formed using normal *planar-process technology.

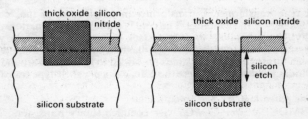

a Coplanar process with nitride etch   *b* Nitride plus silicon etch

copper loss *Syn. for* I^{2}R loss. *See* heating effect of a current; dissipation.

core (1) The ferromagnetic portion of the magnetic circuit of an electromagnetic device. A simple *ferrite core* is a solid piece of ferromagnetic material suitably shaped into a cylinder, torroid, etc. A *laminated core* is composed of laminations of ferromagnetic material insulated from each other; *eddy currents are thus reduced. A *wound core* is one constructed from strips of ferromagnetic material wound spirally in layers.

(2) *Syn.* core store. A static nonvolatile *random-access memory that consists of an array of rings of *ferrite material strung on a grid of wires. The individual rings – *ferrite cores* – are of the order of a millimetre in diameter. Information is stored in the array by causing the direction of magnetization of a core to be either clockwise or anticlockwise, corresponding to the binary digits one or zero. Information is input and output by electronic means using the wire grid to read or write. Core memory has now been superseded by semiconductor memory.

core loss *Syn.* iron loss. The total energy dissipation in the ferromagnetic *core of an inductor or transformer. The energy loss is mainly due to *eddy currents and hysteresis loss (*see* magnetic hysteresis) in the core.

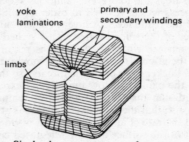

yoke laminations

primary and secondary windings

limbs

Single-phase core-type transformer

core-type transformer A transformer in which most of the *core is enclosed by the windings. The core is made from laminations; usually the yoke is built up from a stack of laminations and the windings are formed around this (*see* diagram). Once the windings have been formed extra laminations are added to form limbs around each winding and thus complete the core. *Compare* shell-type transformer.

cosine potentiometer *See* potentiometer.

COSMOS *Syn. for* C/MOS. *See* complementary transistors; MOS logic circuit.

Cotton balance *Syn.* electromagnetic balance. An absolute means of measuring *magnetic flux density, **B,** in air; it is capable of extremely high accuracy. This method can only be used for fairly strong fields that are uniform over a reasonable volume.

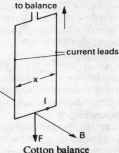

Cotton balance

A long rectangular coil is suspended from an analytical balance with the lower end of the coil suspended in the magnetic flux that is to be measured (*see* diagram). The field is directed horizontally with the lower edge of the coil directed perpendicularly to the lines of **B.** The long sides of the coil experience no vertical force due to **B,** since they are vertical, and act as leads for the lower edge. Provided that these leads are sufficiently long the upper edge of the coil experiences a negligible force due to **B.** Thus the net vertical force, **F,** is just the force on the lower edge of the coil and is given by

$$F = I\int B \mathrm{d}x$$

where I is the current through the coil.

The value of I is measured using a standard resistance and potentiometer and the force **F** by the change in the balance reading when the current is reversed in direction. The value $\int B \mathrm{d}x$ measured is the integrated flux density along the lower edge; the value at any point can be found if the flux-density distribution is known. If the flux is uniform along the length of the lower edge the value of **B** is

$$\int B \mathrm{d}x / x$$

where x is the length of the lower edge.

Magnetic flux densities of about 0·5 tesla have been measured using this method with an accuracy of a few parts in 100 000.

coulomb Symbol: C. The *SI unit of electric *charge, defined as the charge transported through any cross section of a conductor in one second by a constant current of one ampere. Charge, Q, can then be given as

$$Q = \int I dt$$

where I is the current.

coulombmeter *Syn.* coulometer. An instrument that measures electric charge by the amount of material deposited electrochemically. *See also* electrochemical equivalent.

Coulomb's law The mutual force, F, between two electrostatic point charges, q_1 and q_2, that results from the interaction of the electrostatic fields surrounding them is given by

$$F = q_1 q_2 / 4\pi \varepsilon r^2$$

where r is the distance between the charges and ε the *permittivity of the medium. Coulomb's law thus relates electrical and mechanical phenomena. *See also* Ampere's law.

coulometer *See* coulombmeter.

counter (1) A device that detects and counts individual particles and photons. The term is applied to the detector and to the instrument itself. A single particle or photon produces electrons or ions in the detector and these are usually multiplied to produce a pulse of current or voltage for each ionizing event. These pulses are then electronically counted.

(2) *Syn.* digital counter. Any electronic circuit that counts electronic pulses.

In both cases, the average rate of occurrence of events counted is the *count rate*. The *counter lag time* is the delay between the primary event and the occurrence of the count. The *resolving time* is the minimum time between the occurrence of successive primary events that can be successfully counted. In the case of gas-filled radition detectors the inherent resolving time is equal to the dead time.

counter/frequency meter An instrument that can be used as a counter or frequency meter by counting the number of events or periods occurring in a given time. It contains a frequency standard, usually a *piezoelectric oscillator. The time between events may also be counted by comparing the number of standard pulses occurring in the same time as a given number of cycles of the frequency standard.

counter lag time *See* counter.

count rate *See* counter.

coupling The interaction between two circuits so that energy is transferred from one to the other. In *common-impedance coupling* there is an impedance common to both circuits (Figs. *a, b*). The impedance may be a capacitance (*capacitive coupling*), a capacitance and a resistance (*resistance-capacitance coupling*), an inductance (*inductive coupling*), or a resistance (*direct coupling). The impedance may be a part of each circuit or connected between the circuits. In *mutual-inductance coupling* the circuits are coupled by the mutual induc-

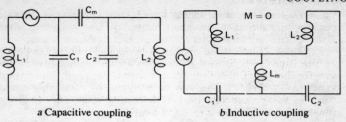

a Capacitive coupling   b Inductive coupling

tance, M, between the coils L_1 and L_2 (Fig. c). The coils used are often those of a transformer. The use of two separate coils between amplifier stages rather than a transformer is termed *choke coupling*. *Mixed coupling* is a combination of mutual-inductance coupling and common-impedance coupling.

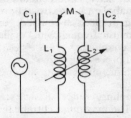

c Mutual-inductance coupling

The *coupling coefficient, K,* is defined as

$$K = X_m / \sqrt{(X_1 X_2)}$$

where X_m is the reactance common to both circuits and X_1 and X_2 are the total reactances, of the same type as X_m, of the two circuits. For Fig. a:

$$K = L_m / \sqrt{[(L_1 + L_m)(L_2 + L_m)]}$$

For Fig. b:

$$K = -C_m / \sqrt{[(C_1 + C_m)(C_2 + C_m)]}$$

For Fig. c:

$$K = M / \sqrt{(L_1 L_2)}$$

The current in the secondary circuit depends on the degree of coupling and the frequency. *Critical coupling* occurs when $KQ = 1$, where Q is the *Q-factor of the circuit. A single peak occurs at the resonant frequency of the circuit and the current has its optimum value. *Overcoupling* occurs when $K > 1/Q$; the current has two side peaks with a dip at the resonant frequency. *Undercoupling*, when $K < 1/Q$, produces a smaller central peak than the optimum.

Band-pass *filters often employ overcoupling, in order to pass a narrow band of frequencies, followed by undercoupling, to compensate for the central dip. In a *tuned circuit, the bandwidth passed varies with frequency. This may be overcome by employing mixed coupling using a capacitance with the mutual inductance to give a constant bandwidth for a range of frequencies.

Cross coupling is unwanted coupling between communication channels, circuits, or components, particularly those with a common power supply. The removal of unwanted signals, especially those due to cross coupling, is called *decoupling*. It is usually achieved using a series *inductance or a shunt *capacitor.

Coupling may also be achieved using an electron beam in one circuit in order to influence a second circuit. Such *electron coupling* is usually unidirectional, the second circuit having little influence on the first. It is utilized in electron-coupled *oscillators, where the load has little effect on the frequency of oscillation.

coupling coefficient *See* coupling.

CPU *Abbrev. for* central processing unit.

crest factor *See* peak factor; pulse.

crest value *See* peak value.

critical coupling *See* coupling.

critical current density The maximum current that can flow in metal conductors which are in contact with a semiconducting substrate without affecting the long-term reliability of the device. At current densities higher than the critical value *metal migration* can occur. Electron scattering at high current densities can cause physical movement of portions of the metal and this leads to an increased resistance and catastrophic failure.

critical damping *See* damped.

critical field (1) *See* avalanche breakdown. (2) *See* magnetron.

CRO *Abbrev. for* cathode-ray oscilloscope.

Crookes dark space *Syns.* cathode dark space; Hittorf dark space. *See* gas-discharge tube.

Crookes tube A low-pressure discharge tube used for studying cathode rays.

crossbar exchange An automatic telephone exchange in which the necessary connections between telephone lines and exchanges are established by means of a series of crossbar switches in order to establish a desired telephone link.

The crossbar switch consists of a number of horizontal selector bars and a number of vertical holding bars to which the telephone lines are connected. Contact is made between lines at the crossing points. Contact fingers on the appropriate selector bar are brought into contact with each of the vertical holding bars; the required contact is formed by means of a small electromagnet on the appropriate holding bar which traps the contact finger. The selector bar then returns to its original position. The crossbar switch has the advantage that other connections to other holding bars can be made without disturbing existing connections whereas the Strowger switch can only be used for one connection at a time.

cross coupling *See* coupling.

crossed-field microwave tube *Syn.* M-type microwave tube. *See* microwave tube.

cross modulation *See* modulation.

cross neutralization *See* neutralization.

crossover The point at which two conductors, insulated from each other, cross paths.

crossover area *See* cathode-ray tube.

crossover distortion *See* distortion.

crossover frequency *See* crossover network.

crossover network. *Syn. for* loudspeaker dividing network. *See* loudspeaker.

crossover network *Syn.* dividing network. A type of filter circuit that divides the frequency range passed between two paths. Frequencies above a specified value pass through one path and those below that value through another. The frequency at which the output passes from one channel to the other is the *crossover frequency* and at that frequency the outputs are equal. Such networks are widely used with loudspeakers to separate the bass and treble components.

crosstalk Interference due to cross *coupling between adjacent circuits or to intermodulation (*see* modulation) of two or more carrier channels, producing an unwanted signal in one circuit when a signal is present in the other. It is common in telephone, radio, and many other data systems.

Crosstalk is classified as *near-end* and *far-end crosstalk* and in speech communication systems as *intelligible* and *unintelligible crosstalk*. Near-end crosstalk is measured at the input or sending terminal. Intelligible crosstalk in a communication system is crosstalk that can be understood by a listener and has a greater interfering effect than unintelligible crosstalk because it diverts his attention. Unintelligible crosstalk cannot be understood by the listener and is often classed as miscellaneous noise.

CRT *Abbrev. for* cathode-ray tube.

cryotron A cryogenic (very low temperature) switching device, depending on *superconductivity. It either consists of a superconductive wire surrounded by a coil or is in thin-film form of two insulated crossing strips made of superconductive material with different critical field curves, such as tin and lead. The presence of a current in the coil or one of the strips changes the superconductivity of the other element and hence switches it off or on.

crystal-controlled oscillator *See* piezoelectric oscillator.

crystal-controlled transmitter A transmitter in which the *carrier frequency is produced by a *piezoelectric oscillator.

crystal counter A radiation *counter that detects and counts subatomic particles. A particle or photon striking a crystal with a potential difference across it produces electron-ion pairs that increase its conductivity. This results in current pulses that may then be counted. *See also* scintillation counter.

crystal cutter A means of cutting gramophone records using a *piezoelectric crystal. The electrical signals from the recording system cause mechanical displacements in a piezoelectric crystal used as a cutting stylus.

crystal detector A *detector that has either a semiconductor junction or a crystal in contact with a metal as its rectifying element.

crystal filter A *filter that uses *piezoelectric crystals to provide its resonant or antiresonant circuits.

crystal growth furnace A furnace that is specially designed to produce a particular temperature profile needed to grow large single crystals from molten material in a controlled manner.

crystal loudspeaker A *loudspeaker in which the sound waves are produced by the mechanical vibrations of a *piezoelectric crystal.

crystal microphone A type of microphone containing a *piezoelectric crystal in the form of two plates separated by an air gap (*see* diagram). Sound pressure variations cause displacements of the crystals and corresponding e.m.f.s are produced across them. Greater sensitivity is obtained using a separate diaphragm that is mechanically coupled to the centre of the crystal but this latter construction is more directive and has a poorer frequency response than the former.

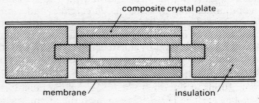

composite crystal plate

membrane / insulation

Crystal microphone

crystal oscillator *See* piezoelectric oscillator.
crystal pick-up *See* pick-up.
crystal puller *See* liquid-encapsulated Czochralski.

crystal rectifier *Syn. for* semiconductor diode. *See* diode.

CTD *Abbrev. for* charge-transfer device.

Curie point *Syns.* magnetic transition temperature; ferromagnetic Curie temperature. *See* ferromagnetism.

Curie's law *See* paramagnetism.

Curie–Weiss law *See* paramagnetism.

current Symbol: I; unit: ampere. The rate of flow of electricity. A *conduction current* is a current flowing in a conductor due to the movement of electrons or ions through the material, usually under the influence of an applied field. The net current is the algebraic sum of the charges. *Compare* displacement current.

A *unidirectional current* is one that always flows in the same direction in a circuit. A unidirectional current of more or less constant magnitude is a *direct current. Compare* alternating current.

current amplifier *Syn. for* power amplifier. *See* amplifier.

current average *See* alternating current.

current balance *See* Kelvin balance.

current density Symbol: j or J. A vector quantity that is equal to the ratio of the current to the cross-sectional area of the current-carrying medium. The medium may be either a radiation beam or a conductor. The current density is defined either at a point or as the *mean current density*.

current feedback *Syn.* series feedback. *See* feedback.

current limiter *See* limiter.

current mode logic (CML) *See* emitter-coupled logic.

current regulator *See* regulator.

current relay *See* relay.

current saturation *See* saturation current.

current sensing *See* charge-coupled device.

current transformer *Syn.* series transformer. *See* transformer; instrument transformer.

cut-off *Syn.* black-out point. The point at which the current flowing through an electronic device is cut off by the control electrode. In a *transistor, the cut-off point is the minimum base current at which the device conducts; in a valve it is the minimum negative grid voltage (the *grid base*) required to stop the current. In a *cathode-ray tube the cut-off bias is the bias voltage that just reduces the electron-beam current to zero. In all cases the values are dependent on the conditions at the other electrodes, which must be specified.

cut-off frequency (1) The frequency at which the attenuation of a passive network changes from a small value to a much higher value; this is the *theoretical cut-off frequency*. The effective cut-off frequency is that frequency where the *insertion loss between two specified impedances has risen by a stated amount compared to the value at a reference frequency. An active network has the same cut-off frequency as a passive one with the same inductances and capacitances. The term is also applied to the limiting frequency of a *filter.

(2) In a *field effect transistor used for microwave applications, that frequency at which the current gain of the FET becomes zero.

(3) *See* waveguide.

cut-off voltage (of a television *camera tube) *See* target voltage.

cut-out A switch, especially a protective device that is operated, usually automatically, under fault conditions.

cutter *See* recording of sound.

C-V curves *See* MOS capacitor.

cycle (1) An orderly set of changes regularly repeated.

(2) One complete set of changes in the values of a recurring variable quantity, such as an alternating current.

cyclic current *Syn. for* mesh current. *See* network.

cyclotron An *accelerator that produces a beam of high-energy particles. The particles travel within two semicircular hollow metal electrodes (known from their shape as *dees*). A unipolar magnetic field, *H,* at right angles to the plane of the dees, causes the particles to execute circular orbits within the dees, the orbital radius being proportional to particle velocity and magnetic field. Particles are accelerated by a radiofrequency field as they pass between the dees and the orbital radius thus increases. The transit time, *t,* within a dee is given by

$$t = \pi m / He$$

for a particle of mass *m* and charge *e,* and is constant until the relativistic mass change causes it to become out of synchronization with the accelerating field.

cylindrical winding A type of winding in which a coil is helically wound, either as a single layer or in multiple layers. The axial length of the coil is usually several times its diameter. *Compare* disc winding.

Czochralski method *See* liquid-encapsulated Czochralski.

D

damped Denoting the progressive diminution of a free oscillation due to an expenditure of energy. *Damping* is used to mean both the cause of the energy loss – friction, eddy currents, etc., – and the progressive decrease in amplitude. The amount of damping is termed *critical* if the system just fails to ocillate (*see* diagram). Greater or lesser degrees of damping lead to *overdamping* and *underdamping* respectively. The *damping factor* of underdamped oscillations is the ratio of the amplitude of any one of the damped oscillations to that of the following one. The natural logarithm of the damping factor, the *logarithmic decrement,* is sometimes quoted.

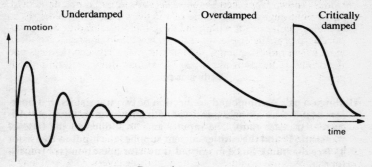

Damped systems

Instrument damping is widely used in indicating instruments to bring the indicator to rest quickly. A device incorporated in an indicating instrument to provide damping is a *damper;* one that is overdamped so that the needle goes to its true position without wavering is a *dead-beat instrument.*

damper *See* damped.

damping *See* damped.

damping factor *Syn.* decrement. *See* damped.

Daniell cell A *cell in which two different electrolytes are separated by a porous partition. The positive electrode is copper immersed in copper sulphate solution; the negative one is zinc amalgam immersed in dilute suphuric acid or sometimes zinc sulphate solution. The cell must be dismantled when not in use to prevent the electrolytes from diffusing into each other. E.m.f.: $\sim$1.08 volts.

105

daraf Symbol: F^{-1}. The reciprocal *farad, used as a unit to measure the reciprocal of capacitance, i.e. *elastance.

dark conduction Low-level conduction that occurs in a photosensitive material when it is not illuminated. If such material is used as a photocathode in devices such as *photocells, it gives rise to a *dark current*.

dark current *See* dark conduction.

dark resistance *See* photocell.

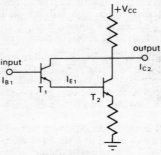

Darlington pair

Darlington pair A compound connection of two transistors that operates as if it were a single transistor with an extremely high forward-current transfer ratio. The input signal is applied to the base of transistor T_1 and the emitter current supplies the input to transistor T_2 (*see* diagram). Since in normal transistor operation (*see* transistor)

$$I_c = \alpha I_e$$

where α (the alpha current factor) is near unity, then the collector current I_{c1} is approximately equal to the emitter current I_{e1} and

$$I_{e1} = \beta_1 I_{b1}$$

β_1 is the *beta current gain factor and I_{b1} the base current. In transistor T_2 the collector current I_{c2} is given by

$$I_{c2} = \beta_2 I_{b2} = \beta_1 \beta_2 I_{b1}$$

If the transistors are matched, i.e.

$$\beta_1 = \beta_2$$

the overall beta current gain of the combination is given by the square of the value for a single transistor.

106

d'Arsonval galvanometer A direct-current galvanometer in which the current to be measured passes through a small rectangular coil suspended between the poles of a permanent horseshoe magnet. The magnetic field produced in the coil reacts with the field of the magnet producing a torque and causing the coil to rotate about the vertical axis in the field. D'Arsonval movement is used in most forms of galvanometer since it combines a high degree of sensitivity with low resistance and high damping.

dart leader stroke *See* lightning stroke.

dashpot A means of providing mechanical damping (*see* damped) in an electrical device. It consists of a piston in a cylinder containing fluid: the friction between the piston and the fluid provides the damping.

DAT *Abbrev. for* digital audio tape. A sound reproduction system that uses magnetic tape to store digitally recorded audio-frequency signals. The signals are sampled, pulse code modulated, and stored as 16-bit digital words as in *compact disc systems, but unlike compact disc the DAT recorders can both record signals and replay pre-recorded tapes.

The DAT system uses a miniature rotating tape head to record slanted tracks on a slow-moving tape. The information is packed on the tape with a much higher density than in conventional tape recording systems and demands extremely high accuracy in manufacture. The tape used is similar to that used in *video recorders. DAT recorders normally use a 48 kHz sampling rate both for recording and replay, but can also replay at the 44.1 kHz sampling rate used by CD systems and can record at 32 kHz as used by DBS (*direct broadcast by satellite) systems. The recorder switches automatically to the appropriate sampling frequency.

Development of DAT systems has drawn heavily on the technologies involved in the development of CD systems and video recording systems to produce a system that uses a robust miniature transportable sound system.

data processing The automatic or semiautomatic organization of numerical data in a desired manner. A data-processing system is any system that can receive, transmit, or store data; many systems can also perform mathematical operations upon the data and tabulate or indicate the results.

Analog and digital *computers are examples of data-processing systems but the term usually refers to systems that perform a limited number of functions automatically, particularly systems that are used to control the operation of other systems.

Day modulation A means of doubling the use of a radio channel by transmitting two *carrier waves in *quadrature, each separately modulated with different signals.

DBR laser *Short for* distributed Bragg reflector laser (*syn. for* distributed feedback laser). *See* injection laser.

DBS *Abbrev. for* direct broadcast by satellite.

d.c. *Abbrev. for* direct current.

d.c. amplifier *Short for* direct-coupled amplifier.

d.c. voltage Informal term for *direct voltage,* i.e. a unidirectional voltage that is substantially constant.

dead (of a conductor or circuit) At earth potential. One not at earth potential is termed *live.*

dead-beat instrument *See* damped.

dead time The time interval immediately following a stimulus during which an electrical device does not respond to another stimulus. A correction for the dead time must be applied to the observed count rate in a *counter to allow for events occurring during this period.

de Broglie waves A set of waves that are associated with a moving particle and represent its behaviour in certain situations, as when a beam of particles undergoes *diffraction. The wavelength is given by

$$\lambda = h/mv$$

where h is the Planck constant and m and v the mass and velocity of the particle.

debug *See* bug.

debunching The spreading of electrons in an electron beam or in a velocity-modulated tube that results from their mutual repulsion. The angle of spread of the electron beam is the *divergence angle. See also* klystron.

Debye length *Syn.* Debye-Hückel length. In a medium such as a *semiconductor containing fixed charges and mobile charges, the distance required for a significant change in mobile carrier population under equilibrium conditions, in which the neutral equilibrium values of charge carrier density are either increased or reduced. The Debye length is a result of the screening effect around a fixed charge such as a donor impurity due to electrostatic attraction between it and the mobile carriers, causing them to cluster around the site of the impurity ion and mask its presence. Hence the electric field surrounding the ion declines much more rapidly than it would in the case of the unscreened ion. Fig. *a* illustrates the concept of Debye length for a positive ion. The ion has a charge $+q$ and is surrounded by a cloud of mobile carriers – electrons in this case. The cloud is about a Debye length in radius and contains an integrated charge of $-q$.

In the case of a semiconductor where both electrons and holes are present, the general form for the Debye length is given by

$$L_D = \sqrt{[kT\varepsilon/q^2(n_0 + p_0)]}$$

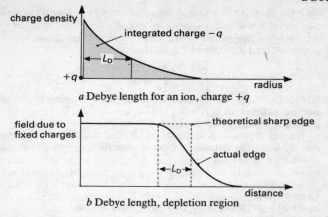

a Debye length for an ion, charge $+q$

b Debye length, depletion region

where k is the Boltzmann constant, T the thermodynamic temperature, ε the permittivity of the material, q the value of the fixed charges, and n_0 and p_0 the neutral equilibrium numbers of electrons and holes respectively.

From this, in an *extrinsic semiconductor where the charge carriers are predominantly due to the presence of impurities an approximate value can be derived as

$$L_{De} \simeq \sqrt{[kT\varepsilon/q^2(N_D - N_A)]}$$

for n-type material, where N_D and N_A are the numbers of donor and acceptor impurities respectively.

For *intrinsic material the value is

$$L_{Di} \simeq \sqrt{[kT\varepsilon/2n_iq^2]}$$

where n_i is the intrinsic carrier density for one type of carrier.

The Debye length can be used as a scaling factor to derive a curve that describes the manner in which the carrier density changes from its neutral equilibrium value to near zero, for example at the edge of a *depletion region, and hence is a measure of the departure in practice from the sharp edge to the depletion region that is often assumed (Fig. *b*).

deca- Symbol: da. A prefix to a unit, denoting a multiple of 10 of that unit.

decade scaler *See* scaler.

decay time *See* storage tube; pulse.

deci- Symbol: d. A prefix to a unit, denoting a submultiple of 10^{-1} of that unit.

decibel Symbol: dB. A dimensionless unit used to express the ratio of two powers, voltages, currents, or sound intensities. It is ten times the common logarithm of the power ratio. Thus if two values of power, P_1 and P_2, differ by n decibels then

$$n = 10 \log_{10}(P_2/P_1)$$
$$\text{i.e. } P_2/P_1 = 10^{n/10}$$

If P_1 and P_2 are the input and output powers, respectively, of an electric network then if n is positive, i.e. $P_2 > P_1$, there is a gain in power; if n is negative there is a power loss.

The *bel,* symbol: B, is equal to 10 decibels. Since it is inconveniently large the decibel is used in practice. If two power values differ by N bels then

$$N = \log_{10}(P_2/P_1)$$

Compare neper.

decoder *Syn.* pulse detector. *See* pulse modulation.

decoupling *See* coupling.

decrement *Syn. for* damping factor. *See* damped.

dee *See* cyclotron.

de-emphasis *See* pre-emphasis.

deep depletion *See* MOS capacitor.

deep level transient spectroscopy (DLTS) A technique used to detect energy levels due to *traps in the forbidden band of a semiconductor. The general technique is to use either photons, electrons, or an applied electric field to excite carriers into these levels, and to measure either the change in capacitance across the semiconductor, or transient currents produced as the equilibrium state in the material is re-established. The technique is very useful in establishing activation energies, concentrations, and capture cross sections of electron traps.

deep ultraviolet exposure *See* photolithography.

defect conduction Conduction in a *semiconductor due to the presence of *holes in the *valence band. The presence of the holes is due to an imperfection in the crystal lattice of the semiconductor. *See* hole conduction.

definition *See* television.

deflection defocusing *See* cathode-ray tube.

deflection plates *See* electrostatic deflection.

deflection sensitivity *See* cathode-ray tube.

deformation potential An electric potential caused by mechanical deformation of the crystal lattice of *semiconductors and conductors. *See* piezoelectric effect.

degeneracy Symbol: *g.* A condition that arises when an atomic or molecular system with a number of possible quantized states (*see* quantum theory) has two or more distinct states of the same energy.

The number of degenerate states each possessing that energy level is the *statistical weight*. Certain semiconductors, for example, exhibit degeneracy in either the valence or conduction bands, in which holes or electrons of the same energy level have different effective masses (*see* direct-gap semiconductor; Gunn effect).

degenerate semiconductor A *semiconductor that has the Fermi level (*see* energy bands) located either in the valence or in the conduction band, giving rise to essentially metallic properties over a wide range of temperature.

degeneration *Syn. for* negative feedback. *See* feedback.

degradation failure *See* failure.

degree Celsius *Syn.* degree centigrade. Symbol: °C. *See* kelvin.

dekatron A type of multielectrode cold-cathode *scaling tube that has ten sets of electrodes that function in turn. As voltage pulses are received, a glow discharge moves from one set of electrodes to the next. A visual display of counts in the decimal system is provided using the tubes in cascade. The tubes may also be used for switching. *Compare* digitron.

delay The time interval between the propagation of a signal and its reception; the time taken for a pulse to traverse any electronic device or circuit. In a switching *transistor, for example, it is the time between the application of a pulse to the input and the appearance of a pulse at the output. Excessive pulse rise and fall times can reduce the speed of operation of a switching circuit and lead to undesirable delay. A known delay may be introduced into a system deliberately, by means of suitable *delay line.

delay distortion *See* distortion.

delayed automatic gain control *Syn. for* biased automatic gain control. *See* automatic gain control.

delayed-domain mode *See* Gunn diode.

delayed sweep *See* time base.

delay equalizer A network or filter that compensates for the effects of delay *distortion and thus maintains the wavefront of a transmitted wave.

delay line Any circuit, device, or transmission line that introduces a known delay in the transmission of a signal. Coaxial cable or suitable L-C (inductance-capacitance) networks may be used to provide short delay times but the attenuation is usually too great when longer delay times are required.

Acoustic delay lines are often employed when a longer delay is needed. The signals are converted to acoustic waves, usually by means of the *piezoelectric effect. They are then delayed by circulation through a liquid or solid medium before being reconverted into electrical signals. Fully electronic *analog delay lines* are now being provided by *charge-coupled devices. *Shift registers and charge-coupled devices may be used for *digital delay lines*.

Dellinger fade-out *See* fading.

delta voltage *See* voltage between lines.

demodulation *See* modulation.

demodulator *See* detector.

demultiplexer *See* multiplex operation.

depletion layer A *space charge region in a *semiconductor that has a net charge due to insufficient mobile charge carriers. Depletion layers are inevitably formed at the interface between two dissimilar conductivity types of semiconductor, in the absence of an applied voltage (*see* p-n junction), and at the interface of a metal and a semiconductor (*see* Schottky barrier). The width of the depletion layers increases when reverse bias is applied; the *depletion-layer capacitance* is the capacitance associated with a given depletion layer, which effectively acts as a dielectric when depleted of mobile carriers. Reversed biased p-n junctions or Schottky diodes can therefore be used as voltage capacitors (*see* varactor). A depletion layer can also form at the surface of a semiconductor of given conductivity type under the influence of an electric field (*see* MOS capacitor).

depletion-layer photodiode *See* photodiode.

depletion mode A means of operating *field-effect transistors in which increasing the magnitude of the gate bias decreases the current. *Compare* enhancement mode.

depletion-mode device *See* field-effect transistor.

deposition The application of a material to a base (such as a substrate) by means of vacuum, electrical, chemical, screening, or vapour techniques. *See also* metallization.

derating Reducing the maximum performance ratings of electronic equipment or devices when operated under unusual or extreme conditions. This ensures an adequate safety margin.

derived units *See* SI units.

de Sauty bridge A four-arm *bridge used for the direct comparison of capacitances (*see* diagram). The capacitors are charged or discharged using the key K; if no response is observed from the ballistic galvanometer G, then

$$R_1 C_1 = R_2 C_2$$

descum *See* scumming.

Destriau effect *See* electroluminescence.

destructive read operation (DRO) *See* read.

detector (1) *Syn.* demodulator. A circuit, apparatus, or circuit element that is used in communications to demodulate the received signal, i.e. to extract the signal from a carrier with minimum distortion. A *linear detector* produces an output proportional to the modulating signal; a *square-law detector produces an output proportional to the square of the modulating signal.

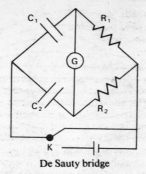

De Sauty bridge

(2) Any device used to detect the presence of a physical property or phenomenon, such as radiation.

detune To adjust the frequency of a *tuned circuit so that it differs from the frequency of the applied signal.

deviation (1) *Syns.* variation; error. The difference between the observed value of a measurement and the true value. In automatic control systems it is the difference between the instantaneous value and the desired value.

(2) *See* frequency modulation.

deviation distortion *See* distortion.

deviation ratio *See* frequency modulation.

device An electronic part that contains one or more active elements, such as a transistor, diode, or integrated circuit.

DFB laser *Short for* distributed feedback laser. *Syn.* distributed Bragg reflector laser. *See* injection laser.

diacritical current The current required in a coil to produce a flux equal to one half of the saturation flux.

diagnostic routine *See* program.

diamagnetism A very weak effect that is common to all substances and is due to the orbital motion of electrons around the nucleus of atoms. Diamagnetism is independent of the temperature of the material.

If a substance is placed in a magnetic flux density B, each individual electron experiences a force due to B since it is a moving charge. The orbits and velocities of the electrons are changed in order to produce a magnetic flux density that opposes B (*see* electromagnetic induction). Each orbital electron therefore acquires an induced magnetic moment that is proportional to B and is in the opposite sense; hence the sample has a negative magnetic *susceptibility.

Diamagnetism causes a reduction of magnetic flux density within a sample; this can be represented schematically (*see* diagram) by a

113

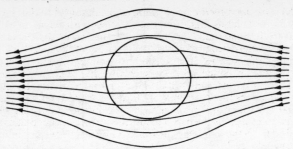

Change in magnetic flux density in a diamagnetic substance

separation of the lines of magnetic flux as they pass through the material. If a diamagnetic substance is placed in a nonuniform magnetic field it tends to move from the stronger to the weaker region of the field. A bar of diamagnetic material placed in a uniform magnetic field tends to orientate itself with the longer axis at right angles to the field. Purely diamagnetic materials include copper, bismuth, and hydrogen.

Certain materials have a permanent molecular magnetic moment and in such substances the diamagnetism is totally masked by the magnetic properties arising from this permanent moment. *See* paramagnetism; ferromagnetism; antiferromagnetism; ferrimagnetism.

diametrical voltage *See* voltage between lines.

diaphragm (1) A vibrating membrane used in sound *transducers, such as microphones, loudspeakers, and the human ear.

(2) A porous partition in an electrolytic cell that separates the electrolytes while allowing ions to pass through.

diaphragm relay *See* relay.

dichroic mirror A mirror used in a colour *television camera to reflect one designated visible frequency band and transmit all others.

Dicke's radiometer An instrument that measures microwave noise power precisely by comparing it with the noise from a standard source in a waveguide.

die *See* chip.

dielectric A solid, liquid, or gaseous material that can sustain an electric field and act as an insulator. A *perfect dielectric* is one in which no energy is lost from an electric field applied across the dielectric. An *imperfect dielectric* is one in which *electric hysteris losses occur. The displacement, *D*, lags behind the applied field, *E*, resulting in a typical hysteris curve (*see* magnetic hysteresis) and in energy losses from the applied field, which appear in the form of heat.

dielectric constant *See* permittivity.

dielectric heating A method of heating an insulating material, such as a plastic, in which a high-frequency alternating electric field is applied to the material. The periodic alternation of the electric field causes an alternating *dielectric polarization of the specimen, which results in the heating effect. Dielectric heating is usually effected by placing the specimen between specially shaped metal *applicators,* which form the electrodes of a capacitor, and applying the alternating field across them. A single applicator is sometimes used, applied to one surface of the specimen, when local heating only is required. The *heating depth* is the depth below the surface at which the heating effect due to the single applicator is observed. *Compare* induction heating.

dielectric isolation A method of isolating individual regions in an integrated circuit, particularly a *bipolar integrated circuit, by surrounding the region with insulating material (dielectric) rather than with isolating diffusions. There are many different methods used to achieve dielectric isolation.

dielectric loss angle *See* dielectric phase angle.

dielectric phase angle The difference between the phase angle of the alternating (sinusoidal) voltage applied across a dielectric material and the phase angle of the resulting alternating current. The difference between the dielectric phase angle and 90° is the *dielectric loss angle.* The cosine of the dielectric phase angle (or the sine of the dielectric loss angle) is the *dielectric power factor.*

dielectric polarization *Syn.* electric polarization. Symbol: P; unit: coulomb per metre squared. An effect observed in a dielectric in the presence of an applied electric field. The electrons in each particular atom are displaced in the direction opposite to the field, and the nucleus in the direction of the field. (The centre of gravity of the atom remains fixed.) Each atom acquires a dipole moment (*see* dipole) parallel to and in the same direction as the field. The polarization is defined as the electric dipole moment per unit volume. It is related to the electric field, E, by the relation

$$P = \chi_e / \varepsilon_0 E$$

where χ_e is the electric *susceptibility and ε_0 the *permittivity of free space. χ_e is a tensor and approximately independent of electric field strength in most practical circumstances. In a uniform isotropic medium P and E are parallel and χ_e is a scalar constant.

dielectric power factor *See* dielectric phase angle.

dielectric strain *See* displacement.

dielectric strength *Syn.* disruptive strength. Unit: volts per metre. The maximum electric field that can be sustained by a dielectric before *breakdown occurs.

difference transfer function *See* feedback control loop.

differential amplifier An *amplifier that has two inputs and produces an output signal that is a function of the difference between the inputs. An ideal differential amplifier produces an output signal of zero when the inputs are identical. In practice a small positive or negative signal may occur. The *common-mode rejection ratio* is a measure of the ability of a differential amplifier to produce a zero output for like inputs.

differential analyser A device for solving differential equations, usually a type of analog *computer although a mechanical device is sometimes used.

differential capacitor A variable capacitor having two sets of fixed plates and one set of moving plates. As the moving plates rotate between the fixed plates, the capacitance to one set of plates is increased while that to the other is decreased.

differential galvanometer A type of *galvanometer that gives a deflection that is a function of the difference between two currents. The currents are passed in opposite directions through two identical coils. The difference between the currents determines the magnitude and direction of deflection.

differential relay *See* relay.

differential resistance The resistance of a device or component part measured under small signal conditions.

differential winding Two or more coils or two windings of a single coil arranged so that when carrying a current their magnetomotive forces are in opposition.

differentiator *Syn.* differentiating circuit. A circuit that gives an output proportional to the differential, with respect to time, of the input. *Compare* integrator.

diffraction A phenomenon occurring when electromagnetic waves or beams of charged particles, such as electrons, encounter either an opaque object or a boundary between two media. The beams are not propagated strictly in straight lines but are bent at the discontinuity. This effect is due to the wave nature of electromagnetic radiation and the *de Broglie waves associated with the charged particles. Interference between the diffracted waves produces a *diffraction pattern* of maxima and minima of intensity; the diffraction pattern produced depends on the size and shape of the object causing the diffraction and the wavelength of the incident radiation. It can be used to investigate crystal structures or surface structures.

Electron and *X-ray diffraction* are employed to investigate crystal structures, the diffraction pattern produced being dependent on the spacing of the crystal planes. Electron diffraction is used to assess the structure of a crystalline semiconductor material near the surface. *Low-energy electron diffraction* uses a beam of low-energy electrons incident normally to the surface and the diffraction patterns from the back scattered electrons are detected. *Reflection high-ener-*

116

gy electron diffraction uses a high-energy electron beam at a very small grazing angle of incidence. In this case the forward scattered electrons produce the diffraction pattern. Electron diffraction patterns obtained from transmission *electron microscopy can be used to provide information about crystalline materials throughout their bulk but this latter is a time-consuming method and very difficult to interpret.

X-ray diffraction is employed to detect imperfections in semiconductor crystals by means of *X-ray topography*, using a slit and photographic plate to record a 'map' of the slice on a photographic film. Although lateral resolution is of the order of several micrometres the technique is rapid and nondestructive.

diffused junction A junction between two different conductivity regions within a semiconductor formed by *diffusion of the appropriate impurity atoms into the material.

diffusion The process of introducing selected impurity atoms into designated areas of a *semiconductor in order to modify the properties of that area. The semiconductor is heated to a predetermined temperature in a gaseous atmosphere of the desired impurity. Impurity atoms that condense on the surface diffuse into the semiconductor material in both the vertical and horizontal directions. The numbers of impurity atoms and distance travelled at any given temperature is well-defined according to *Fick's law. The interface between two different conductivity regions within a semiconductor is a diffused junction.

Early diffused devices were formed by performing nonselective diffusions over the entire semiconductor surface; any unwanted regions of the semiconductor were etched away (as in *mesa transistors) and the junctions were formed below the surface and parallel to it. Modern techniques use the *planar process of selective diffusion into well-defined areas of semiconductor. The edge of the junction is perpendicular to the surface and the device may be formed along the surface of the material.

Double diffusion is a method of forming diffused junctions in which successive diffusions of different impurity types are made into the same well-defined region of semiconductor. The temperature and diffusion time are adjusted to produce the desired impurity concentration. This technique is used if very precise distances between junctions are required, as in *D/MOS circuits, since the geometry is defined by the diffusion process itself and errors caused by misalignment of successive photographic masks are eliminated.

diffusion coefficient *See* Fick's law.

diffusion length *See* semiconductor.

diffusion potential *Syn. for* built-in field. *See* p-n junction.

digital circuit A circuit that responds to discrete values of input voltage and produces discrete values of output voltage. Usually two voltage levels only are recognized, as in binary *logic circuits. *Compare* linear circuit.

digital computer *See* computer.

digital counter *See* counter.

digital delay line *See* delay line.

digital gate *See* gate.

digital inverter *See* inverter.

digital recording *See* compact disc system.

digital voltmeter (DVM) A voltmeter that displays the values of the voltage as numbers. The input is usually supplied as an analog signal and this is repetitively sampled by the voltmeter, which displays the instantaneous values. The output may be supplied in punched card or *paper tape form, the voltmeter functioning as a type of analog-to-digital converter.

digitizer *See* analog/digital converter.

digitron *Syn.* Nixie tube. A type of cold-cathode *scaling tube that has several cathodes (usually ten) shaped into the form of characters (usually the digits 0 to 9). As voltage pulses are received the cathode required is selected by a switching connection to one side of the power supply and a glow discharge illuminates the character. These tubes are widely used for display purposes in *calculators, *counters, etc. *Compare* dekatron.

DIL package *Short for* dual in-line package.

diode Any electronic device that has only two electrodes. There are several different types of diode, their voltage characteristics determining their application. Diodes are most commonly used as rectifiers; those used for other purposes have special names, as with the Gunn diode.

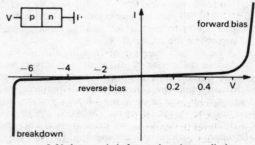

I–V characteristic for semiconductor diode

There are two types of simple diode. A *semiconductor diode* consists of a simple *p-n junction. Current flows when a forward bias is applied to the diode (*see* diode forward voltage) and increases exponentially (*see* diagram). A straight-line approximation of this forward characteristic allows a resistance value (the *forward slope resistance*) to be calculated from the slope of the straight line. Reverse bias produces only a very small leakage current until *breakdown occurs. A *valve diode* is a *thermionic valve that has an anode and cathode and also passes current only in the forward direction.

See also Gunn diode; IMPATT diode; light-emitting diode; photodiode; tunnel diode; varactor; Zener diode.

diode cut-off *See* charge-coupled device.

diode drop *See* diode forward voltage.

diode forward voltage *Syns.* diode drop; diode voltage. The voltage across the electrodes of a diode when current flows. The current increases exponentially with voltage (*see diagram at* diode) and therefore the voltage is substantially constant over the range of currents in common use: a typical value is about 0·7 volts at 10 milliamps, making diodes very useful as *catching diodes. The diode may also function as a *voltage reference diode* when it is used to provide a reference voltage, equal to the diode forward voltage, across its terminals.

diode laser *Syn. for* injection laser.

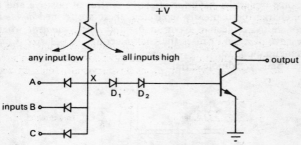

DTL NAND circuit

diode transistor logic (DTL) A family of integrated logic circuits in which each input signal comes through a *diode and the output is taken from the collector of an inverting transistor (*see* inverter). The basic circuit is a NAND gate (*see* diagram). If any of the inputs is at the low logic level the corresponding input diode is forward biased and can conduct current away from the diodes D_1 and D_2. The potential at point X is therefore at a value determined by the *diode forward voltage and represents the low logic level. It is insufficient to forward bias D_2 and no current can flow to the base of the transistor. The transistor is therefore 'off' and the collector voltage is at the high logic level. If all the inputs are high, all the input diodes are reverse biased and cannot conduct current. The potential at point X is therefore high. Both D_1 and D_2 are forward biased and current flows to the base of the transistor, which turns on and saturates. The collector voltage falls to the low logic level.

The speed of the DTL circuit is slow compared to *emitter-coupled logic circuits because the output transistor is operated in saturated mode: *carrier storage at the collector junction causes a delay in the switching time between logic levels. DTL circuits have been largely replaced by *transistor-transistor logic circuits.

diode voltage *See* diode forward voltage.

DIP *Abbrev. for* dual in-line package.

diplexer A device that allows the simultaneous transmission of two signals in the same circuit or channel without interaction.

dipole (1) *Syns.* electric dipole; doublet. A system of two equal and opposite charges very close together. A uniform field produces a torque that aligns the dipole along the field without translation. The product of one of the charges and the distance between them is the *dipole moment* (symbol: *p*). Dipole moment is related to the electric field strength, *E,* and the torque, *T,* by

$$T = p \times E$$

Some molecules have the effective centres of positive and negative charges permanently separated; these are termed *dipole molecules*. Some molecules have an *induced dipole moment* when the presence of a field causes the charge centres to polarize. *Compare* magnetic moment.

(2) *Short for* dipole aerial.

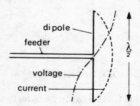

Voltage and current distribution in half-wave dipole

dipole aerial An *aerial commonly used for frequencies below 30 megahertz. It consists of a centre-fed open aerial excited in such a way that the standing wave of current is symmetrical about the mid-point of the aerial. There are several different types of aerial: a *half-wave dipole* has a length equal to half the wavelength, λ (*see* diagram); a *full-wave dipole* has a length of one wavelength; a *folded dipole* consists of two parallel half-wave dipoles separated by a small fraction of the wavelength, connected at their outer ends, and fed at the centre of one of the dipoles; a *multiple folded dipole* consists of more than two parallel half-wave dipoles.

dipole molecule *See* dipole.

dipole moment *Syn.* electric moment. *See* dipole.

direct-access memory *See* random-access memory.

direct broadcast by satellite (DBS) A method of broadcasting that uses a *communications satellite in geostationary orbit as the main transmitter. The signal to be broadcast is transmitted from its point of origin on the earth to the satellite where it is received, amplified, and retransmitted to cover a wide area. It is detected directly by individual receivers using a suitable dish antenna tuned to the DBS signals.

direct-coupled amplifier (d.c. amplifier) An amplifier in which the output of one stage is fed directly to the input of the next or through a resistance. Such an amplifier is used for direct-current amplification and the misnomer *direct-current amplifier* is sometimes used.

direct coupling *Syn.* resistance coupling. *Coupling between electronic circuits or devices, such as amplifier stages, that is not frequency dependent; resistive coupling is a form of direct coupling.

direct current (d.c.) A unidirectional current of substantially constant value. *See also* current.

direct-current Josephson effect *See* Josephson effect.

direct-current restorer A device that restores the d.c. component or low-frequency component to a signal that has had its low-frequency components removed by a circuit element with high impedance to direct current. The device may also be used to add d.c. or low frequency to a signal lacking these components.

Direct-current restoration is used in *television receivers to reconstruct the original video signal. It is required either to restore the d.c. component to the received signal as in *alternating-current transmission or to correct for the presence of an unwanted spurious d.c. component.

direct-current telegraphy *See* telegraphy.

direct-current transmission A method of transmission used in television in which the direct-current component of the luminance signal (*see* colour television) is directly represented in the transmitted signal. *Compare* alternating-current transmission.

direct feedback *Syn. for* positive feedback. *See* feedback.

direct-gap semiconductor Solutions of Schrödinger's equation for an electron moving through a periodic potential that varies with the spacing of atoms in a crystal lattice only exist for certain values of energy, which give the allowed *energy bands; other values – forbidden bands – cannot exist. The difference in energy between the highest electron energy in the first allowed band – the *valence band – and the second allowed band – the *conduction band – is the energy gap, E_g. A simplified energy band diagram is shown in

121

DIRECT-GAP SEMICONDUCTOR

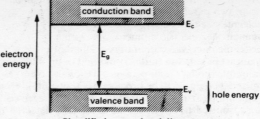

a Simplified energy band diagram

Fig. *a*. When the allowed solutions are plotted as a function of the wave vector, *k*, associated with the moving electron, the values of allowed solutions vary with *k* and the band edge is not flat as shown in Fig *a*.

A *direct-gap semiconductor* is one in which the minimum value of energy $E(k)$ in the conduction band occurs at the same value of *k* as the maximum value of $E(k)$ in the valence band. An *indirect-gap semiconductor* is one in which the minimum value of energy in the conduction band occurs at a different value of *k* than the maximum value in the valence band. Detailed energy bands as a function of *k* for silicon and gallium arsenide are shown in Fig. *b*. Silicon is an indirect-gap semiconductor and gallium arsenide a direct-gap semiconductor.

The wave vector *k* is related to the momentum *p* of the electron. For a free electron, neglecting the potential energy due to the crystal lattice,

$$k = (h/2\pi)p$$

where *h* is the *Planck constant. The solutions $E(k)$ therefore are a function of the electron momentum.

If an electron acquires sufficient energy to allow it to be excited into a higher energy state the law of conservation of momentum must be obeyed. In a direct-gap semiconductor *direct transitions* can occur when a valence electron with maximum energy acquires energy just equal to E_g. Electrons can make a direct transition by absorbing a photon of energy, or by emitting a photon to cross in the reverse direction. This is essential for semiconductor *injection lasers or *light-emitting diodes. In an indirect-gap semiconductor an electron cannot be excited directly across the forbidden gap with an energy E_g; a change in momentum is also required. An *indirect transition* can occur however when a *phonon of wave vector *K* is created during the energy absorption and then

$$k_{absorbed} = k_c + K \simeq 0$$

122

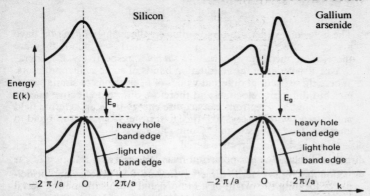

b Energy diagrams for indirect-gap (left) and direct-gap semiconductors

The minimum energy required for an indirect transition is given by

$$E_g + E_{ph}$$

The phonon energy, E_{ph}, is typically very small compared to E_g (Fig. *c*).

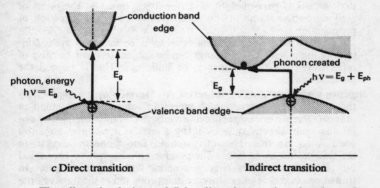

c Direct transition   Indirect transition

The allowed solutions of Schrödinger's equation that form the valence band of a semiconductor consist of a number of sub-bands determined by the crystal structure. *Degeneracy of the sub-bands at the top of the valence band occurs at the value $k = 0$ (Fig. *b*). The bands can be considered as occupied by *holes that have different velocities but the same energy. The effective masses must therefore be different. The effective mass, m^*, is defined as

123

$$m^* = h^2/[4\pi^2(\partial^2 E/\partial k^2)]$$

The holes in a wider sub-band with a smaller value of $\partial^2 E/\partial k^2$ have a greater effective mass than those in a narrower band; they are therefore termed *heavy holes* and *light holes*, respectively. Degeneracy can also occur in the conduction bands of some semiconductors. Where these occur at different values of k, electrons of similar energies but different velocities and therefore different effective masses will be present. Electrons can acquire energy from an external field and interact with the crystal lattice to transfer from one sub-band to another. This results in the *Gunn effect.

directional aerial *See* directive aerial.

directional relay *Syn. for* polarized relay. *See* relay.

direction finding *Syn.* radio direction finding. The practice and principle of locating the origin of a radio signal. A discriminating aerial and some form of receiver is required. *Automatic direction finding* carries out the process automatically using either a rotating *directive aerial or two such aerials at right angles. The rotating aerial is often used in conjunction with a *cathode-ray tube (CRT) as an indicator to display strength of signal against direction. The direction of maximum strength is the bearing of the radio source. When a CRT is used in this way the term *cathode-ray direction finding* is sometimes applied. *Frame direction finding* employs a loop aerial with a polar figure-of-eight directional response. At the point of zero signal the frame of the aerial points along the direction of propagation.

  Direction finding in mountainous areas or in urban areas containing many very tall buildings can be subject to error because of reflections from the mountains or buildings. This is termed the *mountain effect*.

directive aerial *Syn.* directional aerial. An *aerial that is a more effective transmitter or receiver of energy in some directions than in others (*compare* omnidirectional aerial). A common method of obtaining such directivity is by using a *passive aerial* and an *active aerial* in conjunction. The active aerial is one connected directly to the transmitter or receiver. The passive aerial is not so connected but influences the directivity by reacting with the active aerial. In transmission there is an induced e.m.f. produced in the passive aerial; in reception the reaction between the aerial elements results from their mutual inductance. A passive aerial placed behind an active aerial is called a *reflector;* one placed in front is a *director. See also* aerial array.

direct lightning surge *See* lightning stroke.

directly heated cathode *See* thermionic cathode.

director *Short for* director aerial. *See* directive aerial.

direct ray The path along which a radiowave travels in the minimum possible time between a transmitting and receiving aerial.

direct slice writing *See* electron beam lithography.

direct stroke *See* lightning stroke.

direct transition *See* direct-gap semiconductor.

direct voltage *See* d.c. voltage.

direct-voltage accelerator *See* accelerator.

direct wave The portion of a transmitted wave that travels along the path of a *direct ray. It may suffer from tropospheric refraction. *See also* ground wave. *Compare* indirect wave.

disc (1) *Colloq*. A gramophone record.

(2) *See* disk.

discharge (1) To remove or reduce the electric charge on a body such as a capacitor.

(2) The passage of an electric current or charge through a medium, often accompanied by luminous effects. *See* breakdown; gas-discharge tube; arc; spark.

(3) The conversion of chemical energy into electricity in a *cell by drawing current from it.

discharge lamp *See* gas-discharge tube.

discharge tube *See* gas-discharge tube.

discharge-tube rectifier A *gas-discharge tube that may be used as a rectifier, the electrodes being so arranged that the discharge current in one direction is greater than that in the opposite direction upon reversal of the applied voltage. *See also* mercury-arc rectifier.

discriminator (1) A circuit that converts a frequency-modulated or phase-modulated signal into an amplitude-modulated signal. *See* modulation.

(2) A circuit that selects signals with a particular range of amplitude or frequency and rejects all others.

disc thermistor *See* thermistor.

disc winding A type of winding for transformers consisting of flat coils in the form of a disc. Disc windings are usually employed for high-voltage applications. *Compare* cylindrical winding.

disintegration voltage The voltage at which the cathode of a gas-filled tube starts to disintegrate due to bombardment by positive ions. The voltage at which *cathode sputtering starts.

disk *Short for* magnetic disk. *See* moving magnetic surface memory.

dislocation An imperfection in the structure of a crystal. When dislocations occur in the crystalline structure of a semiconductor they can have serious deleterious effects. They introduce unwanted energy levels in the forbidden band (traps), they can alter the etching properties of the material, and can seriously change the electrical properties of devices. For example the values of source-drain current and threshold voltage of *field-effect transistors are strongly dependent on the dislocation density in the semiconducting substrate. The dislocation density in any particular crystal depends on the material used, the purity, and the method of production. Perfect or near perfect small crystals can be produced, but larger crystals are more difficult to produce without dislocations. Large virtually dislocation-free silicon crystals are now being produced, but the dislocation density in large gallium arsenide crystals is significant. Dislocation density can be determined by etching the crystal in a solvent that preferentially etches at dislocations, and then counting the etched pits. Dislocation density is therefore sometimes termed *etch pit density.*

displacement *Syns.* electric displacement; dielectric strain; electric flux density. Symbol: D; unit: coulombs per metre squared. A vector quantity defined as

$$D = \varepsilon_0 E + P$$

where E is the electric field strength, P the *dielectric polarization, and ε_0 the permittivity of free space. In a dielectric medium the total charge within any given closed surface consists of any free charges, ρ_e, in the surface together with the apparent charge density, $-\operatorname{div} P$, due to the polarization. The volume charge density then becomes $\rho_e - \operatorname{div} P$ and *Gauss's theorem relating to a dielectric is written as

$$\operatorname{div} E.\mathrm{d}\tau = (1/\varepsilon_0)\!\int(\rho_e - \operatorname{div} P)\mathrm{d}\tau$$

It can therefore be shown that the displacement D is given by

$$\operatorname{div} D = \rho_e$$

In a vacuum P is zero and Gauss's theorem is written as

$$\operatorname{div} E = \rho_e$$

displacement current The rate of change of electric flux in a dielectric with respect to time ($\partial D/\partial t$) when the applied electric field changes. No motion of charge carriers is involved apart from the setting up of electric dipoles (*see* dielectric polarization). A displacement current gives rise to magnetic effects similar to those of a conduction current and these effects form the basis of Maxwell's electromagnetic theory of light.

display The visual presentation of information as on the screen of a *cathode-ray tube or by using *scaling tubes in cascade, as in a *digital voltmeter.

disruptive discharge A sudden large increase in current in an insulator, semiconductor, or gas when the material breaks down under the influence of an applied electric field.

disruptive strength *See* dielectric strength.

dissipation *Syn.* loss. A loss of power due to the tendency of electronic circuits and components to resist the flow of current. In a resistive circuit the power dissipated is equal to I^2R, where I is the current and R the resistance. This is I^2R *loss*. In an inductor or capacitor the *dissipation factor* is the cotangent of the *phase angle, α, or the tangent of the loss angle, δ. In low-loss components it is almost equal to the power factor, $\cos\alpha$, and can be given approximately by $\sigma/2\pi f\varepsilon$, where σ is the electrical conductivity, ε is the permittivity of the medium, and f the frequency.

Dissipation causes free oscillations to be *damped and removes the sharpness of cut-off in *filters. High-frequency industrial heating is made possible because of dissipation of eddy-current energy in conductors (*see* induction heating) and displacement-polarization energy in dielectrics (*see* dielectric heating).

A network that is designed to absorb power is a *dissipative network,* as compared to a network that attenuates by impedance reflection. All networks provide some dissipation since entirely loss-free components cannot be made.

dissipation factor *See* dissipation.

dissipative network *See* dissipation.

dissociation constant *See* Ostwald's dilution law.

dissymmetric quadripole *See* quadripole.

dissymmetric transducer *See* transducer.

distortion The extent to which a system or component fails to reproduce accurately at its output the characteristics of the input. The modification of a waveform by a transmission system or network involves the introduction of features not present in the original input or the suppression or modification of features that are present. Distortion is a significant problem in telecommunication systems. There are several different types of distortion.

Amplitude distortion occurs when the ratio of the root-mean-square value of the output to the r.m.s. value of the input varies with the amplitude of the input, both waveforms being sinusoidal. If harmonics are present in the output waveform only the fundamental frequency is considered.

Attenuation distortion occurs when the gain or loss of the system depends on frequency.

Crossover distortion occurs in *push-pull operation when the transistors are not operating in the correct phase with each other.

Delay distortion is a change in the waveform because of the variation of the delay with frequency.

Deviation distortion occurs in frequency-modulated receivers that have an inadequate bandwidth or nonlinear discriminator.

Harmonic distortion is due to harmonics not present in the original waveform.

Intermodulation distortion results from spurious combination-frequency components in the output of a nonlinear transmission system when two or more sinusoidal voltages, applied simultaneously, form the input. Intermodulation distortion of a complex waveform arises from intermodulation (*see* modulation) within the waveform.

Nonlinear distortion is produced in a system when the instantaneous transmission properties depend on the magnitude of the input. Amplitude, harmonic, and intermodulation distortion are all results of nonlinear distortion.

Phase distortion occurs when the phase change introduced is not a linear function of frequency.

Optical distortion of an image is seen in electronic systems, such as *cathode-ray tubes, television picture tubes, etc., and in facsimile transmission. It is due to errors in the electron-lens focusing systems.

Aperture distortion of an image occurs in a scanning system when the scanning spot has finite dimensions rather than infinitely small dimensions.

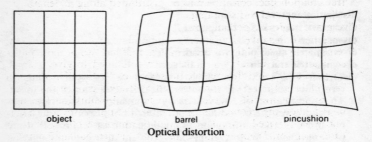

object   barrel   pincushion
Optical distortion

Barrel and *pincushion distortion* are seen when the lateral magnification is not constant but depends on image size (*see* diagram). Barrel distortion occurs when the magnification decreases with object size, pincushion distortion when it increases with object size.

Coma is a plumelike distortion of the spot occurring when the focusing elements of the electron gun are misaligned.

Keystone distortion is due to the length of the horizontal scan line varying with the vertical displacement of the line. It is most pronounced when the electron beam is at an acute angle to the screen and results in a trapezoidal image instead of a rectangular one. It can be removed using suitable transmitter circuits.

Origin distortion is a flattening of the waveform where it crosses the zero line in a gas-focused cathode-ray tube employing electrostatic deflection. It occurs at low deflecting voltages when there is a nonlinear relationship between the angular deflection and the deflecting voltage.

Trapezium distortion is a trapezoidal pattern on the screen of a cathode-ray tube instead of a square one and occurs when the deflecting voltage applied to the plates is unbalanced with respect to the anode.

distributed Bragg reflector laser (DBR laser) *Syn. for* distributed-feedback laser. *See* injection laser.

distributed capacitance (1) The capacitance of an electrical system, such as a transmission line, considered to be distributed along its length.

(2) The capacitance between individual turns on a coil or between adjacent conductors. The distributed capacitance in a coil lowers the inductance of the coil and may be represented by a single capacitor across the terminals.

distributed feedback laser (DFB laser) *Syn.* distributed Bragg reflector laser. *See* injection laser.

distributed floating-gate amplifier *See* charge-coupled device.

distributed inductance The inductance of an electrical system, such as a transmission line, considered to be distributed along its length.

distribution control *See* scanning.

divergence angle *See* debunching.

diversity gain *See* diversity system.

diversity system A communication system that has two or more paths or channels. The outputs of these are combined to give a single received signal and thus reduce the effects of *fading. The *diversity gain* is the gain in reception achieved by using a diversity system.

Frequency diversity employs independent transmission channels on neighbouring frequencies. *Space diversity* employs several receiving aerials spaced several wavelengths from each other. In both cases each aerial supplies its own receiver; the demodulated outputs are then combined. *Polarization diversity* uses aerials that are arranged to receive oppositely polarized waves.

divider A circuit that reduces the number of pulses or cycles by an integral factor.

dividing network *See* crossover network.

D-layer *Syn.* D-region *See* ionosphere.

D/MOS MOS circuits or transistors that are fabricated using double *diffusion. Regions of different conductivity type are formed by successive diffusion of different impurities through the same opening in the oxide layer. D/MOS devices are short-channel high-performance devices that were originally developed for microwave applications. They have a very precise channel length that is determined by the double diffusion rather than the inherently less precise method of *photolithography.

The speed of operation of an MOS transistor is determined by the channel length; for high operating speeds short channel lengths are required. Ordinary MOS transistors formed by a single diffusion (Fig. *a*) suffer from *punch-through at short channel lengths since the depletion layer associated with the reverse-biased p-n$^+$ drain

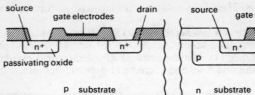

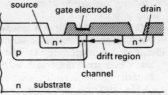

a Cross section of n-channel
MOSFET

b Cross section of n-channel
D/MOS transistor

junction spreads rapidly through the p-region as the drain voltage is increased. In the D/MOS device an n⁻ substrate is used. The p-regions are formed by diffusion (*see* planar process) followed by an n⁺ source/drain diffusion in which the source diffusion is made into the p-regions. An n⁺-p-n⁻-n⁺ structure is produced (Fig. *b*) in which an n⁻ region, termed the *drift region,* separates the p-region from the n⁺ drain. The drain junction therefore becomes a reverse-biased p-n⁻ junction and the associated depletion layer is almost entirely contained within the n-region.

The *breakdown voltage of a D/MOS device is determined by the width of the drift region up to a theoretical maximum determined by the characteristics of the n⁻ semiconductor (*see* avalanche breakdown; depletion layer). High-voltage MOS transistors can be formed using relatively wide drift regions. Breakdown voltages up to 300 volts with drift regions of about 25 micrometres have been produced. Devices with relatively short drift regions are produced as integrated circuits for low-power high-speed (up to microwave frequency) applications.

Epitaxial D/MOS transistors are formed in an n⁻ epitaxial layer grown on a p-type substrate. Individual transistors on a chip may then be isolated by performing extra p-type isolating diffusions. *See also* V/MOS.

domain (1) *See* ferromagnetism.
(2) *See* Gunn effect.

dominant mode *See* mode.

donor *Short for* donor impurity. *See* semiconductor.

donor level *See* semiconductor.

dopant *See* doping.

doping The addition of a particular type of impurity to a *semiconductor in order to achieve a desired n-conductivity or p-conductivity: donor impurities are added to form an n-type semiconductor and acceptor impurities a p-type. The impurity added is the *dopant*.

doping compensation The addition of a particular type of impurity to a *semiconductor in order to compensate for the effect of an impurity already present.

doping level The amount of doping necessary to achieve the desired characteristic in a semiconductor. Low doping levels (p, n) give a high-resistivity material; high doping levels (p$^+$, n$^+$) give a low-resistivity material. *See also* semiconductor.

doping profile *Syn. for* impurity profile. *See* Fick's law.

Doppler effect The change in the apparent frequency of a source of electromagnetic radiation (or sound) when there is relative motion between the source and the observer. The observed frequency f_o is given by

$$f_o = f(c - v_o)/(c - v_s)$$

where c is the velocity of light or sound, v_o is the velocity of the observer, and v_s is the velocity of the source.

The effect is utilized in *Doppler navigation,* which is a navigation system (in a moving object) that operates by ground reflection. *Doppler radar* employs the Doppler effect to distinguish between fixed and moving targets: the measurement of the change in the frequency of the reflected wave is used to determine the velocity and direction of the moving target.

Doppler navigation *See* Doppler effect.

Doppler radar *See* Doppler effect; radar.

dot generator A test generator used with a television receiver to adjust the convergence of the *picture tube. A pattern of evenly spaced dots or small squares is produced on a dark background and the dynamic focusing (*see* colour picture tube) is adjusted until a satisfactory image is formed on the screen.

dot matrix tube *See* colour picture tube.

double amplitude *Syn. for* peak-to-peak amplitude. *See* amplitude.

double-base diode *See* unijunction transistor.

double-beam cathode-ray tube *See* cathode-ray tube.

double-current system A telegraph system that reverses the direction of the electric current in order to effect transmission of the signals. *Compare* single-current system.

double detection reception *Syn. for* double superheterodyne reception. *See* superheterodyne reception.

double diffusion *See* diffusion.

double drift device *See* IMPATT diode.

double-ended *See* single-ended.

double heterostructure laser *See* injection laser.

double image *See* ghost.

double modulation Multiple *modulation involving two carriers only.

131

double-pole switch A switch that can operate simultaneously in two independent circuits. *Compare* single-pole switch.

double-sideband transmission Transmission of both sidebands generated when a *carrier wave is amplitude modulated, but not of the carrier itself. Double-sideband suppressed carrier systems are not often used because of the difficulties encountered when reintroducing the carrier at the receiver.

double superheterodyne reception *Syn.* double detection reception. *See* superheterodyne reception.

doublet *See* dipole.

drain The electrode of a *field-effect transistor through which *carriers leave the interelectrode space.

DRAM *Abbrev. for* dynamic random access memory. *See* dynamic memory.

D-region *Syn. for* D-layer. *See* ionosphere.

drift The variation with time of any electrical property of a circuit or apparatus. Drift often occurs during warm up or when the device is nearing the end of its useful life. In a voltage regulator or reference standard the variation of output voltage with respect to time is the *drift rate*.

drift mobility *Syn.* carrier mobility. Symbol: μ. The average velocity of excess *minority carriers in a *semiconductor per unit electric field. In general, *holes and *electrons have different mobilities.

drift rate *See* drift.

drift region *See* D/MOS.

drift space (1) A region in an *electron tube that is free of electric or magnetic fields.

(2) *See* klystron.

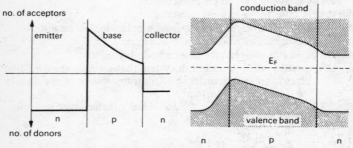

Variation of doping levels and energy bands in drift transistor

drift transistor *Syn.* graded-base transistor. A transistor in which the impurity concentration in the *base varies smoothly across the base region. The *doping level is high at the emitter-base junction drop-

132

ping to a low doping level (therefore high resistivity) at the base-collector junction. The impurity profile and associated energy levels of an n-p-n transistor are shown in the diagram. The bending of the conduction bands, required to maintain the constancy of the *Fermi level throughout the material, assists the passage of electrons through the base and thus reduces the base recombination current. The high-frequency response of these transistors is good.

drift tube *See* linear accelerator.

drift velocity The average velocity of electrons in a conductor or of ions in a gas, in the direction of applied electric field.

driven aerial *Syn. for* active aerial. *See* directive aerial.

driver A circuit or device that provides the input for another circuit or controls the operation of that circuit.

driving impedance *See* motional impedance.

driving-point impedance (1) The ratio at the input terminals of a network of the root-mean square (rms) value of the applied sinusoidal voltage to the rms value of the resulting current between the terminals.

(2) *See* motional impedance.

driving potential *See* photocell.

DRO *Abbrev. for* destructive read operation. *See* read.

droop *See* pulse.

drop-in The presence of a spurious *bit in the information recorded at the input or output of a digital *computer due to faulty reading of the input/output signals. Drop-in is most commonly caused by dust particles or other irregularities of the magnetic medium in magnetic tape or disk input/output devices. *Drop-out* occurs when the reading head fails to read a bit.

drop-out *See* drop-in.

dropper *See* dropping resistor.

dropping resistor. *Syn.* dropper. A resistor introduced into a circuit to provide a voltage drop across its terminals and hence reduce the voltage in the circuit.

dry battery *See* battery.

dry cell *See* cell.

dry etching *Syn.* plasma-assisted etching. *See* etching.

dry flashover voltage *See* flashover voltage.

dry joint A faulty soldered joint that has a high resistance because of a residual oxide film.

DTL *Abbrev. for* diode transistor logic.

D-type flip-flop *See* flip-flop.

dual-channel sound A technique used in television receivers to separate the sound and video signals after the common first detector stage. Separate intermediate-frequency stages are employed for each signal.

133

dual in-line package (DIP; DIL package) A standard form of package used for integrated circuits. It consists of a ceramic or plastic casing containing a *leadframe. The frame is used to form the connections to the bonding pads of the integrated circuit and is connected to output pins arranged in two parallel lines at opposite sides of the package. The number of pins available varies from eight with smaller circuits to about 48 for large circuits. *See also* leadless chip carrier; pin grid array; tape automated bonding.

dubbing The combining of two sound signals into a composite recording. At least one source of sound will have been prerecorded.

dull emitter *See* thermionic cathode.

dummy antenna *See* artificial aerial.

duplexer A two-channel multiplexer (*see* multiplex operation) that uses a *transmit-receive (TR) switch so that one aerial may be used for both reception and transmission. The switch protects the receiver from the high power of the transmission. Duplexers are commonly used in *radar, the TR switch operating in the time between transmission of the pulse and reception of the return echo.

duplex operation Simultaneous operation of a communications channel in both directions. *Half-duplex operation* occurs when the operation is limited to either direction but not both directions at once. *Compare* simplex operation.

Dushman's equation *See* Richardson's equation.

dust core A magnetic core that is made from a powdered material such as *ferrite. Such cores have a very low eddy-current loss at high frequencies.

duty A statement of the operating conditions and their durations to which a device is subjected, including rest and de-energized periods. *Uninterrupted duty* is the operation of a device without any off-load (*see* load) periods. *Continuous duty* is uninterrupted duty that continues for an indefinite time. *Intermittent duty* has on-load (*see* load) periods alternating with off-load periods. When the on-load periods are small in comparison with off-load periods the intermittent duty is termed *short-time duty; periodic duty* occurs if the load conditions are regularly recurrent. Operation at loads and for durations that are both subject to wide variation is *varying duty*. The *duty cycle* is a group of variations of load with time. The ratio of the on-load period under specified conditions to the sum of the on-load and off-load periods is the *duty ratio*.

duty cycle *See* duty.

duty factor (of a pulse train) *See* pulse.

duty ratio *See* duty.

DVM *Abbrev. for* digital voltmeter.

dynamic Denoting any electrical device, circuit, or apparatus in which the electrical parameters are constantly changing. The term can be applied to those devices that operate with alternating currents and voltages, especially those with a marked frequency dependence. It is also used to describe components, such as varactors, in which an electrical property varies with the operating conditions, e.g. if the reactance is a function of the applied voltage. Any device or circuit in which the signals decay over a period of time unless regenerated, is also termed dynamic. The MOS random access memory is a particular example of this type of device. Any component, circuit, or device that is not dynamic or that is operated with essentially constant electrical conditions is described as *static*.

dynamic characteristic *See* characteristic.

dynamic current injection *See* charge-coupled device.

dynamic focusing *See* colour picture tube.

dynamic impedance *Syn.* dynamic resistance. The impedance at resonance of a parallel *resonant circuit. It is purely resistive by definition (*see* resonant frequency).

dynamic memory A solid-state memory in which the stored information decays over a period of time. The decay time can range from milliseconds to seconds depending on the nature of the device and its physical environment. The memory cells must undergo refresh operations sufficiently often to maintain the integrity of the stored information. MOS random-access memory and CCD memory are both dynamic memories.

dynamic operation *See* MOS integrated circuit.

dynamic range The range over which an active electronic device can produce a suitable output signal in response to an input signal. It is often quoted as the difference in decibels between the noise level of the system and the level at which the output is saturated (the *overload level*).

dynamic resistance (1) The resistance of any electronic device under normal operating conditions. Many devices exhibit a variation of resistance with frequency.

(2) *See* dynamic impedance.

dynamic sensitivity *See* phototube.

dynamo An electromagnetic *generator that produces either alternating or direct current. It consists of a plane coil that is made to rotate in a uniform magnetic field of flux density B. The coil, of area A, rotates with angular velocity ω; an angle θ is subtended by the normal to the plane of the coil and B. If time $t = 0$ is chosen so that $\theta = \omega t$, then the e.m.f. induced in the coil is given by:

$$V = A\omega B \sin \omega t$$

This is an alternating voltage, of period ω, that is maximal when the plane of the coil lies in the direction of the magnetic field.

Alternating current will flow if the two ends of the coil are connected to a pair of slip rings, R and R' (Fig. *a*). The ends of the coil can be connected to a commutator by means of semicircular seg-

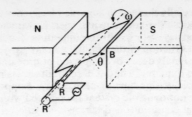

a Slip-ring operation of a dynamo

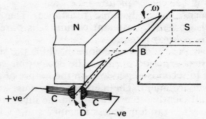

b Carbon-brush operation of a dynamo

ments, D, making contact with a pair of carbon brushes, C (Fig. *b*). Then the voltage in each terminal always has the same sign since each segment moves to the next brush as the e.m.f. changes sign. This produces a direct current in a load connected across the brushes. The current is not steady however since the induced voltage is alternating. The fluctuations about the mean value are known as *ripple*.

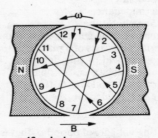

c 12-pole drum armature

The ripple may be reduced using a drum armature. This has the coil wound around a drum-shaped armature so as to produce sever-

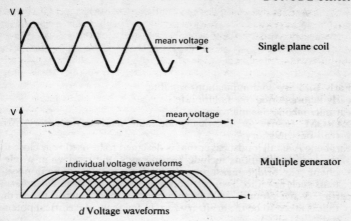

d Voltage waveforms

al plane coils symmetrically spaced around the drum (Fig. *c*); each vertical conductor forms a pole. The brushes are placed opposite each pole and are connected in pairs. The effect is to superimpose a number of voltage waveforms all varying slightly in phase (Fig. *d*). The output voltage has a higher mean value and less magnitude of ripple than that of the single coil. The period of the ripple voltage, which corresponds to the time interval at which successive conductors occupy the same position, is also much higher.

dyne The unit of force in the obsolete CGS system of units. One dyne equals 10^{-5} newton.

dynode *See* electron multiplier.

dynode chain *See* electron multiplier.

E

Early Bird *See* communications satellite.

early failure period *See* failure rate.

ear microphone *See* microphone.

EAROM *Abbrev. for* electrically alterable read-only memory. *See* read-only memory.

earphone A small loudspeaker that is designed to be used very close to the ear. Applications include hearing aids, the receiver in a telephone, a.c. bridge measurements, and use with reproduction systems such as radio. Two earphones used together form a *headset*.

earth *U.S. syn.* ground. (1) A large conducting body, such as the earth, that is taken to be the arbitrary zero in the scale of electrical potential.

(2) A connection, which may be accidental, between a conductor and the earth. An effective earth for electrical equipment is formed by a wire connected to a cold water pipe. Connection may also be made to an *earth electrode,* i.e. a large copper plate buried in moist soil.

(3) The point or portion in an electric circuit or device that is at zero potential with respect to earth.

(4) To connect an electric circuit or device to earth.

earth bus *See* bus.

earth capacitance The capacitance between any circuit or equipment and a point at *earth potential.

earth current (1) A current that flows to earth, especially one that results from a fault in a system.

(2) Any current flowing in the earth. Particular earth currents are associated with ionospheric disturbances. Buried cables sometimes have their lead sheaths corroded by a direct earth current.

earth electrode *Syn.* earth plate. *See* earth.

earth fault A fault that occurs when a conductor is accidentally connected to earth or when the resistance to earth of an insulator falls below a specified value.

earth plane *See* ground plane.

earth plate *Syn. for* earth electrode. *See* earth.

earth potential *Syn.* zero potential. The potential of a large conducting body, such as the earth, taken to be the arbitrary zero in the scale of electrical potential.

earth-return circuit A circuit composed of one or more conductors in parallel that connect two points in a *telecommunication system and that is completed through earth at the two points.

E/B *Abbrev. for* electrode per bit operation. *See* CCD memory.
e-beam lithography *Short for* electron-beam lithography.
e-beam resist *See* electron-beam lithography.
E-bend *Syns.* E-plane; edgewise bend. *See* waveguide.
EBIC *Abbrev. for* electron-beam induced current analysis.

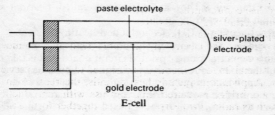

paste electrolyte

silver-plated electrode

gold electrode

E-cell

E-cell A solid-state timing device consisting essentially of a thimble-shaped electrolytic cell with a central gold electrode, an outer silver-plated electrode, and a paste electrolyte consisting of a suitable silver salt (*see* diagram). When a current passes, silver is lost from the outer electrode, which is made the anode, and is deposited at the same uniform rate on the central gold cathode. After a given time, determined by the elements of the external circuit, the current reverses and the cathode becomes the anode. When the cathode is 'deplated' the reverse current ceases and the process is once more reversed.

An E-cell is a useful and versatile timing device since it can be made very small, is robust, and can be used for a wide range of time intervals simply by altering the external circuit. The desired time interval is usually determined by charging and discharging a capacitor.

e.c.g. *Abbrev. for* electrocardiograph or electrocardiogram.
ECH *Abbrev. for* eddy-current heating. *See* induction heating.
echo (1) A communications wave that has been reflected or refracted and that has sufficient magnitude and delay to distinguish it from the direct wave. In radio an echo is heard. In television it appears as a *ghost on the screen; it may or may not be heard simultaneously.

(2) The portion of a transmitted radar signal reflected back to the receiver.

Echo *See* communications satellite.
echo sounding A system based on the same principles as *radar but using sound waves instead of radiowaves. *See* sonar.
ECH set *Short for* eddy-current heating set. A radiofrequency generator that is used to produce the alternating magnetic field used for *induction heating.
ECL *Abbrev. for* emitter-coupled logic.

139

eddy current A current induced when a conductor is subject to a varying magnetic field. Energy is dissipated by eddy currents (*eddy-current loss*), usually appearing as heat, and becomes significant in high-frequency applications. This effect is utilized in *induction heating. Eddy currents in a moving conductor react with the magnetic field to produce retardation of the motion and are used to produce damping. They are sometimes called *Foucault currents* (although discovered by Joule).

eddy-current heating (ECH) *See* induction heating.

eddy-current loss *See* eddy current.

edge connector A track on a *printed circuit board that is taken to one edge of the board to form a connector. Each board has several edge connectors, which may be plugged into a suitable socket allowing connections to be made to the circuit on the board.

edge effect Deviation from parallel in the lines of force representing the electric field at the edges of parallel-plate capacitors. This results in a field that is nonuniform at the edges.

edge profile The shape of the semiconductor edges produced after *etching away portions of the slice to form *mesas, particularly after wet etching. The shape of the edge profile is a function of the crystalline structure of the semiconductor, the crystal orientation, and the particular etchant used. Gallium arsenide for example has a very different crystal structure to silicon and therefore has etch characteristics that are different. Specific crystalline planes can have etch rates that are significantly different, due to anisotropy of the crystalline structure. Materials with larger anisotropy show greater differences in etch rates. The shape of the resulting profile for a given crystal orientation can be predicted if the relative etch rates are known, and the predicted shape represented on a *Wulff plot*. If the dissolution rates in given directions are represented by vectors, and the normals to these vectors plotted on a polar diagram, the resulting diagram gives an envelope of normals corresponding to the shape of the edge.

edge profile modification *See* lift-off.

edgewise bend *Syn. for* E-bend. *See* waveguide.

Edison cell *Syns.* nickel-iron cell; Ni-Fe cell. A secondary *cell in which the plates are impregnated steel grids. The positive plate is a nickel/nickel hydrate mixture and the negative one is iron oxide. Potassium hydroxide is the electrolyte. The cell is lighter and more durable than lead cells and is capable of delivering heavy currents. E.m.f.: $\sim$ 1.3 volts.

Edison effect *Obsolete syn. for* thermionic emission.

EDS *Abbrev. for* energy dispersive spectroscopy. *See* electron microprobe.

e.e.g. *Abbrev. for* electroencephalograph or electroencephalogram.

effective resistance *Syn.* alternating-current resistance. The resistance to alternating current of a conductor or other circuit element. It is measured as the power in watts dissipated as heat divided by the current in amperes squared. It includes the resistance to direct current and resistance due to *eddy currents, *hysteris, and the *skin effect.

effective value *See* root-mean-square value.

EFM *Abbrev. for* eight to fourteen modulation. *See* compact disc system.

EHF *Abbrev. for* extremely high frequency. *See* frequency band.

e.h.t. or **EHT** *Abbrev. for* extra high tension. Usually applied to the high-voltage supply for cathode-ray tubes or television picture tubes.

EIA *Abbrev. for* Electronic Industries Association. *See* standardization.

Einstein photoelectric equation *See* photoelectric effect.

Einthoven galvanometer *Syn.* string galvanometer. A *galvanometer that has a tightly strung conducting thread between the poles of a strong electromagnet. A current passed through the thread causes it to be deflected at right angles to the magnetic field. The deflection is observed with a high-power microscope. The instrument is highly sensitive and can detect currents of 10^{-11} ampere.

elastance Symbol: S; unit: farad^{-1}. The reciprocal of *capacitance.

elastic recoil detection analysis (ERDA) A technique used for surface analysis of a material up to depths of one micrometre. The technique is similar to *Rutherford back scattering (RBS), but uses a beam of energetic heavy charged particles rather than the alpha particles used in RBS. The ERDA technique is used for detecting and measuring the concentration of light elements in the material, including hydrogen.

elastoresistance A change in the resistance of a material when it is stressed within its elastic limit.

E-layer *Syns.* E-region; Heaviside layer; Kennelly-Heaviside layer. *See* ionosphere.

electret A substance that is permanently electrified and has oppositely charged extremities: it is the electrical analogue of a permanent magnet. Electrets have been used in *electrometers and in *capacitor microphones.

electrically alterable read-only memory (EAROM) *See* read-only memory.

electric axis The direction in a crystal of maximum *conductivity. It is the X-axis of a *piezoelectric crystal.

electric charge *See* charge.

electric conduction *See* conduction.

electric constant *Syn. for* permittivity of free space. *See* permittivity.

electric controller *See* automatic control.

electric current *See* current.

electric degree One 360th part of an alternating current cycle. Currents or voltages arising in different parts of a circuit may be represented as vectors; the *phase difference is the angle, expressed in electric degrees, between the vectors.

electric dipole *See* dipole.

electric dipole moment *See* dipole.

electric displacement *See* displacement.

electric eye *Colloq.* A *photocell or *photovoltaic cell, particularly when used in burglar alarm systems, lifts, etc.

electric field The space surrounding an electric charge or varying magnetic field in which another electric charge experiences a perceptible mechanical force. The charge acted upon by the field is assumed to be sufficiently small so that the electrical conditions are not altered by it. *See also* electric field strength; Coulomb's law.

electric field strength Symbol: E; unit: volts per metre. The strength of an *electric field at a point, measured in terms of the mechanical force per unit charge. The force, F, given by *Coulomb's law, is thus related to field strength by:

$$F = eE$$

where e is the electron charge.

The potential difference between two points separated by distance s is given by

$$dV = -E.ds$$

or in more general terms

$$E = -i\partial V/\partial x - j\partial V/\partial y - k\partial V/\partial z$$

where i, j, and k are unit vectors along the x-, y-, and z-axes, respectively.

electric flux Symbol: Ψ; unit: coulomb. The quantity of electricity displaced across a given area in a dielectric. It is given by the scalar product $D.dS$ of the electric *displacement, D, and an element dS of area. *Gauss's theorem states that

$$\int D.dS = \int \rho_e d\tau$$

where ρ_e is the volume electric charge density in a small volume element $d\tau$. The total flux through an area surrounding a charge q is thus equal to q. This flux is unaltered by the presence of a dielectric medium.

electric flux density *See* displacement.

electric heating The production of heat from electric energy. Methods used include passing a current through a *resistance, use of an electric *arc, *induction heating, and *dielectric heating.

electric hysteris loss Electrical loss in a dielectric meterial due to internal forces in the material produced by a varying electric field. The loss usually appears in the form of heat.

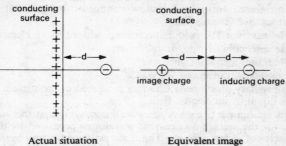

Electric image

electric image *Syn.* image charge. A concept devised by Kelvin for solving electrostatic problems associated with a point charge in the vicinity of a conducting surface. It can be shown that under certain conditions the charges induced, by the point charge, on the surface of the conductor have electrical effects identical with those that would be produced by an imaginary point charge located at a particular point relative to and below the surface. This imaginary point charge is known as the electric image of the first charge. The force exerted on the first charge, due to the induced surface charges, may be calculated as if it were due to the image and is called the *image force*. In the case of an infinite plane surface the image charge is equal and opposite in sign to the inducing charge and at a distance behind the plane equal to the distance of the actual charge in front of it (*see* diagram).

electric intensity *Obsolete syn. for* electric field strength.

electric interference Interference caused by electrical apparatus (other than radio stations) and creating disturbances to a signal.

electricity The phenomena associated with static or dynamic electric charges, such as electrons.

electric moment *Syn. for* dipole moment. *See* dipole.

electric polarization *See* dielectric polarization.

electric potential Symbol: V; unit: volt. The work required to bring a unit positive charge from infinity to a point in an *electric field. The potential is one volt when one joule is needed to transfer a charge of one coulomb.

electric recording A sound recording system that uses the amplified signals from one or more suitable microphones. *See* recording of sound.

electric screening The surrounding of apparatus with an electrical conductor to prevent interference from unwanted electrical disturbances. The shield is usually a *Faraday cage,* which is an earthed wire screen completely surrounding the apparatus. No electric field can be produced inside the hollow conductor by external electric charges. *See* Gauss's law.

electric spectrum The colour spectrum produced by an electric *arc.

electric susceptibility *See* susceptibility.

electric transducer *See* transducer.

electroacoustic transducer *See* transducer.

electrocapillary phenomena Changes produced in the surface tension of a liquid by an electric field.

electrocardiograph (e.c.g.) A sensitive instrument that measures and records the voltage and current waveforms produced by the heart muscles of living animals. The trace produced is an *electrocardiogram.*

electrochemical equivalent The mass of any ion that is deposited from an electrolyte solution by a current of one ampere flowing for one second. *See also* coulombmeter.

electrochemical series *See* electromotive series.

electrode A device that emits, collects, or deflects electric charge carriers. It is usually in the form of a solid plate, a wire, or a grid that controls current into and out of an electrolyte, gas, vacuum, dielectric, or semiconductor. Liquid mercury electrodes are also used. The *electrode current* is the current that flows through a specified electrode, such as a *collector, *grid, or *drain.

electrode a.c. resistance *See* slope resistance.

electrode characteristic *See* characteristic.

electrode current *See* electrode.

electrode differential resistance *See* slope resistance.

electrode dissipation The heat dissipated in a particular electrode, usually the anode, as a result of bombardment by electrons, ions, or radiation from other electrodes.

electrode drop The potential difference across an electrode caused by its resistance.

electrode efficiency The ratio of the actual mass deposited from solution in an electrolytic *cell to the theoretical mass.

electrode per bit operation (E/B) *See* CCD memory.

electrodeposition *Deposition by electrolysis.

electrode potential *See* electromotive series.

electrodynamic instrument An instrument in which the interaction of magnetic fields produced by currents in a system of movable and fixed coils produces a torque. These instruments operate with both direct and alternating currents.

electrodynamics The study of the mechanical forces acting upon and between currents.

electrodynamometer An electrodynamic instrument that contains two coils, one fixed and one movable, and produces a deflection equal to the square of the current passing through the coils, which are connected in series. It is frequently used as a standard for current or voltage measurements.

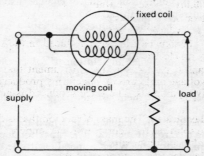

Single-phase electrodynamic wattmeter

For use as a *wattmeter, the fixed coil is connected in series with the load and the movable coil shunted across a high noninductive resistance (*see* diagram). Transformer coupling to the current circuit may be used to isolate the instrument from the main current. The couple produced is then proportional to the wattage.

electroencephalograph (e.e.g.) A sensitive instrument that measures and records the voltage waveforms produced by the brain. The trace produced is an *electroencephalogram*.

electroendosmosis *See* electro-osmosis.

electrogen A photoelectric molecule.

electroless plating Any plating process performed without the presence of an external applied voltage (*compare* electroplating). The techniques are particularly appropriate for plating nonconducting or semiconducting surfaces with a noble metal such as gold. The surface to be plated is immersed in an appropriate solution. *Immersion plating* proceeds by the chemical replacement of a low electrode potential metal on the surface by the required higher electrode potential metal. The reaction ceases once a thin layer has formed. *Autocatalytic plating* involves the continuous chemical reduction of the more noble metal to form a coating on the base metal. The coating metal may act as a catalyst for the reduction. Both techniques are complex.

145

electroluminescence *Syn.* Destriau effect. The emission of light by certain phosphorescent substances under the influence of an applied electric field. The phenomenon is utilized in the *luminescent lamp:* this consists of a slab of dielectric material within which phosphorescent power is dispersed and which is sandwiched between transparent plate electrodes. Light is produced when voltages of 400 to 500 volts are applied across the plates.

electrolysis The production of chemical change, usually decomposition, when current flows through an electrolyte.

electrolyte A substance that conducts electricity when in solution or when molten because of its dissociation into *ions. *Strong electrolytes* are compounds, such as mineral acids, that are completely dissociated into ions when in solution. *Weak electrolytes* are compounds that are only partially dissociated in solution. Weak solutions of such compounds are better conductors than strong solutions because of their greater degree of dissociation (*see* Ostwald's dilution law).

electrolytic capacitor Any *capacitor in which the dielectric layer is formed by an electrolytic method. The capacitor does not necessarily contain an electrolyte. When a metal electrode, such as an aluminium or tantalum one, is operated as the anode in an electrolytic cell a dielectric layer of the metal oxide is deposited. The capacitor is formed using either a conducting electrolyte as the second electrode or a semiconductor, such as manganese dioxide. The electrolyte used is either in liquid form or in the form of a paste, which saturates a paper or gauze. Tantalum capacitors containing a nonliquid electrolyte are usually in bead form. Electrolyte capacitors have a high capacitance per unit volume but suffer from high leakage currents.

electrolytic cell *See* cell.

electrolytic dissociation The reversible separation of certain substances in solution into oppositely charged electrolytic ions. The positively charged ions are *cations, the negatively charged ones *anions. Some compounds, such as sodium chloride, are completely dissociated whereas others, such as acetic acid, are only partially dissociated. *See also* electrolyte.

electrolytic meter An instrument, such as a *coulombmeter, that measures electric charge by the amount of material deposited or gas liberated electrolytically. It can be used as an energy meter with a constant voltage supply; it is then calibrated in joules or kilowatt-hours.

electrolytic photocell An electrolytic cell, constructed from certain materials such as selenium electrodes in selenium dioxide solution, that is sensitive to light when a small external d.c. voltage is applied across it. The cell has a linear sensitivity of about one milliampere per lumen of luminous flux.

electrolytic polarization A reduction in the potential difference across an electrolytic system caused by polarization in the electrolyte.

electrolytic polishing *Syn.* electropolishing. Polishing of a metal by making it the anode of an electrolytic cell and passing current through the cell. Under suitable conditions the protuberances dissolve preferentially leaving a smooth lustrous surface.

electrolytic rectifier A *rectifier that has two dissimilar electrodes in an electrolyte. Suitable combinations of electrodes and electrolyte produce a system that conducts much more readily in one direction than the other. A typical combination is the *tantalum rectifier* that contains tantalum and lead electrodes and a dilute solution of sulphuric acid as the electrolyte.

electrolytic separation The separation of isotopes by electrolysis, based on the fact that different isotopes of an element are liberated at an electrode at different rates.

electrolytic tank A device that can be used to predict the behaviour of certain physical systems, such as a system of electrodes, by immersing a metal scale model of the system in a poorly conducting electrolyte. When suitable voltages are applied to the model the boundary conditions of the electrodes are determined by plotting lines of equipotential in the electrolyte.

electromagnet A device that is magnetized only when an electric current flows in it. It consists of a helical winding around which a magnetic field is produced when current flows through the winding. A ferromagnetic *core is almost invariably used: the design of the core is a major factor in determining the magnetic flux density. The winding around which the field is produced is the *field coil;* the current used to produce the field is the *field current.*

electromagnetic balance *See* Cotton balance.

electromagnetic deflection *Syn.* magnetic deflection. The use of *electromagnets to deflect an electron beam. The most common use is in the *cathode-ray tube in which two pairs of deflection coils produce vertical and horizontal deflections.

electromagnetic focusing *Syn.* magnetic focusing. *See* focusing.

electromagnetic induction The production of an electromotive force in a conductor when there is a change in magnetic flux through the conductor. The laws of electromagnetism may be expressed as follows.

(i) When a moving conductor cuts the flux of a magnetic field or when a changing magnetic field crosses a conductor an *induced electromotive force* is produced across the conductor.

(ii) *Faraday–Neumann law*: if a conductor cuts a magnetic flux, ϕ, the induced potential difference, V, is propotional to the rate of change ($d\phi/dt$) of flux.

(iii) *Lenz's law*: the induced potential difference is in such a direction as to oppose the change that produces it:

147

$$V = -\mathrm{d}\phi/\mathrm{d}t$$

If a current in a circuit varies, the associated magnetic flux also changes in direct proportion causing a back e.m.f. This is *self-inductance*, and the back e.m.f. is given by

$$V = -L\mathrm{d}I/\mathrm{d}t$$

where I is the current and L is the coefficient of self-inductance, also called *self-inductance*, which is measured in henrys.

The change in flux associated with a varying current can also link with another circuit and produce an e.m.f. in it. This is *mutual inductance*. The induced e.m.f. in a second circuit is given by

$$V_2 = -M\mathrm{d}I_1/\mathrm{d}t$$

where M is the coefficient of mutual inductance, also called *mutual inductance*, which is measured in henrys. In an ideal mutual inductance between two circuits with self-inductance L_1 and L_2,

$$M^2 = L_1 L_2$$

electromagnetic lens *Syn.* magnetic lens. An *electron lens consisting of an arrangement of coils that focuses an electron beam electromagnetically. *See* focusing. *Compare* electrostatic lens.

electromagnetic radiation Energy that is radiated by a charged particle undergoing acceleration. Transverse sinusoidal electric and magnetic fields are propagated at right angles to each other and to the direction of motion, the instantaneous values of the fields being related to the charge and current densities by *Maxwell's equations. These equations define the fields as *electromagnetic waves* propagated through free space with a constant velocity c, where c is the velocity of light and is equal to $2 \cdot 997\,924\,58 \times 10^8$ metres per second.

Just as moving charged particles have associated wavelike features (*see* de Broglie waves) so electromagnetic radiation has a wave/particle duality: it may also be considered as a stream of particles (*photons) that move at the velocity of light, c, and have zero rest mass. Although wave motion is sufficient to explain the properties of reflection, refraction, and interference, *quantum theory, which is concerned with the particulate nature of electromagnetic waves, must be used to explain phenomena, such as the *photoelectric effect, that occur when radiation and matter interact.

The characteristics of the radiation depend on the frequency, ν, of the waves. The frequency, wavelength λ, and velocity are related by

$$c = \nu\lambda$$

The photons have energy E related to the frequency ν by

$$E = h\nu$$

148

where h is the *Planck constant. Energy is exchanged between radiation and matter by absorption and emission in discrete amounts, termed *quanta, the energy of each quantum being $h\nu$.

The total range of possible frequencies is defined as the electromagnetic spectrum (*see* Table 10, backmatter). *Radiowaves have the lowest frequencies; progressively higher frequencies are associated with infrared radiation, light, ultraviolet radiation, *X-rays, through to *gamma rays at the highest frequencies.

electromagnetic relay *See* relay.

electromagnetic spectrum The entire range of frequency of *electromagnetic radiation. *See* Table 10, backmatter.

electromagnetic strain gauge *Syn.* variable inductance gauge. *See* strain gauge.

electromagnetic units (emu) *See* CGS system.

electromagnetic wave *See* electromagnetic radiation.

electromagnetism (1) The magnetic phenomena associated with moving electric charges, the electric phenomena associated with varying magnetic fields, and the magnetic and electric phenomena associated with electromagnetic radiation.

(2) The study of these phenomena.

electromechanical apparatus Any device or apparatus that converts electrical signals into related mechanical movement, or vice versa. Electromechanical devices, such as calculators or counters, are now being replaced by fully electronic devices; the greatest use for electromechanical devices is for sound-reproduction apparatus, such as record cutters and players.

electrometer A device that measures potential difference or electric charge by means of the mechanical forces that exist between bodies that possess electrostatic charges. *See* attracted-disc electrometer; quadrant electrometer; string electrometer; electroscope.

electromotive force (e.m.f. or emf) Symbol: E; unit: volt. The property of a source of electrical energy that causes a current to flow in a circuit. The algebraic sum of the potential differences in a circuit equals the e.m.f., which is measured by the energy liberated when unit electric charge passes completely round the circuit. A battery of e.m.f. E will supply a current I to an external resistance R:

$$E = I(R + r)$$

where r is the internal resistance of the battery. The term 'electromotive force' strictly applies to a source of electrical energy but is sometimes misused as being equivalent to *potential difference.

electromotive series *Syn.* electropositive series. If a metal is placed in a molar solution of one of its salts a potential, termed the *electrode potential,* is developed between the metal and the solution. The electromotive series is the arrangement of chemical elements in order of their electrode potentials (*see* Table 13, backmatter). The standard

149

reference is the *hydrogen electrode,* which is given arbitrarily the value zero. This electrode consists of gaseous hydrogen at a pressure of one atmosphere in contact (by means of a platinum electrode) with an acidic solution containing one molar hydrogen ions at 25°C. Elements that come above hydrogen in the electromotive series and tend to give up electrons to acquire a positive charge are *electropositive*; those below it, including the halogens, tend to acquire a negative charge and are *electronegative.*

The electromotive series implies the relative ability of the elements to form positive ions in solution and in this context it is often referred to as the *electrochemical series.* Elements higher in the table form positive ions more readily than those lower in the series and will therefore displace lower elements from solution. Zinc, for example, will displace copper from solution because it is higher in the series. If a strip of zinc is placed in a solution of copper ions, copper will be deposited and zinc ions will form in the solution.

electromyograph A sensitive instrument, such as an *electrocardiograph, that measures and records the current waveforms generated by contractions in muscles. The trace produced is an *electromyogram.*

electron A stable elementary particle that has a negative charge, e, of 1.602×10^{-19} coulomb, mass m of 9.109×10^{-31} kg, and spin 1/2. It is the natural unit of electric charge. Electrons are constituents of all atoms, moving around the nucleus in several possible or 'allowed' orbits (*see also* energy levels). They also exist independently. They are primarily responsible for electrical conduction in most materials (*see* energy bands). Electrons moving in one direction under the influence of an electric field constitute an electric current, the direction of conventional current flow being opposite to the direction of motion of the electrons. Electrons were first discovered as *cathode rays by Sir J. J. Thomson in 1897.

Electrons are liberated by various effects: in *gas-discharge tubes by the ionization of gas molecules; by heating metal filaments (*thermionic emission); by the action of light, ultraviolet radiation, X-rays, or gamma-rays on matter (*photoelectric effect); by the application of an intense electric field at the surface of a metal (*field emission); by the scattering of X-rays or gamma-rays (Compton effect); in *beta decay, when they are spontaneously emitted as *beta particles from certain radioactive elements.

Electrons experience an electric force F_E in the presence of an electric field E; moving electrons experience a transverse force F_H in the presence of a magnetic field. These have values

$$F_E = Ee$$
$$F_H = ev \times B$$

where B is the magnetic flux density vector acting at right angles to v, the velocity vector. These forces are utilized in the *focusing of electron beams. The electric force also accelerates the electrons.

Electrons interact with matter to produce effects dependent on the velocity of the electrons and the state of matter involved. They may undergo elastic scattering, producing deflection and localized heating due to the energy lost, or inelastic collisions, losing discrete amounts of energy and producing various phenomena: with a gas, excitation and light emission or ionization occurs; with solids and liquids, effects such as the production of X-rays, fluorescence, and secondary electron emission are observed.

An electron beam also has wavelike properties, similar to those of electromagnetic radiation, the wavelength being given by

$$\lambda = h/mv$$

where h is the Planck constant and mv the electron's momentum (*see also* de Broglie waves).

The antiparticle of the electron is the *positron,* which has a positive charge and a mass equal to that of the electron.

electron affinity (1) Symbol: A or E_a; unit: electronvolt. The energy released when an electron becomes attached to an atom or molecule to form a negative ion. Many atoms or molecules have a positive electron affinity, i.e. the negative ion is more stable than the neutral species.

(2) Symbol: χ; unit: electronvolt. The difference in energy between the bottom of the conduction band, E_c, in a semiconductor and the vacuum level, i.e. the energy of free space outside the semiconductor. *Compare* work function.

electron beam A beam of *electrons that are usually emitted from a single source, such as a *thermionic cathode.

The *electron-beam voltage* at a point in an electron beam is the average, with respect to time, of the voltage between that point and the electron-emitting surface. The *electron-beam d.c. resistance* is the ratio of the electron-beam voltage and the direct-current output of the electron beam. *See also* electron gun.

electron-beam d.c. resistance *See* electron beam.

electron-beam device Any device that utilizes one or more electron beams as an essential part of its operation. Several electrodes may be present to form, control, and direct the electron beam. The deflected beam is made to strike a fluorescent screen in measuring instruments such as the *cathode-ray tube. *See also* electron microscope; klystron.

electron-beam induced current analysis (EBIC) A technique that is used to detect crystal defects in a semiconductor slice. The technique requires the presence of a p-n junction, Schottky barrier, or MOS capacitor to be successful. An electron beam is scanned across the sample, which leads to the generation of electron-hole pairs. The charges generated are collected by the diode and the resulting current detected. Any defect or inhomogeneity that affects the production or recombination of the electron-hole pairs will also affect the current detected. Use of a *cathode-ray tube synchronized with the scanning electron beam produces an image of the area scanned. EBIC can be easily added to many scanning *electron microscopes.

electron-beam lithography (e-beam lithography) A method of *lithography that uses energetic electrons instead of light to expose the *resist. It is used to form patterns on photomasks for use in *photolithography. It can also be used to form patterns directly on slices – *direct slice writing*. This latter application is often used to form small features on slices since greater resolution is possible using an electron beam than with photolithography.

Special resists – *e-beam resists* – are used. These are not usually sensitive to ordinary light and can therefore be handled under normal lighting conditions. Both positive and negative resists are available. Unlike photolithography no mask is used in e-beam lithography. The electron beam is scanned across the slice under computer control in order to expose the resist in the desired pattern. Two main scanning methods are used: *raster scanning*, in which the beam is moved backwards and forwards across the slice in a rectangular pattern and the beam switched on and off as appropriate, and *vector scanning*, in which the beam is moved, under computer control, to trace out the pattern directly. The scanning method is chosen according to the resolution required and the type of pattern to be produced. Different machines are required for the two methods.

Because of the extremely high resolution required in direct slice writing only a very small portion of the slice can be exposed at a time. After exposure of one field, another portion of the slice must be moved into position. Final alignment within the field is performed by the use of *alignment marks* on the slice which are scanned by the beam and monitored by a detector. Sophisticated machines are capable of producing different exposure patterns at different locations on the slice. This allows diagnostic patterns – *plug bars* – to be placed in a few positions on the slice.

electron-beam voltage *See* electron beam.

electron binding energy *See* ionization potential.

electron capture *See* recombination processes.

electron charge *See* charge; electron.

electron coupling *See* coupling.

electron density (1) The number of electrons per unit mass of a material.

(2) The number of electrons per unit volume of a material. *See* carrier concentration.

(3) *Syn.* equivalent electron density. The product of the ion density in an ionized gas and the ratio of the electron mass to the mass of a gas ion.

electron diffraction *See* diffraction.

electronegative *See* electromotive series.

electron emission The liberation of *electrons from the surface of a material. *See* emission.

electron gas The concept that the *free electrons in a solid, liquid, or gas may be treated as a gas and compared with a real gas dissolved in the appropriate material. Such an electron gas obeys a totally different energy distribution than a real gas, obeying *Fermi–Dirac statistics rather than the Maxwell–Boltzmann statistics obeyed by an ordinary gas. It also has a vanishingly small specific heat.

electron gun A device that produces an *electron beam and forms an essential part of many instruments, such as *cathode-ray tubes, *electron microscopes, and *linear accelerators, all of which need electron beams for their operation.

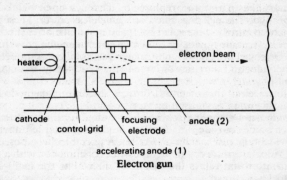

Electron gun

It consists of a series of electrodes (*see* diagram) usually producing a narrow beam of high-velocity electrons. Electrons are released from the indirectly heated *thermionic cathode, the intensity being controlled by variation of the negative potential of the cylindrical control grid surrounding the cathode. The control grid has a hole in front to allow passage of the electron beam. The electrons are accelerated by a positively charged *accelerating anode* before being focused by the focusing electrode and then further accelerated by the second anode.

electron-hole pair *See* hole. *See also* semiconductor.

electron-hole recombination Syn. for band-to-band recombination. *See* recombination processes.

electronic device A device that utilizes the properties of electrons (or ions) moving in a vacuum, gas, or semiconductor.

electronic efficiency The ratio of the output power delivered to the load of an *oscillator or *amplifier at the desired frequency to the average power input to the device.

electronic energy levels *See* energy levels.

electronic-gap admittance *See* gap admittance.

Electronic Industries Association (EIA) *See* standardization.

electronic instrument *See* electrophonic instrument.

electronic mail A means of transmitting and receiving text using a network of computers and the telephone system. A central computer acts as a clearing house for messages: it is accessed by other computers using the telephone lines and can receive and store messages or transmit stored messages on demand. A service is provided by British Telecom on Prestel and also on Telecom Gold. *See also* videotext.

electronic memory A *memory that has no moving parts and in which the read and write operations are entirely electronic. The earliest forms of electronic memory employed thermionic valves in the memory cells. These were replaced by the much smaller and cheaper ferrite *cores, which consumed very much less power. These in turn have been largely superseded by *solid-state memories that are extremely versatile high-speed low-power devices: the growth of microelectronic techniques has made possible the production of semiconductor memories of large memory capacity and extremely small physical size.

Other electronic-memory devices include *magnetic-bubble memories and *Josephson memories.

electronic news gathering A recording system used in *television in which scenes outside the television studio are recorded directly on to *videotape rather than on to film. A portable *televison camera and videotape recorder are used, often in conjunction with a mobile transmitter that relays the recording directly to the main control centre.

electronics The study, design, and use of devices that depend on the conduction of electricity through a vacuum, gas, or semiconductor. *Electrons and mobile *holes are the most important forms of charge carriers in electronic devices but *ions also play a part. Semiconductor devices are assuming an ever-increasing role in modern electronics.

electronic switch An electronic device, such as a transistor, that is used as a switch. These devices are usually operated as high-speed switches where a very fast response is required, as for example in a computer.

electronic tuning The use of a coupled electron beam (*see* coupling) to change the operating frequency of a system. The velocity, intensity, or configuration of the electron beam may be altered to provide the desired frequency change.

electron indicator tube *See* magic eye.

electron lens A device that is used for *focusing an electron beam using either magnetic or electrostatic fields (*see* electromagnetic lens; electrostatic lens). The two types, which are analagous to optical lenses used with light beams, can be combined in instruments such as electron microscopes. Most electron-beam devices, such as cathode-ray tubes or camera tubes, use an electron lens to provide a sharp narrow beam.

electron microprobe A method of analysing the composition of semiconductor material below the surface, using X-rays emitted by the material in response to stimulation by a beam of electrons. The instrumentation is essentially a scanning *electron microscope (SEM) combined with an X-ray fluorescence detector. The X-ray detectors are mounted on a normal SEM and the electron beam of the SEM excites atoms in the material. Both the normal SEM picture and the X-ray spectra are produced simultaneously.

There are two types of X-ray detector used: *energy dispersive spectroscopes* (EDS) detect all wavelengths simultaneously whereas *wavelength dispersive spectroscopes* (WDS) use a crystal to direct X-rays of a given wavelength only onto the detector. The EDS detector is faster and more sensitive than the WDS method but in some cases the X-ray spectra can overlap and the WDS method yields significantly better resolution. The method is not very sensitive to low atomic number elements, which tend to emit *Auger electrons rather than X-rays. The technique is nondestructive. *X-ray fluorescence* (XRF) is a very similar technique but an X-ray beam is used to produce the X-ray output rather than an electron beam. XRF produces poor lateral resolution because of the lack of focusing of the X-ray beam but it does have very good sensitivity. Once again low atomic number elements are not detected because of the emission of Auger electrons rather than X-rays.

electron microscope An instrument that uses a beam of electrons to investigate a sample in order to achieve a higher magnification than is possible with an optical microscope. There are two main types of electron microscope.

In the *transmission electron microscope* (TEM) the electron beam is focused by an *electromagnetic lens or sometimes an *electrostatic lens and has an energy of 50–100 kilovolts (Fig. *a*). A sharply focused image in one plane can only be obtained by using monoenergetic electrons. To avoid energy losses in the beam the sample must be extremely thin (<50 nanometres) so that the scattered electrons that form the image are not changed in energy, and the image therefore appears two-dimensional. A resolution of 0·2–0·5 namometres is possible.

The *scanning electron microscope* (SEM) has lower resolution and magnification but produces a seemingly three-dimensional image,

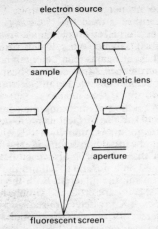

a Transmission electron microscope

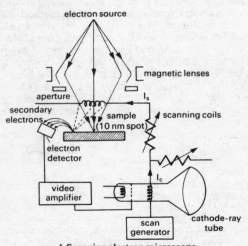

b Scanning electron microscope

with great depth of field, from a sample of any convenient size or thickness (Fig. *b*). The sample is scanned by the electron beam, the numbers of resulting secondary electrons being proportional to the

156

geometry and other properties of the sample. These electrons are converted, by means of an electron detector, *scintillator, and *photomultiplier, into a highly amplified signal that is used for intensity modulation of the beam of the display cathode-ray tube. The resolution is about 10–20 nm.

The *scanning transmission electron microscope* (STEM) has been produced in order to combine the high resolution of the transmission instrument with the three-dimensional image of the scanning type.

Both the TEM and the STEM can be used to produce diffraction patterns from thin samples of crystalline material such as *semiconductors, rather than the usual image. This can be used to analyse the crystal structure of such materials. *See also* diffraction.

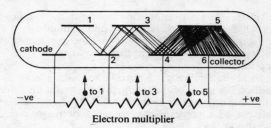

Electron multiplier

electron multiplier An electronic tube in which current amplification is achieved by means of *secondary emission of electrons. Primary electrons are released from the cathode by some means, such as the *photoelectric effect. These are accelerated by a high potential applied to the first anode. The anodes, termed *dynodes,* are made from a good secondary emitter and on impact a greater number of electrons is produced. These are then accelerated in turn by the second and subsequent anodes, to each of which an increasing positive potential is applied (*see* diagram). A large output pulse is produced at the final anode – the collector – which is operated at a very high potential. The chain of anodes is called a *dynode chain. See also* photomultiplier.

electron optics The study of the behaviour of *electron beams under the influence of magnetic and electrostatic fields in a vacuum or very low pressure gas: the analogue of light beams passing through refractive media. The applied fields form *electron lenses that are used for *focusing or defocusing the electron beam.

electron-ray tube *See* magic eye.

electron scanning *See* scanning.

electron-stream potential *Syn. for* electron-beam voltage. *See* electron beam.

electron synchrotron *See* synchrotron.

electron tube An electronic device in which conduction between two electrodes takes place in an envelope that is sealed or continuously exhausted and contains a gas or a vacuum. Tubes that are evacuated to a high vacuum are termed *hard tubes;* those that contain traces of a gas are *soft tubes.* The concentration of gaseous atoms in soft tubes is less than that of a *gas-filled tube but is sufficient to cause some modification to the characteristics compared with those of hard tubes. Such tubes frequently contain more than two electrodes. *See also* valve; thermionic valve.

electronvolt Symbol: eV. The energy acquired by one electron when passing freely through a potential difference of one volt. The unit is extensively used in atomic physics. One eV equals 1.602×10^{-19} joule.

electron voltaic effect A phenomenon similar to the *photovoltaic effect in which electrons striking the photocathode of a photocell cause electron emission. At low energies the gain increases rapidly with voltage to a maximum and then decreases.

electro-optical effect *See* Kerr effects.

electro-optical shutter *Syn. for* Kerr cell. *See* Kerr effects.

electro-optics The study of the interactions between the refractive indices of some transparent dielectrics and the electric fields in which they are placed. Changes in the optical properties of dielectrics are produced. *See* Kerr effects.

electro-osmosis *Syn.* electroendosmosis. The movement of an electrolyte through a fine tube or membrane under the influence of an electric field. Under conditions of pressure equality on each side of the partition, the volume of liquid transferred depends upon the nature of the electrolyte: it is independent of the area and thickness of the partition, being approximately proportional to the *resistivity of the solution. A contact potential difference (*see* contact) between the membrane and electrolyte has been postulated to explain the effect.

electrophonic effect The perception of sound when the human body is subjected to alternating currents of particular frequencies and magnitudes.

electrophonic instrument *Syns.* electrosonic instrument; electronic instrument. A musical instrument that creates sound by electrical or electronic means. There are various types, one of which is the *Moog synthesizer.* It contains many sine-wave and square-wave generators, which produce waves singly or in complex groups, together with a large number of *filters. Individual circuits may be plugged in or out externally and the instrument may be set and played with an extremely wide variety of different sounds. Other electrophonic instruments include the Hammond organ and the Compton organ. Electrophonic instruments are never completely successful when used to imitate musical instruments but a wide variety of unusual sounds can be produced.

The term is also used to refer to conventional instruments, such as the guitar or piano, in which the sound is amplified (and possibly distorted) by electrical means.

electrophoresis The movement of colloidal particles in a liquid under the influence of an electric field. Positively charged particles migrate to the cathode (*cataphoresis*) and negatively charged ones to the anode (*anaphoresis*).

electrophorus An early form of simple electrostatic charge generator. A flat dielectric plate ('cake') is positively charged by friction. A varnished metal plate with an insulated handle is placed on the cake and momentarily earthed, taking with it and induced negative charge when it is removed. The process can be repeated until the original positive charge leaks away.

electroplating The application of a metal to the surface of another material by electrolysis. The technique is widely used to produce a protective or decorative surface on metal objects. It is also used in the fabrication of integrated circuits where relatively thick layers of metal are required. *See also* light-assisted plating.

electropneumatic Denoting control systems that contain both electronic and pneumatic elements.

electropolar Having magnetic poles or permanent positive and negative charges, as in a magnet or an electret.

electropolishing *See* electrolytic polishing.

electropositive *See* electromotive series.

electropositive series *See* electromotive series.

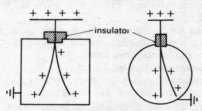

Gold-leaf electroscopes

electroscope An electrostatic instrument that detects small potential differences and electric charges. The most common form is the *gold-leaf electroscope,* which consists of a pair of gold leaves hanging side by side from an insulated metal support enclosed in a draught-proof case. If a charge is supplied to the support the leaves separate due to their mutual repulsion. A more precise form replaces one of the gold leaves by a rigid metal plate (*see* diagram).

Lauritzen's electroscope utilizes metallized quartz fibre as the sensitive element. The *pith-ball electroscope* utilizes two pith-balls suspended by silk threads.

An instrument that is capable of accurate quantitative measurement is called an *electrometer.

electrosonic instrument *See* electrophonic instrument.

electrostatic actuator *See* actuator.

electrostatic adhesion Adhesion between two substances or surfaces due to the presence of opposite charges, which attract each other.

electrostatic charge-storage tube *See* storage tube.

electrostatic deflection The use of electrostatic fields produced between two metal electrodes for deflecting an electron beam. The electrodes used are called *deflection plates* and two pairs of plates, at right angles to each other, are used in electron-beam devices, such as *cathode-ray tubes, to provide deflection in two orthogonal directions. *Compare* electromagnetic deflection.

electrostatic field The electric field associated with charged particles at rest. It is the region, around a distribution of electrostatic charge, in which a stationary charged particle would experience a force. *See also* Coulomb's law.

electrostatic focusing *See* focusing.

electrostatic generator *See* generator.

electrostatic induction The production of a charge distribution on a conductor when in the vicinity of another charged body under the influence of the associated electric field. If the conductor is in the vicinity of a positively charged body the region nearest to the charged body becomes negatively charged; the more remote regions become positively charged. The reverse effect is seen if the body is negatively charged.

electrostatic lens An *electron lens consisting of an arrangement of electrodes that focuses an electron beam electrostatically. *See* focusing. *Compare* electromagnetic lens.

electrostatic loudspeaker *See* loudspeaker.

electrostatic memory A type of *memory that stores information in the form of electrostatic charges. *See* storage tube.

electrostatic precipitation A method of precipitating solid or liquid particles from a gas. An electrostatic field is applied across the gas between two electrodes, one of which is earthed (usually the positive one). The particles collect on the earthed electrode.

electrostatic printing A type of printing in which a pattern of electrostatic charges is produced on the surface of the paper. A fine dark powder is applied to the paper and adheres to the electrostatic charges; it is then fixed by heat. *See also* Xerography.

electrostatics The study of electric charges at rest and their associated phenomena.

electrostatic separation A method of separating fine powders of different *permittivities. The powders undergo different deflections when placed in the intense electrostatic field between two highly charged electrodes.

electrostatic units (esu) *See* CGS system.

electrostatic voltmeter *See* voltmeter.

electrostatic wattmeter *See* wattmeter.

electrostriction A change in the dimensions of a body under the influence of an electric field in a medium of relative permittivity different from its own. Forces of extension or compression may be produced. In a nonhomogeneous field the body will also tend to move: one of higher relative permittivity than its surroundings tends to move into the region of higher field strength and vice versa.

electrothermal instrument *See* thermal instrument.

electrovalent bond *Syn.* ionic bond. *See* valency.

electrovalent crystal *See* ionic crystal.

element A substance that consists entirely of atoms of the same atomic number. Over 100 different elements have been identified (*see* periodic table). Elements are the basic substances from which compounds are built up by chemical combination: this excludes disruptive processes in which the atomic nuclei comprising the element are disturbed.

elementary particle Any of the particles of matter that cannot be subdivided into smaller particles. Elementary particles are described by a set of quantum numbers describing their intrinsic properties, such as *charge and *spin. The stable particles are the *electron, *proton, *photon, and neutrino. The *neutron is also stable when bound in an atomic nucleus. The electron is the natural unit of electric charge.

Elementary particles can be classified in groups according to the type of nuclear interaction to which they are subject. These groups are the hadrons (including the neutron and proton), the leptons (including the electron and neutrino), and the photon, which is in a group by itself.

ELSI *Abbrev. for* extra large scale integration. *See* integrated circuit.

e.m.f. (or **emf**) *Abbrev. for* electromotive force.

emission The liberation of electrons or electromagnetic radiation from the surface of a solid or liquid, usually electrons from a metal. The outer electrons of the atoms in a metal (conduction electrons) move in a random manner among the lattice atoms with no net forces on them in the bulk of the material. Electrons near the surface of the material with directions of motion out of the surface can leave the surface, but then experience a force directing them back to the metal as the metal is left positively charged. The charge on the metal can be considered as an *electric image located the same distance inside the metal as the electron is outside it. The force on the electron

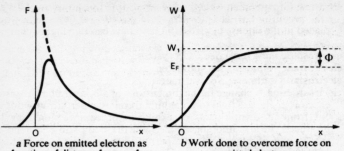

a Force on emitted electron as function of distance from surface

b Work done to overcome force on emitted electron

varies with distance x from the surface (Fig. *a*). As the electron moves out from the surface, work (W) is done to overcome this force (Fig. *b*), where

$$W = \int_0^x F \, dx$$

The value W_1 represents a potential barrier that must be overcome by the electron. Electrons can only escape if their energies are greater than W_1. W_1 is related to the *work function, Φ, by

$$\Phi = W_1 - E_F$$

where E_F is the *Fermi level.

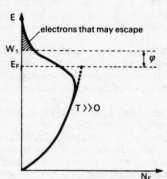

c Number of electrons as a function of electron energy

Normally only a few electrons will have velocities, due to their thermal energy, large enough to escape (Fig. *c*). Emission occurs when sufficient energy is given to the electrons to allow them to escape (as in *photoemission, *thermionic, or *secondary emission) or the potential barrier is distorted by the presence of an intense electric field (as in *field emission).

emissivity Symbol: ε. When radiation is produced by the thermal excitation of atoms, molecules, etc., the emissivity is defined as the ratio of the power per unit area radiated from the surface to that radiated by a black body at the same temperature.

emitter (1) *Short for* emitter region. The region of a bipolar junction transistor from which carriers flow, through the emitter junction, into the *base. The electrode attached to this region is the *emitter electrode. See also* transistor; semiconductor.
  (2) *Short for* emitter electrode.

emitter-coupled logic (ECL) *Syn.* current mode logic (CML). A family of integrated *logic circuits so called because a pair of transistors coupled by their emitters forms a fundamental part of the circuit. The basic ECL gate has simultaneously the function required and its complement.

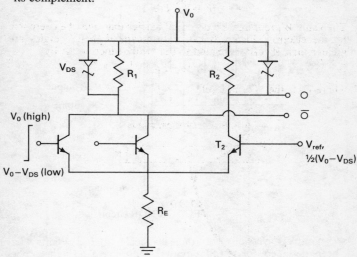

b Low-voltage ECL gate

A simple OR/NOR circuit is shown in Fig. *a*. Input is via the transistors $T_{1a,b,c}$; these are emitter-coupled to transistor T_2 and form a *long-tailed pair with it. This is an excellent *differential amplifier. An *emitter-follower buffer forms the output stage. Transistor T_2 has a fixed bias applied to its base with magnitude halfway

163

between a logical 1 and a logical 0. If a logical 0 is applied to all three input transistors then current flows through T_2 causing a voltage drop across R_2. This in turn produces a logical 0 at the OR output and a logical 1 at the NOR output. If any one of the input transistors $T_{1a,b,c}$ has a logical 1 applied, current flows through that transistor producing a voltage drop across R_1 and the outputs are hence reversed, i.e. a logical 1 occurs at the OR output. Typical values of applied voltages are $-1\cdot55$ volts (logical 0), $-0\cdot75$ volts (logical 1), $-1\cdot15$ volts (fixed bias).

The transistors are operated in nonsaturated mode and the *delay is exceedingly short (approximately one nanosecond) making ECL circuits inherently the fastest type of logic circuit.

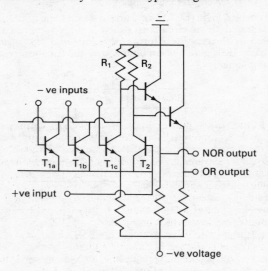

a ECL OR/NOR circuit

Simpler versions of the classical ECL circuits have been designed for VLSI circuits; these have a higher *packing density and operate with lower voltage swings. Fig. *b* shows a simple low-voltage ECL gate in which the emitter-follower transistors are replaced by *Schottky clamped-load resistors R_1 and R_2. The fixed reference bias applied to transistor T_2 is generated 'on-chip' rather than being supplied externally. The total difference betwen the 'high' and 'low' logic levels is equal to the forward bias of the Schottky diode, V_{DS}.

An alternative form of higher packing density ECL circuit uses ECL circuits connected in series (gated); this allows a more complex logic function to be implanted on a smaller area of chip. This method of *series-gated* or *cascoded* circuit design is also widely used in FET circuitry.

emitter electrode *See* emitter.

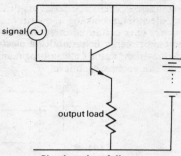

signal

output load

Simple emitter follower

emitter follower An amplifier that consists of a bipolar junction *transistor with *common-collector connection, the output being taken from the *emitter (*see* diagram). The transistor is suitably biased so that it is nonsaturated and conducting. The emitter voltage thus has a constant value relative to the base at all times, and the emitter follows the signal applied to the base. The voltage gain of the amplifier is therefore nearly unity but the current gain is high. The amplifier is often used as a *buffer and is characterized by a high input impedance and low output impedance.

The *thermionic-valve analogue is the *cathode follower* and the FET analogue is the *source follower* although neither of these is as efficient a unity-gain buffer amplifier as the emitter follower, the voltage gain, particularly of the source follower, being further from unity.

emitter region *See* emitter.

emu *Abbrev. for* electromagnetic unit. *See* CGS system.

enabling Activating a particular circuit or group of circuits from a larger set of circuits in order to effect their operation. An *enabling pulse* is often used to select the desired circuit. Examples include selecting a particular input stage from several inputs in parallel, or selecting a particular semiconductor *chip from an array of several similar chips.

encephalograph *Short for* electroencephalograph.

enclosed arc An arc struck between carbon electrodes placed within a transparent or translucent enclosure. The enclosure is designed so that the flow of air reaching the arc is restricted and the arc burns in an atmosphere containing the products of combustion. *Compare* open arc.

encoder A device, such as a *rotary encoder, that produces an output in a desired coded form.

endfire array *Syn.* staggered aerial. *See* aerial array.

energy bands In a single atom, according to quantum theory, the orbiting electrons can only have certain discrete energies. The atom has a number of associated *energy levels and the electrons occupy the lower levels and obey the *Pauli exclusion principle, i.e. not more than two electrons can occupy each level.

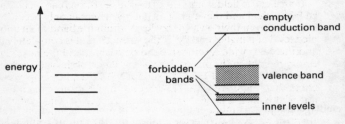

a Energy levels in a free atom　　　　Energy bands in a solid

In a solid the situation is more complicated. As atoms come closer together the narrow energy levels become bands of allowed energies (Fig. *a*), each band consisting of a large number of closely spaced energy levels capable of containing two electrons. In general, within a band the number of levels is approximately equal to the number of atoms in the crystal. The bands of permissible energy are called the *allowed bands*. Within any allowed energy bands the number of constituent discrete energy levels varies with the energy.

The energy bands may either be filled or empty depending on whether they correspond to filled or empty levels in the free atoms. The energy regions between the bands are called *forbidden bands*. The band of energies of the *valence electrons in a solid is called the *valence band*. Inner electrons, i.e. electrons close to the nucleus, are not greatly affected when atoms are brought together to form a crystal. They are tightly bound to the nuclei and cannot move through the crystal.

The theory of these energy bands in solids depends on quantum mechanics. In general, the Schrödinger equation is solved for an electron moving in a varying electric potential $U(k)$, the periodicity of which is created by the spacing of the ions in the crystal lattice. The allowed solutions give the allowed bands of energy and the energies for which there are no solutions are the forbidden bands. In some materials, such as gallium arsenide, the allowed solutions exhibit *degeneracy.

The possible solutions are plotted on the reciprocal lattice of the crystal lattice under consideration. The solutions for $k = 1, 2, \ldots n$ are enclosed in zones, known as *Brillouin zones*. The *reciprocal lattice* is a theoretical lattice defined by a', b', and c', which are related to sides a, b, and c of the unit cell of the crystal by:

$$a' = [b \times c]/[a.(b \times c)]$$
$$b' = [c \times a]/[a.(b \times c)]$$
$$c' = [a \times b]/[a.(b \times c)]$$

If an electric field is applied to a solid, electrons can be accelerated by the field and they thus gain energy. This can only occur if they can move from their own energy level within the band to an unoccupied level at a higher energy. If the valence band is completely filled and there is a wide forbidden band between it and the next highest empty band then the material is an *insulator. If the valence band is not completely filled or if it overlaps with a higher empty energy band, then there are vacant levels that the electrons can enter. The material is then a good *conductor of electricity. This is the situation occurring in most metals.

In *semiconductors there is a small forbidden gap between the valence band and a higher empty band. At absolute zero, all the electrons occupy the valence band and the material acts as an insulator. As the temperature is increased some electrons gain enough thermal energy to escape from the valence band and cross the forbidden gap into the *conduction band, leaving the valence band unfilled. Thus conduction is possible in both bands and the conductivity rises with temperature.

At absolute zero all the electrons in a solid occupy the lower energy levels and the valence band is filled to a certain energy, while no higher levels are occupied. This level of maximum energy is called the *Fermi level* E_F. At temperatures above absolute zero some electrons gain thermal energy and the distribution of electrons in the given energy levels can be derived by *Fermi–Dirac statistics. The probability of finding an electron at a particular energy is its *Fermi function*, $F(E)$. Fig. *b* shows the function for a metal at absolute zero and a temperature above absolute zero. The Fermi level is then defined as the energy at which the probability of finding an electron is $\frac{1}{2}$. The Fermi level is approximately constant with temperature. In semiconductors the electrons are distributed between the valence band and the conduction band and the Fermi level lies in the forbidden gap between the two (Fig. *c*). In an insulator the Fermi level lies in the valence band. The electronic distribution

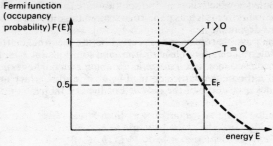

Fermi function
(occupancy
probability) $F(E)$

b Energy distribution in a metal

curves on Fig. *c* show how the number of electrons in each band varies with energy.

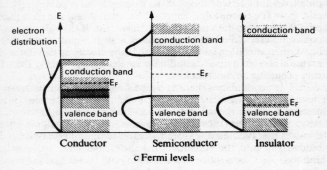

c Fermi levels

See also direct-gap semiconductor.

energy component (1) (of a current or voltage) *See* active current or active voltage.

(2) (of the volt-amperes) *See* active volt-amperes.

energy dispersive spectroscopy (EDS) *See* electron microprobe.

energy levels The possible values of energy of an atom or molecule. *Quantum theory dictates that only certain fixed values are possible and these may only change by an integral multiple of some fixed amount. An electron orbiting the nucleus of the hydrogen atom may only occupy certain orbits of different energy. These allowed states of energy of the atom are called *electronic energy levels*. The *ground state* is the lowest possible energy level, higher states being called *excited states*. The total energy of an atom is composed of kinetic energy, as it translates through space, and the electronic energy.

168

A molecule also has contributions to its energy from vibrations within it and from its rotation. The vibrational and rotational energies are also confined to certain discrete values. The differences between successive *vibrational energy levels* are smaller than between electronic energy levels, *rotational energy levels* having even smaller differences. Each electronic energy level has several associated vibrational levels, which in turn have numbers of rotational levels. Changes in the energy level result from the emission or absorption of *photons of radiation.

enhancement mode A means of operating *field-effect transistors in which increasing the magnitude of the *gate bias increases the current. *Compare* depletion mode.

enhancement-mode device *See* field-effect transistor.

ENIAC Acronym from *E*lectronic *N*umerical *I*ntegrator *A*nd Computer. One of the first large-scale *digital computers to use electronic techniques. It was built by the University of Pennsylvania and employed about 18 000 valves.

epitaxial growth *See* epitaxy.

epitaxial layer *See* epitaxy.

epitaxial transistor *See* planar process.

epitaxy *Syn.* epitaxial growth. A method of growing a thin layer of material upon a single-crystal substrate, such as silicon, so that the lattice structure of the layer is identical to that of the substrate. The material, which may be the same as the substrate or a different one, is usually deposited from a gaseous mixture (*see* vapour phase epitaxy). The technique is extensively used in semiconductor technology when a layer (the *epitaxial layer*) of different conductivity to the substrate is required. *See also* liquid phase epitaxy; MOCVD; molecular beam epitaxy.

E-plane *Syn. for* E-bend. *See* waveguide.

epoxy resin A type of synthetic resin that is widely used for encapsulating electronic devices because of its properties of high strength and low shrinkage.

equal energy source A source of energy in which the emitted energy is equally distributed throughout the entire frequency range of the source's spectrum.

equalization A means of reducing *distortion in a system by introducing networks that compensate for the particular type of distortion over the required frequency band. For example, if a loudspeaker system has a poor response to bass frequencies, *bass boost* (amplification of the bass frequencies with respect to the treble frequencies) may be introduced into the system to compensate.

In a telephone system equalization is also required to compensate for *group delay*, i.e. the relative delay between the signal and the carrier wave when travelling along the telephone line. Group delay equalizers are systems that compensate for the phase shift and pulse distortion introduced when signals at different frequencies experience different delay relative to each other and the carrier signal. Equalization is also needed to compensate for particular telephone lines that may have slightly different line characteristics. *Adaptive system equalization* automatically equalizes for the line characteristics of the different lines.

In radio communication systems *adaptive radio microphone equalization is used to compensate for variations in the received signal strength.

equalizer (1) A network that provides *equalization.

(2) *Syn.* equipotential connection. A low-resistance connection made between two points in an electrical machine to ensure that the points are always at the same potential.

equipotential connection *See* equalizer.

equivalent circuit An arrangement of simple circuit elements with the same characteristics as a more complicated circuit or device, under specified conditions; it can be used to predict the behaviour of the more complicated system.

equivalent network An electrical network that may replace another network without materially affecting the conditions in other parts of the system, but usually at one particular frequency only.

equivalent resistance The value of total *resistance that, if placed at a point in a circuit, would dissipate the same power as the total of smaller resistances in the circuit.

equivalent sine wave A sine wave that has the same *fundamental frequency and *root-mean-square value as a particular periodic wave.

erase To remove stored information from a location in *storage tubes, *memories, *magnetic tape, etc.

erasing head *See* magnetic recording.

ERDA *Abbrev. for* elastic recoil detection analysis.

E-region *Syn. for* E-layer. *See* ionosphere.

erg The unit of work in the obsolete CGS system of units. One erg equals 10^{-7} joule.

error *See* deviation.

Esaki diode *See* tunnel diode.

esu *Abbrev. for* electrostatic unit. *See* CGS system.

etching Chemical erosions of selected portions of a surface in order to produce a desired pattern on the surface. The technique is widely used in microminiature electronics. *Wet etching* utilises liquids, such as acids or other corrosive chemicals, as the etching agent. The etching process takes place by means of chemical reactions at the surface of the material. This process is limited both by the rate at which the chemical reactions can take place, and by the rate at which the products of the chemical reaction can be removed. These factors are a function of the nature of the liquid used and the temperature at which the etching is carried out. Etching in which the main limiting factor is the reaction rate is known as *reaction rate limited, surface limited* or *kinetically limited* etching. Where the removal rate of the chemical products is the predominant controlling factor the etching is known as *diffusion limited* or *mass transport limited* etching. In certain applications the etching proces can be electrically aided by making the slice to be etched the anode (*anodic etching*) or cathode (*cathodic etching*) of an electrolytic cell.

Dry etching or *plasma assisted etching* uses either chemical or physical reactions between a low-pressure plasma or glow discharge and the surface to be etched. It has several advantages compared to wet etching, but the etching process itself is very complex, and the results can be greatly affected by small variations in the process parameters. Dry etching is capable of patterning smaller geometries than wet etching, lateral etch rates close to zero can be produced under certain conditions, and smooth *edge profiles can be produced when needed for metal crossovers. Certain semiconductors, particularly gallium arsenide, have no suitable liquid etchants that can produce deep narrow features, and dry etching is particularly important in these cases.

Plasma etching is any process in which a plasma generates reactive species that then chemically etch material in direct proximity to the plasma. If the chemical reactions are enhanced by the kinetic energy of the ions in the plasma the process is described as *kinetically assisted chemical reaction. Reactive ion etching* is similar to plasma etching, but uses only kinetically assisted chemical etching. The applied voltage drop is mainly at the surface of the slice. *Reactive ion beam etching* separates the slices from the plasma by a grid that accelerates the ions created in the plasma towards the slice. The ion energy is higher, and some of the etching is due to physical reactions.

Sputter etching uses energetic ions from the plasma to physically blast (sputter) atoms from the surface. No chemical reactions are involved. *Ion milling* is also a purely mechanical method that uses a roughly collimated beam of energetic ions to erode a surface by bombardment. Ion milling, unlike other dry etching procedures, can be used at angles other than perpendicular to the slice.

etch pit density *Syn. for* dislocation density. *See* dislocation.

Ettinghausen effect The development of a very small transverse temperature gradient in a conductor when carrying current in a magnetic field. The temperature gradient is established in a direction perpendicular to both the magnetic field and the current.

eutectic bond A thermometallurgical bond used to provide contact between semiconductor chips. It is formed by heating a eutectic mixture, such as gold and silicon, and then allowing it to cool. A

171

eutectic mixture is a mixture that solidifies as a whole when cooled without change in composition.

even larger scale integration (ELSI) *Colloq.* Extra large scale integration. *See* integrated circuit.

E wave *Syn. for* TM wave. *See* mode.

excess conduction Conduction in a *semiconductor due to the presence of electrons that are not required to complete the chemical bonding of the semiconductor and are therefore able to conduct charge through the semiconductor. These *excess electrons* are usually supplied by a donor impurity (*see* semiconductor).

excess voltage *See* overvoltage.

exchange forces *See* ferromagnetism.

excitation (1) The addition of energy to an atom or molecule, transferring it from the ground state to a higher *energy level.

(2) The application of a signal to the *base or control electrode of a *transistor or *valve.

(3) The application of voltage to an oscillating crystal.

(4) The application of radiofrequency pulses to a tuned circuit.

(5) The application of current to the winding of an *electromagnet in order to produce a magnetic flux. The applied current is called the *exciting current.*

excited state *See* ground state; energy levels.

exciting current *See* excitation.

exciton *See* semiconductor.

exclusive OR gate *See* logic circuit.

expanded sweep *See* time base.

expanded-sweep generator *See* time base.

expander (1) *Short for* volume expander. *See* volume compressor.

(2) *Short for* gate expander.

exploring coil *Syn.* search coil. *See* flip-coil.

exposure tower *Colloq.* Photolithographic equipment.

external memory *See* memory.

extinction potential *See* gas-discharge tube.

extra large scale integration (ELSI) *See* integrated circuit.

extrapolated failure rate *See* failure rate.

extrapolated mean life *See* mean life.

extremely high frequency (EHF) *See* frequency band.

extrinsic photoconductivity *See* photoconductivity.

extrinsic semiconductor A semiconductor in which impurities or imperfections determine the charge-carrier concentration. *See also* semiconductor. *Compare* intrinsic semiconductor.

eyelet-construction mica capacitor *See* mica capacitor.

Eyring equation *See* Arrhenius equation.

F

Fabry-Perot modes *See* injection laser.

facsimile transmission *Syn.* picture telegraphy. A method of transmitting any kind of graphic material to produce a pictorial likeness of the object. The system employs *facsimile scanning:* the subject copy is scanned, so providing a successive analysis from which electrical signals are produced. These signals are transmitted to a receiver. The *facsimile bandwidth* is the range of frequencies that is required for adequate transmission of the copy. A form of *carrier wave is often used for facsimile transmission, the derived electrical signals – the *facsimile baseband* – being used to modulate (*see* modulation) the carrier before transmission.

The *facsimile receiver* detects and amplifies the incoming signals then applies them to a light source. The light source is brightness-modulated by the receiver signal in order to record the picture elements. *Picture inversion* can be produced by introducing a 180° phase shift into the detected signal in order to reverse the black and white shades of the original copy.

One form of facsimile transmission uses telephone lines to transmit the data to the receiver. The transmitted signals are produced in digital form and *pulse code modulation is used. The system, popularly known as *FAX*, is used for high-speed conveyance of documents and is predicted as the postal service of the future.

Facsimile telegraph is a telegraph system for the transmission of pictures. *Phototelegraphy* is a facsmile telegraph system that has special regard to half-tone reproduction.

fader A device that maintains an electrical signal at a constant level while one signal is being *faded out,* i.e. smoothly reduced in amplitude, and another *faded in,* i.e. smoothly increased in amplitude.

fading Variations in signal strength at a receiver due to variations in the transmission medium. Destructive interference between two waves travelling by two different paths to the receiver is the most common cause of fading; this is termed *interference fading.*

Amplitude fading occurs when all transmitted frequencies are attenuated approximately equally, resulting in a smaller received signal. *Selective fading* occurs when some frequencies are more attenuated than others, resulting in a distorted received signal. *Dellinger fade-out* is a complete loss of the received signal, which may last for minutes or even hours following a burst of hydrogen particles from an eruption associated with a sun spot. This causes the formation of a highly absorbent D-layer of the *ionosphere lower than the regular E- and F-layers.

failure Termination of the ability to perform its required function by a device, component, circuit, or any part or subsystem that can be separately tested. The *failure mechanism* is the physical, chemical, metallurgical, or other process resulting in a failure. The predicted or observed result of a failure mechanism on a particular item in relation to the operating condition at the time of failure is the *failure mode*. The predicted or observed result on the function of the item or of related items is the *failure effect*.

Failure is classified according to cause, suddenness, and degree. There are various causes of failure:

misuse failure is attributable to the application of stresses beyond the stated capabilities of the item;

inherent weakness failure is attributable to a weakness inherent in the item itself when subjected to stresses within its stated capabilities;

primary failure is not caused either directly or indirectly by the failure of another item whereas *secondary failure* is;

wear-out failure results from deterioration processes or mechanical wear, the probability of occurrence increasing with time.

Failure can be either sudden or gradual, i.e. unanticipated by prior examination or anticipated. It can also be partial, complete, or intermittent. A failure that is both sudden and complete is termed *catastrophic failure;* one that is both gradual and partial is called *degradation failure. See also* failure rate.

failure rate The number of *failures of an item per unit measure of life (cycles, time, etc., as appropriate). For any particular item the failure rate will be based on the results of *life tests and one or more of the following will be quoted.

The *observed failure rate* is the ratio of the total number of failures in a single population to the sum of the times during which each item has been subjected to stress conditions. It is associated with particular and stated time intervals and stress conditions. The criteria for what constitutes a failure should be stated.

The *assessed failure rate* is determined as a limiting value of the confidence interval with a stated probability level based on the same data as the observed failure rate of nominally identical items. The following conditions apply: the source of the data should be stated; results may be combined only when all conditions are similar; it should be stated whether one- or two-sided intervals are being used; the upper limiting value is usually used for failure-rate statements and the assumed underlying distribution should be stated.

The *extrapolated failure rate* is an extension of the observed or assessed failure rate by defined extrapolation or interpolation for durations or stress conditions different from those applying to the conditions of the assessed rate.

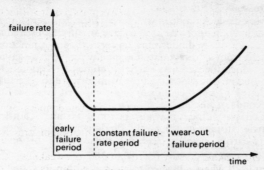

Change of failure rate with time

The failure rate of an item varies during its lifetime (*see* diagram). In the *early failure period* the failure rate decreases rapidly. To avoid an early failure in use a manufacturer will operate and test an item in a process, known as *burn-in,* that stabilizes its characteristics. The failure rate increases rapidly in the *wear-out failure period* due to deterioration processes. The *useful life* of an item is the period from a stated time during which, under stated conditions, an item has an acceptable failure rate. For items showing a failure rate pattern as in the diagram, the useful life corresponds to the *constant failure-rate period,* when failures occur at an approximately uniform rate.

fall-time (1) The time required for a *logic circuit to change its output from a high level (logical 1) to a low level (logical 0).

(2) *Syn. for* decay time. *See* pulse.

fan-in The maximum number of inputs acceptable by a *logic circuit.

fan-out Within a given family of *logic circuits, the maximum number of inputs to other circuits that the output of a given circuit can drive.

farad Symbol: F. The *SI unit of *capacitance. A capacitor has a capacitance of one farad when a charge of one coulomb increases the potential difference between its plates by one volt. The farad is too large for most practical purposes, the submultiples microfarad, nanofarad, and picofarad being more convenient.

Faraday cage *See* electric screening.

Faraday constant Symbol: F. A fundamental constant having the value $9.648\,670 \times 10^4$ coulombs per mole. It is the quantity of electricity that is equivalent to one mole of electrons and that can deposit or liberate one mole of a univalent ion. It the product of electron charge and Avogadro constant.

175

Faraday dark space *See diagram at* gas-discharge tube.

Faraday–Neumann law *See* electromagnetic induction.

far-end crosstalk *See* crosstalk.

fast-recovery diode A diode in which very little *carrier storage occurs and that may therefore be used to give an ultrahigh speed of operation. Fast-recovery diodes may be formed as p-n junction diodes fabricated from a *direct-gap semiconductor, such as gallium arsenide, in which the minority-carrier lifetimes are very much smaller than in silicon. The switching time from forward to reverse bias is of the order of 0.1 nanoseconds or less compared to about one to five nanoseconds in silicon.

Fast-recovery diodes may also be formed from *Schottky diodes. The carrier storage in a Schottky diode is negligible since they are majority-carrier devices, and a similar speed of operation is achieved. *Compare* charge-storage diode.

FATFET

FATFET A planar *field-effect transistor (FET) that has a long gate length (*see* diagram) and is used to measure *drift mobility. The device is so called because of its geometry.

fat zero A significant charge packet used in *charge-coupled devices that corresponds to a zero input sample. Typically a fat zero corresponds to about 10% of the maximum charge packet and is used to minimize distortion of the signal due to surface trapping states in the semiconductor. A *skinny zero* is a zero charge level that contains no deliberately introduced background charge level.

fault A defect in any circuit or device that interferes with it or prevents it from operating normally.

fault current A *current that may flow through a circuit or device as a result of a fault, such as a defect in the insulation. The current may take the form of an *arc back, *short circuit, electrical *surge, current to earth, etc.

FAX *Colloq. for* facsimile transmission.

FCC *Abbrev. for* Federal Communications Committee. *See* standardization.

feedback The process of returning a fraction of the output energy of an energy-converting device to the input. The circuit that transmits the feedback signal to the input is the *beta circuit;* the circuit containing the active device, which generates the output signal, is the *mu circuit.*

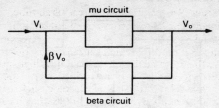

a Voltage feedback circuit

In the case of an active device, such as a *transistor, that introduces a gain A in the absence of feedback, *voltage feedback* is employed if a fraction β of the output voltage is returned to the input (Fig. *a*). The effective output voltage is given by

$$v_o = A(v_i + \beta v_o)$$

The overall gain of the combination is then

$$v_o/v_i = A/(1 - \beta A)$$

If β is negative the feedback voltage opposes the input voltage and the process is termed *negative feedback*. The overall gain of the device is reduced but there is a corresponding reduction in the amount of *noise and *distortion in the output. If the term $(-\beta A)$ is made large compared with unity the overall gain reduces to $1/\beta$ and is independent of the elements in the mu circuit. An amplifier operated in this manner is very stable and is independent of minor variations in the operating conditions.

If β is positive the feedback voltage reinforces the input voltage and the process is termed *positive feedback*. The overall gain of the device is increased and if the factor $(-\beta A)$ becomes equal to or greater than unity the output voltage becomes independent of any input signal and oscillations occur. The point at which the term $(-\beta A)$ just becomes unity for any given circuit is termed the *singing point*. The overall combination can then be considered to have an effective *negative resistance (*see* oscillator).

Current feedback is a form of feedback in which a fraction of the current output to the load is fed back to the input (Fig. *b*). The effective output current is given by

$$i_o = A(i_i + \beta i_o)$$

and the overall gain is given by

$$i_o/i_i = A/(1 - \beta A)$$

An analysis similar to that of voltage feedback can then be applied.

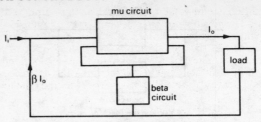

b **Current feedback circuit**

In a multistage amplifier, feedback may be applied to each individual amplifier stage (*local feedback*) or across the composite device (*multistage feedback*). The phase of the feedback at the input is maintained in the correct relationship to the input by introducing a reactance in the feedback circuit. *Capacitive feedback* uses one or more capacitors and *inductive feedback* uses a self-inductance or a mutual inductance.

Feedback is also used in control systems when a fraction of the controlled parameter is fed back in order to produce any necessary correcting signals (*see* feedback control loop).

feedback control loop A method of control, used for many different types of control system, in which a portion of the output derived from a system is fed back to the input circuit in order to control the output signal in a desired manner (*see* diagram).

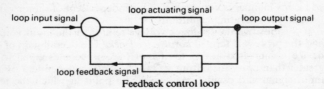

Feedback control loop

The external signal applied to the loop is the *loop input signal;* the controlled signal that is output by the loop is the *loop output signal.* The portion of the signal fed back to the input circuit is the *loop feedback signal.* The feedback signal is mixed with the input signal and produces a *loop actuating signal,* which is used to produce the controlled output. The transmission path between the loop input and the loop output is the *through path;* the path between the loop actuating signal and the loop output is the *forward path.*

The *forward transfer function* is the mathematical relationship between the actuating signal and the controlled loop output. The value of the loop actuating signal is determined by the *loop error,* which is the difference in value between the actual output and the desired value. The *actuating transfer function* is the relationship between the loop input signal and the loop actuating signal. The *difference transfer function* is the relationship between the loop error and the loop input signal. The loop feedback signal and the loop input signal are mixed in a suitable manner to maintain the desired transfer functions. In many systems the loop error is used as the loop actuating signal.

feedback oscillator *See* oscillator.

feeder (1) The part of a radio system that conveys the radiofrequency energy from the transmitter to the aerial or from the aerial to the receiver.

(2) An electric line that conveys electrical energy from a generating station to a point of a distributing network without being tapped at any intermediate points. *See* transmission line.

feedthrough A contact on a printed circuit board that connects one layer of interconnections with the next layer, passing through the insulating material that separates them. Up to 12 layers of interconnections have been mounted on a single board, although only a double-sided board is commonly used. In an *integrated circuit containing multilayer interconnections, feedthroughs can be used to make contact between one layer of interconnections and the next.

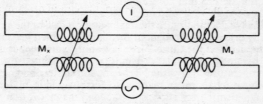

Felici balance

Felici balance A type of alternating-current *bridge that is used to determine the mutual inductance between the windings of an inductor. The unknown inductance M_x is compared to a known standard inductance M_s (*see* diagram). At balance, when there is no response from the indicating instrument, I,

$$M_x = -M_s$$

femto- Symbol: f. A prefix to a unit, denoting a submultiple of 10^{-15} of that unit.

Fermi–Dirac statistics A system of quantum statistics that is used to describe the behaviour of solids in terms of a *free electron model. In this model the most weakly bound electrons of the constituent atoms are considered to behave as a gas subject to certain conditions: the electrons are free to move in any direction through the solid, they do not interact with each other, and are subject to the *Pauli exclusion principle.

The probability of an energy level of energy E being occupied by an electron is given by the *Fermi–Dirac distribution function,* f(E):

$$f(E) = 1/[e^{(E - \mu)/kT} + 1]$$

where μ is a constant and a function of temperature, k is the Boltzmann constant, and T the thermodynamic temperature.

The free electron model is used to explain a number of important physical properties of metals.

Fermi function *See* energy bands.

Fermi level Symbol: E_F. The maximum electronic energy level that is occupied by an electron in a solid at a temperature of absolute zero. At higher temperatures some electrons are excited into higher energy states. The Fermi level then corresponds to the value of energy at which the Fermi–Dirac distribution function has a value ½. *See also* energy bands.

ferrimagnetism An effect observed in certain solids, notably garnets and ferrites, in which the magnetic properties change at a certain critical temperature known as the *Néel temperature.* Ferrimagnetism occurs in materials that have a permanent molecular magnetic moment associated with unpaired electron spins. At temperatures above the Néel temperature thermal agitation causes the spins to be randomly orientated throughout the material, which becomes paramagnetic: it obeys the Curie-Weiss law approximately but is characterized by a negative Weiss constant (*see* paramagnetism).

At temperatures below the Néel temperature ferrimagnetic materials behave in a similar manner to ferromagnetic materials (*see* ferromagnetism): they show spontaneous magnetization within a domain structure and *magnetic hysteresis, but the spontaneous magnetization observed is less than that of ferromagnetic materials and does not correspond to full parallel alignment of the individual magnetic moments.

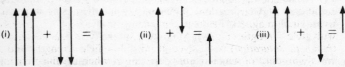

Possible arrangements of magnetic moments on two sublattices

Néel explained this behaviour by suggesting that the interatomic exchange forces are antiferromagnetic (*see* antiferromagnetism) in nature but that the magnetic moments of the sublattices containing the antiparallel spins are unequal, causing a net magnetization. Some possible arrangements are shown in the diagram: (i) unequal numbers of identical moments; (ii) unequal moments; (iii) two identical moments on each sublattice plus one unequal moment on one. The net small spontaneous magnetization causes the material to behave as a weak ferromagnetic material.

Ferrimagnetic materials are technically important because they are usually insulators and hence have low *eddy-current losses in radiofrequency applications, while exhibiting substantial magnetic moments at room temperature, although less than those of ferromagnetics.

ferrite A low-density ceramic material with composition $Fe_2O_3 \cdot XO$, where X is a divalent metal, such as cobalt, nickel, manganese, or zinc. These magnetic materials have very low eddy-current loss and cores made from them are used for high-frequency circuits and as *dust cores in computers.

ferrite core *See* core.

ferroelectric crystals Crystals that exhibit electrical properties analogous to certain magnetic properties, such as ferromagnetism. In an alternating electric field very large values of the piezoelectric and dielectric constants are developed, usually in one particular direction within a certain temperature range. These crystals are particularly useful as detectors of vibrations.

ferromagnetic Curie temperature *Syn. for* Curie point. *See* ferromagnetism.

ferromagnetism A phenomenon observed in certain solids in which the magnetic properties change abruptly at a certain characteristic temperature known as the *Curie point*. Below the Curie point the solid exhibits ferromagnetic properties. Above this temperature the thermal energy of the atoms is sufficient to produce magnetic properties typical of *paramagnetism: the *susceptibility obeys the Curie-Weiss law approximately, the value of the Weiss constant being close to that of the *Curie point and a few degrees higher.

The chief ferromagnetic elements are iron, cobalt, and nickel; there are also many ferromagnetic alloys based on these materials. Ferromagnetic materials are characterized by a large positive susceptibility: very large values of magnetization are produced by relatively small magnetic fields, the magnetization varying nonlinearly with field strength (Fig. *a*). Maximum intensity of magnetization (*magnetic saturation*) is achieved at fairly low field strengths and a certain amount of magnetization is retained when the magnetizing field is removed, i.e. the materials exhibit *magnetic hysteresis.

Ferromagnetism was first explained by Weiss, who suggested that spontaneous magnetization occurs within ferromagnetic materials due to large interatomic forces acting between neighbouring atoms

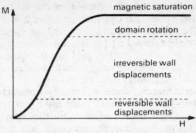

a Typical magnetization curve of a virgin ferromagnetic specimen

in the crystal lattice. Below the Curie point these forces can overcome the thermal effects and tend to produce an ordered state. The interatomic forces were discovered by Heisenberg and are known as *exchange forces*. Weiss also postulated that the groups of atoms are organized into tiny bounded regions called *domains*. In each individual domain the magnetic moments of the atoms are aligned in the same direction. The domain is thus magnetically saturated and behaves like a magnet with its own magnetic moment and axis. In an unmagnetized sample the domains are randomly orientated so that the magnetization of the specimen as a whole is zero. The existence of domains have been verified experimentally using *Bitter patterns and by the *Barkhausen effect.

The magnetic moments of the atoms arise from the *spin of electrons in an unfilled inner shell. In any stable material in the absence of an applied magnetic field, the detailed arrangements of the magnetic moments result from the interaction between the various forces operating within the sample. It is always such as to produce the minimum energy possible. In ferromagnetic materials this minimum energy state occurs when the electron spins of the atoms within a domain are arranged in parallel.

Magnetization of the domains is much harder along certain directions relative to the crystal axes than in others: more energy is required to magnetize the domains lying along these directions. This anisotropy energy is least for small domains. To form the boundaries between the domains however also requires energy because of the exchange forces between neighbouring atoms and this tends to increase domain size. The domain size is determined by a compromise between these competing forces. The boundaries are called *Bloch walls* and extend over a finite number of atoms, each of whose spins are slightly displaced relative to that of its neighbours (Fig. *b*). The energy state is also affected by the degree of crystalline perfection, the existence of strains and impurities affecting significantly

182

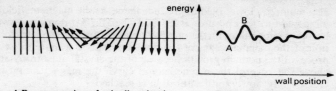

b Representation of spin direction in a Bloch wall with 180° rotation	*c* Energy as a function of Bloch wall position

the ferromagnetic behaviour. It can be shown that in any particular material the energy state of the domains is least when the Bloch wall intersects as many dislocations as possible. A typical energy curve as a function of wall position is shown in Fig. *c*. A virgin specimen would have a wall located at the minimum energy position, marked A.

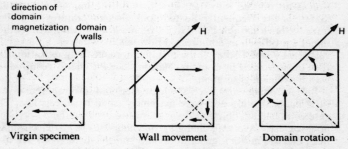

d Representation of magnetization of a ferromagnetic sample

When a magnetic field is applied to a ferromagnetic material the characteristic shape of the magnetization curve (Fig. *a*) is explained by consideration of the domain behaviour. At small values of magnetic field the net effect is to displace the Bloch walls over a few atoms, away from the minimum energy state; thus those domains with spins parallel or nearly parallel to the field grow at the expense of the others (Fig. *d*). If the field is removed the walls tend to move back to the minimum energy state and for small values of applied field the magnetization changes are small and reversible. At larger values of applied field the wall excursions are sufficiently large so that an energy maximum, marked B on Fig. *c,* is passed through and the change becomes irreversible. A single crystal with few dislocations allows much greater reversible wall excursions than a polycrystalline material with many strains and impurities present; it also requires a lower field to produce them. As the magnetic field is increased further a position is reached when further domain growth becomes impossible; further magnetization is only possible by rotation of the magnetic axes of the domains. This is a more difficult process than domain growth because of the crystal anisotropy and above the knee of the magnetization curve the magnetization increases only slowly until saturation is reached.

Ferromagnetic materials are classified as either *hard* or *soft*. Hard materials have a low relative permeability, very high coercive force, and are difficult to magnetize and demagnetize; soft materials have a high relative permeability, low coercive force, and are easily magnetized and demagnetized.

Hard ferromagnetics, such as cobalt steel and various ferromagnetic alloys of nickel, aluminium, and cobalt, retain a high percentage of their magnetization and have a relatively high hysteresis loss (*see* magnetic hysteresis). They are most suitable for use as *permanent magnets, as used in *loudspeakers. A high degree of dislocation is introduced into their structure during manufacture. Hard materials are frequently heated to high temperatures and then quenched in a suitable liquid to introduce strains. Alternatively they may be produced as compressed powders in which each particle is sufficiently small so as to be a single domain; magnetization can then only proceed by domain rotation since the energy required for wall movement is so great.

Soft ferromagnetics, such as silicon steel and soft iron, retain very little magnetization and have extremely small hysteresis loss. The ease of magnetization and demagnetization makes them very suitable for uses involving changing magnetic flux, as in electromagnets, electric motors, generators, and transformers. They are also useful for *magnetic screening. Their properties are enhanced by careful manufacture, as by heating and slow annealing, in order to achieve a high degree of crystal purity.

Although the large magnetic moment at room temperatures makes soft ferromagnetic materials extremely useful for magnetic circuits, most ferromagnetics are very good conductors and suffer energy loss from *eddy currents produced within them. The ideal material for magnetic circuits would be a ferromagnetic insulator. There is also an additional energy loss due to the fact that magnetization does not proceed smoothly but in minute jumps (*see* Barkhausen effect). This loss is known as *magnetic residual loss* and depends purely on the frequency of the changing flux density, not on its magnitude. *See also* antiferromagnetism; ferrimagnetism.

FET *Abbrev. for* field-effect transistor.

fetron A junction *field-effect transistor mounted in a suitable package so that it may directly replace a valve in a circuit without any modification to the circuit.

fibre-optics system An optical system that uses one or more glass or perspex fibres as a *light guide* or for transmitting optical images. The fibres have polished surfaces coated with a material of suitable refractive index. Light entering one end within a certain solid angle undergoes total internal reflection at the surface and is transmitted through the fibre. Provided that the critical angle of the material is not exceeded the fibres can be curved.

Parallel fibres fused together can be used to form the screens of *cathode-ray tubes and image intensifiers, thus avoiding any loss of definition as the light passes from the inner phosphor layer through the glass screen. Fibre-optics systems are now used for telecomunication purposes. The electrical signals from a transmitter are used to modulate a *laser beam, which is then transmitted through the fibres. Many more signals can be carried through a bunch of fibres than through a conventional cable of the same diameter. Low-loss fibres several kilometres in length have been produced.

Fick's law If a concentration gradient of mobile impurity atoms (or ions) exists in a semiconductor there will be a flow of such atoms (or ions) from the region of high concentration to the region of low concentration. This effect is utilized in producing a desired impurity profile in a particular specimen of semiconductor.

At normal temperatures the impurity atoms are immobile until heated to a high temperature. A concentration gradient is formed by heating the semiconductor wafer in a gaseous atmosphere of the impurity atoms so that a high concentration exists at the surface. Under such conditions the impurity atoms diffuse into the semiconductor according to *Fick's first law:*

$$f = -D(\partial N/\partial x)$$

where f is the particle flux through a plane parallel to the surface, in numbers per square metre per second. D is a constant of proportionality – the *diffusion coefficient* – and $\partial N/\partial x$ is the concentration gradient. The diffusion coefficient, D, is given by

$$D = \mu kT/e$$

where k is the Boltzmann constant, T the thermodynamic temperature, e the electron charge, and μ the mobility of the impurity atom.

Since N is also a function of time during the diffusion process, *Fick's second law* can be derived from the first law and is given as:

$$\partial N/\partial t = D(\partial^2 N/\partial x^2)$$

Diffusion also takes place in the y- and z-directions, parallel to the plane of the surface, as soon as a concentration gradient is established below the surface. Diffusion flow will continue until the impurity concentration becomes uniform across the specimen; in practice the process is stopped by reducing the temperature when the desired profile has been achieved.

field (1) The region in which a physical agency exerts its influence. Typical examples are electric and magnetic fields resulting from the presence of charge or magnetic dipoles. These are *vector fields*. Such a field may be pictorially represented by a set of curves, referred to as *lines of force or of flux. The *field density* and direction at a point represent the strength and direction of the field at that point.

(2) (in computing) A set of symbols treated together as a unit of information. The symbols need not necessarily belong to the same *word.

(3) *Syn.* raster. *See* television.

field coil *See* electromagnet.

field control A method of adjusting the voltage of a generator or the speed of a motor by varying the current in the field coil.

field current *See* electromagnet.

field density *See* field.

field-effect transistor (FET) A unipolar multielectrode semiconductor device in which current flows through a narrow conducting *channel* between two electrodes and is modulated by an electric field applied at a third electrode. The regions in the transistor are known as the *source, drain,* and *gate;* the narrow channel connects the source and drain regions and the modulating signal is applied to the gate. Application of a suitable bias across the transistor causes charge carriers to flow from source to drain. Devices with n-type source and drain regions and hence an n-type channel are referred to as *n-channel* devices, those with p-type regions as *p-channel* devices.

Field-effect transistors differ from junction transistors in that the FET is a unipolar device (current is carried by majority carriers only) and conduction takes place in the plane of the substrate wafer parallel to the semiconductor junctions. The junction transistor is a bipolar device (majority and minority carriers both contribute to conduction) and current flows across the junctions.

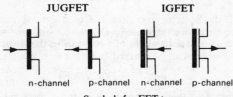

a Symbols for FET types

There are two main types of FET: the *junction FET* (JFET, JUGFET), in which the conducting channel forms part of the structure of the device, and the *insulated-gate FET* (IGFET), in which the channel is formed in use by the action of the gate voltage. The graphical symbols are shown in Fig. *a*.

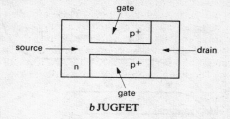

b JUGFET

The junction FET consists of a wafer of semiconductor material flanked by two highly doped layers of the opposite conductivity type (n^+ or p^+), forming the source, drain, and gate regions (Fig. *b*). If a positive voltage is applied to the drain, electrons in the source are attracted into it and flow through the channel between the source and drain. As the drain voltage increases, *depletion layers associated with the p^+–n junctions increase in size and reduce the cross-sectional area of the channel, thus increasing the resistance of the device. When the depletion layers first meet, the drain voltage is called the *pinch-off voltage*, V_P (Fig. *c*). At drain voltages above pinch-off, the depletion layers extend further into the drain region but remain essentially constant in the channel (Fig. *d*). The current therefore remains substantially constant until the voltage reaches a value at which *breakdown occurs.

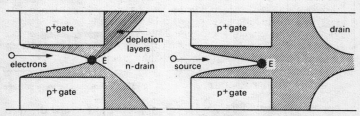

c Drain voltage just at pinch-off   *d* Drain voltage above pinch-off

Applying a negative voltage to the gate will also increase the size of the depletion layers and hence the pinch-off condition will be reached at a lower drain voltage. A family of characteristics will thus be generated for different values of gate bias (Fig. *e*). The gate voltage is therefore used to modulate the channel conductivity. The pentode valve generates a similar set of characteristics and has been replaced by the FET for many applications (*see* fetron). *MESFETs are a form of JFET in which the gate electrode is replaced by a Schottky barrier. Most gallium arsenide FETs are MESFETs.

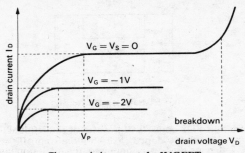

e Characteristic curves of a JUGFET

The basic structure of an insulated-gate FET is shown in Fig. *f*. A wafer of semiconductor material has two highly doped regions of

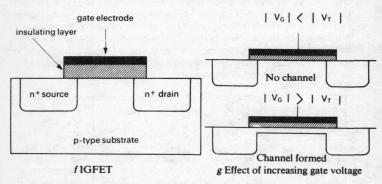

f IGFET   *g* Effect of increasing gate voltage

opposite polarity diffused into it to form the source and drain regions. An insulating layer is then formed on the surface between these regions and a conductor deposited on top of this layer to form the gate electrode. This device has several alternative names and abbreviations in common use: metal insulator silicon field-effect transistor (*MISFET, MIST*); metal oxide silicon field-effect transistor (*MOSFET, MOST*). MOS transistors have a silicon dioxide layer as the insulator and are a special case of MIS transistors. Most IGFETs are MOS devices.

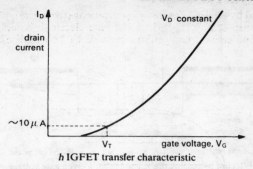

h IGFET transfer characteristic

If a positive voltage is applied to the gate electrode of an n-channel device, a depletion layer is formed in the substrate below the insulating layer. As the bias is increased, the depletion layer spreads into the semiconductor until a point is reached at which the semiconductor becomes inverted (i.e. of opposite conductivity type) at the surface. Beyond this point the depletion region ceases to grow (*see also* MOS capacitor). The inverted layer constitutes a narrow channel connecting source and drain through which current can flow (Fig. *g*). The effect of increasing the bias further is to increase the number of carriers in the inversion layer and hence the conductivity of the channel. The gate voltage at which strong inversion takes place is the *threshold voltage, V_T* (Fig. *h*). Applying a small positive voltage to the drain attracts electons from the source and current then flows through the conducting channel. The channel acts as a resistor and the drain current, I_D, is approximately proportional to the drain voltage V_D (*see* Figs. *i, j* and equation (2) below).

As the drain voltage is increased the channel depth is reduced near the drain until it eventually reaches a pinch-off point at which the channel depth is just zero. If the magnitude of the drain voltage is further increased the current remains essentially constant, since the channel remains constant; the depletion region near the drain increases in size as with junction FETs. A p-channel device operates in a similar manner to an n-channel device but negative gate and drain voltages are applied.

Field-effect transistors fall into two categories depending on the transfer characteristics of the devices (Fig. *k*): *depletion-mode devices* are those in which conduction takes place with zero gate bias; *enchancement-mode devices* are those in which a voltage must be applied to the gate before conduction can occur. All junction FETs are depletion-mode devices. Ideally, all insulated-gate FETs, as described above, are enhancement-mode devices but when an oxide is formed on the surface of a semiconductor a quantity of positive

189

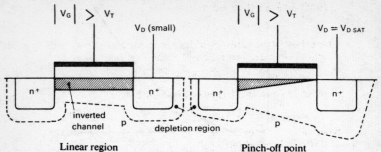

i Effect of increasing drain voltage for a given gate voltage

charge inevitably exists in the interface. In n-channel devices the presence of this positive charge can result in the presence of a spontaneous inversion layer even with zero gate bias. In p-channel devices the effect is to increase the magnitude of the threshold voltage.

Practical IGFETs can therefore be either enhancement- or depletion-mode devices. Enhancement-mode devices are simpler to use in logic circuits, since they are 'off' at zero gate bias and hence simpler to use as switches. n-channel devices are preferred due to the greater mobility of electrons compared to holes, which leads to a higher gain. Reliable manufacture of n-channel enhancement devices is now routine and these have replaced the p-channel devices, which were formerly easier to manufacture.

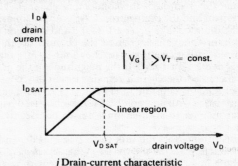

j Drain-current characteristic

Field-effect transistors may be broadly described as square-law devices, i.e. the output current I_{DS} varies with the square of the input voltage V_{GS}. (*Compare* junction transistor where there is an exponential dependence.) The basic device equations are as follows:

190

Junction FET (beyond pinch-off):

$$I_{DS} = I_{DSS}[1 - (V_{GS}/V_P)]^2 \qquad (1)$$

where I_{DSS} is the saturation drain current at $V_G = 0$ and V_P is the pinch-off voltage. The pinch-off condition is

$$V_{DS} = V_{GS} - V_P$$

Insulated-gate FET (below pinch-off):

$$I_{DS} = k[2(V_{GS} - V_T)V_{DS} - V_{DS}^2] \qquad (2)$$

where k is the gain factor and V_T is the threshold voltage.

IGFET (beyond pinch-off):

$$I_{DS} = k(V_{GS} - V_T)^2 \qquad (3)$$

The pinch-off condition is

$$V_{DS} = V_{GS} - V_T$$

The *mutual conductance, $\partial I_{DS}/\partial V_G$, of both types of FET is proportional to the square root of the drain current, i.e.

$$g_m = \partial I_{DS}/\partial V_G \propto \sqrt{I_{DS}}$$

(*Compare* junction transistors: $g_m \propto I_c$.)

The input impedance of an FET is always very high. In the junction devices the input is through a reverse-biased diode; in the insulated-gate version the input impedance is purely capacitive, the gate being separated from the channel by an insulator. (*Compare* junction transistors where the input is through a forward-biased diode.)

Depletion-mode devices   **Enhancement-mode devices**

n-channel

p-channel

JUGFETs + IGFETs   **IGFETs only**
k Transfer characteristics of FETs

191

Junction FETs are invariably used as discrete devices in circuits in which their square-law and/or high input impedance characteristics are required. These include high input impedance *amplifiers and square-law *mixers. They are also used as bidirectional switches.

Insulated-gate FETs are also used in similar discrete device applications but their main use is in *MOS integrated circuits. MOSFETS have been claimed to have several advantages over junction transistors as the basic component in integrated circuits. There are fewer steps in processing and there is a smaller area per device. They therefore have a greater functional density (devices per unit area) and lower cost than junction transistors. Other advantages include their use as high-value resistors and their low power dissipation.

field emission *Syns.* autoemission; cold emission. A type of *emission in which the presence of a large external accelerating electric field reduces the potential barrier at the surface of the emitter (*see* Schottky effect) and allows electrons to escape from the surface. The potential barrier is shown in the diagram as the work done, W, for an electron to escape.

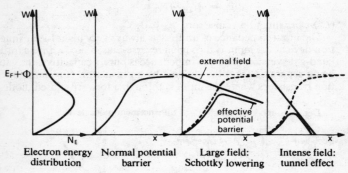

| Electron energy distribution | Normal potential barrier | Large field: Schottky lowering | Intense field: tunnel effect |

The distortion of the potential barrier at sufficiently large values of the field results in an effective narrowing of the barrier and allows the *tunnel effect to operate: electrons with energies around the Fermi level (*see* energy bands) may also be emitted. The current density j has been shown to vary with the electric field E as:

$$j = CE^2 \exp(-D/E)$$

where C and D are approximately constant. The very intense fields required are of the order of 10^{10} volts per metre and are usually only obtained at sharp points on the emitter surface.

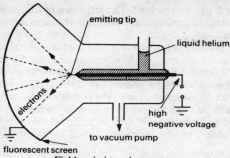

Field-emission microscope

field-emission microscope An instrument for studying the surface of a solid, usually a metal, by causing it to undergo *field emission. A simple microscope is shown in the diagram. A high voltage is applied in a vacuum between the single crystal solid tip and the curved fluorescent screen. The electrons emitted from the tip form an image on the screen; the different intensities of the image represent areas on the tip that have different work functions. Vibrations of the metal atoms limit the resolution and the tip is usually held at liquid helium or hydrogen temperature. The structure of alloys, the behaviour of impurities in metals, and the effect of adsorption at the metal point can be studied.

field-enhanced emission An increase in *photoemission and *secondary emission in the presence of a strong electric field at the surface of the emitter.

field frequency *Syn.* raster frequency. *See* television.

field ionization A process similar to *field emission but one in which electrons are emitted from an atom in the presence of an intense electric field and are captured by the surface of a nearby metal, resulting in ionization of the atom. The electrons are normally prevented from leaving the atom by a potential barrier equal to the ionization potential of the atom. This barrier may be distorted by an electric field in a similar manner to that occurring in field emission, allowing electrons to escape.

field-ion microscope An instrument for studying the surface of a solid, usually a metal, by subjecting it to *field ionization. The form is identical to the *field-emission microscope but the voltage is applied in the opposite direction. Low-pressure helium is allowed into the microscope and helium ions, formed at the surface of the tip, are accelerated to the fluorescent screen to form the image. Atomic vibrations, which affect the resolution, are minimized by holding the tip at liquid helium or hydrogen temperature; individual atoms

193

of the metal can be resolved. The structure of alloys, the structure and behaviour of surfaces, and adsorption on the metal surface can be studied.

field magnet The magnet used to provide the magnetic field in magnetic circuits and machines. An *electromagnet is most commonly used but small electrical machines may use a permanent magnet.

filament A thread-like body of metal or carbon, particularly the conductor of an incandescent lamp, the cathode of a thermionic valve, or the electrode used to heat an indirectly heated cathode.

filamentary transistor *Obsolete syn. for* unijunction transistor.

filament cathode *Syn. for* indirectly heated cathode. *See* thermionic cathode.

filament getter *See* getter.

filament reactivation Replenishment of the thorium at the surface of a thoriated tungsten filament, achieved by flashing (*see* flash) the exhausted filament.

fill-and-spill injection *Syn.* potential equilibration injection. *See* charge-coupled device.

film An electonic coating with a minimal thickness dimension. Thin films have thicknesses in the range one nanometre to one micrometre; thick films range from 10 to 100 micrometres.

film resistor A type of resistor that uses a thin layer of resistive material deposited on an insulating core. For low-power applications film resistors are more stable than composition resistors and except for very high precision requirements are smaller and less expensive than accurate wire-wound resistors.

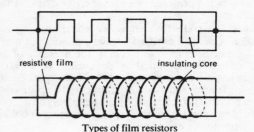

resistive film   insulating core

Types of film resistors

Resistive materials used are crystalline carbon, boron-carbon, and various metallic oxides or precious metals. Film resistors usually have a continuous uniform film applied in a particular pattern to the core, the film thickness determining the resistance (*see* diagram). Higher resistances are obtained by using a spiral pattern of film on the core, the tighter the spiral the higher the resistance. High-power applications are limited by the film resistor's 200°C maximum operating temperature but below this limit the resistance achieved for a given physical size is higher than that of the corresponding wire-wound resistor.

Flat thin- or thick-film resistors are also used in *integrated circuits. These may be produced in a form suitable for hybrid integrated circuits or as an integral compenent in a fully integrated circuit.

Continuously adjustable film resistors are also produced; they may be either linear of circular and are actuated by a lead screw.

Type of filter	Pass band(s)	Attenuation band(s)
low pass	$0 - f_c$	$f_c - \infty$
high pass	$f_c - \infty$	$0 - f_c$
band pass	$f_1 - f_2$	$0 - f_1,\ f_2 - \infty$
band stop	$0 - f_1,\ f_2 - \infty$	$f_1 - f_2$

filter An electrical network that will transmit signals with frequencies within certain designated ranges (*pass bands*) and suppress signals of other frequencies (*attenuation bands*). The frequencies that separate the pass and attenuation bands are the cut-off frequencies, which have the symbols f_c if there is only one cut-off frequency or f_1 and f_2 if more than one. Filters are classified according to the ranges of their pass or attenuation bands as *low-pass, high-pass, band-pass* and *band-stop filters;* the four main classifications with their corresponding frequency limits are shown in the table.

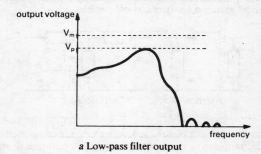

a Low-pass filter output

An ideal filter would transmit the pass band without attenuation and completely suppress the attenuation band, with a sharp cut-off profile. Practical filters however do attenuate the pass band, due to absorption, reflection, or radiation, which results in loss of signal power; neither do they completely suppress the attenuation bands. A typical curve of output voltage with frequency is shown in Fig. *a* for a simple low-pass filter: V_p is the peak voltage and V_m is the
195

maximum voltage of an ideal filter. The *filter attenuation* is defined as the loss in signal power in decibels or nepers through the filter; the *filter discrimination* is the difference between the minimum value of *insertion loss in an attenuation band and the maximum value in a pass band.

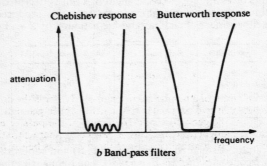

b **Band-pass filters**

The components of a practical filter may be arranged to give the desired output curve. For example, *Chebishev* (or *Tchebyshev*) and *Butterworth filters* are band-pass filters with different output characteristics (Fig. *b*). Butterworth filters have a flat response in the pass band whereas Chebishev filters have some variation of the residual response in the pass band but have a more rapid increase of attenuation giving a sharper cut-off profile.

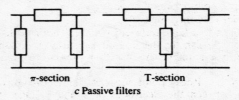

c **Passive filters**

Filters are *active* or *passive* according to their components. Active filters contain active components, such as operational amplifiers, that introduce some gain into the signal combined with suitable R-C feedback circuits to give them the desired frequency-response characteristic. Most passive filter networks are constructed from impedances arranged in shunt and in parallel (L-C networks). Two basic arrangements are used: *π-sections* and *T-sections* (Fig. *c*). Composite networks are built up from these basic sections and the arrangement is termed a *ladder network* because of the alternation of shunt and parallel sections. Another type of configuration is the *lattice filter* in which the impedance elements are arranged in a bridge network (Fig. *d*).

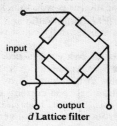

input

output
d **Lattice filter**

The bandwidth of a band-pass or band-stop filter is the difference in hertz between two particular frequencies whose geometric mean equals that of the geometric mid-frequency of the pass or attenuation band. Frequencies exhibiting a particular characteristic, such as the point at which the response is three decibels below the peak value, are usually chosen.

filter attenuation *See* filter.

filter discrimination *See* filter.

firing *See* gas-discharge tube.

first ionization potential *See* ionization potential.

fixed-cycle operation *See* synchronous computer.

fixed-point representation A method of representing numbers in which the decimal or binary point is fixed in a given location and the number contains a constant predetermined number of digits. *Compare* floating-point representation.

fixed store *See* read-only memory.

flash To subject to a large transient voltage. Thoriated filaments are flashed to replenish the thorium on the surface; carbon filaments are flashed in an atmosphere of carbon to produce a uniform cross section.

flash arc *U.S. syn.* Rocky point effect. A sudden disruptive transient discharge between the electrodes of a high-voltage *thermionic valve that usually but not invariably results in the destruction of the valve.

flashback voltage The peak inverse voltage across a *gas-discharge tube needed to produce ionization.

flashover A disruptive discharge in the form of an *arc or *spark between two electrical conductors or between a conductor and earth. The occurrence of an arc is termed *arcover* and of a spark *sparkover*.

flashover voltage The voltage between two conductors at which *flashover just occurs. The *dry flashover voltage* is the voltage at which

197

flashover occurs when the conductors are separated by a clean dry insulator; the *wet flashover voltage* is the voltage at which flashover occurs when the clean insulator is wet, thus simulating rain.

flat-band capacitance *See* MOS capacitor.

flat tuning Tuning with a substantially equal response to a range of frequencies.

flatwise bend *Syn. for* H-bend. *See* waveguide.

F-layer, F_1-layer, F_2-layer *Syn.* Appleton layer. *See* ionosphere.

Fleming's rules Mnemonics in frequent use by practical electricians for the directional relation between current, motion, and magnetic field in a dynamo or electric motor. In each case the thu*M*b represents the direction of *M*otion, the *F*irst finger represents the *F*ield, and the m*I*ddle finger the current, *I*. When these three fingers are held at right angles, the right hand represents the relation in a dynamo (*right-hand rule*) and the left hand represents the relation in an electric motor (*left-hand rule*).

flexible resistor A type of wire-wound *resistor in which the wire is wound round a flexible insulating core.

flicker Visual perception of brightness fluctuations of less than 25 to 30 per second, i.e. less than the persistence of vision. The threshold for perceiving flicker is dependent on the brightness of the observed light and the angle it subtends to the optic axis. In *television flicker prevents complete continuity of the images. In *colour television flicker can result from either unwanted variations in the luminance signal – *luminance flicker* – or in the chrominance signal – *colour flicker.*

flicker noise *See* Schottky noise.

flip-chip A semiconductor *chip with thickened and extended *bonding pads enabling it to be flipped over and mounted upside down on a suitable substrate, such as a *thin-film or *thick-film circuit. Extension of the bonding pads is achieved by depositing metallic pellets on top of the lead areas in small holes in the overlying oxide layer.

flip-coil The classical means of measuring the value of *magnetic flux density, *B,* at a point in air. The flip-coil consists of a number of turns, N, wound on a small former of area A. If the coil is placed in a magnetic field, with its axis parallel to the direction of *B,* the *magnetic flux linking the coil is NAB. If the coil is removed very quickly ('flipped') to a point where the flux density is zero, the flux change is just NAB; this can be measured using either a *ballistic galvanometer or a *fluxmeter.

Flip-coils can be constructed with different values of NA (turns × area): the appropriate coil can be chosen for the field to be measured, so as to produce a suitable instrument deflection. Accuracy of about one per cent can be achieved using a fluxmeter and about 0·1 per cent with a ballistic galvanometer.

The coil may also be used to investigate the magnetic flux distribution of a magnetic field. It is used in a similar way but the measuring instrument does not need to be calibrated absolutely. The comparative measurements can be plotted to produce a graphical representation of the flux distribution. When used in this way the coil is often termed an *exploring coil*.

An alternative way of using a flip-coil to measure magnetic flux density is to rotate it rapidly in the field. An alternating voltage, V, is produced given by

$$V \propto NAB\omega$$

where ω is the angular velocity of the coil. As this method does not depend on a single throw measurement, less sensitive measuring instruments may be used.

flip-flop *Syn.* half-shift register. A bistable *multivibrator circuit that usually has two inputs corresponding to the two stable states. It is so called because application of a suitable input pulse causes the device to 'flip' into the corresponding state and remain in that state until a pulse on the other input causes it to 'flop' into the other state.

Flip-flops are widely used in computers as counting and storage elements and several types have been developed. Flip-flops as described above are *unclocked* and are triggered directly by the input pulses. *Clocked flip-flops* have a third input to which a clock pulse is applied. The output state of the device is determined by the state of the inputs at the moment a clock pulse is applied. The basic types of flip-flops are described below.

A *D-type flip-flop* ('D' stands for delay) is a clocked flip-flop whose output is delayed by one clock pulse: if a logical 1 appears at the input, a logical 1 will appear at the output one clock pulse later.

Input		Output	
R	S	Q	$\overline{Q}$
0	0	no change, same as previous state	
1	0	1	0
0	1	0	1
1	1	indeterminate	

clock input — R / S — Q / $\overline{Q}$

Clocked R-S flip-flop

An *R-S flip-flop* is a flip-flop whose inputs are designated R and S. The outputs corresponding to the various input combinations are shown in the table. Logical 1s should not be allowed to appear on the inputs together.

A *J-K flip-flop* is a flip-flop whose inputs are designated J and K. These devices are almost invariably clocked and their outputs are the same as the R-S type except when logical 1s appear together at the inputs. In these circumstances the device changes state.

An *R-S-T flip-flop* has three inputs designated R, S, and T. The R and S inputs produce outputs as described above. Application of a pulse to the T input causes the device to change state.

A *T flip-flop* has only one input. Application of a pulse to this input causes the device to change state.

floating Denoting a circuit or device that is not connected to any source of potential.

floating battery *See* battery.

floating-carrier modulation *Syn.* controlled-carrier modulation. A type of *amplitude modulation in which the amplitude of the *carrier wave does not remain constant but is automatically varied in a manner dependent upon the amplitude of the modulating wave, which is averaged over a short time period. The modulation factor therefore remains substantially constant.

floating-diffusion amplifier *See* charge-coupled device.

floating-gate amplifier *See* charge-coupled device.

floating-gate PROM *Syn.* optically erasable memory. *See* read-only memory.

floating-point representation A method of representing numbers by means of a predetermined number of significant digits – the *mantissa* – together with a decimal or binary multiplier – the *exponent:* the number x can be written as

$$x = y \times n^z$$

where y is the mantissa, n is either 10 or 2, and z is the exponent in integer form. *Compare* fixed-point representation.

flooding *See* storage tube.

flooding gun *See* storage tube.

floppy disk *See* moving magnetic surface memory.

fluorescence *See* luminescence.

fluorescent lamp A type of lamp in which light is generated by fluorescence (*see* luminescence). A common form of fluorescent lamp consists of a *gas-discharge tube containing a low-pressure gas, such as mercury, with the inner surface of the tube coated with a *phosphor. When a current passes through the tube the ultraviolet radiation produced strikes the phosphor, which then emits visible radiation. Another type of lamp, the usual sodium vapour or mercury vapour street lamp, does not have a fluorescent coating; electrons in the discharge excite the atoms of vapour and these atoms fluoresce as they decay from the excited states thus produced.

fluorescent screen A type of screen used in various electronic devices, such as *cathode-ray tubes, *image converters, or storage cathode-ray tubes, to convert the electron beam into a visible image. These screens consist of an array of many small-diameter (two to three micrometre) phosphor crystals that emit light when bombarded by high-energy radiation, such as X-rays, or by electrons.

flutter An unwanted type of *frequency modulation in high-fidelity sound-reproduction systems resulting in audible variations of pitch above about 10 hertz. *Compare* wow.

flux A measure of the strength of a field of force through a specified area. *See* electric flux; magnetic flux.

fluxmeter An instrument that measures changes in magnetic flux. The most usual type is the *Grassot fluxmeter,* which consists essentially of a moving-coil *galvanometer that is designed so that the restoring couple on the moving coil is negligibly small and electromagnetic damping is large. The galvanometer is used in conjunction with an exploring coil (*see* flip-coil) of known area. A change in the magnetic flux cutting the exploring coil causes an induced current in the galvanometer coil and hence the latter is deflected. The angle of deflection is directly proportional to the change in magnetic flux through the exploring coil. The instrument is calibrated empirically using a magnetic flux standard.

flyback (1) *See* time base. (2) *See* sawtooth waveform.

flying-spot scanner A device that produces a video signal from an object, such as a film, by scanning the object with a spot of light, which is then focused on a *photocell to produce corresponding electrical signals. The moving (or 'flying') spot of light is normally produced on the screen of a high-intensity cathode-ray tube used as a light source. Mechanical scanning of the object has also been employed, using a single point source of light, with a suitably perforated rotating disc between it and the object.

flywheel effect The continuation of oscillations in an *oscillator during the intervals between exciting pulses. It results from electrical inertia, which is analogous to mechanical inertia of a flywheel.

flywheel time base *See* time base.

f.m. (or FM) *Abbrev. for* frequency modulation.

FM receiver A *radio or *television receiver that detects frequency-modulated signals (*see* frequency modulation).

focus error signal *See* compact disc system.

focusing The process or a method of making a beam of radiation or particles converge. In an electron-beam device, such as a *cathode-ray tube, three principal methods of focusing the beam are used.

In *electrostatic focusing* two or more electrodes at different potentials are used to focus the electron beam. The electrostatic fields set up between the electrodes cause the beam to converge; the focusing effect is controlled by varying the potential of one of the electrodes, termed the *focusing electrode*. The electrodes are usually cylindrical, mounted coaxially with the electron tube, and are used in conjunction with deflection plates in instruments such as cathode-ray tubes.

In *electromagnetic focusing* the action of a magnetic field is used to make the electron beam converge. The field is produced by passing direct current through a *focusing coil,* the focusing effect being controlled by varying the current through the coil. The coil is usually one of short axial length that surrounds the tube and is coaxial with it. Electrons of different energies converge at different points along the beam axis. Thus if the electron beam is not monoenergetic, longitudinal spreading of the nominal convergence point results. Electromagnetic focusing is used in conjunction with deflection coils in instruments such as cathode-ray tubes.

In *gas focusing* a small amount of an inert gas, such as argon or helium, is introduced into the tube. The gas becomes ionized along the path of the electron beam and the mutual attraction between the electrons and the positive ions causes a narrowing of the electron beam. A loss of focus occurs at high-frequency operation of the tube due to the relatively high inertia of the positive ions. The life of these gas-focused tubes is comparatively short since the positive ions bombard the cathode and soon destroy it; gas focusing is therefore little used.

focusing coil *See* focusing.

focusing electrode *See* focusing.

focus servo system *See* compact disc system.

foil capacitor A capacitor in which the electrodes are metal foil. The term is most commonly applied to *paper capacitors but some polystyrene or polyester *film capacitors use foil electrodes and one form of tantalum *electrolytic capacitor uses a tantalum foil as one of the electrodes.

folded dipole *See* dipole aerial.

forbidden band *See* energy bands.

forced oscillations Oscillations produced in a circuit that is acted upon by an external driving force, such as oscillations in a resonant circuit coupled to a fixed-frequency oscillator. The resulting oscillations have two components: a transient component, whose frequency is determined by the natural frequency of the circuit and decays rapidly, and a steady component, whose frequency equals that of the external driving force.

If a circuit is acted upon by an external voltage

$$V = V_0 \cos\omega t$$

applied at $t = 0$, then the steady state solution is given by

$$I = (V_0/Z) \cos(\omega t - \phi)$$

where Z is the impedance and is given by

$$Z^2 = R^2 + (\omega L - 1/\omega C)^2$$

and ϕ, the phase angle, is given by

$$\phi = \tan^{-1}[(\omega L - 1/\omega C)/R]$$

The phase differs from that of the applied voltage except when

$$\omega L - 1/\omega C = 0$$

i.e. when

$$\omega = \omega_0 = 1/\sqrt{(LC)}$$

where ω_0 is the natural frequency of the circuit. This is the resonance condition and the current is then maximal. *See also* resonant frequency. *Compare* free oscillations.

force factor The ratio of the force required to block the movement of an electromechanical *transducer to the corresponding current in the electrical system.

form factor The ratio of the *root-mean-square value of an alternating quantity (such as current or voltage) to the half-period mean value, for a half-period beginning at a zero point. A simple sine wave has a form factor equal to $\pi/2\sqrt{2}$ (i.e. 1·111).

FORTRAN *See* programming language.

forward bias *Syn.* forward voltage. *See* forward direction.

forward blocking state *See* pnpn device; silicon-controlled rectifier.

forward current *See* forward direction.

forward direction The direction in which an electrical or electronic device has the smaller resistance. A voltage applied in a forward direction is the *forward bias;* it produces the larger current, known as the *forward current. Compare* reverse direction.

forward path *See* feedback control loop.

forward slope resistance *See* diode.

forward transfer function *See* feedback control loop.

forward voltage *Syn. for* forward bias. *See* forward direction.

forward wave *See* travelling-wave tube.

Foster–Seeley discriminator *See* frequency discriminator.

Foucault current *See* eddy current.

Fourier analysis A mathematical method of analysing complex waveshapes or signals into a series of simple harmonic functions, the frequencies of which are integral multiples of the fundamental frequency. In the case of an arbitrary periodic phenomenon, u, of period T, provided that certain conditions – Dirichlet conditions – are satisfied the periodic process may be represented by the *Fourier series* given by

$$u = F(t)$$

where

$$F(t) = \Sigma_{n=-\infty}^{n=+\infty} a_n e^{in\omega t}$$

where ω is equal to $2\pi/T$, i is the square root of -1, and n is an integer; a_n is the nth coefficient and is given by

$$a_n = (1/T)\int_0^T F(t)\, e^{-in\omega t}\, dt$$

The Fourier series may alternatively be written as a series of sines and cosines or as a series involving the phase angles of the various components.

As the period, T, becomes infinitely large so that $1/T$ tends to zero, the Fourier series in its limiting form becomes an integral – the *Fourier integral.* The values of the Fourier series or the Fourier integral are determined by the physical conditions of the phenomenon under consideration.

Fourier analysis is widely used in the electrical communication field. A slightly different representation is commonly used in which the Fourier integral is written as

$$F(t) = \int_{-\infty}^{+\infty} g(\omega)\, e^{i\omega t}\, d\omega$$

where the function

$$g(\omega) = 1/2\pi\int_{-\infty}^{+\infty} F(t)\, e^{-i\omega t}\, dt$$

is called the *Fourier transform* of the function $F(t)$. Similarly $F(t)$ is also the Fourier transform of the function $g(\omega)$.

Fourier integral *See* Fourier analysis.

Fourier series *See* Fourier analysis.

Fourier transform *See* Fourier analysis.

four-layer diode *Syn.* Schockley diode. *See* pnpn device.

four-phase CCD *See* charge-coupled device.

four point probe A method of measuring the sheet resistance of slices of semiconductor. Four equally spaced probes are used – the two outer probes supply a small current I and the two inner probes are connected to a high-impedance voltmeter to measure the potential difference V developed. In the ideal case of a homogeneous semi-infinite material the sheet resistance R_s is given by

$$R_s = 2\pi s V/I$$

where s is the spacing between the probes. In practice, the material is not infinitely thick; the conductive layer is usually very thin relative to the probe spacing, and a correction factor of $\pi/\log_e 2$ must be applied. The lateral extent of a slice is not infinite and a second correction factor must be used – the value of this factor is a function of the slice size and positioning of the probe.

four-terminal network *See* quadripole.

four-terminal resistor A standard resistor that has four terminals. Two are used to connect the resistor to the current source and the other two to connect it to a measuring instrument. This arrangement ensures that the potential drop across the resistor is not affected by contact resistances at the terminals.

four-wire circuit A circuit that consists of two pairs of conductors and that forms a simultaneous two-way communication channel between two points of a telecommunication system. One pair of conductors forms the 'go' channel and the other the 'return' channel. In the case of a *phantom circuit two pairs or groups of conductors form the circuit. A circuit that operates in the same manner as a four-wire circuit, although not necessarily containing four conductors, is a *four-wire type circuit*. An example is a circuit that contains two wires and that has a different frequency band for each direction of transmission. *Compare* two-wire circuit.

frame (1) The complete picture in television.

(2) One cycle of a number of pulses that regularly recur in a pulse train used in pulse-train communications.

frame aerial *See* loop aerial.

frame direction finding *Syn.* loop direction finding. *See* direction finding.

frame/field transfer device *See* solid-state camera.

frame frequency *Syn.* picture frequency. *See* television.

franklin Symbol: Fr. A unit of charge in the obsolete CGS electrostatic system of units. One franklin equals $3 \cdot 336 \times 10^{-10}$ coulomb.

free electron An electron that is not bound to a specific atom or molecule and is therefore free to move when influenced by an applied electric field. *See also* energy bands; semiconductor; Fermi-Dirac statistics.

free-electron paramagnetism *See* paramagnetism.

free field A *field in which any boundary effects are negligible in the region of interest.

free-field calibration Calibration of a microphone in which the open-circuit voltage produced by a certain value of sound pressure in the free wave is determined. The presence of the microphone upsets the *free field that existed before the introduction of the microphone, causing difficulties in determining the pressure value. At higher frequencies, when the dimensions of the microphone are comparable with the wavelength, the actual pressure on the pressure-sensitive portion of the microphone can approach twice that of the actual pressure in the free wave. At lower frequencies, where the dimensions are small compared to wavelength, the pressures are substantially equal.

free oscillations Oscillations arising in a circuit under the influence of internal forces, such as a capacitor discharging through a resistance and inductance, or of a constant external force, such as a direct voltage. Both these conditions are analogous to a mechanical vibrating system being displaced from the neutral point.

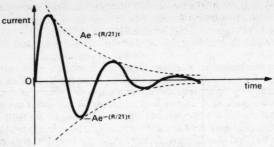

Decay of free oscillations

The oscillations decay gradually, depending on the amount of damping in the circuit, with a frequency, f, termed the natural frequency, which is approximately equal to $(LC)^{-2}$, when the resistence R in the circuit is small (*see* diagram). The amplitude of the current is given by

$$i = A\, e^{-(R/2L)t} \sin\omega t$$

where the angular frequency ω is equal to $2\pi f$, and A is a constant determined by the initial conditions. The maxima of successive oscillations lie on the curve

$$i = A\, e^{-(R/2L)t}$$

and the amplitudes (i_m, i_{m+1}) of successive maxima of the same sign decrease by a constant ratio. It can be shown that

$$\log_e(i_m/i_{m+1}) = \pi/Q$$

where Q, the *Q-factor of the circuit, is given by

$$Q^2 = L/R^2C$$

Compare forced oscillations.

free space The region used as an absolute standard and characterized by an absence of gravitational and electromagnetic fields. Free space was formerly referred to as a vacuum. The electric constant and the magnetic constant are the formally defined values of the *permittivity of free space and the *permeability of free space, respectively. The velocity of light in free space is constant and is the maximum possible value.

F-region *Syn. for* F-layer. *See* ionosphere.

frequency Symbol: ν or f; unit: hertz. The number of complete oscillations or cycles of a periodic quantity occurring in unit time. The frequency is related to the angular frequency ω by the relation $\omega = 2\pi\nu$. The frequency of a periodic quantity, such as an alternating current, is given by the number of times the quantity passes through its zero value in the same sense in unit time.

The frequency of *electromagnetic radiation is related to the wavelength, λ, by the equation $\nu = c/\lambda$, where c is the velocity of light.

frequency analyser *See* wave analyser.

Wavelength	Band	Frequency
1 mm — 1 cm	extremely high frequency ; EHF	300 — 30 GHz
1 cm — 10 cm	superhigh frequency ; SHF	30 — 3 GHz
10 cm — 1 m	ultrahigh frequency ; UHF	3 — 0.3 GHz
1 m — 10 m	very high frequency ; VHF	300 — 30 MHz
10 m — 100 m	high frequency ; HF	30 — 3 MHz
100 m — 1000 m	medium frequency ; MF	3 — 0.3 MHz
1 km — 10 km	low frequency ; LF	300 — 30 kHz
10 km — 100 km	very low frequency ; VLF	30 — 3 kHz

a Radiofrequency bands

frequency band A particular range of *frequencies that forms part of a larger continuous series of frequencies. The internationally agreed radiofrequency bands are shown in Table *a. Microwave* frequencies, ranging from VHF to EHF bands (i.e. from 0·225 to 100 gigahertz), are usually subdivided into bands designated by letters. These are not internationally agreed but the commonly used subdivisions are shown in Table *b*. Limits for the bands may differ slightly from those shown. The entire electromagnetic spectrum is shown in Table 10, backmatter.

Band	Frequency (GHz)	Wavelength (cm)
P	0.225 – 0.390	133.3 – 76.9
L	0.390 – 1.550	76.9 – 19.3
S	1.55 – 5.20	19.3 – 5.77
X	5.20 – 10.90	5.77 – 2.75
K	10.90 – 36.00	2.75 – 0.834
Q	36.0 – 46.0	0.834 – 0.652
V	46.0 – 56.0	0.652 – 0.536
W	56.0 – 100.0	0.536 – 0.300

b Microwave bands

frequency bridge An alternating-current *bridge whose balance point is dependent on the frequency at which the measurement is carried out.

frequency changer *Syn.* conversion transducer. (1) A device that converts alternating current of one frequency to alternating current of another frequency. The *conversion gain ratio* of a frequency changer is defined as the ratio of signal power available at the output to that available at the input; the *conversion voltage gain* is the ratio of the output voltage to the input voltage.

(2) *See* mixer.

frequency compensation Modification of a circuit or device to produce a flat response to a particular range of frequencies.

frequency control *See* automatic frequency control.

frequency deviation *See* frequency modulation.

frequency discriminator A *discriminator that selects input signals of constant amplitude and produces an output voltage proportional to the amount that the input frequency differs from a fixed frequency. Frequency discriminators are commonly used in *automatic frequency control systems (when the output is used to correct the frequency) and in *frequency-modulation systems to convert the frequency-modulated signals to amplitude-modulated signals. The design of frequency discriminators is such that noise due to amplitude variations in the received signal is almost completely eliminated. The best known type is the *Foster-Seeley discriminator*. This device employs a tuned reactance, in which the output is changed through a 90° phase angle, followed by a limiter.

frequency distortion *Syn. for* attenuation distortion. *See* distortion.

frequency diversity *See* diversity system.

frequency divider A device that produces an output signal whose frequency is an exact integral submultiple of the input frequency.

frequency-division multiplexing A form of *multiplex operation in which each user of the system is assigned a different frequency band. The transmitted signal contains several *carrier waves each of a different frequency and separately modulated with a different input signal. At the receiver a number of tuned circuits are used to separate the different carrier frequencies.

frequency doubler A *frequency multiplier that produces an output signal with a frequency twice that of the input signal.

frequency meter An instrument that is used to measure the frequency of an alternating current. The frequency of an electromagnetic wave is commonly measured using a *cavity resonator.

frequency-modulated cyclotron *See* synchrocyclotron.

frequency-modulated radar *See* radar.

frequency modulation (f.m. or FM) A type of *modulation in which the frequency of the *carrier wave is varied above and below its unmodulated value by an amount proportional to the amplitude of the signal wave and at the frequency of the modulating signal, the amplitude of the carrier wave remaining constant (*see* diagram). If the modulating signal is sinusoidal then the instantaneous amplitude, e, of the frequency-modulated wave may be written:

$$e = E_m \sin[2\pi F t + (\Delta F/f)\sin 2\pi f t]$$

where E_m is the amplitude of the carrier wave, F the unmodulated carrier wave frequency, ΔF the peak variation of the carrier-wave frequency from F due to modulation, and f is the modulating signal frequency. ΔF is called the *frequency swing* and the maximum value (ΔF_{max}) of the frequency swing for which the system has been designed is the *frequency deviation*. The *deviation ratio* is defined as $\Delta F_{max}/f_{max}$; the *modulation index*, β, is given by $\Delta F/f$.

Frequency modulation has several advantages over *amplitude modulation, the most important being the improved *signal-to-noise ratio. *Compare* phase modulation.

frequency multiplier A circuit or device that produces an output signal whose frequency is an exact integral multiple of the frequency of the input. Particular cases of frequency multipliers are frequency doublers and triplers.

One type of frequency multiplier uses a nonlinear amplifier, usually a *class B or *class C amplifier, which produces an output containing many harmonics of the input. The required harmonic is abstracted using a *filter, often in the form of a tuned circuit. Another type has a *multivibrator triggered by the input signal to produce oscillations that are an exact integral multiple of the frequency of the input.

frequency overlap *See* colour television.

frequency pulling (1) *See* pulling. (2) *See* magnetron.

frequency pushing *See* magnetron.

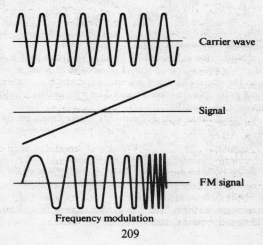

Frequency modulation

frequency range The range of frequencies at which a circuit or device operates normally. The frequency range at which a particular device is useful depends on the particular operating conditions for the device.

frequency relay *See* relay.

frequency-response characteristic The variation with frequency of the *transmission loss or gain of any apparatus, circuit, or device. A series of tests at different frequencies designed to determine the frequency-response characteristic is termed a *frequency run.*

frequency run *See* frequency-response characteristic.

frequency-selective relay *See* relay.

frequency selectivity The ability of any circuit or device to differentiate between signals at different frequencies or between a signal at a particular frequency and interference at different frequencies, and to select the desired signal.

frequency spectrum (of electromagnetic waves) *See* Table 10, backmatter.

frequency standard, primary A very stable and precise *oscillator that may be calibrated against national standard frequencies and used as a laboratory standard.

frequency swing *See* frequency modulation.

frictional electricity *Syn.* triboelectricity. The phenomena associated with electrostatic charge produced by friction between two dissimilar materials, such as glass and silk or ebonite and catskin. *Frictional machines,* such as the Wimshurst machine, were machines designed to produce electricity by friction: they are now obsolete.

fringe area A region round a broadcasting transmitter in which satisfactory reception of the broadcast signal is not always obtained.

front porch *See* television.

front-wall photovoltaic cell *See* photovoltaic cell.

frying (1) *Noise that is caused by external sources and is invariably present in the recording of sound. It results in a noise, resembling frying, in the audio output of a sound-reproduction system. *Surface noise* on a gramophone record caused by irregularities in the recording surface can contribute to frying.

(2) A frying noise that is heard in a telephone receiver and is caused by an excessive current being passed through the carbon granules of the transmitter or receiver.

f.s.d. (or **FSD**) *Abbrev. for* full-scale deflection.

full-scale deflection (f.s.d.; FSD) The maximum value of the measured quantity for which a measuring instrument is calibrated.

full-wave dipole *See* dipole aerial.

full-wave rectifier circuit A *rectifier circuit that rectifies both the positive and negative half-cycles of the single phase a.c. input and deliv-

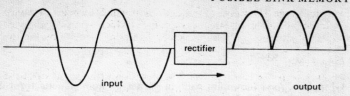

Full-wave rectifier circuit

ers them as unidirectional current to the load (*see* diagram). *Compare* half-wave rectifier circuit.

functional packing density *See* packing density.

function generator (1) A *signal generator that produces various specific waveforms for test purposes over a wide range of frequencies.

(2) A unit in an analog computer that produces an output signal corresponding to the value of a specified function of the independent variable input.

fundamental *Short for* fundamental frequency.

fundamental frequency (1) The frequency of a sinusoidal component of a periodic quantity that has the same *period as the periodic quantity.

(2) The lowest frequency present in a complex vibration. *See also* harmonic.

fuse A short length of easily fusible wire that is used to protect electric circuits or devices by melting ('blowing') at a specific current and thus breaking the circuit. The *fuse current rating* is the maximum value of current that the fuse will conduct without melting. The frequency and voltage at which a fuse is designed to operate are specified by the *fuse frequency rating* and *fuse voltage rating*. The *fuse characteristic* is the relation between the current through the fuse and the time taken for the fuse to operate.

fusible-link memory *See* read-only memory.

G

G³ *Abbrev. for* gadolinium gallium garnet. A nonmagnetic form of garnet that can be grown as single crystals and used as a nonmagnetic substrate material for solid-state magnetic circuits, particularly *magnetic bubble memory.

gain Unit: decibel. A measure of the ability of an electronic circuit, device, or apparatus to increase the magnitude of a given electrical input parameter. In a power *amplifier the gain is the ratio of the power output by the amplifier to the power input to it. For a voltage *amplifier that supplies relatively little power to the load, the gain is the ratio of the voltage developed across a specified load impedance to the input voltage. For a *directive aerial the gain is the ratio of voltage generated in the direction of maximum sensitivity to that produced by the same signal in an omnidirectional aerial.

gain control *Syn.* volume control. A circuit or device that varies the amplitude of the output signal of an *amplifier.

gain region *Syn.* active layer. *See* injection laser.

galactic noise *Syn.* Jansky noise. *See* radio noise.

gallium arsenide Symbol: GaAs. A *direct-gap 3-5 *semiconductor that has a relatively large band gap and high carrier mobility. The relatively high carrier mobility allows the semiconductor to be used for high-speed applications and because of the large energy gap it has a high resistivity that allows easier isolation between different areas of the crystal. The conduction band is a two-state conduction band, i.e. solutions of Schrödinger's equation exhibit double degeneracy in the conduction band. Some electrons therefore are 'hot' electrons, i.e. they have small effective mass and higher velocity; this results in the *Gunn effect (*see also* direct-gap semiconductor).

Gallium arsenide has several advantages compared to silicon for certain applications. It can be operated at microwave frequencies where silicon devices cannot function; gallium arsenide devices are more radiation tolerant than silicon devices, making them attractive for some space and military applications; as a direct-gap semiconductor it can be used for optical components such as *light-emitting diodes and *lasers, giving the potential for fabricating integrated opto-electronic circuits.

Gallium arsenide processing differs substantially from silicon processing. Diffusion of impurities into the material is extremely difficult, unlike silicon, therefore *epitaxy or *ion implantation must be used to produce areas of different conductivity type. There is no stable native oxide for gallium arsenide, which means that the widely used *MOS devices cannot be formed on gallium arsenide. The most widely used device is the *MESFET, a GaAs junction field-effect transistor employed in high-speed *logic circuits.

Gallium arsenide is used for microwave devices such as *Gunn or *IMPATT diodes, in monolithic microwave *integrated circuits (MMICs) for use in *phased array radar or *direct broadcast by satellite, and in high-speed logic circuits. *High electron mobility transistor (HEMT) logic circuits are being developed with operating speeds approaching those of the superconducting *Josephson junctions. Most gallium arsenide devices are unipolar, but the *heterojunction bipolar transistor (HJBT) is a bipolar device with potential for digital applications.

galvanic cell *Obsolete syn. for* cell.

galvanic current *Obsolete* Direct current, such as that produced by a cell.

galvanoluminescence *See* luminescence.

galvanomagnetic effect Any of various phenomena arising when a current is passed through a conductor or semiconductor in the presence of a magnetic field. The effects include the *Ettinghausen effect, *Hall effect, *magnetoresistance, and the *Nernst effect.

galvanometer An instrument that detects and measures small currents. The most commonly used type is the *moving-coil galvanometer,* in which a small coil is suspended between the poles of a permanent magnet. When the current to be measured is passed through the coil a couple acts on the coil forcing it to rotate in the magnetic field. Before passing the current the suspension is adjusted so that, in most types, the coil sets along the direction of the magnetic field; some galvanometers have the coil set perpendicular to the field. When a current, i, is passed through the coil, the couple and hence the angular deflection, θ, is proportional to $iB\cos\theta$, where B is the magnetic flux density. In practice the poles of the magnet are shaped and have a soft iron core between them (Fig. a) so that the magnetic field is radial and always parallel to the plane of the coil whatever the deflection; thus $\cos\theta = 1$ and $\theta \propto i$. The suspension is used to provide the necessary torsional control; it may be in the form of a strip of phosphor bronze or a light hair spring. The coil is wound on a metal former in which *eddy currents are induced; these make the instrument deadbeat (*see* damped).

The deflection may be observed by means of a light nonmagnetic pointer that moves over a scale or more usually by using a small mirror attached to the suspension. A beam of light is reflected by the mirror and the spot of light moves across a linear scale. This arrangement is sometimes called a *mirror galvanometer.* Greater sensitivity may be obtained by causing the reflected light spot to fall on a suitable *photocell. This is known as a *photoelectric galvanometer.* The moving-coil galvanometer is virtually independent of the

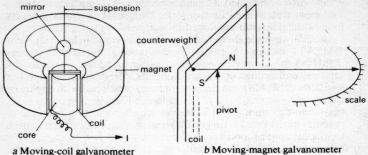

a Moving-coil galvanometer   b Moving-magnet galvanometer

earth's magnetic field, which is very weak compared to the field of the magnet.

The *moving-magnet galvanometer* is now little used because it has the disadvantage that it must be very carefully aligned along the magnetic meridian and is easily disturbed by external magnetic fields. It consists of a small magnet, often in the form of a needle, pivoted at the centre of a fixed coil, which is large in comparison to the magnet. The magnet is under the influence of a uniform magnetic field, usually that of the earth, and is set up so that in the zero current situation it is in the plane of the coil (Fig. *b*). A long light nonmagnetic pointer, usually of aluminium, is attached perpendicularly to the magnet so that it swings over a scale. When a current, i, is passed through the coil the magnet is forced to rotate and the angular deflection, β, is measured. The earth's magnetic field provides the necessary retarding force. At equilibrium $i \propto \sin \beta$. The *tangent galvanometer is one form of moving-magnet galvanometer; a more sensitive form is the *astatic galvanometer.

Transient direct currents are measured using a *ballistic galvanometer. Alternating currents may be measured with either a *thermogalvanometer* or a *vibration galvanometer* depending on their frequency.

Currents of frequency less that 1000 hertz are detected with a vibration galvanometer. This is usually a moving-coil (or sometimes a moving-magnet) galvanometer with light damping, which vibrates about the zero position when the current passes. Optimum response is obtained by tuning the instrument to the detected frequency. This is achieved by adjusting the tension in the wire supporting the coil. The vibration galvanometer may also be used as the indicating instrument in a.c. bridge circuits instead of the more usual telephone receiver.

Currents of frequency above 1000 hertz are measured with a thermogalvanometer, which consists of a resistance wire through which the current is passed. The subsequent rise in temperature in the wire (*see* heating effect of a current) is measured with a thermocouple.

Larger currents are measured using an *ammeter. Most ammeters are shunted galvanometers: a resistor is connected in *parallel with the galvanometer to reduce its sensitivity. The resistor is known as a *galvanometer shunt.*

galvanometer constant The multiplying factor that must be applied to the scale reading of a galvanometer in order to give the value of the current in amps.

galvanometer shunt *See* galvanometer.

gamma rays (γ-rays) Very high frequency electromagnetic radiation that is emitted spontaneously by certain radioactive elements in the course of a nuclear transition or can be produced in nuclear reactions, as in the annihilation of an elementary particle and its antiparticle. The wavelength of gamma rays emitted by radioactive substances is characteristic of the radioisotope involved and ranges from about 4×10^{-10} to 5×10^{-13} metre. Although gamma rays at one time lay at the extreme low-wavelength end of the electromagnetic spectrum, modern high-voltage generators can now produce X-rays of much shorter wavelength than that of most gamma rays.

Gamma rays are distinguished from *alpha and *beta particles by their greater depth of penetration and their nondeflection in electric and magnetic fields. The depth of penetration is controlled by their energy, which depends on the wavelength. The energy of gamma rays is determined by measuring the maximum energy of photoelectrons that they produce or their diffraction by certain crystal lattices.

ganged circuits Two or more circuits with variable elements that are mechanically coupled so that they can be operated simultaneously by a single control.

ganging oscillator An oscillator that has a constant output over a wide frequency range and whose frequency may be rapidly adjusted. It is used for testing the accuracy of adjustment of ganged tuned circuits over their tuning range.

gap (1) *Syn. for* forbidden band. *See* energy bands; semiconductor.

(2) An air gap in a magnetic circuit that increases the inductance and saturation point. An air gap is a necessity where moving parts are involved.

(3) The space between the electrodes in any electron tube. A spark gap is a special arrangement of the electrodes so that a *spark occurs if the voltage exceeds a predetermined value. A spark gap is often used to divert high-voltage surges and thereby protect a device.

gap admittance The admittance of a circuit at a gap between electrodes. The *circuit-gap admittance* is the admittance in the absence of electron conduction across the gap; the *electronic-gap admittance* is the difference between the admittances with electron conduction and with no electron conduction.

gap length *See* magnetic recording.

gas amplification *See* gas multiplication.

gas breakdown A type of *breakdown that occurs in a *gas-filled tube when the voltage reaches a given value. Ions in the gas are accelerated by the field and reach high kinetic energies. Little recombination of ions occurs because of the high energies but further ions are produced by collisions between the ions and molecules of the gas; a multiplication effect thus occurs causing rapid breakdown of the gas. The process is analogous to *avalance breakdown in a semiconductor.

gas cell *See* gas electrode.

gas-discharge tube A *gas-filled tube in which the presence of gaseous molecules contributes significantly to the characteristics of the tube. Normally a gas is a poor electrical conductor but a sufficiently high electric field causes *ionization of molecules and atoms in the immediate vicinity of the electrodes; these gaseous ions are attracted to the charged electrodes and a small current – the *preconducting current* – flows. At sufficiently high values of applied potential, below the breakdown voltage, an external ionizing event causes the current between the electrodes to change in a manner dependent on the operating conditions (*see* gas-filled radiation-detection tubes).

If a sufficiently high potential difference is applied to the tube *gas breakdown occurs and a large current flows across the tube. The potential difference across the tube drops to a relatively small value and the discharge is self-sustaining. The *threshold current* is the value of current at which the discharge becomes self-sustaining; the process of establishing the discharge is termed *firing*. The minimum voltage required for the discharge to continue is the *maintaining voltage;* the voltage at which the discharge ceases completely is the *extinction potential* (*see also* reignition voltage).

Current flows as a result of a multiple collision of ions within the tube: collisions between ions cause excitation and further ionizations, light being produced when excited atoms and ions return to the ground state. Recombination between positive and negative ions also results in a small amount of light emission.

The phenomena observed depend on the pressure of gas in the tube. At pressures near atmospheric pressure a *spark passes between the electrodes. As the pressure is reduced *glow discharge* occurs. At relatively high pressures the mean free path of ions is small; a *positive glow* is observed near the anode and a *negative glow* near the cathode, the centre of the tube being dark. As the gas pressure is reduced the mean free path increases and the positive glow extends across the tube until the whole tube is filled. The colour of the glow discharge is brilliant and characteristic of the gas in the tube. *Glow-*

discharge tubes are frequently used for luminous signs and for lighting purposes. A *starter electrode is often used in a glow-discharge tube in order to initiate the discharge.

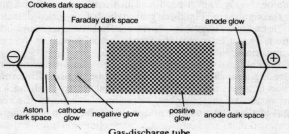

Gas-discharge tube

A further reduction in pressure causes a change in the glow pattern (*see* diagram). Additional dark regions appear in which electrons, produced by ionization and excitation, have insufficient energy for excitation and the probability of recombination is low. The largest potential drop in the tube is across the *Crookes dark space*. Striations in the positive glow are caused by alternate ionizations and recombinations in the tube.

At very low pressures the Crookes dark space fills the tube and very few collisions occur. The high kinetic energies of the ions cause *secondary emission of electrons (originally known as *cathode rays*) from the cathode, and at sufficiently high fields secondary emission of positive ions in the form of *anode rays* from the anode. With very high fields *X-rays are emitted from the anode as a result of bombardment by high-energy electrons. This phenomenon was utilized in early *X-ray tubes.

If the electrodes are placed relatively close together an *arc forms across them. The heat generated by the arc causes *thermionic emission of electrons from the negative electrode and the current density is therefore very high. The *arc discharges* occur over a very wide range of pressures and the light associated with them is rich in ultraviolet frequencies. *Arc-discharge tubes* are used as ultraviolet lamps and as *rectifiers (*see also* mercury-arc rectifier). *Misfire* of an arc-discharge tube or glow-discharge tube is a failure to establish a discharge when the correct conditions exist.

The *Townsend discharge* is a luminous discharge that occurs at lower current densities than the glow discharge; the voltage across the tube is a function of the current density. At low current densities the potential falls uniformly across the tube and the luminous region extends across the tube.

gas electrode An electrode that absorbs or adsorbs a gas so that when in contact with an electrolyte the gas effectively acts as the electrode. A *gas cell* is one that contains a gas electrode.

gas-filled radiation-detection tubes Gas-filled tubes that operate by virtue of the ionization produced in them when charged particles pass through them. The three main types of gas-filled detectors are the *ionization chamber, *proportional counter, and *Geiger counter. The difference in the three types of tube is in their operating conditions.

A typical tube consists of a gas-filled chamber with a central electrode that is insulated from the chamber walls, with a voltage V applied between the wall and the central electrode. If an ionizing event occurs in the chamber, releasing N ion pairs, the resulting output is a function of the voltage, V, as shown in the diagram.

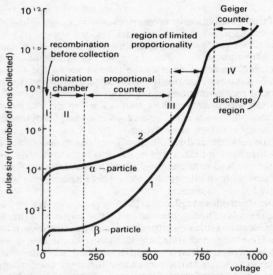

Dependence of pulse size on voltage in gas-filled radiation detectors

At very low voltages some of the ion pairs are lost due to recombination within the chamber (region I) but as the voltage increases recombination loss becomes negligible and all the charge is collected at the electrodes (region II). This region is known as the saturation or ionization chamber region.

As the voltage is further increased gas multiplication occurs, the multiplication factor for a given voltage being independent of the initial ionizing event. The output pulse size is therefore proportional to the number of ion pairs produced by the ionizing event, and therefore to the energy of the radiation producing that event. This region (region III) is known as the region of proportionality.

At higher voltages still, the strict proportionality of pulse size breaks down (region of limited proportionality) until the pulse size becomes independent of the initial ionization, due to avalanche in the gas. The pulse size is limited by the characteristics of the external circuit and the chamber, and a plateau region is reached (region IV). This is termed the Geiger-Müller region.

gas-filled relay *See* thyratron.

gas-filled tube An *electron tube that contains a gas or vapour, such as mercury vapour, in sufficient quantity so that the electrical characteristics of the tube are determined entirely by the gas, once ionization has taken place. The *gas ratio* of such a tube is the ratio of the *ionization current in the tube to the electron current required to produce it.

gas focusing *See* focusing.

gas laser *See* laser.

gas maser *See* maser.

gas multiplication *Syn.* gas amplification. (1) The production of additional ions by ions produced in a gas under the influence of a sufficiently strong electric field.

(2) The ratio of the total ionization to the initial ionization as a result of the above process.

gas noise *Noise arising in a *gas-filled tube due to the random ionization of the gas molecules in the tube. It is a form of white noise, i.e. it has a relatively flat frequency spectrum; *gas-discharge tubes are therefore used as white-noise generators for testing purposes.

gas phototube *See* phototube.

gas ratio *See* gas-filled tube.

gassing The evolution of gas in the form of small bubbles from one or more electrodes during electrolysis. Gassing occurs in an accumulator towards the end of the charging period.

gate (1) *Short for* gate electrode. An electrode or electrodes in a *field-effect transistor, an *MOS capacitor, *MOS integrated circuits, a *charge-coupled device, and a *silicon controlled rectifier.

(2) *Digital gate.* A digital circuit that has two or more inputs but only one output. The conditions applied to the inputs determine the voltage level at the output. The output is switched between two or more discrete values. Digital gates are widely used in *logic circuits.

(3) *Analog gate.* A *linear circuit or device that produces an output signal only during a specified interval of the input signal. During this interval the output is a continuous function of the input

219

signal. Analog gates are widely used in *radar and electronic control systems.

(4) An electric signal or trigger that allows a circuit or device to operate. The most common method of gating is to employ a *clock.

gate expander An array of diodes connected to the input stage of a *diode-transistor logic gate and used to extend the possible number of inputs to that gate.

gating *See* receiver.

gauss Symbol: Gs or G. The unit of magnetic flux density in the obsolete CGS electromagnetic system of units. One gauss equals 10^{-4} tesla.

Gauss's theorem The electric field strength, *E,* over a closed surface containing electric charges is given by

$$\int E \cdot dS = \Sigma q / \varepsilon_0$$

where dS is a small element of area on the surface S, ε_0 is the *permittivity of free space, and Σq the total charge enclosed within the surface.

If the volume enclosed by the surface contains a distributed charge density, ρ_e, Gauss's theorem becomes

$$\int E \cdot dS = (1/\varepsilon_0) \int \rho_e d\tau$$

where dτ is a small volume element. Since

$$\int E \cdot dS = \int \text{div } E \, d\tau$$

then

$$\text{div } E = \rho_e / \varepsilon_0$$

In a dielectric medium the total charge density includes an apparent charge density due to polarization of the atoms or molecules within the medium and Gauss's theorem becomes

$$\text{div } D = \rho_e$$

where *D* is the electric *displacement. This is one of *Maxwell's equations.

The two main conclusions that can be drawn from Gauss's theorem are firstly that the electric field inside a hollow conductor containing no charge is zero, the enclosed space being at the same potential as the conductor: electrical apparatus can therefore be shielded from the effects of external electric fields by surrounding it with an earthed conductor; secondly that any excess static charge on a conductor must reside on the outer surface.

Geiger counter *Syn.* Geiger-Müller counter. A gas-filled tube that is used to detect ionizing radiation, especially *alpha particles, and to count particles. It has a thin wire anode mounted coaxially inside a cylindrical cathode with a potential difference, slightly lower than

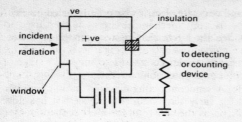

a **Geiger counter** circuit

that required to produce a discharge, maintained across the electrodes (Fig. *a*). When radiation, such as an alpha particle, enters through a thin window the gas along its path becomes ionized. The ions are accelerated by the field and produce an avalanche, which is quickly quenched. The resulting current pulse is amplified and registered by a detector (typically a loudspeaker) or a counting device.

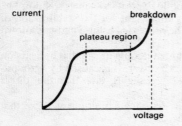

b Current-voltage characteristic

The output characteristic of a Geiger counter shows a large plateau region (Fig. *b*) where the output is substantially constant for a wide range of voltage; this is the operating region of the tube. The tube is also independent of the energy of the incident radiation in this region. A typical counter has a dead time, following a discharge, of about 200 microseconds. Geiger counters are useful for count rates up to about 2000 counts per second. *See also* gas-filled radiation-detection tubes.

Geissler tube A type of *gas-discharge tube that demonstrates the luminous effects of an electrical discharge through a gas at low pressure. It has a characteristic central capillary section that concentrates the glow. It is used to examine the spectra of gases and is also useful for spectrometer calibrations.

generating station *See* power station.

generation rate (of electron-hole pairs) *See* semiconductor.

generator A machine that converts mechanical energy into electrical energy. An electromagnetic generator (i.e. a *dynamo) has a coil that is made to move so as to cut lines of magnetic flux. An *electrostatic generator,* such as a *Van de Graaff generator or a Wimshurst

221

machine, has equal and opposite electric charges produced by electrostatic induction or friction; mechanical energy is then used to separate the charges. *See also* alternating-current generator; induction generator.

geostationary orbit *See* communications satellite.

geosynchronous orbit *See* communications satellite.

germanium Symbol: Ge. A semiconducting element, atomic number 32, extensively used for early semiconductor devices. It has now been displaced by silicon as the substrate for most semiconductor devices.

getter A material that has a strong chemical affinity for other materials. Such substances can be used to remove undesirable elements or compounds that are present in small quantities in an environment. Uses include the introduction of barium, often in filament form as a *filament getter,* into a sealed vacuum system in order to remove residual gases, and the introduction of phosphorus into oxide layers on silicon in order to remove mobile impurities such as sodium. The latter application is particularly important for the stabilization of MOS *field-effect transistors. The process of using a getter is known as *gettering.*

ghost *Syn.* double image. An unwanted double image that appears on the screen of a television or radar receiver. It is caused by *ground-reflected waves arriving at the receiver slightly later than the *direct waves.

giga- Symbol: G. A prefix to a unit, denoting a multiple of 10^9 of that unit: one gigahertz equals 10^9 hertz.

gilbert Symbol: Gb. The unit of magnetomotive force in the obsolete CGS electromagnetic system of units. One gilbert equals 0·7958 amperes (or ampere-turns).

glass electrode A thin-walled glass bulb filled with a buffer solution into which dips a platinum wire; the wire is thus in electrical contact with the inner wall of the bulb. The device is used to measure the pH of a solution. The potential between the glass surface and an external solution varies with the hydrogen ion concentration, i.e. with the pH of the solution. The pH is found by measuring the potential difference between the glass bulb (platinum contact) and the external solution.

glassivation *See* passivation.

glass tube *See* glass valve.

glass valve *Syn.* glass tube. An *electron tube in which the envelope is made entirely from glass without any different base material. The leads are taken in through special metal-glass seals.

glow discharge *See* gas-discharge tube.

glow-discharge microphone *See* microphone.

glow-discharge tube *See* gas-discharge tube.

222

glow lamp A *gas-discharge tube operated in the glow-discharge region, using the glow discharge as the source of light. The colour of the glow depends on the gas used in the tube: a *neon lamp,* containing neon, emits a red glow. Since the voltage across the tube remains substantially constant for a range of operating currents, glow lamps are frequently used as voltage regulators.

glow switch A *switch that is formed from a glow-discharge tube containing heat-sensitive contacts. When a glow discharge is produced the heat generated by it is sufficient to cause the contacts to operate.

gold-bonded diode A *point-contact diode formerly used in computer logic circuit applications when moderately fast switching was required. It is formed by pressing a preformed gold whisker against an n-type germanium chip and using pulses of current to alloy the two together. It is thought that a *p-n junction is formed at the interface.

gold-leaf electroscope *See* electroscope.

goniometer An instrument that measures angles, such as the angle between two radiofrequency waves (*radiogoniometer*); the direction of propagation of a wave may also be measured.

g parameters *See* transistor parameters.

graded-base transistor *See* drift transistor.

Gramme winding *See* ring winding.

gramophone *Syn.* record player. An instrument that converts the undulations of the grooves of a gramophone record into sound. A stylus moves along the grooves and vibrates in sympathy with the modulations in the groove. A *transducer (the *pick-up), which may be either an electromagnetic or piezoelectric type, converts the vibrations into electrical signals; these are amplified and used to drive a loudspeaker (*see also* reproduction of sound). Stereophonic records have groove modulations in two directions (laterally and vertically), detected by two transducers driving two loudspeakers.

graphecon *See* storage tube.

graphical symbols Symbols that represent the various types of components and devices used in electronics, telecommunications, and allied subjects. The graphical symbols recommended by the British Standards Institute are shown in Table 1, backmatter.

graphic equalizer A *tone control in which the frequency range is divided into bands. The signal in each band may be adjusted by sliding contacts, the positions of which indicate frequency response.

graphic instrument *Syns.* recording instrument; recorder; grapher; chart recorder. A measuring instrument that presents the quantity being measured in the form of a graph. The graph is produced either in ink on a suitable paper chart or on the screen of a *cathode-ray oscillograph. A permanent record may be produced by photographing the screen in the latter case.

graphic panel A master control panel in automatic and remote control systems in which coloured block diagrams are used to show the relation and functioning of the different parts of the control system. It may also have the controls and recording instruments mounted on it in their correct relative positions.

grass *Syn.* picture noise. *Colloq.* *Noise that occurs in radar receivers due to irregularities in the *time base of the display. These may be due to random fluctuations in the time-base generator or to *electric interference. It results in noise on the picture that resembles grass. *See also* lawnmower. *Compare* snow.

Grassot fluxmeter *See* fluxmeter.

gravity cell A type of primary *cell that contains two different electrolytes kept apart by their different densities.

green gun *See* colour picture tube.

grenz rays Soft X-rays produced when electrons are accelerated through 25 kilovolts or less. Grenz rays are produced in many types of electronic equipment, such as colour television sets, but have an extremely low penetrating power.

grid (1) An electrode that has an open structure, such as a mesh or a plate with a hole in it, thus allowing an electron beam to pass through it. *See* thermionic valve.

(2) The nationwide high-voltage transmission line system that interconnects many electricity power stations. It transmits voltages of up to 400 kilovolts. Voltages as high as 735 kV are used in some countries.

grid base (of a thermionic valve) *See* cut-off.

grid bias The potential applied to the grid of a thermionic valve that determines the portion of the characteristic curve at which the valve will operate, or that modifies the cut-off value of the valve. *Automatic grid bias* uses a resistor in either the grid or cathode circuits of the valve to supply the grid bias. The voltage across the resistor is determined by either the grid or cathode current respectively and in turn supplies the potential to the grid. *See also* thermionic valve.

grid control ratio *Syn. for* control ratio. *See* thyratron.

grid emission The emission of electrons or ions from the grid of a thermionic valve.

grid leak A high resistance connected between the grid and cathode of a thermionic valve that prevents an accumulation of charge on the grid and may also be used to develop the *grid bias.

grid modulation *See* amplitude modulation.

grid stopper *See* parasitic oscillations.

ground *U.S. syn. for* earth.

ground absorption A loss of energy that occurs during transmission of radiowaves due to absorption in the ground.

ground capacitance *U.S. syn. for* earth capacitance.

ground clutter *See* ground return.

ground current *U.S. syn for* earth current.

grounded-base connection *See* common-base connection.

grounded-collector connection *See* common-collector connection.

grounded-emitter connection *See* common-emitter connection.

ground electrode *U.S. syn. for* earth electrode. *See* earth.

ground noise *Syn. for* random noise. *See* noise.

ground plane *Syn.* earth plane. A sheet of conducting material that is adjacent to a circuit and is at earth potential. It may be used to provide a low-impedance earth at any point in the circuit. For example, one side of a double-sided printed circuit board may be used as the ground plane. Connections to it from the printed circuit on the other side of the board are easily effected through the board from any desired point in the circuit.

ground plate *U.S. syn. for* earth electrode. *See* earth.

ground potential *U.S. syn. for* earth potential.

ground ray *See* ground wave.

ground-reflected wave A radiowave that travels between a transmitting and a receiving aerial situated above the ground and that undergoes at least one reflection from the ground. It can be affected by the properties of the ground and can also undergo refraction in the troposphere. *See also* ground wave.

ground reflection Reflection of a transmitted radar wave by the ground before it reaches the target.

ground return Echoes received by a radar receiver due to reflections of the radar wave by the ground or by objects on the ground. Ground return from extraneous sources can lead to *ground clutter* on the radar screen. This type of noise can obscure the target.

ground state The lowest possible *energy level of a system. If a system has a configuration such that its energy is greater than its ground-state energy, it is said to be in an *excited state*. Many phenomena, including luminescence and semiconduction, depend upon systems being in an excited state.

ground wave *Syn.* ground ray. A radiowave that travels between a transmitting and a receiving aerial situated above the earth. It has two main components: the *space wave, which includes the *direct wave and the *ground-reflected wave, and the *surface wave. *Compare* indirect wave.

group operation The operation of all the poles of a multiple switch or circuit-breaker by a single mechanism.

Grove cell A two-fluid primary *cell in which a zinc rod immersed in dilute sulphuric acid forms the negative element, which is separated from the positive element by a porous partition. The positive element consists of a platinum plate immersed in fuming nitric acid. E.m.f.: 1·93 volts.

grown junction A *p-n junction that is formed in a single crystal of semiconductor material while the crystal is being grown from a melt. The amount and type of impurities added to the semiconductor are varied in a controlled manner as the crystal grows. A *grown-diffused junction* is produced by *diffusion of impurities into the semiconductor after a grown junction has been formed, in order to produce the precise doping profile required.

guard band A frequency band that is left vacant during broadcasting to minimize mutual interference between two neighbouring frequency bands.

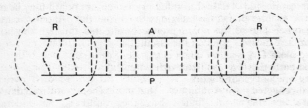

Principle of guard ring

guard ring A device used to ensure uniform fields and to define the sensitive volume in absolute electrometers and standard capacitors. It consists of a metal plate surrounding and coplanar with a smaller plate; a narrow air gap separates the two. The diagram shows the lines of force between a plate A surrounded by a guard ring RR, and a parallel earthed plate P. The variations in field at the edges affect only the guard ring RR.

A similar device in the form of an auxiliary electrode is commonly used in semiconductor devices and vacuum tubes.

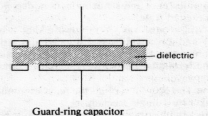

Guard-ring capacitor

guard-ring capacitor A standard capacitor that uses a *guard ring to reduce the edge effect (*see* diagram). The *guard-well capacitor* is a special type of guard-ring capacitor used for capacitance below 0·1 picofarads. In this type the guard ring forms a well on to which a pyrex disc is mounted in order to locate the electrode assembly accurately.

Gudden–Pohl effect The transient luminescence observed in a phosphor, previously exposed to ultraviolet radiation, when an electric field is applied to it.

guide *Short for* waveguide.

Guillemin effect *See* magnetostriction.

Guillemin line A network that produces pulses with very sharp rise and fall times so that they are almost square.

Gunn diode *Syn.* transferred electron diode. A negative-resistance microwave oscillator that operates by means of the *Gunn effect. It is a diode formed from a sample of low-resistivity n-type gallium arsenide that produces coherent microwave oscillations when a large electric field is applied across it. The most fundamental form of operation of a Gunn diode is *transit time mode* in which the output frequency of the oscillations is determined by the time taken for a domain of high-energy low-mobility carriers to drift across the length of the semiconductor. This mode has several disadvantages and low efficiency.

Delayed domain mode of operation is a more efficient and useful operating mode. The diode is connected to an external tuned circuit that determines the operating frequency of the diode. The field across the diode is biased by a radiofrequency (r.f.) field so that the threshold voltage is only exceeded during the positive cycles of the r.f. fields. Domains can only form during the positive cycles and therefore the output current pulses are forced to occur at an externally determined frequency.

Limited space charge accumulation mode is an operating mode that produces an output frequency much greater than the transit time frequency. The frequency of the r.f. bias, determined by the external tuned circuit, is high and therefore stable domains do not have time to form. The field within the device remains above the threshold value for most of the operating cycle because stable domains are not present, and the electrons experience a negative resistance as they flow across the diode.

Gunn effect *Syn.* negative differential resistance effect. An effect that occurs when a large d.c. electric field is applied across a short sample of n-type gallium arsenide. At values above the threshold value, typically several thousand volts per cm, coherent microwave oscillations are generated.

The effect results from the *degeneracy of the energy levels in the conduction band and is due to the charge carriers of different mobilities forming bunches, known as *domains*. These migrate along the potential gradient at a rate determined by the carrier mobility. Some of the low-energy high-mobility carriers as they gain energy from the applied field change into a higher-energy low-mobility state, i.e. a velocity reduction in response to an increase in electric field is observed. Charges will accumulate to form a domain of low-

227

mobility electrons with an apparent mass of approximately twenty times that of the high-mobility carriers. As the domain forms, nearby regions in the semiconductor become depleted of carriers and an effective dipole results. Much of the applied field is applied across the dipole and the field in the rest of the material falls below the threshold value. Thus, once one stable domain has formed no others will form while it still exists. The domain drifts towards the anode where it produces a current spike, and the process begins again. The result is that the domains produce microwave current at the output.

gyrator A component that does not obey the reciprocal theorem, i.e it reverses the phase of signals transmitted in one direction but does not affect signals transmitted in the opposite direction. Gyrators are usually used at microwave frequencies, forming part of a waveguide. They may be entirely passive or may contain active elements.

H

halation *Syn.* halo. A glow observed around the spot on the screen of a *cathode-ray tube. It is caused by total internal reflections within the thickness of the glass.

half-adder *See* adder.

half-duplex operation *See* duplex operation.

half-power point The point on a characteristic curve that corresponds to operation at a power intensity of half the maximum value.

half-shift register *See* flip-flop.

half-wave dipole *See* dipole aerial.

half-wave rectifier circuit A *rectifier circuit that delivers unidirectional current to the load only during alternate half-waves of the

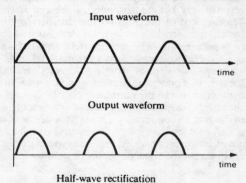

Half-wave rectification

single-phase alternating-current input (*see* diagram). Usually the positive half-cycles are rectified.

half-wave voltage doubler A *voltage doubler that operates only during the input wave half-cycle.

Hall coefficient *See* Hall effect.

Hall effect An effect observed when a current-carrying conductor is placed in a magnetic field whose direction is perpendicular to both the direction of current and of magnetic field.

    The force due to the magnetic flux density, B, on charge carriers of charge e and drift velocity v is equal to $ev \times B$. This force displaces the carriers and sets up a nonuniform charge distribution giving rise to an electric field, E. At equilibrium the force on the

229

charge carriers, $e\mathbf{E}$, due to this field just balances that due to the flux density:

$$e\mathbf{E} + e\mathbf{v} \times \mathbf{B} = 0$$

The drift velocity, $\mathbf{v}$, is related to the current density, $\mathbf{j}$, by the equation

$$\mathbf{j} = ne\mathbf{v}$$

where n is the number of charge carriers per unit volume. Thus the electric field is related to the vector product of the magnetic flux density and the current density by

$$\mathbf{E} = -R_{\mathrm{H}}(\mathbf{j} \times \mathbf{B})$$

R_{H} is the *Hall coefficient* and is equal to $1/ne$.

Measurement of the Hall coefficients for various materials have shown that in some cases the direction of the Hall field is reversed, implying that these materials have a positive Hall coefficient. This effect, sometimes termed the *Suhl effect*, indicates that the current is carried by positively charged carriers in these materials, i.e. by *holes. Measurement of the Hall coefficient therefore determines whether the charge carriers are electrons or holes and the value of the carrier concentration. The electric field results in a potential difference – the *Hall voltage* – across the material.

The above analysis applies strictly to metals and *degenerate semiconductors. In nondegenerate semiconductors additional factors are introduced due to the thermal energy of the carriers, asymmetry of the energy bands, and the effect of two or more conduction bands contributing to conduction. Direct measurement of the Hall coefficient still enables the type of carrier and its concentration to be determined, however, provided that one type of carrier predominates. Thus in an n-type semiconductor ($n \gg p$)

$$R_{\mathrm{H}} \simeq -r/en$$

and in a p-type semiconductor ($p \gg n$)

$$R_{\mathrm{H}} \simeq +r/ep$$

where n and p are the electron and hole concentrations, respectively, and r is a function of mean free time between carrier collisions and is thus determined by the scattering mechanism.

The *Hall mobility*, μ_{H}, is defined as the product of R_{H} and conductivity σ. In the case of metals and degenerate semiconductors the Hall and *drift mobilities are equal. In most nondegenerate semiconductors the difference is between 10 and 20%. Both the Hall and drift mobilities are commonly used to assess the quality of a material: mobility is reduced by increased concentrations of impurities in a material.

Hall-effect device A device that utilizes the *Hall effect in order to measure the charge carrier concentration in a metal or semiconductor or to detect magnetic fields. Measurement of the Hall voltage produced in a flat plate or wafer – a *Hall probe – of known dimensions can be used to detect the presence of a magnetic flux density if the Hall coefficient of the material is known. Under given conditions the carrier concentration in the wafer can be deduced from the Hall coefficient, which is obtained from the measured Hall voltage.

Hall mobility *See* Hall effect.

Hall probe A convenient means of measuring magnetic flux density, *B,* by utilizing the *Hall effect. A small sample of a suitable semiconductor, such as indium antimonide or indium arsenide, is used in which the Hall voltage developed across it varies almost linearly with the magnetic flux density. This method is less accurate than other means of measuring *B* (*compare* Cotton balance) but can be used in nonuniform fields and can be carried out quickly and easily.

Hall voltage *See* Hall effect.

Hallwacks effect *Photoemission from a metal when the incident radiation is in the form of ultraviolet rays. It is an obsolete name for *photoelectric effect.

halo *See* halation.

ham radio *Colloq.* Noncommercial (amateur) radio communication between licensed individuals. The frequencies used are restricted to internationally agreed values in order to prevent interference with commercial broadcast transmission and shipping or aircraft communication.

handset *See* telephony.

hard disk *See* moving magnetic surface memory.

hard ferromagnetic material *See* ferromagnetism.

hard tube *Syns.* high-vacuum tube; hard valve. *See* electron tube.

hard valve *Syn. for* hard tube. *See* electron tube.

hardware The physical units, devices, and circuits that make up a computer system. *Compare* software.

hardwiring *See* read-only memory.

hard X-rays *See* X-rays.

harmonic An oscillation of a periodic quantity, present in a complex vibration, having a frequency that is an integral multiple of the *fundamental frequency. An oscillation having a frequency that is an integral submultiple of the fundamental is termed a *subharmonic*. In practice the fundamental need not necessarily be present.

harmonic analyser A device that analyses a periodic function from its graph in terms of the Fourier series (*see* Fourier analysis) corresponding to the function.

harmonic distortion *See* distortion.

231

harmonic generator A signal generator that produces *harmonics of an input fundamental frequency. These devices may take the form of an oscillator controlled by a tuning fork.

harmonic oscillator *See* oscillator.

Harris flow *See* space charge.

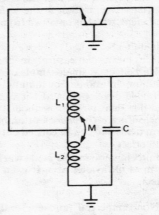

Hartley oscillator

Hartley oscillator A type of *oscillator consisting of a transistor with *common-emitter connection and a parallel *resonant circuit between the emitter and collector (*see* diagram). Ignoring the resistance of the coil the resonant frequency, ω_0, is given by

$$\omega_0{}^2 \simeq C/(L_1 + L_2 + 2M)$$

Hartley oscillators may also be crystal-controlled (*see* piezoelectric oscillator, Fig. *d*).

Hartree diagram *See* magnetron.

Hartshorn bridge An alternating-current *bridge that measures mutual inductance. Direct comparison of a mutual inductance with a standard mutual inductance is virtually impossible because effects, such as the self- and mutual capacitances of the coils, produce voltage components in the secondary circuits in phase with the primary current. This problem is overcome in the Hartshorn bridge by the use of a variable resistor, of resistance *r,* common to both the primary and secondary circuits of the bridge (*see* diagram). The standard inductance and the resistance are varied until a balance is achieved with no response registered by the indicating instrument I; assuming that the secondaries are connected in antiphase, then

232

$$M_1 = M_2$$
$$r = \pm(\rho_1 - \rho_2)$$

where ρ_1 and ρ_2 are the resistances of the mutual inductances M_1 and M_2.

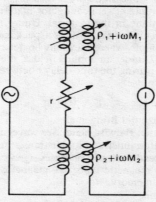

Hartshorn bridge

Hay bridge A four-arm *bridge used for the measurement of large inductance (*see* diagram). At balance, when there is no response on the indicating instrument, I,

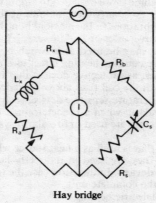

Hay bridge
233

$$L_x = R_a R_b C_s / (1 + \omega^2 C_s^2 R_s^2)$$
$$R_x = L_x \omega^2 C_s R_s$$

where ω is the angular frequency.

Haynes–Schockley experiment An experiment by which the carrier *drift mobility in semiconductors is measured. Excess carriers are produced in a sample of semiconductor material and diffuse outwards and recombine in the material. Under the influence of an applied electric field they also drift as a 'package' towards the negative end of the sample, where they may be detected. With a sample of known length, l, under an applied field E, the drift mobility, μ, is calculated by measuring the time delay t between the applied pulse and the detected pulse:

$$\mu = l / Et$$

HB *Abbrev. for* horizontal Bridgeman.

H-bend *Syns.* H-plane; flatwise bend. *See* waveguide.

HBT *Abbrev. for* heterojunction bipolar transistor.

head A device that reads, records, or erases signals stored on a suitable medium, such as magnetic tape. *See* magnetic recording; moving magnetic surface memory.

headset *See* earphone.

hearing aid A complete sound-reproduction system that increases the sound intensity at the ear. Modern hearing aids use a small crystal microphone, battery-powered amplifier, and earpiece. The development of integrated circuits has allowed hearing aids to become very small and light and they may be hidden in a spectacles frame. Good sound quality at sufficient power output for most cases is obtained with volume and tone controls available to the wearer. Special types of hearing aids include those designed to amplify a specific frequency band only and those in which the output fom the amplifier is used to produce vibrations in the mastoid bone behind the ear, thus bypassing a defective outer or middle ear.

heat coil A coil that may be used as a switch by the sensing of the temperature of the coil (*see* heating effect of a current). Such coils are most often used as protective devices that open a circuit when the current through the coil rises above a predetermined value.

heat detector A temperature-sensitive device, such as a thermocouple, that operates an alarm when a predetermined rise in temperature occurs. Heat detectors are used in fire-alarm systems or in burglar-alarm systems.

heater Any resistor that is used as a heat source when subjected to an electric current. They are used in indirectly heated cathodes and domestic heating devices. In the latter case the term applies both to the element and the complete unit.

heating depth *See* dielectric heating.

heating effect of a current *Syn.* Joule effect. If a current, I, is passed through a circuit of resistance R, electrical energy is dissipated as the current flows from a higher to a lower potential. The rate of dissipation of energy equals I^2R. This is known as I^2R *loss*. The energy appears as heat; if all the electrical energy is converted into heat energy, Q, then

$$Q = I^2Rt$$

where t is the time for which the current flows. The temperature does not increase indefinitely since a stable state is reached when the rate of emission from the surface of the resistor just equals the rate of generation of heat.

heating time The time a device takes to heat up to a steady temperature. The term is usually applied to the cathode heating time of a *thermionic valve. *See also* warm-up.

heat sink A device that is employed to dispose of unwanted heat in a circuit and prevent an excessive rise in temperature. Heat sinks are particularly useful for protecting transistors in power applications.

Heaviside layer (or **Heaviside–Kennelly layer**) *Syn. for* E-layer. *See* ionosphere.

heavy hole *See* direct-gap semiconductor.

hecto- Symbol: h. A prefix to a unit, denoting a multiple of 100 of that unit.

height control A control in a *television or *radar receiver that adjusts the overall size of the frame scan.

Heil tube An early type of *klystron.

Helmholtz coils An arrangement of two identical cylindrical coils

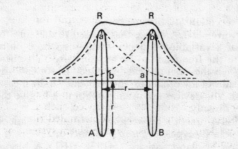

aa	magnetic field due to coil A
bb	magnetic field due to coil B
RR	resultant magnetic field

Helmholtz coils

mounted coaxially a distance r apart. When r just equals the radius of the coils a uniform calculable magnetic field is produced between them when the same current flows through both coils (*see* diagram). This arrangement was first used in the Helmholtz galvanometer but is now commonly employed when a precise uniform magnetic field is required over a substantial volume.

Helmholtz galvanometer An obsolete form of galvanometer. It operates in a similar manner to the *tangent galvanometer but employs *Helmholtz coils instead of a single coil to provide a uniform magnetic field in which the magnet is deflected.

HEMT *Abbrev. for* high electron mobility transistor.

henry Symbol: H. The *SI unit of *inductance and *permeance. The self- or mutual inductance of a closed loop is one henry if a current of one ampere gives rise to a magnetic flux of one weber.

heptode *Syn.* pentagrid. A *thermionic valve that has seven electrodes, i.e. five grids between the anode and cathode. The most common use of the device is as a *mixer.

hertz Symbol: Hz. The *SI unit of frequency. It is the frequency of a periodic phenomenon that has a period of one second. It replaced the cycle per second (c.p.s.).

Hertzian waves *Obsolete* Electromagnetic waves of radiofrequency.

heterodyne Denoting the production of *beats in a radio receiver by the combination of the received wave with a locally generated wave of slightly different frequency. *See* beat reception.

heterodyne analyser *See* wave analyser.

heterodyne interference *Syn.* whistle. A type of interference occurring in *superheterodyne reception. It consists of a high audiofrequency note caused by *beats between two carrier waves with almost equal frequencies.

heterodyne reception *See* beat reception.

heterodyne wavemeter A device that is used to measure frequency. It consists of a variable-frequency oscillator whose output is combined with the frequency to be measured so that *beats are produced. When the desired beat frequency is obtained by adjusting the frequency of the oscillator, the frequency to be measured is then calculated. The devices are usually calibrated to give the unknown frequency directly.

heterojunction A junction between two dissimilar *semiconductors of opposite polarity types. *See also* p-n junction.

heterojunction bipolar transistor (HJBT; HBT) A bipolar junction *transistor in which the emitter is formed from a wide-gap semiconductor (i.e. one with a wide energy *gap) and the base from a narrow-gap semiconductor. The effect of using this emitter-base *heterojunction gives several advantages compared to the usual homostructure of bipolar transistors. It was first conceived by

Schockley in 1948 in his original patent application for the bipolar transistor, but viable devices have only recently been produced with the development of *molecular beam epitaxy and *metal-organic chemical vapour deposition techniques. Heterojunctions are also used for double heterostructure *injection lasers and *light-emitting diodes.

Heterojunctions utilize the variations in energy gap between the two sides of the junction in addition to the electric fields in order to control the distribution and flow of both *electrons and *holes. In the case of the bipolar junction transistor, where both types of charge carrier contribute to the current flow through the transistor, the ability to control the forces acting on electrons and holes so that they act independently of each other offers enormous practical advantages.

The simplified energy levels for an n-p-n transistor with a wide-gap n-type emitter under forward voltage conditions are shown in Fig. *a*. An additional potential barrier ΔE_g exists in the valence band and inhibits the flow of holes from the base region to the emitter. The *beta-current gain factor is given by

$$\beta_{max} \equiv I_n / I_p$$

under ideal conditions where the effects due to electron-hole recombination in the base and electron trapping by acceptor impurities are negligible.

If the potential barrier heights at the emitter-base junction are given by qV_n and qV_p for electrons and holes respectively, and the doping levels of emitter and base are given by N_e and P_b, then for nondegenerate semiconductors the injected electron current density is given by

$$J_n = N_e v_{nb} \exp\left(- qV_n / kT\right)$$

and the injected hole current density is given by

$$J_p = P_b v_{pe} \exp(-qV_p / kT)$$

where v_{nb} and v_{pe} are the mean velocities of electrons at the emitter end of the base and of holes at the base end of the emitter respectively; k is the Boltzmann constant and T the thermodynamic temperature.

The beta-current gain factor is given by

$$\beta_{max} = J_n / J_p$$

In a conventional homojunction transistor, where the barrier heights are equal, the ratio of J_n / J_p gives

$$\beta_{max} \propto N_e / P_b$$

In the heterojunction transistor the ratio gives

237

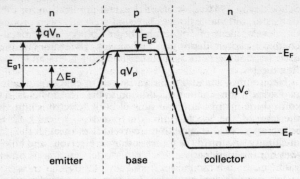

a Simplified energy levels of a wide-gap emitter n-p-n transistor

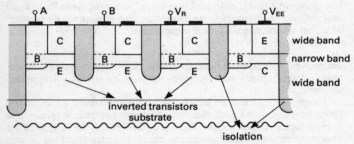

b Double heterostructure ECL input stage

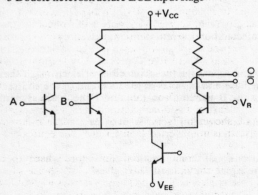

c ECL input stage–circuit

$$\beta_{max} \propto (N_e/P_b) \exp(\Delta E_g/kT)$$

where $\Delta E_g = q(V_p - V_n)$

To obtain high values of β_{max} in a conventional homojunction transistor, it is necessary to have $N_e >> P_b$, i.e. a high doping level in the emitter and a low doping level in the base; this leads to high base resistance. Using a heterojunction structure and choosing the semiconductor materials so that the energy-gap differences are several times the value of kT, large values of β_{max} may be achieved almost regardless of the doping ratio. The base doping level may therefore be increased to give much lower values of base resistance. This has advantages both for high-frequency operation and for digital switching operation.

The maximum oscillation frequency of a bipolar transistor is inversely proportional to the square root of the base resistance, R_b, and therefore reductions in R_b lead to an increase in the operating frequency. For transistors used as switches in integrated *logic circuits, the speed of operation is dependent on a number of factors, including the load resistance and capacitance of the circuit, the base resistance and collector capacitance of the transistor, and the transit time of injected carriers across the base region. Reducing the base resistance can lead to significant improvements in switching speed – up to eight times – and also reduces the power dissipated in the base region.

Double heterojunction transistors have been developed in which the collector is also a wide-gap material. In integrated logic circuits such as *transistor-transistor logic, the transistors are operated in saturation in part of the switching cycle. During this part of the operation, the switching speed is reduced and power dissipation increased due to injection of holes into the collector. A wide-gap collector material can be used to suppress this phenomenon.

Suitable design and selective doping of the wide-gap materials can lead to the production of transistors with smaller collector areas than emitter areas. There is a consequent reduction in collector capacitance, also leading to a reduced switching time. This is most easily achieved in integrated circuits using an 'inverted' transistor design in which the collector regions are at the surface of the slice, as in I^2L circuits. Part of a double heterostructure for emitter-coupled logic (ECL) is shown in Fig. b. Three of the transistors are inverted while the fourth is noninverted (emitter up). The equivalent circuit is shown in Fig. c.

A number of different materials have been used to achieve HJBTs. 3–5 *semiconductors offer the widest choice of suitable energy-gap materials. GaAlAs on GaAs provides a suitable choice, especially when combined with photoelectronic devices, such as *injection lasers or *phototransistors, on the same chip. Where silicon is the required substrate material, oxygen doped *polysilicon on silicon gives a reasonably efficient energy-gap difference for heterojunction implementation.

heterojunction photodiode *See* photodiode.

heteropolar generator A type of electromagnetic *generator, such as a dynamo, in which the direction of the induced voltage in each conductor is reversed when that conductor successively traverses the magnetic field: i.e. the direction of the magnetic field sensed by the conductor is reversed each time the conductor cuts the field. In order to produce direct current from the generator some form of commutator must be used.

heterostructure laser *See* injection laser.

Heusler alloys Alloys that exhibit ferromagnetic properties although they contain no ferromagnetic material. They contain manganese, aluminium, and zinc or copper.

hexagon voltage *Syn.* mesh voltage. *See* voltage between lines.

hexode A *thermionic valve that has six electrodes, i.e. four grids between the anode and cathode. The most common use of the device is as a frequency changer.

HF *Abbrev. for* high frequency. *See* frequency band.

hi-fi Acronym from *high fi*delity. *See* recording of sound; reproduction of sound.

high-density packaging The packaging of a given set of electrical components in a very small volume through the application of miniaturization techniques, as by the use of *integrated circuits.

high electron mobility transistor (HEMT) A type of *gallium arsenide *field-effect transistor that is designed to allow electron flow to occur in an undoped layer of GaAs so that the electron mobility is not limited by impurity scattering. The device uses a layer of doped AlGaAs adjacent to a layer of undoped GaAs. Donors in the AlGaAs supply electrons that migrate into the undoped GaAs. Discontinuity between the energy gaps in the two materials causes the electrons to remain in the GaAs, but very close to the heterojunction because of the electrostatic attraction from the donor atoms. Electron flow therefore occurs in the undoped layer, and at high mobility. The electrons are sometimes referred to as a two-dimensional electron gas (2DEG or TEG) and the device is thus sometimes known as a *TEGFET.

high-electron-velocity camera tube *Syn.* anode-voltage-stabilized camera tube. *See* camera tube. *See also* iconoscope.

high fidelity *See* recording of sound; reproduction of sound.

high frequency (HF) *See* frequency band.

high-frequency resistance *See* skin effect.

high-level injection *See* injection.

high-level programming language *See* programming language.

high logic level *See* logic circuit.

high-pass filter *See* filter.

high-pressure Czochralski *See* liquid-encapsulated Czochralski.

high-pressure puller *See* liquid-encapsulated Czochralski.

high recombination-rate contact A contact between two semiconductors or between a semiconductor and a metal that has a substantially constant charge-carrier concentration whatever the current density at the contact. This is because the excess minority charge recombination rate is high so that the excess charges generated as the current flows disappear very quickly.

high tension (HT) *Syn. for* high voltage, especially when it refers to the voltage supply to the anode of a *thermionic valve, usually in the range 60 to 250 volts.

high-vacuum tube *Syn. for* hard tube. *See* electron tube.

high-velocity scanning *See* scanning.

high-voltage test A test of the insulation of any electronic device. A voltage in excess of the normal operating voltage is applied across the insulation to ensure that it does not break down.

hi-lo Read diode *See* IMPATT diode.

Hittorf dark space *Syn. for* Crookes dark space. *See* gas-discharge tube.

Hittorf's principle If electrodes are arranged in a gas at a given pressure, discharge between them does not necessarily occur between the closest points of the electrodes. It will occur between points of potential difference equal to or greater than the threshold voltage required, although the threshold voltage depends on both the electrode separation and gas pressure (*see* Paschen's law).

HJBT *Abbrev. for* heterojunction bipolar transistor.

H-network *Syns.* H-pad; H-section. *See* quadripole.

holding anode *Syn.* keep-alive arc. *See* mercury-arc rectifier.

holding beam *See* storage tube.

holding current *See* pnpn device; silicon-controlled rectifier.

hole An empty energy level in the *valence band of a semiconductor due to an electron being lost from the band by thermal excitation or being trapped by an acceptor impurity (*see* semiconductor). The total current resulting from electrons in a filled valence band is zero:

$$j = nev = e\sum_{i=1}^{n} v_i = 0$$

where j is the current density, n and e the electron density and charge, and v the average velocity of electrons in the valence band. If the jth electron is excited to the conduction band or trapped by an impurity then

$$e\sum_{i=1, i \neq j}^{n} v_i = -ev_j$$

241

The net effect of the mobile electrons occupying all but the jth energy level is therefore equivalent to the effect of a single positive electronic charge 'occupying' the vacant level and hence termed a hole. The hole velocity is equal to the velocity of an electron in the same energy level.

When an electron is thermally excited into the conduction band leaving behind a hole in the valence band an *electron-hole pair* is generated. Relaxation of a conduction electron into an unoccupied energy level in the valence band results in band-to-band recombination (*see* recombination processes).

Under the influence of an electric field holes can drift through the material by a process of continuous exchange with adjacent electrons (*see* hole conduction). They therefore act as mobile positive charge *carriers in the valence band and are the majority carriers in p-type semiconductors.

A hole at or near the top of the valence band has positive effective mass and is mathematically equivalent to a *positron. *See also* direct-gap semiconductor.

hole capture *See* recombination processes.

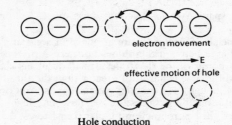

electron movement

E

effective motion of hole

Hole conduction

hole conduction Conduction of electricity in a semiconductor in which a *hole is propagated through the crystal lattice under the influence of an electric field. An adjacent electron moves under the influence of the field and fills the vacancy, leaving a corresponding vacancy behind. The effective movement of the hole due to a process of continuous exchange is equivalent to the movement of a positive charge in the same direction (*see* diagram). Holes move in the direction of positive field.

hole current The current in a *semiconductor associated with the movement of *holes through the material.

hole density *See* carrier concentration.

hole injection *See* injection.

hole trap A site in a semiconductor crystal, such as a donor impurity or lattice defect, that can trap *holes. *See* semiconductor.

homeostasis The condition of a system that is in dynamic equilibrium: the inputs and outputs are balanced so that the system gives the appearance of a steady-state system.

homing beacon *Syn.* locator beacon. *See* beacon.

homojunction A junction between two regions of opposite polarity types within a semiconductor. *See* p-n junction.

homopolar generator An electromagnetic *generator in which the voltage induced in the conductors always has the same sense with respect to the conductors.

homostructure laser *See* injection laser.

honeycomb coil A coil in which the turns are wound in a criss-cross fashion in order to reduce the distributed capacitance.

hook-up *Colloq.* A connection, usually of a temporary nature, between any electric or electronic circuits.

horizontal blanking *See* blanking.

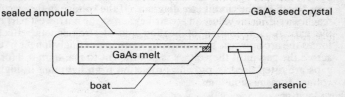

Schematic diagram of horizontal Bridgeman growth method

horizontal Bridgeman (HB) A method of growing *gallium arsenide crystals. The process uses a boat in which the crystal is grown. The boat containing either pure gallium or polycrystalline gallium arsenide is sealed into a long quartz ampoule filled with an inert gas. Pure arsenic is placed in the neck of the ampoule and a seed crystal of gallium arsenide is placed at one end of the boat (*see* diagram). Heaters are used to cause the arsenic to become gaseous and the material in the boat to melt. A temperature profile is created to cause the melt to be held at the melting point of gallium arsenide and the seed crystal region at just below the solidification point. The gaseous arsenic reacts with the gallium in the boat. Once the reaction is complete the ampoule and heaters are moved relative to each other so that the temperature front is moved slowly along the length of the boat. Crystal growth occurs from the seed crystal following the temperature front, to produce a large single crystal with a cross-sectional shape matching the shape of the boat. Gallium arsenide produced by this method is sometimes referred to as *boat grown* gallium arsenide.

horizontal hold *See* television receiver.

horizontal polarization (1) Polarization of an electromagnetic wave in which the electric field vector is parallel to the geographic horizon. (2) The horizontal arrangement of a *dipole aerial.

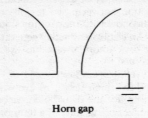

Horn gap

horn gap A device that protects circuits from excessive transient voltages. It is a spark gap, in the shape of a horn, that provides a temporary *earth for the circuit (*see* diagram). If the voltage in the circuit reaches a sufficient value, an arc strikes across the narrowest part of the gap. A combination of heating and electromagnetic effects forces the arc upwards so that it follows an increasingly longer path across the gap until the voltage is insufficient to maintain it. Horn gaps are often used to provide protection from lightning in high-voltage transmission lines.

horn loudspeaker *See* loudspeaker.

horseshoe magnet A magnet that is in the shape of a horseshoe so that the poles are close together.

hot-carrier diode *See* Schottky diode.

hot cathode *See* thermionic cathode.

hot-cathode stepping tube A type of multielectrode thermionic valve with two or more stable paths that the electron beam can follow. A suitable input signal can cause the beam to step in a desired sequence from one position to another.

hot-cathode tube *Colloq.* A thermionic valve.

hot electron A conduction electron in a solid that has an energy such that the electron is not in thermal equilibrium with the crystal lattice, i.e. it is more than a few kT above the *Fermi level, where k is the Boltzmann constant and T the thermodynamic temperature of the lattice. A *hot hole* is a mobile hole that is similarly not in thermal equilibrium with the lattice.

hot spot A small portion of an electrode, such as an arc spot (*see* arc), that is at a higher temperature than the rest of the electrode.

hot-wire ammeter *See* thermoammeter.

hot-wire gauge *See* Pirani gauge.

hot-wire instrument A thermal instrument that utilizes the elongation by heat of a strip or wire carrying an electric current.

hot-wire microphone A device that utilizes the change in resistance of a hot wire when exposed to sound waves. The change in resistance of the wire is equal to that resulting from a steady draught of velocity equal to the maximum velocity of the simple harmonic motion of the sound waves.

There are two factors contributing to the observed resistance drop: a steady drop depending on the particle velocity at the wire, and therefore on the sound intensity, and an oscillatory change due to the sinusoidal variations in amplitude of the sound wave.

howl An unpleasant high-pitched audiofrequency tone heard in receivers and caused by unwanted electric or *acoustic feedback.

howler (1) An oscillator that produces a high-pitched audiofrequency tone in order to attract attention. It is used by telephone exchanges to attract attention to a telephone whose receiver has been left off.

(2) An audiofrequency oscillator that is used as a *null-point detector in certain bridges. The oscillator produces a high-pitched audiofrequency tone that either disappears or reaches a maximum intensity when the balance point is achieved.

H-pad *Syn. for* H-network. *See* quadripole.

h parameter *Short for* hybrid parameter. *See* transistor parameters.

H-plane *Syn. for* H-bend. *See* waveguide.

H-section *Syn. for* H-network. *See* quadripole.

HT *Abbrev. for* high tension.

hum A low-pitched audiofrequency droning noise heard in audiofrequency systems. It usually originates from the mains supply and occurs at frequencies that are harmonics of the mains-supply frequency.

hum modulation Unwanted *modulation of a signal by hum arising in an audiofrequency system. It results in *noise or *distortion of the output.

hunting Fluctuation of a controlled signal about its desired value that results from overcorrection by a control device. The stable desired value is never actually attained.

H wave *Syn. for* TE wave. *See* mode.

hybrid computer *See* computer.

hybrid integrated circuit *See* integrated circuit.

hybrid junction A junction between *transmission lines, either waveguides or coaxial lines. An ideal junction has no direct coupling between arms 1 and 4 or between arms 2 and 3 (*see* diagrams). If arm 1 is excited power flows in arm 4 as a result of reflections in arms 2 and 3.

The two main types of hybrid junction are the *ring junction,* in which all 4 arms are coplanar (Fig. *a*), and the *T-junction,* in which arm 4 is perpendicular to the plane containing the other three arms (Fig. *b*). If the decoupled arms (1 and 2) are independently matched in impedance and the other two arms (3 and 4) are terminated in

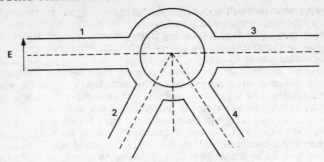

a Waveguide ring junction (E-bend)

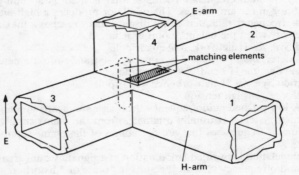

b Waveguide T-junction

their characteristic impedances, Z_0, then all four arms are matched at their inputs.

hybrid parameter *See* transistor parameters.

hybrid-π equivalent circuit *See* transistor parameters.

hybrid ring junction *Syn. for* ring junction. *See* hybrid junction.

hybrid-T junction *Syn. for* T-junction. *See* hybrid junction.

hydrogen electrode *See* electromotive series.

hydrogen thyratron *See* thyratron.

hydrophone A *transducer that produces electrical signals in response to water-borne sound waves.

hysteresis A delay in the change of an observed effect in response to a change in the mechanism producing the effect. The best-known example of hysteresis is *magnetic hysteresis.

246

hysteresis distortion Distortion introduced by nonlinear hysteresis in one or more elements of a system.

hysteresis factor *See* magnetic hysteresis.

hysteresis loop *See* magnetic hysteresis.

hysteresis loss (1) *See* magnetic hysteresis. (2) *See* electric hysteresis loss.

hysteresis meter An instrument that detects and measures *magnetic hysteresis.

Hz *Abbrev. for* hertz.

IC *Abbrev. for* integrated circuit.

iconoscope A high-electron-velocity photoemissive *camera tube in which the target is formed on a thin mica plate. One side of the plate is covered by a *mosaic* of very many small areas of photoemissive material and the other side by a thin metallic *signal electrode* from which the output signal is obtained (*see* diagram). The optical image

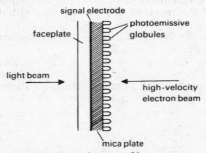

Section of mosaic target of iconoscope

produced on the mosaic causes *photoemission of electrons from each elemental area. The mica acts as the dielectric for what is effectively an array of elemental capacitors, in which a pattern of positive charges is produced as a result of the photoemission. A high-velocity electron beam is used to scan the target area and discharge each capacitor through the signal electrode. The magnitude of the resulting current is a function of the charge on each target area and hence of the illumination producing the charge pattern.

The *orthicon* operates in a similar manner to the iconoscope but employs a low-velocity electron scanning beam.

ideal bunching *See* velocity modulation.

ideal crystal A theoretical crystal that has a perfect and infinite crystal structure and therefore contains no impurities or defects.

ideal diode equation *Syn. for* Schockley equation. *See* p-n junction.

ideal transducer A hypothetical transducer that produces the maximum possible signal power output under specified input and load conditions; i.e. power loss in the transducer is negligible.

ideal transformer *Syn.* perfect transformer. *See* transformer.

idiochromatic crystal *See* intrinsic crystal.

idle component (1) (of current) *See* reactive current.
  (2) (of voltage) *See* reactive voltage.
  (3) (of volt-amperes) *See* reactive volt-amperes.

IEC *Abbrev. for* International Electrotechnical Commission. *See* standardization.

IEE *Abbrev. for* Institute of Electrical Engineers. *See* standardization.

IEEE *Abbrev. for* Institute of Electrical and Electronics Engineers, Inc. *See* standardization.

IF *Abbrev. for* intermediate frequency. *See* superheterodyne reception.

iff *See* logic circuit.

IF strip The part of the receiver in *superheterodyne reception that is concerned with intermediate-frequency amplification. It consists of the intermediate-frequency amplifier together with its physical environment.

IGFET *Abbrev. for* insulated-gate field-effect transistor. *See* field-effect transistor.

ignition coil *Syn.* spark coil. An *induction coil in which the voltage pulses output by the secondary coil are sufficiently large to produce sparks. The sparks may be used to ignite an inflammable mixture, such as the gases in an internal-combustion engine.

ignitor *Syns.* ignitor electrode; ignitor rod. An auxiliary electrode in an arc-discharge tube (*see* gas-discharge tube) that is used to control the arc across the tube. It usually consists of a semiconducting rod with one end immersed in a mercury-pool cathode. When a current passes through the ignitor, local heating of the mercury pool allows the arc to strike, and the arc discharge continues as long as the ingnitor current continues. An arc discharge initiated by an ignitor is termed an *ignitor discharge* and occurs at a voltage lower than that required for an arc discharge in the absence of an ignitor. The *ignitor voltage drop* is the voltage across the tube that is necessary for the ignitor discharge at a specified ignitor current. Ignitors are used in gas-discharge rectifiers, such as the *ignitron, and in *switching tubes.

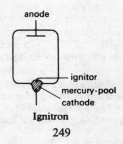

Ignitron

ignitron A type of mercury-arc rectifying tube in which the discharge is initiated by a subsidiary *ignitor. The tube contains an anode and a mercury-pool cathode. A rod of semiconductor, such as silicon carbide or boron carbide, is immersed in the cathode (*see* diagram). When a current is passed through the ignitor rod to the mercury, local heating of the mercury forms a hot spot, which allows the arc to strike across the tube at a lower anode voltage than would otherwise be necessary.

An *ignitron rectifier* is a rectifier that contains an ignitron as the main rectifying element; a *thyratron is usually included to control the firing time of the ignitron.

I²L *Abbrev. for* integrated injection logic. *Syn.* merged transistor logic (MTL). A family of *bipolar integrated logic circuits that are very compact and provide a very high functional packing density combined with the possibility of high-speed operation. I²L is essentially a development derived from *diode transistor logic (DTL). The input diodes and transistors of DTL are combined in I²L into a single multicollector n-p-n transistor, and the current source is provided by a p-n-p transistor rather than a load resistor. The p-n-p transistor is formed in the same area of chip as parts of the multicollector transistor allowing a very compact structure. The multicollector structure provides the isolation between logic gates provided in DTL by the input diodes of the following gates.

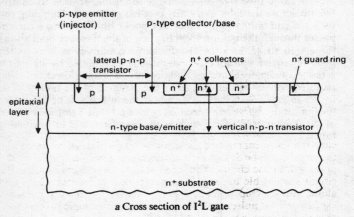

a Cross section of I²L gate

A typical structure is shown in Fig. *a* and the equivalent circuit in Fig. *b*. The p-n-p transistor is arranged laterally on the chip; the p-type emitter is common to a large number of gates and is termed the *injector*. The n-p-n transistor is a vertical transistor, which is inverted compared to the usual method of fabrication. The n-type epitaxial layer forms the base of the p-n-p transistor and the emitter of the n-p-n transistor and is common to all the gates. An n^+ *guard ring surrounds each p-type p-n-p collector/n-p-n base region containing the n^+ n-p-n collectors, in order to provide isolation between individual gates. The *fan-out is determined by the number of multiple collectors.

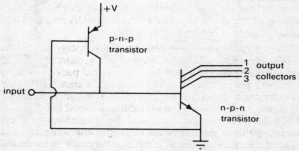

b Equivalent circuit of single I^2L gate

A basic 3-input NOR gate is shown in Fig. *c*. If any of the collectors of gates 1, 2, or 3 is at the low logic level, current from the p-n-p transistor flows to that collector (indicated by the dashed line) and the transistor T_4 is 'off'. The collectors of T_4 are therefore at the high logic level. If all the inputs are high T_4 is 'on' and current from the p-n-p transistor flows through T_4 and the collector voltages are low. The p-n-p transistors are not shown but are indicated by the label 'injection current'.

The difference between the high and low logic levels is determined by the forward voltage of the base-emitter junction of the multicollector transistor and the circuits can operate with a total voltage swing of about 0·7 volts. The power consumed by the circuits is a linear function of the speed at which they are operated and the circuits can be designed to optimize the speed and power at any point within the circuit. The small voltage change renders the I^2L circuit susceptible to stray noise pulses or interference when used alone, and on-chip input and output buffer circuits are normally used to convert pulses from *transistor-transistor logic to those suitable for I^2L and vice versa.

Schottky I^2L (STL) is a form of I^2L in which the collectors are formed as *Schottky diodes. Excessive *carrier storage at the collector junction is reduced and the speed *of operation is therefore increased. The total voltage swing is also reduced due to the

251

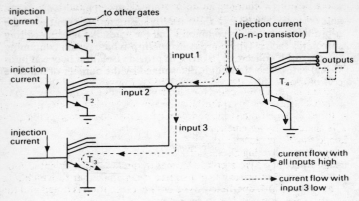

injection current

to other gates

T_1

input 1

injection current

T_2

input 2

injection current

T_3

input 3

injection current (p-n-p transistor)

outputs

T_4

→ current flow with all inputs high

---→ current flow with input 3 low

c Basic NOR gate

characteristically small forward voltage of the Schottky diode. The speed of operation of the STL circuits may also be increased using Schottky diodes across the base-collector junction to prevent the output transistors from going too far into the saturated mode, and hence keeping charge-storage times to a minimum.

I^2L has found many varied applications. It is fabricated using standard bipolar techniques and other types of circuit, such as light-emitting diode drivers, operational amplifiers, and oscillators, can be easily produced on the same chip allowing great flexibility.

image *Short for* electric image.

image attenuation constant *Syn.* image attenuation coefficient. *See* image transfer constant.

image charge *See* electric image.

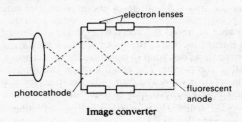

electron lenses

photocathode

fluorescent anode

Image converter

image converter *Syn.* image tube. An *electron tube that converts an image outside the visible part of the spectrum, such as an infrared image, into a visible image. The 'nonvisible' image is focused on a photosensitive cathode and produces electrons. These are attracted

252

to a positively charged fluorescent anode screen, and focused on it by an *electron-lens system (*see* diagram). A visible image is produced on the screen. Image converters have many applications and are used in X-ray intensifiers, infrared telescopes and cameras, electron telescopes, and microscopes.

image dissector A type of television *camera tube, little used now, in which electrons from each portion of a photosensitive plate are focused in turn on a collector electrode in order to produce a video signal. It differs from the *iconoscope or *image orthicon camera tubes in which the static charge pattern on the photosensitive plate is scanned by an electron beam.

image force *See* electric image.

image-force lowering *See* Schottky effect.

image frequency *Syn.* second-channel frequency. An unwanted input frequency that arises from a source other than that to which a receiver used in *superheterodyne or *beat reception is tuned and that causes spurious signals to be output.

In superheterodyne reception an unwanted response is generated in the intermediate-frequency amplifier. The spurious signals are termed *image interference* and occur particularly when a nonlinear mixer element is used. The image frequencies are those frequencies such that the difference between them and the local-oscillator frequency falls within the dynamic range of the intermediate-frequency amplifier. The characteristics of the particular receiver determine the input conditions that cause image interference. The image frequencies can be equal to the desired input signal frequency, equal to the intermediate frequency, or to twice the local-oscillator frequency.

In beat reception the image frequencies are those frequencies such that *beating with the beat-frequency oscillator produces the same frequency as beating with the desired signal.

In either case the *image ratio* is the ratio of the image-frequency amplitude to the desired signal amplitude when identical outputs are produced.

image impedances The values Z_{i1} and Z_{i2} of *impedance that simultaneously satisfy the following conditions: if Z_{i1} is connected across one pair of terminals of a *quadripole then the impedance between the second pair is Z_{i2}; if Z_{i2} is connected across the second pair of terminals then the impedance across the first pair is Z_{i1}.

image interference *See* image frequency.

image orthicon A low-electron-velocity *camera tube in which light from a scene is focused on a *photocathode consisting of a light-sensitive material deposited on a thin sheet of glass. Electrons are emitted from the photocathode in proportion to the intensity of the light and are focused on to a target consisting of a thin glass disc with a fine mesh on the photocathode side of the disc (*see* diagram).

The impact of electrons from the photocathode causes *secondary emission of electrons from the target that is greater than, but proportional to, the original electron density from the photocathode. The secondary electrons are collected by the mesh screen and drained off to a power supply. The target is left with a positive static charge pattern corresponding to the original light image.

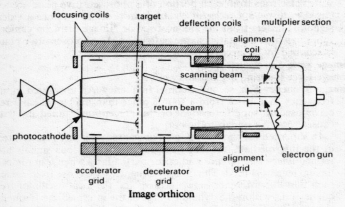

Image orthicon

The reverse side of the disc is scanned with a low-velocity electron beam produced from an *electron gun. Positively charged areas of the target are neutralized by electrons from the beam, which is consequently intensity-modulated by the original picture information. The electron beam is reflected by the target glass and travels back towards the electron gun. It is collected by an electrode that surrounds the electron gun and acts as the final aperture for the scanning beam. This electrode acts as the first dynode of an *electron multiplier section from which the video ouput is obtained.

The image orthicon has high sensitivity, the spectral sensitivity is close to that of the human eye, and the speed of response is relatively fast.

image phase constant *Syn.* image phase-change coefficient. *See* image transfer constant.

image potential The potential at a point just outside the surface of a material due to an *electric image.

image ratio *See* image frequency.

image transfer constant *Syn.* image transfer coefficient. Symbol θ. A complex quantity, given by $(\alpha + i\beta)$, of a quadripole network terminated in its *image impedances. It is half the natural logarithm of the complex ratio of the steady-state *volt-amperes input to output of the network:

$$\theta = \frac{1}{2}\log_e(E_1 I_1 / E_2 I_2)$$

where E_1, I_1 and E_2, I_2 are the voltages and currents at the input and output terminals, respectively.

The real part of the image transfer constant (α) is the *image attenuation constant;* the imaginary part (iβ) is the *image phase constant. Compare* propagation coefficient.

image tube *See* image converter.

immersion plating *See* electroless plating.

impact ionization Ionization of an atom or molecule due to the loss of orbital electrons following a high-energy collision. In a semiconductor electron-hole pairs can be generated if the electrons have sufficient energy to enter the conduction band.

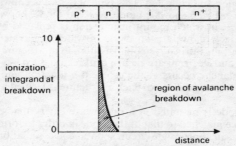

a Typical IMPATT diode: Read diode

IMPATT diode Acronym from *IMP*act ionization *A*valanche *T*ransit *T*ime. A semiconductor diode that acts as a powerful source of microwave power. When a p-n junction is reverse-biased into *avalanche breakdown, it exhibits negative resistance at microwave frequencies and may be used as a *negative-resistance oscillator. The differential current is out of phase with the differential voltage due to two effects: following a voltage increment the current builds up with a delay time, t_A, characteristic of the avalanche; the terminal current increment is further delayed by a time t_t(the transit time) during which the carriers are collected by the electrodes. The diode is usually formed so that the current is delayed by half a cycle with respect to the voltage.

Although any p-n junction diode will exhibit IMPATT mode operation, typical devices used consist of an avalanching region together with a drift region in which no avalanche occurs. Examples of such diodes are shown in the diagrams. Fig. *a* shows a *Read diode* with structure p$^+$-n-i-n$^+$. Although both holes and electrons are produced by the avalanche breakdown, only the electrons are given

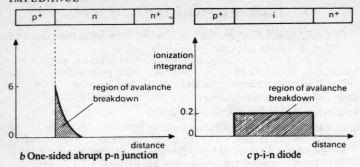

b One-sided abrupt p-n junction

c p-i-n diode

a drift region and collected. This type of device is known as a *single drift device*. Fig. *b* shows a modified Read diode, known as a *hi-lo Read diode*, in which the intrinsic region is replaced by an n-type region. Fig. *c* shows a p-i-n diode in which the avalanche occurs throughout the intrinsic region. An alternative structure where both the holes and electrons are given a drift region is a p^+-p-n-n^+ structure. Avalanche occurs at the centre p-n junction and both holes and electrons are collected. Such a structure is termed a *double drift device*.

The diode is mounted in a microwave cavity that can have its impedance matched to that of the diode in order to form a resonant system. In an appropriate circuit spontaneous oscillation will occur.

impedance (1) Symbol: Z; unit: ohm. A measure of the response of an electric circuit to an alternating current. The current is opposed by the capacitance and inductance of the circuit in addition to the resistance. The total opposition to current flow is the impedance, which is given by the ratio of the voltage to the current in the circuit.

The alternating current is given by

$$I = I_o \cos\omega t$$

where I_o is the peak current and ω is the angular frequency. When *reactance, due to the capacitance and inductance, is present in a circuit, the voltage will be out of phase with the current and is given by

$$V = V_o \cos(\omega t + \phi)$$

where ϕ is the *phase angle. In a circuit containing resistance, R, capacitance, C, and inductance, L, the voltage is given by

$$V_o \cos(\omega t + \theta) = IR + L(\mathrm{d}I/\mathrm{d}t) + 1/C \int I \mathrm{d}t$$

Solving this equation shows that the current is equal to

256

$$I = V_o \cos(\omega t + \phi)/\sqrt{[R^2 + (\omega^2 L^2 - 1/\omega^2 C^2)]}$$

Since the impedance is the ratio of current to voltage then

$$Z = \sqrt{[R^2 + (\omega^2 L^2 - 1/\omega^2 C^2)]}$$
$$= \sqrt{[R^2 + X^2]}$$

where X is the reactance. Z is thus a complex quantity whose magnitude or modulus $|Z|$ is equal to the vector sum of R and X. The *complex impedance* can thus be given by

$$Z = R + iX$$

where i is equal to $\sqrt{-1}$. The real part, the resistance, represents a loss of power due to dissipation. The imaginary part, the reactance, indicates the phase difference between the voltage and current. It is either positive or negative depending on whether the current lags or leads the voltage, respectively. In a circuit containing only resistance, or in a resonant circuit, the current and voltage are in phase and Z is purely resistive. In a circuit containing only reactance the current and voltage are out of phase and Z is purely imaginary, i.e. there is no dissipation in the circuit.

The complex impedance is the ratio of the complex voltage to the complex current, given respectively as

$$V = V_o \exp[i(\omega t + \phi)]$$
$$I = I_o \exp(i\omega t)$$

Hence the complex impedance can be expressed as

$$Z = |Z|\exp(i\phi)$$
$$Z = |Z|(\cos\phi + i \sin\phi)$$

(2) *See* impedor.

impedance coupling *See* coupling.

impedance matching The matching of the *impedances of the parts of a system to ensure optimum conditions for transfer of power from one part of the system to another.

The maximum power is transferred from an *amplifier to a load if the load impedance is made the *conjugate impedance of the amplifier output impedance.

A nonuniform *filter will transmit a wave undisturbed provided that the *iterative impedance of each filter section is made equal. If the iterative impedances of the sections are different, loss of power output occurs due to reflections of the wave occurring at the junctions. A small section terminated in its *image impedances, where one impedance equals the iterative impedance of the filter sections and the other the load impedance, may be used to match the load to the filter.

Reflections of wave power in a *transmission line are eliminated by making the load impedance equal to the generator output impedance, and the line impedance of the transmission line equal to both the above impedances. Transmission lines of differing line impedances may be joined in a system using a *quarter-wavelength line.

impedor *Syn.* impedance. Any circuit element, such as a resistor, capacitor, or inductance, that is used mainly for its *impedance.

imperfect dielectric *See* dielectric.

imperfection Any deviation in the structure of a crystalline solid compared to that of an *ideal crystal.

impregnated cable A cable with paper insulation in which the insulating properties of the paper tapes are improved by impregnating them with an insulating compound, such as oil.

impregnated carbon The carbon in an *arc lamp that is mixed with other materials in order to produce a coloured arc. The materials used are suitably chosen to produce an arc of a particular colour.

impregnated cathode A cathode that is made of porous tungsten impregnated with a barium salt so that as the barium evaporates from the surface of the cathode it is continuously replenished from the bulk of the material.

impulse *Short for* impulse voltage or current. *See* impulse voltage.

impulse current *See* impulse voltage.

impulse excitation *See* shock excitation.

impulse flashover voltage *See* impulse voltage.

impulse generator *Syn.* surge generator. An electronic device that produces a single pulse, typically one representative of the *surges generated by lightning in a transmission line. A typical impulse generator operates by charging and discharging one or more capacitors.

impulse noise *See* noise.

impulse puncture voltage *See* impulse voltage.

impulse ratio for flashover (or **puncture**) *See* impulse voltage.

impulse voltage A unidirectional voltage that rises rapidly to a maximum value and then falls to zero more or less rapidly without any appreciable superimposed oscillations. An *impulse current* is a unidirectional current with similar characteristics. Impulses are usually unwanted, being generated by fault conditions in electrical equipment or apparatus or by procedures such as switching on or off. A common source of large impulse voltages is a *lightning stroke.

A typical waveshape is shown in Fig. *a.* The maximum voltage, V, of the impulse is the peak value. The *wavefront* is the rising portion OA and the *wavetail* is the falling portion ABC. The duration of the wavefront is the time, T_1, required for the voltage to rise from zero to the peak value. The time to half value of the wavetail, T_2, is the time required for the impulse to rise from zero, reach the peak value, and then decay to half the peak value. The waveshape in Fig. *a* is described as a T_1/T_2 impulse. The average rate of increase of voltage with time is the *steepness* of the wavefront. A typical impulse voltage that results from lightning is a 1/50 wave, where T_1 and T_2 are measured in microseconds; standard testing equipment uses a wave with these proportions produced by a surge generator.

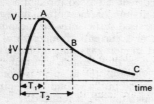
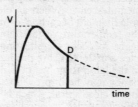

a Waveshape of impulse voltage

b Waveshape of chopped impulse voltage

An impulse voltage or current frequently causes *flashover or puncture in electrical apparatus. If this occurs the impulse voltage collapses rapidly and is described as a *chopped impulse voltage* (Fig. b). Flashover or puncture can occur either during the wavefront or during the wavetail. The actual value of the impulse voltage at which flashover or puncture occurs during the wavefront is the *impulse flashover voltage* or the *impulse puncture voltage*. If flashover or puncture occurs on the wavetail the peak value of the impulse voltage is quoted. The *time to flashover* or the *time to puncture* is the time interval between the beginning of the impulse voltage or current and the instant at which the wave is chopped (point D in Fig. b).

In an insulator used in a.c. power transmission systems, the impulse voltage at which flashover or puncture occurs in the insulator is in general different to the voltage of the a.c. power at which failure occurs. The *impulse ratio for flashover* (or *puncture*) is the ratio of the impulse flashover voltage (or puncture voltage) to the peak value of the alternating voltage at which flashover or puncture occurs. Typical values for commonly used insulators are between 1·5 and 1·2.

impurities Foreign atoms in a semiconductor that are either naturally occurring or deliberately introduced. Impurities have a fundamental effect on the type and amount of conductivity of the semiconductor. *Impurity diffusion* is the deliberate diffusion of impurities into selected regions of a semiconductor in order to produce the desired characteristics. The energy levels due to the impurities are *impurity levels*. The presence of impurities affects the mobilities (both *Hall and *drift mobilities) of charge carriers due to *impurity scattering* between the carriers and the impurity atoms. *See also* diffusion; semiconductor.

inactive interval *See* sawtooth waveform.

incandescence The emission of visible light from a substance at a high temperature. The term is also used to describe the radiation itself. *Compare* luminescence.

incandescent lamp An electric lamp in which light is produced by incandescence. When an electric current is passed through a suitable filament of carbon, osmium, tantalum, or (most often) tungsten the temperature of the filament is raised sufficiently for incandescence to occur (*see* heating effect of a current). Temperatures in excess of 2600°C are produced and an inert gas filling is often used in the lamp to prevent disintegration of the filament. Thermal efficiency may be improved using a *coiled coil:* the filament is wound into a close spiral, which is itself wound into a second close spiral. Thus heat loss, by conduction through the gas, is reduced.

incident current *Syn.* initial current. *See* reflection coefficient.

incremental hysteresis loss *See* magnetic hysteresis.

incremental permeability *See* permeability.

incremental plotter A device that produces a graphic display from data generated by a computer. A smooth curved line is not usually achieved because of the digital nature of the computer output; small increments in the $x-$, $y-$, and diagonal directions are used to approximate the curves.

An incremental plotter is most often used as an *off-line plotter*. The source of data is a previously prepared magnetic tape, disk, or paper tape and the plotter may be located remotely from the computer. An incremental plotter may be linked directly to the computer and used as an *on-line plotter*, i.e. the points are plotted graphically as soon as they are generated.

A *microfilm plotter* is an incremental plotter that produces a graph on photographic film rather than paper.

incremental recorder An *off-line device that records data, as it is generated, in a form suitable for inputing directly to a *computer. Suitable recording media include magnetic tape, paper tape, and punched cards.

index error *Syn.* zero error. An error on the scale of a measuring instrument such that under zero input conditions a reading x is registered by the instrument. In the absence of other errors, a simple correction of $-x$ can be applied to all readings on the instrument.

indicating instrument An instrument that indicates the presence of a variable electrical quantity or measures its value. Examples include the *ammeter, *galvanometer, and *digital voltmeter.

indicator tube An extremely small *cathode-ray tube, often with a screen diameter measured in millimetres. It is used to indicate the value of a varying quantity by altering the size or shape of the image on the screen in accordance with the input voltage. A rectangular voltage pulse, termed an *indicator gate,* may be applied to the grid or cathode circuit of the tube in order to sensitize or desensitize it.

indirect-gap semiconductor *See* direct-gap semiconductor.

indirect lightning surge *Syn.* induced lightning surge. *See* lightning stroke.

indirectly heated cathode *Syns.* filament cathode; unipotential cathode. *See* thermionic cathode.

indirect photoconductivity *Syn.* phonon-assisted photoconductivity. *See* photoconductivity.

indirect ray *See* indirect wave.

indirect stroke *See* lightning stroke.

indirect transition *See* direct-gap semiconductor.

indirect wave *Syns.* indirect ray; reflected wave. The portion of a transmitted wave that does not travel directly from the transmitter to the receiver but is reflected by the *ionosphere. *Compare* ground wave.

induce To cause an electrical or magnetic effect in one circuit or device by altering the condition of a second circuit or body. *See* electromagnetic induction; electrostatic induction.

induced current A current that flows in a conductor as a result of a changing *magnetic flux density, in which the lines of magnetic flux intersect with the conductor. *See also* electromagnetic induction.

induced dipole moment *See* dipole.

induced electromotive force *See* electromagnetic induction.

induced lightning surge *Syn. for* indirect lightning surge. *See* lightning stroke.

induced noise *Noise that appears in any circuit, device, or apparatus as a result of *electromagnetic induction from other nearby circuits.

inductance (1) Unit: henry. A constant that relates the magnetic flux, ϕ, linking a circuit to the current flowing in the circuit or in a nearby circuit. *See* electromagnetic induction.

The *self-inductance* (symbol: L) of a circuit is defined as one henry if the circuit is threaded by a total flux of one weber when a current of one ampere is flowing. The *mutual inductance* (symbol: M or L_{12}) between two circuits is defined as one henry when a flux of one weber is present in one circuit due to a current of one ampere flowing in the second circuit.

(2) *See* inductor.

induction *Short for* electromagnetic induction; electrostatic induction.

induction coil A device that uses *electromagnetic induction to produce a series of pulses of high potential and approximately unidirectional current. It consists of a primary coil of wire with only a few turns wound on an iron core, surrounded by a coaxial secondary coil of many turns and insulated from it (Fig. *a*). When the current in the primary coil is interrupted suddenly, a large e.m.f. is induced in the secondary. When the primary circuit is remade a much small-

261

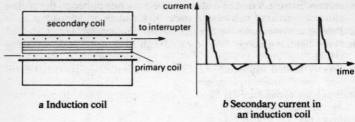

a Induction coil

b Secondary current in an induction coil

er e.m.f. is induced in the secondary in the opposite sense. The relatively high resistance introduced into the primary circuit at break, compared to remake, results in a much smaller *time constant in the primary and consequently in the higher e.m.f. The voltage output of the secondary depends on the sharpness of the break. The performance of the coil depends on the type of interrupter used in the circuit. Vibrating hammers (using the principle of the electric bell) are frequently used for small coils; larger coils may employ mercury breaks. The output from the secondary consists of a series of large pulses corresponding to the breaks in the primary circuit alternating with much smaller inverse pulses at the remake points (Fig. *b*).

induction compass A device that indicates direction. It consists of a small coil that is made to rotate in the earth's magnetic field. The direction indicated by the compass depends on the magnitude of the induced current in the coil.

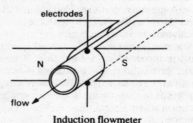

Induction flowmeter

induction flowmeter A device that measures the rate of flow of a conducting liquid. The liquid is made to flow through a tube, length L, placed in a magnetic field of flux density B (*see* diagram). Electrodes are placed on a diameter of the tube and the induced e.m.f., E, across this diameter depends on the rate of flow, v, of the liquid:

$$E = v \times BL$$

induction furnace A device that utilizes the heating effect of an induced current in the secondary of a transformer.

induction generator An alternating-current (a.c.) generator in which a coil or number of coils are made to rotate in an alternating magnetic field produced by a number of electromagnets excited from an alternating-current source. Provided that the rotating coils are driven at a sufficient speed – determined by the a.c. source frequency and the number of pairs of magnetic poles – a.c. power is supplied to a load at the same frequency as the exciting voltage. The induction generator is fundamentally an *induction motor used as an a.c. generator.

This type of a.c. generator cannot be operated independently of any other a.c. source and must be operated in parallel with one or more *synchronous a.c. generators that provide the excitation for it.

induction heating *Syns.* eddy-current heating; r.f. heating. Heating caused by induced *eddy currents in a conducting material when subjected to a varying magnetic field. The varying magnetic field is commonly produced by an alternating current in a coil – the *load coil* – surrounding the load to be heated.

One type of furnace uses induction heating for melting metal: the advantages of this type of heating are that the heat is generated within the metal itself, and the eddy currents set up circulatory movements in the molten metal that have the effect of stirring the melt. *Compare* dielectric heating.

induction instrument An instrument that depends for its operation on the interaction between a varying magnetic field in a fixed winding and the *eddy currents induced by the field in a movable conductor (usually in the form of a disc or cylinder). The resulting torque produces a deflection of the conductor.

induction machine A machine that produces electric charge by electrostatic induction. *See* Van de Graaff generator.

induction microphone A type of *microphone that has a straight-line conductor as the moving element. The conductor moves in a magnetic field, and the current induced in it by electromagnetic induction depends on the sound pressure and causing it to move.

induction motor An alternating-current *motor in which the current in the secondary winding, usually the rotor, is produced by *electromagnetic induction when alternating current is supplied to the primary winding, usually the stator. Mechanical movement results from the torque produced by the interaction between the rotor current and the magnetic field due to the current in the stator.

inductive Denoting any electric circuit, device, or winding having an appreciable self-inductance (*see* electromagnetic induction). In practice it is extremely difficult to produce a total lack of inductance and the term is usually applied to a circuit, etc., in which for a particular application the effect of inductance is not negligible. *Compare* noninductive.

inductive coupling *See* coupling.

inductive feedback *See* feedback.

inductive interference A type of interference in a communication system caused by electromagnetic induction from the electric supply system.

inductive load *See* lagging load.

inductive-output device An electronic device in which the output voltage is produced by electromagnetic induction between the current and the output electrode, without the current carriers being collected by the output electrode.

inductive reactance *See* reactance.

inductive tuning *See* tuned circuit.

inductor *Syns.* inductance; reactance coil. A device or circuit element, usually in the form of a coil, that possesses inductance and is used primarily because of that property. *See also* choke.

inert cell A primary *cell that contains the chemicals and other necessary ingredients in solid form, but will not function until water is added to form an electrolyte.

inertia switch A type of *switch that operates when an abrupt change in its velocity occurs.

information satellite *See* satellite.

information theory An analytical technique that determines the optimum amount of information required (i.e. the necessary and sufficient amount) to transmit a message or solve a specific problem in communication, control, or computer systems. The *information content* of a message is the minimum amount of information required to transmit the message with a desired accuracy, in the absence of *noise. *Information retrieval* is the means of extracting specific information from stored or transmitted data.

infrared image converter *See* image converter.

infrared radiation The portion of the electromagnetic spectrum of radiation (*see* Table 10, backmatter) extending from the limit of the red end of the visible spectrum to the microwave region. The infrared region extends from about 730 nanometres to about one millimetre in wavelength.

   Infrared radiation is fairly long-wave heat radiation, emitted by hot bodies, and is detected by devices such as *bolometers, thermopiles (*see* thermocouple), and *photocells.

inherent weakness failure *See* failure.

inhibiting input An input applied to a digital *gate that prevents any output that might otherwise occur.

initial current *Syn. for* incident current. *See* reflection coefficient.

injection (1) In general, the application of a signal to an electronic circuit or device.

(2) The introduction of excess charge carriers, either electrons or holes, into a *semiconductor material so that the total number present exceeds that at thermal equilibrium. In *low-level injection* the number of excess carriers is small whereas in *high-level injection* the number of excess carriers is comparable to the numbers at thermal equilibrium.

injection efficiency The efficiency of a p-n junction under forward bias, defined as the ratio of the current carried by injected minority carriers to the total current across the junction.

injection laser *Syns.* diode laser; semiconductor laser. A *laser that consists essentially of a *p-n junction diode. Under forward bias conditions electrons from the n-type material cross into the p-type where they form an excess minority carrier concentration. This process is referred to as *injection of electrons. *Recombination of the injected minority carriers occurs both by radiative and nonradiative processes. Photons produced from radiative recombination can interact with donors in the valence band and be absorbed, or can interact with electrons in the conduction band and stimulate the emission of an identical photon, or they can be emitted from the diode as a flash of light (*see* light-emitting diode). At sufficiently large values of minority carrier concentration, the numbers of stimulated photons exceeds that of absorbed photons and optical gain occurs.

Oscillation of the laser is achieved by positive feedback of part of the light to stimulate further emission, and coherence of the light output is also achieved using a resonant cavity. This is usually done by cleaving and polishing the ends of the diode perpendicular to the plane of the junction to form partial reflective surfaces. The diode itself acts as the resonant cavity, with a number of natural resonant frequencies – *Fabry-Perot modes*. The modes are derived from solving Maxwell's equations for the optical electric field; this yields discrete values of the propagation constant for the boundary conditions that apply at the diode junction. For many applications multimode operation is acceptable but for some applications, particularly optical communications, single frequency operation is essential to eliminate interference losses between the different frequencies. The unwanted modes must be filtered out and this is most conveniently done by incorporating a suitable grating into the laser. The light is confined and guided by the structure of the junction itself, which functions as a dielectric waveguide (Fig. *a*). The laser light is produced in a narrow region on the p-side of the junction known as the *gain region* or *active layer*. The thickness of the gain region is determined by the diffusion length of the injected carriers. n_1, n_2, and n_3 are the refractive indexes of the three layers, where

$$n_2 > n_1 \geqslant n_3$$

265

INJECTION LASER

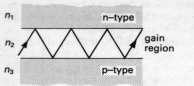

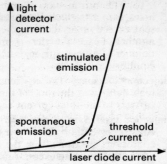

a Injection laser junction, functioning
as dielectric waveguide

b Light output as a function of
diode current

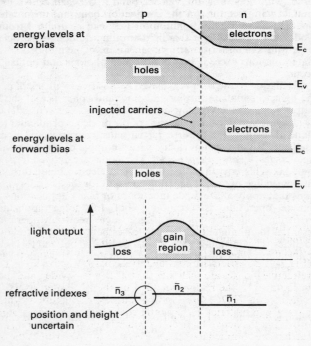

c Homojunction structure

266

The injected carrier concentration is a function of the current across the junction, which is related to the current in the external circuit. At values above the *threshold current* laser action occurs with high values of optical gain (Fig. *b*).

*Direct-gap semiconductors such as gallium arsenide allow direct transitions from the conduction to the valence bands and vice versa, and radiative recombination plays a major part in such transitions. These semiconductors are therefore suitable for production of injection lasers. Indirect-gap semiconductors such as silicon are not suitable because of the extra absorption losses in stimulating indirect transitions and it is virtually impossible to stimulate laser action in such materials.

Early injection lasers were *homostructure lasers*, i.e. a p-n *homojunction diode was used. Disadvantages of the homostructure were the very high threshold current required to stimulate laser action and the poor optical confinement, resulting in extra losses, due to the very small differences in refractive index between different layers within the crystal. The gain region was of the order of the diffusion length, and nonuniform gain across it was experienced because of the exponential decrease in injected electron concentration with distance from the junction (Fig. *c*).

Heterostructure lasers use a *heterojunction diode as the basis for laser action; the discontinuities in the energy bands at the junction prevent the injected electrons crossing back into the wider gap material (Fig. *d*). *Single heterostructure lasers* use a simple diode. The most common configuration is a *double heterostructure laser* in which a thin layer of narrow energy gap material is 'sandwiched' between layers of wide gap semiconductor (Fig. *e*). The injected carriers are confined to the middle layer, which can be made narrower than the diffusion length to give a more uniform gain across the active layer; the larger differences in refractive indexes between the different materials provide an efficient dielectric waveguide with good optical confinement.

A range of different materials and modifications to the basic structure have been produced to provide lasers of different wavelengths and to optimize the operation. *Separate confinement heterostructure (SCH) lasers* have five layers of semiconductor material (Fig. *f*); the carriers are confined to the narrow active layer, but the refractive indexes are such that the waveguide is wider, and the light is confined to the region bounded by the two outer layers. *Large optical cavity (LOC) lasers* have a four-layer structure in which the narrow gap material in the centre consists of a thin layer of p-type providing the active layer and a wider layer of n-type (Fig. *g*). The carriers are confined to the active layer of p-type whereas the optical confinement is across the width of both n- and p-layers.

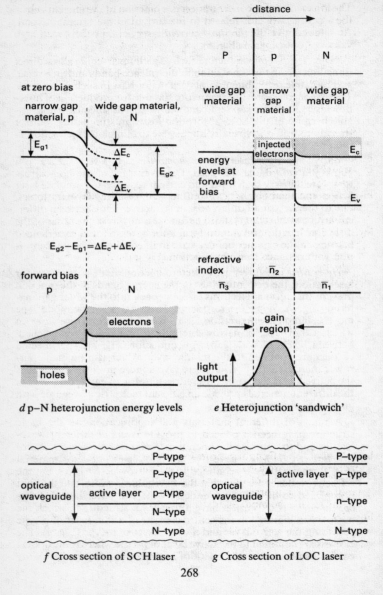

d p–N heterojunction energy levels

e Heterojunction 'sandwich'

f Cross section of SCH laser

g Cross section of LOC laser

268

Distributed feedback (DFB) lasers are constructed with a periodic variation of the refractive index within the optical waveguide, rather than cleaved ends to the cavity, in order to obtain the necessary feedback of the light. Such periodic variations are generally produced by corrugating the interface between the two dielectric layers.

Injection lasers differ from other types of laser in that the initiating excitation is due to minority carriers rather than light, and multimode light output is produced. Injection lasers are small, robust, cheap, and flexible and have an extensive range of applications, particularly in the fibre optics communications field, machine control systems, and compact disc systems.

injector *See* I²L.

in opposition *See* opposition.

in parallel *See* parallel.

in phase *See* phase.

in-phase component (1) (of a current or voltage) *See* active current; active voltage.

(2) (of the volt-amperes) *See* active volt-amperes.

input (1) The signal or driving force applied to a circuit, device, computer, machine, or other plant.

(2) The terminals at which this signal is applied.

(3) To apply as an input signal or driving force.

input impedance The *impedance of a circuit or device presented at its input.

Data bearing medium	Device	
	input	output
punched cards	card reader	card punch
paper tape	tape reader	tape punch
magnetic tape	magnetic tape unit	
characters on paper	optical character reader	line printer
graphs, diagrams	visual display unit	
		incremental plotter

input/output (I/O) Denoting those operations, devices, and data-bearing media that are used to pass information into or out of a *computer. Most forms of data-bearing medium are involved in both input and output processes and some in common use are shown in the table with the input and output devices.

in series *See* series.

insertion gain *See* insertion loss.

insertion loss The loss of power in a load that occurs when a network is inserted between the load and the generator supplying the load.

269

Insertion gain occurs when a gain rather than a loss of power results. The loss or gain is usually expressed as the ratio of the power delivered to the load after insertion of the network, to the power delivered to the load before insertion; it is measured in *nepers or *decibels. The value usually depends not only on the network parameters but also on the load and generator impedances.

instantaneous automatic gain control A fast-acting automatic gain control in *radar systems that reduces the *clutter by responding very rapidly to variations in the mean clutter level.

instantaneous carrying-current The peak value of current that may be carried instantaneously by a switch, circuit-breaker, or similar device at rated voltage under specific conditions.

instantaneous frequency The rate of change of phase of any oscillating electric variable, measured in radians per second divided by 2π. Particular applications are in *frequency and *phase modulation.

instantaneous power The value, at the output terminals of a circuit, of the rate at which power is transmitted from that circuit to the next portion of the system.

instantaneous sampling *See* sampling.

instantaneous value The value, at a particular moment, of any quantity that varies with time.

Institute of Electrical and Electronics Engineers, Inc. (IEEE) *See* standardization.

Institute of Electrical Engineers (IEE) *See* standardization.

instruction set The repertoire of arthmetic and logical operations of a computer. All programs exist in a computer as sequences of instructions drawn from the instruction set.

instrument damping *See* damped.

instrument rating The limits, designated by the manufacturer, between which an instrument can operate without damage. This is not necessarily the same as the *full-scale deflection.

instrument sensitivity The response of a measuring instrument expressed as the ratio of the magnitude of the physical response to the magnitude of the quantity measured. It is often stated directly as, for example, divisions per volt.

instrument shunt *See* shunt.

instrument transformer A *transformer that is used in conjunction with a measuring instrument. It utilizes the current-transformation (*current transformer*) or voltage-transformation (*voltage transformer*) properties of a transformer. The primary winding forms part of the main circuit and carries the current or voltage to be measured. The secondary winding is connected to the measuring instrument, such as an ammeter or voltmeter. Instrument transformers are used to extend the dynamic range of a.c. instruments and to isolate the instruments from circuits operating at high voltages.

270

insulate To support or surround a conductor with insulating material so that the current through it is confined to a desired path.

insulated-gate field-effect transistor (IGFET) *See* field-effect transistor.

insulated system A system of distribution of electric energy that is nowhere connected to *earth under normal conditions. Such a system has the advantage that a fault to earth does not prevent the system from being used, but the disadvantage that a fault condition can cause the build-up of static charges and give the user an unpleasant electric shock. In practice an earthed system is most often used.

insulating barrier A screen of insulating material that is fitted to certain electrical apparatus, such as switch gear or fuses. It prevents the formation of an arc between the apparatus and some other point, which may be an operator or another part of the apparatus, and thus prevents damage.

insulating resistance The resistance between two electrical conductors or systems of conductors that are normally insulated from each other. It is usually of the order of megohms but in the case of cables it is expressed as megohms per kilometre.

insulation Material that is used to insulate an electrical conductor. It is also the process of insulating a conductor.

insulator A material, such as glass or ceramic, that has a very high resistance to electric current so that the current flow through it is usually negligible. An insulator is used to prevent the loss of electric charge or current from a conductor. *See also* energy bands.

integrated circuit (IC) *Syn.* microcircuit. A complete circuit that is manufactured as a single package. A *hybrid integrated circuit* consists of several separate component parts attached to a ceramic substrate and interconnected either by wire bonds or a suitable metallization pattern. The individual parts are unencapsulated and may consist of diffused or thin-film components or one or more monolithic circuits. It is much smaller than a circuit made from discrete packaged components and once fabricated an individual component cannot be altered without destroying the entire circuit.

In a *monolithic integrated circuit* all the circuit components are manufactured into or on top of a single *chip of silicon. The individual parts of the circuits are not separable from the complete circuit once formed and some devices, such as *charge-coupled devices, cannot be made from discrete components. Interconnections between the various parts of the circuit are made by a pattern of conducting material on the surface of the circuit.

*Bipolar integrated circuits are based on bipolar junction *transistors and are formed using bipolar technology. *MOS integrated circuits are based on insulated-gate *field-effect transistors and *MOS capacitors. The development of self-aligned gate technology

271

and overlapping silicon-gate technology for MOS technology has allowed extremely complex MOS circuits to be produced.

The complexity of circuits that may be produced on a single chip of silicon is described by the numbers of parts that form the circuits: *small-scale integration* (SSI) describes fairly simple circuits; *medium-scale integration* (MSI) denotes circuits of medium complexity, such as a decade *scaler. More complex circuits are most often used in computers and can be described by the number of *bits that can be stored in a memory circuit of similar complexity: *large-scale integration* (LSI) denotes circuits of complexity up to 16 kilobits; *very large scale integration* (VLSI) describes circuits with a capability between 16 kilobits and one megabit; *extra large scale integration* (ELSI) refers to circuits containing more than one megabit.

integrated injection logic *See* I²L.

integrated Schottky logic (ISL) *Syn. for* Schottky transistor logic (STL). *See* I²L.

integrating array *See* solid-state camera.

integrating frequency meter *Syn.* master frequency meter. An instrument that allows a check to be made on a source of alternating voltage by integrating the number of cycles that occur in a specified time interval. Comparison with the number of cycles that should occur in the same time interval indicates whether the prescribed frequency has been maintained.

integrating wattmeter *See* watt-hour meter.

integrator A device that performs the mathematical operation of integration so that the output of the device is substantially the integral with respect to time of the input to it.

A *capacitance integrator* utilizes a capacitor, usually in series with a resistor, to perform integration. The voltage across the capacitor C when a direct current I flows into it is given by

$$V = (1/C)\int I dt$$

The capacitor thus integrates the current with respect to time.

intelligent terminal *See* terminal.

intelligibility A function of a speech communication system, such as a telephone system, in which it is more important for the received information to be intelligible than for the input to be flawlessly reproduced. The intelligibility is measured by the *syllable articulation score,* which is the percentage of correctly received monosyllabic nonsense words uttered in an uncorrelated sequence.

intelligible crosstalk *See* crosstalk.

Intelsat Acronym from *In*ternational *Tele*communications *Sat*ellite Consortium. *See* communications satellite.

intensity *Short for* magnetic intensity or electric intensity (both obsolete terms). *See* magnetic field strength; electric field strength.

intensity modulation *Syn.* z-modulation. The variation in brilliance of the spot on the screen of a cathode-ray tube in accordance with the magnitude of an input signal.

intensity of magnetization *See* magnetization.

interaction space A region in an electron tube that roughly corresponds to the interelectrode space and in which the electrons interact with an alternating magnetic field.

interactive Allowing continuous two-way communication between the user of an *on-line peripheral device, such as a *terminal, and a *computer. Interactive operation enables a user at a remote location to send and receive information to and from a computer quickly, and to modify the operation of a *program during its execution following the production of intermediate results or interrogation. *See also* time sharing. *Compare* batch processing; real-time operation.

interactive videotext *See* videotext.

interchangeability (of electronic components and devices) *See* standardization.

interconnecting feeder *See* trunk feeder.

interconnection (1) Any method of providing an electrical path between any of the materials (metals, semiconductors, etc.) that combine to form a circuit.

(2) Connections between and external to any functional item that forms a circuit or system of circuits. Functional items include component parts, devices, subassemblies, and assemblies.

Compare intraconnection.

interconnector *See* trunk feeder.

interdigital magnetron *See* magnetron.

interdigitated capacitor *See* monolithic capacitor.

interelectrode capacitance The capacitance between specified electrodes of an electronic device (such as the base and emitter of a transistor) that may form a small capacitor within the device. The operation of such devices can be significantly affected by the existence of interelectrode capacitances.

interference A disturbance to the signal in any communication system caused by unwanted signals. A common cause of interference in radio reception is the operation of electrical machinery and apparatus, particularly commutating machines and apparatus containing gas-discharge tubes. Television signals frequently suffer serious interference from motor-vehicle ignition systems.

Man-made interference such as that described above can usually be eliminated by fitting special devices (*suppressors*) to the offending apparatus, but interference arising from natural causes, such as changes in the atmosphere, is not easily prevented. *See* crosstalk; heterodyne interference; hum; image frequency.

interference fading *See* fading.

273

interlaced scanning *See* television.

interline transfer device *See* solid-state camera.

interlock A safety device that allows a piece of apparatus to function only when predetermined conditions are fulfilled.

intermediate frequency (IF) *See* superheterodyne reception.

intermediate-frequency amplifier *See* superheterodyne reception.

intermediate-frequency harmonic interference *Syn. for* image interference. *See* image frequency.

intermittent duty *See* duty.

intermodulation *See* modulation.

intermodulation distortion *Syn.* combination-tone distortion. *See* distortion.

internal discharge *See* ionization discharge.

internal memory *See* memory.

internal photoelectric effect *See* photoconductivity.

internal resistance Symbol: r; unit: ohm. A small resistance possessed by a cell, accumulator, or dynamo. It is given by

$$r = (E - V)/I$$

where E is the e.m.f. generated by the device, V is the potential difference across the terminals, and I is the current. *See also* cell.

international ampere Symbol: A_{int}. The former standard of electric current defined as the constant current that, when passed through an aqueous solution of silver nitrate, will deposit silver at the rate of 0·001 118 00 grams per second. It was replaced (1948) as the standard unit by the *ampere.

$$1\ A_{int} = 0{\cdot}999\ 85\ \text{abamperes}$$
$$1\ A_{int} = 9{\cdot}9985\ \text{amperes}$$

International Electrotechnical Commission (IEC) *See* standardization.

international ohm *See* ohm.

International Radio Consultative Committee (CCIR) *See* standardization.

International Standards Organization (ISO) *See* standardization.

international system A former system of units that expressed the values of electrical quantities in terms of the *international ampere, international *ohm, centimetre, and second. It has now been replaced by *SI units. *See also* CGS system.

International Telecommunication Union (ITU) *See* standardization.

International Telegraph and Telephone Consultative Committee (CCITT) *See* standardization.

international units *See* international system; CGS system.

international volt *See* volt.

interrogating signal *See* transponder.

interrupter A device, such as an *induction coil, that periodically interrupts a continuous current. *See also* make-and-break.

interstage coupling *Coupling between successive stages of a multistage amplifier employing several amplifying stages in cascade. The type of coupling chosen (direct, resistive, etc.) depends on the particular design of amplifier stage used.

intraconnection An electrical connection inseparably associated with circuit elements within a component part. *Compare* interconnection.

intrinsic conductivity The conductivity of a *semiconductor associated with the semiconductor material itself and not contributed by impurities. At any given temperature, equal numbers of electrons and holes are thermally generated and these give rise to the intrinsic conductivity. The numbers of charge carriers are dependent on temperature and the conductivity, σ, is a function of temperature:

$$\sigma = A \exp(E_g / 2kT)$$

where A is a constant that depends on the particular material, E_g is the band-gap energy, k is the Boltzmann constant, and T the thermodynamic temperature.

In an extrinsic semiconductor, the intrinsic conductivity is usually negligible compared to the extrinsic conductivity at normal working temperatures. At sufficiently high temperatures however the numbers of thermally generated carriers become much larger than those contributed by impurities and the semiconductor becomes intrinsic (*see* intrinsic temperature range). In practice, at room temperature, the intrinsic carrier concentration in silicon is about 10^{10} per cm^3. The minimum impurity carrier concentration obtainable in silicon is about 10^{14} per cm^3, i.e. the extrinsic carrier concentration is about 10^4 times greater than the intrinsic value.

intrinsic crystal *Syn.* idiochromatic crystal. A crystal that exhibits photoelectric properties in its pure state.

intrinsic mobility The mobility of charge *carriers in an intrinsic semiconductor. The mobility of electrons is approximately three times that of holes.

intrinsic photoconductivity *See* photoconductivity.

intrinsic semiconductor *Syn.* i-type semiconductor. A pure *semiconductor that has equal concentrations of *electrons and *holes under conditions of thermal equilibrium. Absolute purity is unobtainable in practice and nearly pure materials are termed intrinsic. *See also* semiconductor; intrinsic conductivity; intrinsic temperature range. *Compare* extrinsic semiconductor.

intrinsic stand-off ratio *See* unijunction transistor.

intrinsic temperature range The temperature range in which the electrical properties of a semiconductor are not essentially modified by impurities in the crystal. For a pure sample of *intrinsic semiconductor, the intrinsic temperature range is the whole range of working temperatures. For an extrinsic semiconductor at most temperatures, the impurities contribute most of the charge carriers; at sufficiently high temperatures however the intrinsic carrier concentration rises to such a level that the extrinsic conductivity becomes negligible compared to the *intrinsic conductivity and at that temperature the semiconductor becomes intrinsic.

inverse direction *See* reverse direction.

inverse feedback *Syn. for* negative feedback. *See* feedback.

inverse gain The *gain of a bipolar junction *transistor when it is connected in reverse, that is, so that the *emitter acts as the *collector and the collector as the emitter. The inverse gain is usually less than the gain normally obtained, due to the higher doping level of the emitter compared with the collector. This causes the emitter to have a higher *injection efficiency into the base than has the collector.

inverse limiter *Syn. for* base limiter. *See* limiter.

inverse photoelectric effect *See* photoelectric effect.

inversion The production of a layer of opposite polarity at the surface of a semiconductor under the influence of an electric field, usually an applied one. Spontaneous inversion can occur in the surface of p-type material when it is in contact with an insulating layer due to the presence of positive ions in the insulator. Inversion can only occur when sufficient mobile minority carriers are present in the semiconductor material, otherwise a depletion layer forms. The phenomenon is utilized for formation of the channel of an insulated-gate *field-effect transistor. *See also* MOS capacitor.

inverter (1) A device that converts direct current into alternating current. A common type of inverter is a rotating machine designed for the purpose.

(2) *Syn.* linear inverter. An amplifier that reverses the polarity of a signal, i.e. that introduces a 180° phase shift.

(3) *Syns.* digital inverter; NOT circuit. A *logic circuit that inverts the level of the input signal, i.e. that produces a low output for a high input and vice versa.

inverting transistor A bipolar or MOS transistor used as an analog or digital *inverter. Operation as an analog inverter results from the 180° phase shift introduced by both types of transistor between the input and output. A digital inverter may be formed with a bipolar transistor by applying the input to the base. A high input causes saturation (*see* transistor) with a resulting drop in the collector (output) voltage. With a low input level the saturation ceases and the output therefore rises. An MOS *field-effect transistor is used as a digital inverter essentially by operating it as a switch that connects the output to a point of low potential difference (usually earth potential). A high voltage level input to the gate causes the conducting channel to form across the transistor and therefore the drain (output) potential falls to the low value of the source. A low input level, which must be less than the threshold voltage of the transistor, causes the drain to be isolated from the source. Since the drain is connected to the power supply by means of a dropping resistor the drain potential then rises to the high level.

I/O *Abbrev. for* input/output.

ion An atom, molecule, or group of atoms or molecules that has an electric charge. Negative ions (*anions) contain more electrons than are necessary for electrical neutrality of the atom or group; positive ions (*cations) contain fewer. *See* electrolysis.

ion-beam analysis A method of analysing the surface of a material, such as a thin film or a semiconductor, using a beam of ions. Various techniques have been developed. *See* elastic recoil detection analysis; Rutherford back scattering; soft X-ray appearance spectroscopy.

ion-beam lithography A method of *lithography similar to *electron-beam lithography except that the electron beam is replaced by a beam of ions. The ions are heavier than the electrons used in e-beam lithography and therefore suffer far less scattering within the resist and produce very few low-energy secondary electrons. They can also supply more energy to the resist, which results in greater resist sensitivity and reduced writing time. There are difficulties in producing collimated ion beams and with the effect of the ions on the semiconductor substrate below the resist layer. *Multilevel resists are needed to eliminate the latter effect.

ion burn A defect of the phosphor coating on the screen of a *cathode-ray tube that is caused by bombardment by heavy negative ions present in the beam. Permanent localized deactivation of the phosphor results and appears as a dark area or a rash of darker areas in the image. The use of an *ion trap in the tube can minimize this problem. *Compare* ion spot.

ion density The number of pairs of positive and negative ions per unit volume.

ionic atmosphere The atmosphere, consisting of ions, that surrounds an individual ion in an electrolyte. In the absence of an electric field each anion is surrounded by a symmetrical accumulation of cations and vice versa. When an electric field is applied the anions migrate to the anode and the cations to the cathode. The symmetry of the ionic atmosphere is therefore disturbed with respect to the surrounded ion (*see* diagram) and the ions experience a retarding force due to the migration of the opposite-polarity ionic atmosphere in the opposite direction.

If high-frequency alternating current or a fast-pulse direct current is applied to the electrolyte, the symmetry is little disturbed and a higher conductivity of electrolyte results.

No applied field	Applied field

Ionic atmosphere

ionic bond *Syn. for* electrovalent bond. *See* valency.

ionic conduction The movement of charges through a *semiconductor due to the displacement of ions within the crystal lattice, such movement being maintained by a continuous supply of external energy.

ionic crystal *Syn.* electrovalent crystal. A crystal that is in the form of an array of positively and negatively charged ions, the interatomic forces being of the Coulomb type. The Coulomb attraction between the ions is balanced by the repulsion experienced when the outer electronic shells of each ion approach too closely.

ionic heating Heating of a surface by ion bombardment. Ionic heating is sometimes used for heating the *cathode of a thermionic valve.

ionic mobility The velocity of an ion when moving under the influence of unit electric field. It is measured in metres per second per volt per metre or $m^2 s^{-1} V^{-1}$.

ionic semiconductor A type of semiconductor in which the movement of *ions through the material contributes more to the conductivity than the movement of mobile charge carriers (*electrons and *holes).

ion implantation The technique of implanting *ions into the lattice of a semiconductor crystal by bombarding the surface with ions under controlled conditions. The implanted ions come to rest at random locations within the crystal lattice and cause a significant amount of damage to the lattice. It is therefore necessary to activate the implant, i.e. to follow the implantation with high-temperature annealing to remove the damage to the lattice and allow the implanted atoms to move to occupy lattice sites.

Ion implantation can be reliably performed with high uniformity, and because doping can be performed locally by selective masking it offers great flexibility. The technique is used in the manufacture of integrated circuits and solid-state components, such as transistors, and may be used as an alternative to *diffusion or in conjunction with it.

ionization Any process in which *ions are formed from neutral atoms. Ions are formed spontaneously when an electrolyte dissolves in a suitable solvent. The action of ionizing radiation (X-rays, alpha, beta, or gamma rays, fast electrons, etc.) is required for ions to be formed in a gas.

ionization chamber A *gas-filled radiation-detection tube that is used to detect ionizing radiation. The ionization chamber is a very versatile radiation detector since it can be used to detect and measure a wide range of energies and intensities of radiation.

A typical chamber has a sensitive volume defined using a *guard ring, contained within it, and two electrodes with a voltage applied across them. Parallel-plate chambers consist of two parallel-plate electrodes with the sensitive volume contained between them. Cylindrical ionization chambers have a central electrode in the form of a thin wire surrounded by and coaxial with a circular electrode, with the sensitive volume bounded by the outer electrode. When the gas is ionized by a beam of radiation the ions migrate to the electrodes under the influence of the applied voltage.

In conjunction with a suitable external circuit the ionization chamber can be used as a counter – an *ionization counter* – to count particles, such as alpha or beta particles, a pulse of current being produced by each particle of radiation. The most common application however is as a continuous measuring instrument in which an ionizing current is produced and the size of the current is proportional to the intensity of the ionizing radiation.

The sensitivity of the instrument is proportional to the mass of gas enclosed within the sensitive volume, and the size of instrument depends on the intensity of radiation to be measured. Extremely large chambers have been produced for measuring background radiation levels and very small chambers are used to calibrate high-output beams of X-rays or electrons.

ionization counter *See* ionization chamber.

ionization current Current that results from the movement of *ions through a conducting medium under the influence of an electric field.

ionization discharge *Syn.* internal discharge. Partial discharge of a capacitor caused by the ionization of gas bubbles in the dielectric material. Ionization discharge is a major cause of failure of capacitors used with alternating currents above about 200 volts. Such discharges rarely occur with direct currents.

ionization gauge A pressure gauge that is used to measure extremely low gas pressures down to 10^{-8} mmHg. It consists of a high-vacuum three-electrode thermionic valve fused to the gas system (*see* diagram). Electrons leaving the cathode are accelerated to the grid but fail to reach the anode because of its negative potential. Any gas molecules in the system are ionized by collision with some of the electrons that pass through the grid; the positive ions thus formed migrate to the anode producing an output current that is a function

279

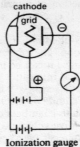

cathode

grid

⊖

⊕

Ionization gauge

of the number of gas molecules present and hence of the gas pressure.

ionization potential *Syns.* electron binding energy; radiation potential. Symbol: I; unit: electronvolt. The minimum energy required to remove an electron from a given atom or molecule to infinity. Originally the ionization potential was defined as the minimum potential through which an electron would fall in order to ionize the atom, and was measured in volts.

The electron usually removed from the atom or molecule is in the outermost orbit, i.e. the least strongly bound electron. Some atoms and molecules may be ionized by the removal of an electron that is not in the outermost orbit, i.e. one that is more strongly bound. The resulting ion will be in an excited state. The ionization potential for removal of the least strongly bound electron is designated the *first ionization potential* (I_1) and results in an ion in the ground state. Ionization by the removal of the second (or subsequent) least strongly bound electron requires a greater ionization potential, termed the second (or third, etc.) ionization potential (I_2, I_3, etc.).

ionizing event Any physical process that produces an ion or group of ions. The physical agent involved in the process, such as a charged particle passing through a gas, is termed an *ionizing agent.*

ionizing radiation Any radiation, such as streams of energetic charged particles (electrons, protons, alpha particles, etc.) or energetic ultraviolet radiation, X-rays, or gamma rays, that produces ionization or excitation of the medium through which it passes.

ion microprobe *See* secondary-ion mass spectroscopy.

ion milling *See* etching.

ionosphere A spherical shell of ionized air that surrounds the earth, extending from about 50 km to over 1000 km, and is used to reflect radiowaves in order to obtain long-distance radio transmission between points on the earth's surface.

The ionosphere consists of several distinct layers or regions that can change in thickness between day and night and also show seasonal and latitude variations. The lowest layer is the *C-layer,* which lies approximately 40 to 60 km above the earth's surface. The *D-layer,* extending from about 60 to 90 km, has a relatively low concentration of electrons and reflects low-frequency radiowaves. The *E-layer* extends from about 90 to 150 km, has a higher concentration of electrons than the D-layer, and reflects medium-frequency radiowaves.

The highest layer is the *F-layer,* which lies approximately from 150 to over 1000 km above the earth. This layer splits during the day into the F_1-*layer* (lower) and F_2-*layer* (higher). It contains the highest concentration of free electrons and reflects high-frequency radiowaves. At night the D- and E-layers become relatively inactive since there is no solar radiation to regenerate ion pairs lost by recombination. The F-layer has a lower density and hence a lower recombination rate of ions and can therefore be used for radio transmission at all times. It is thus the most useful region for long-range radio communication. There is a well-marked ionization maximum for the F-layer, the region above the maximum being the *topside ionosphere.* Radiowaves deflected by the ionosphere are termed *ionospheric waves.*

Waves with wavelengths between about six millimetres and 20 metres lie within the *radio window and are not reflected by the ionosphere but pass straight through it. High-frequency *television transmissions fall within this band and require *communications satellites, usually in geostationary orbit, in order to achieve long-distance television links. Radioastronomy is restricted to wavelengths within the radio window.

ionospheric defocusing *See* ionospheric focusing.

ionospheric focusing A process that results in an enhancement of the field strength at a given receiver due to the focusing that arises from either small-scale or large-scale curvature of the ionospheric layers. *Ionospheric defocusing* results from a reduction of the field strength at a receiver due to the defocusing that arises from the curvature of the layers.

ionospheric wave *Syn.* sky wave. *See* ionosphere.

ion source A device that provides ions, particularly for use in particle *accelerators. A common type consists of a minute jet of a suitable gas that is subjected to electron bombardment. The resulting ions (protons, alpha particles, etc.) are injected into the accelerator.

ion spot (1) A darker area appearing on the screen of a *cathode-ray tube and resulting from decreased luminescence of the screen (not necessarily of the phosphor) after bombardment by heavy negative ions in the beam. Use of an *ion trap can minimize this effect. *Compare* ion burn.

(2) A spurious signal appearing in the output of *camera tubes and *image converters due to alteration of the charge pattern on the target or cathode by ion bombardment.

ion trap A device used in *cathode-ray tubes that attracts heavy ions present in the electron beam and thus prevents them impinging on the phosphor coating of the screen and causing blemishes.

iris *See* waveguide.

I²R loss *Syn.* copper loss. *See* dissipation; heating effect of a current.

iron Symbol: Fe. A metal, atomic number 26, that exhibits *ferromagnetism and has a high tensile strength. It is widely used in electronics as a magnetic material and for screening purposes. It is relatively abundant and therefore has a low cost.

iron loss *See* core loss.

irradiation The exposure of a body or substance to electromagnetic or corpuscular *ionizing radiation.

ISL *Abbrev. for* integrated Schottky logic, *syn. for* Schottky transistor logic. *See* I²L.

island effect An effect, occurring in *thermionic valves when the grid voltage falls below a certain minimum value, in which the emission from the cathode is restricted to a few small areas of the cathode.

ISO *Abbrev. for* International Standards Organization. *See* standardization.

isochronous cyclotron A form of *cyclotron that takes account of the relativistic mass change of the accelerated particles. Above a particular energy (about 15 MeV/proton) the relativistic mass change causes the orbital time (Be/mc) of the particles to fall out of synchronism with the accelerating alternating voltage. This is overcome by causing the magnetic flux density, B, to increase with radius in an appropriate manner so as to balance the increase in the mass, m, and cause the circulation time to remain constant and in synchronism with the accelerating voltage. *See also* synchrocyclotron.

isoelectronic Denoting groups of atoms that exhibit similar electronic properties, such as similar distributions of electrons in the outer orbits.

isolating Disconnecting a circuit or device from an electric supply system, usually by making a circuit open at a time when it carries no current.

isolating transformer A transformer that is used to isolate any circuit or device from its power supply; thus the circuit derives power from the source without a continuous wire connection between them.

isolation diode (1) A *diode that is used in a circuit to allow signals to pass in one direction but to block those in the other direction and therefore prevent damage from surges in the reverse direction.

(2) The diodes formed by the collector-substrate junctions in *bipolar integrated circuits. In order to maintain isolation between parts of the integrated circuits these junctions must always be re-

verse biased. This is achieved by maintaining the potential of the substrate so that none of the diodes becomes forward biased.

Similar diodes are formed in *MOS integrated circuits by the source-substrate junctions, drain-substrate junctions, and, once formed, the channel-substrate junctions. The substrate material must be held at a suitable potential to maintain the isolation between parts of the circuit, as with bipolar circuits.

isolator A device, usually made of a ferrite, that allows microwave energy to pass in one direction with little loss but absorbs power in the reverse direction.

Isoplanar process *Tradename. See* coplanar process.

isotopes Two or more nuclides that have the same *atomic number but differ in nuclear mass due to different numbers of neutrons in the nucleus. Each nuclide is said to be an isotope of the element defined by the given atomic number. The isotopes of a given element are almost identical chemically but have different physical properties. Several different naturally occurring isotopes of most of the elements have been discovered; high-speed bombardment of suitable material by corpuscular radiation is frequently used to prepare artificial radioactive isotopes.

isotropic Denoting a substance in which physical properties, such as magnetic susceptibility, are the same in all directions.

iterative impedance The impedance that when connected to one pair of terminals of a quadripole produces a like impedance at the other pair of terminals. In general a quadripole has two iterative impedances, one for each pair of terminals. If the two iterative impedances are equal, their common value is termed the *characteristic impedance* of the network. *Compare* image impedance.

ITU *Abbrev. for* International Telecommunication Union. *See* standardization.

i-type semiconductor *See* intrinsic semiconductor.

jack plug and socket A type of plug-and-socket connector used when rapid and easy connections between circuits or devices are required. Insertion or removal of the plug can cause one or more switching functions to occur, such as the breaking of a short circuit. The basic construction is shown in the diagram. The plug slides into the sprung socket and is correctly located by means of a groove running round the plug. The contacts are arranged linearly along the length of the plug and socket and are insulated from each other. Two or more contacts may be used. The end of the plug forms one contact and the others lie along the length of the metal casing. Contact is made by means of a central wire or conducting rod. More than one rod can be used if more than two contacts are required, the rods being insulated from the casing and from each other except at the connection point.

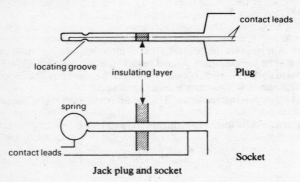

Jack plug and socket

Any kind of wire or cable may be used with these plugs and sockets, although the characteristic impedance of about 600 ohms is not suitable for joining two lengths of coaxial cable. The most common applications of these connectors have been for *patching in manually operated telephone exchanges and in stereo headphones.

jammer *Syn.* jamming transmitter. *See* jamming.

jamming Deliberate interference in communications and radar by means of unwanted signals that are intended to render unintelligible or to falsify the whole or part of the desired signal. A *jammer* is used to produce the jamming signal. Reducing the effects of jamming is termed *antijamming*.

Jansky noise *Syn. for* galactic noise. *See* radio noise.

JEDEC *Abbrev. for* Joint Electron Device Engineering Council. *See* standardization.

JFET *Abbrev. for* junction field-effect transistor. *See* field-effect transistor.

jitter Short-term instabilities in either the amplitude or phase of a signal, particularly the signal on a *cathode-ray tube. It has the effect of causing momentary displacements of the image on the screen, giving it a shaky or 'jittery' appearance. An oscilloscope that is used to measure the amount of jitter is termed a *jitter scope*. Momentary errors of synchronization between the scanner and receiver in *television or *facsimile transmission can cause jitter of the received images. This type of jitter is known as *jitters*.

In hi-fi sound reproduction systems that use a digital recording method, jitter can cause unwanted audible variations in the pitch of the sound output. These can be quite unpleasant. In *compact disc systems, for example, a noise similar to ignition interference is heard. Jitter in digital recording is the equivalent of *wow and *flutter in a system using analog recording.

See also pulse.

J-K flip-flop *See* flip-flop.

Johnson–Lark–Harowitz effect The change in resistivity of a metal or a *degenerate semiconductor due to scattering of the charge carriers by impurity atoms.

Johnson noise *Syn. for* thermal noise. *See* noise.

Johnson–Rahbeck effect An effect observed when a high potential difference of about 200 volts is applied across two plates, one of metal and one of a semiconductor material, such as slate or agate. The two plates are held together as a result of the applied voltage.

Josephson effect An effect that occurs when a sufficiently thin layer of insulating material is introduced into a superconducting material (*see* superconductivity). A superconducting current can flow across the junction, known as a *Josephson junction,* in the absence of an applied voltage. This is the *direct-current Josephson effect*. If the value of the current exceeds a critical value, I_c, determined by the properties of the insulating barrier, current can only flow when a finite voltage is applied. The current-voltage characteristic is shown in the diagram, in which the dashed curve is the current-voltage characteristic in the nonsuperconducting state.

The *alternating-current Josephson effect* occurs when a small direct voltage, V, is applied across a Josephson junction. The superconducting current across the junction becomes an alternating current given by

$$I_s = I_c \sin\omega t$$

where

$$\omega = 2\pi\nu = 2e/hV$$

h is the Planck constant, ν the frequency, and e the electron charge.

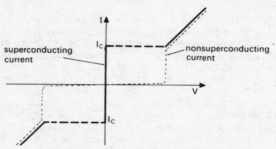

I–V characteristic of Josephson junction

The direct-current Josephson effect is utilized in several devices, particularly the *Josephson memory. The alternating-current Josephson effect is utilized for radiofrequency detection, the determination of h/e, for accurate measurement of frequency, and as a monitor of voltage changes in standard cells or for the comparison of cells at different Standards Laboratories.

Josephson junction *See* Josephson effect.

Josephson memory A cryogenic *memory that consists of an array of Josephson cells, i.e. memory cells containing *Josephson junctions, held at a temperature very close to critical temperature. In the absence of an external magnetic field the cell is superconducting but the presence of a magnetic field destroys the *superconductivity and hence the voltage across the device changes. Information is stored in the form of local variations of a magnetic field; the data is sensed by the voltage across the appropriate cell.

Josephson memories are inherently extremely fast in operation but because of the cryogenic requirements are extremely expensive to operate. The development of materials that exhibit superconductivity at higher temperatures is reducing the difficulties of using this type of memory.

Joshi effect Variations of the current in a gas discharge due to the effect of light on the gas.

joule Symbol: J. The *SI unit of energy, including work and quantity of heat. It is defined as the work done when the point of application of a force of one newton is displaced one metre in the direction of the force. One joule equals one watt-second. *See also* kilowatt-hour.

Joule effect *See* heating effect of a current.

Joule magnetostriction *Syn.* positive magnetostriction. *See* magnetostriction.

JUGFET *Abbrev. for* junction field-effect transistor. *See* field-effect transistor.

jump A transition of an orbital electron from one atomic energy level to another.

jumper A direct electrical connection between two points in a printed circuit that is not a part of the interconnection pattern of the circuit.

junction (1) A contact between two different conducting materials such as two metals, as in a rectifier or thermocouple.

(2) A boundary between two semiconducting regions of differing electrical properties. *See* p-n junction.

(3) A connection formed between two or more conductors of the same type or between sections of transmission lines.

junction box An enclosed distribution panel for the semipermanent connection or branching of electrical circuits.

junction coupling Coupling of a cavity resonator to a coaxial line by direct connection to the coaxial conductor.

junction field-effect transistor (JUGFET, JFET) *See* field-effect transistor.

junction transistor *Short for* bipolar junction transistor. *See* transistor.

K

K band A band of microwave frequencies ranging from 10·9 to 36·0 gigahertz. *See* frequency band.

keep-alive arc *Syn. for* holding anode. *See* mercury-arc rectifier.

keep-alive circuit *See* transmit-receive switch.

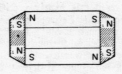

Keepers

keeper *Syn.* armature. A piece of ferromagnetic material, usually soft iron, that is placed across the extremities of one or more permanent magnets when not in use (*see* diagram). The keeper completes the magnetic circuit and neutralizes the demagnetizing effect in the magnet.

Kell factor *See* television.

kelvin Symbol: K. The basic *SI unit of *thermodynamic temperature, defined as 1/273·16 of the thermodynamic temperature of the triple point of water, i.e. the equilibrium point at which pure ice, air-free water, and water vapour can coexist in a sealed vacuum flask. It is related to the *degree Celsius* (°C):

$$t \, °C = T - 273·15 \text{ kelvin}$$

where T is the thermodynamic temperature and 273·15 K is the temperature of the ice point, which is the zero of the Celsius (or centigrade) temperature scale.

The kelvin is also used as a unit of temperature difference, in which case it is identical to the degree Celsius.

Kelvin balance *Syns.* current balance; ampere balance. A type of instrument in which the electromagnetic forces resulting from the passage of a current through a set of coils are balanced against the force of gravity. In one form two coils B, E at each end of a balanced rod are suspended between four fixed coils, A, F and C, D (*see* diagram). When current flows through all six coils, i.e. in the order A, B, C, D, E, F, the resulting electromagnetic forces cause each fixed

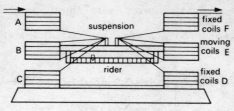

Kelvin balance

coil to displace the balanced arm in the same direction. The arm is rebalanced by moving a rider along a scale on the arm, calibrated to give a reading of the current. If the current is reversed the direction of the displacement is unaltered since the current flow is reversed in all the coils. Thus both direct and alternating currents can be measured.

Kelvin contacts A means for testing or making measurements in electronic circuits or components, particularly when low values are being measured. Two sets of leads are used to each test point, similar with respect to thickness, material, and length; one set carries the test signal and the second set connects with the measuring instrument used. The effect of resistance in the leads is thus eliminated from the measurement.

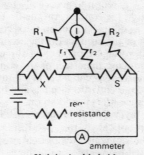

Kelvin double bridge

Kelvin double bridge *Syn.* Thomson bridge. A bridge with six arms that has been developed from the *Wheatstone bridge for measuring low resistances. The bridge is connected as shown in the diagram: X is the low resistance to be measured and S is a standard low resistance. Two sets of variable resistances, R_1, R_2 and r_1, r_2 are varied together until balance is achieved. At balance:

$$R_1/R_2 = r_1/r_2 = X/S$$

289

The method eliminates errors due to contact resistance and the resistance of the leads.

Kelvin effect *Syn.* Thomson effect. *See* thermoelectric effects.

Kelvin–Varley slide

Kelvin–Varley slide A device used in vernier potentiometers to reduce the effect of contact resistance. The device consists essentially of two or more sets of slide wires or resistance coils in cascade, each coil acting as a decade voltage divider for the preceding one (*see* diagram). If the total resistance of the first coil is $11R$, the resistance of the second coil is made $2R$ and it is connected to the first by a pair of sliding contacts that move together and shunt $2R$ of the first coil. The total resistance of the shunt and the shunted portion of the first coil is R, thus the first coil is effectively divided into 10 equal resistances. The second coil therefore acts as a vernier scale.

When a voltage V is required accurately the approximate voltage is set up by adjusting the contacts on the first coil, fine adjustment being achieved by a sliding contact on the second coil. A small error on the positioning of the contact on the second coil has much less effect on the total voltage tapped than the equivalent error on a single coil, thus reducing the effect of the contact resistance of the slider. Further subdivisions can be effected in a similar manner, by subdividing the second coil into 11 parts and bridging with a further coil of total resistance $4R/11$, and so on.

Kennelly–Heaviside layer *Syn. for* E-layer. *See* ionosphere.

Kerr cell *Syn.* electro-optical shutter. *See* Kerr effects.

Kerr effects Two effects in which the optical properties of transparent material are affected by electric or magnetic fields.

The *electro-optical effect* is the effect whereby the direction of polarization of plane-polarized light through a refractive medium is rotated by an electric field applied perpendicularly to the direction of propagation of the light. The *Kerr cell* utilizes this effect. It consists of two parallel plates immersed in a liquid that exhibits a marked Kerr effect. Polarized light passing through the cell can be interrupted by the application of an electric field. *Pockel's effect* is the Kerr effect when it occurs in a piezoelectric material. Pockel's effect can be used for the measurement of distance by a *mekometer*.

Such an instrument can measure distance to an accuracy of 0·05 millimetres in 50 metres.

The *magneto-optical effect* occurs when plane-polarized light is reflected from a highly polished pole face of a strong electromagnet. Slight elliptical polarization of the light beam is produced.

keyboard transmitter *See* teleprinter.

key punch *See* punched card.

keystone distortion *See* distortion.

kilo- Symbol: k. (1) A prefix to a unit, denoting a multiple of 10^3 (i.e. 1000) of that unit: one kilometre equals 10^3 metres.

(2) A prefix used in computing to denote a multiple of 2^{10} (i.e. 1024): one kilobyte equals 2^{10} bytes.

kilogram (or **kilogramme**) Symbol: kg. The *SI unit of mass equal to the mass of the international prototype of the kilogram, which is a piece of platinum-iridium kept at Sèvres, France.

kinetically assisted chemical reaction *See* etching.

kilowatt-hour Symbol: kWh. A unit of work or energy used commercially and defined as the energy produced when one kilowatt of power is expended for one hour. One kWh equals $3·6 \times 10^6$ joules.

Kirchhoff's laws (i) At any point in an electric circuit the algebraic sum of the currents meeting at that point is zero.

(ii) In any closed electric circuit the algebraic sum of the products of current and resistance in each part of the network is equal to the algebraic sum of the electromotive forces in the circuit.

Kirk effect An effect observed in epitaxial power *transistors designed for microwave applications. At high injection levels the current is sufficiently high so that the injected minority carrier concentration in the base becomes comparable to or greater than the base doping concentration. At these high injection levels, the Kirk effect occurs such that the high field region is relocated from the base-epitaxial layer transition region to the epitaxial-substrate interface. The effective width of the base is therefore increased from W_B to ($W_B + W_C$) where W_C is the width of the epitaxial layer. *See* epitaxy.

klydonograph A type of surge-voltage recorder that consists essentially of a photographic film or plate placed between two electrodes. One electrode is earthed and the other is connected to the conductor being investigated. When a voltage surge, such as that due to lightning, passes along the conductor the potential difference established between the electrodes affects the photographic emulsion and a record of the characteristics of the surge is obtained upon development. The photographic surface may be a fixed plate or a moving film. The latter provides a continuous record.

klystron An *electron tube that is used as a microwave amplifier or oscillator. It is a linear-beam *microwave tube in which *velocity modulation is applied to an electron beam in order to produce amplification of a microwave-frequency field.

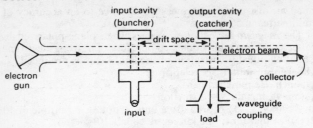

a Two-cavity klystron amplifier

Several variations of the basic klystron exist. A simple two-cavity klystron is shown in Fig. *a*. A beam of high-energy electrons produced from an electron gun is passed through a *cavity resonator excited by high-frequency radiowaves. The interaction between the high-frequency waves and the electron beam produces velocity modulation of the beam. The modulated beam leaving the cavity resonator (the *buncher*) traverses a field-free region (the *drift space*) where bunching occurs as the faster electrons catch up with the slower ones. The periodic current-density variations in the beam due to the formation of the bunches are of the same frequency as the exciting radiowaves. The beam then passes through a second cavity resonator (the *catcher*) placed at a distance x away, where the current-density variations produce a voltage wave in the catcher, which is tuned to the exciting frequency or a harmonic of it.

The magnitude of the output waves depends on the velocity of the electrons; the phase is such that the negative maximum corresponds to the centre of the bunch. Most of the energy of the beam is given to the catcher since many more electrons are retarded by the induced field than are accelerated by it. Voltage amplification is obtained by conversion of the d.c. energy of the original beam into radiofrequency energy in the output circuit.

It can be shown that the optimum condition for power extraction from the beam occurs when

$$\omega t = 2\pi(n + \tfrac{3}{4})$$

where ω is the angular frequency, t the transit time between the two resonators, and n is an integer known as the *mode number*. Since $t = x/v_0$, where v_0 is the initial electron velocity, the transit time may be altered by adjusting the voltage of the electron gun. A collector electrode is used to collect that part of the electron beam leaving the second cavity. Two-cavity klystrons can be made to oscillate if positive feedback to the input cavity is employed.

The most important type of klystron is the *reflex klystron,* used as a low-power oscillator. This type of klystron has only one cavity,

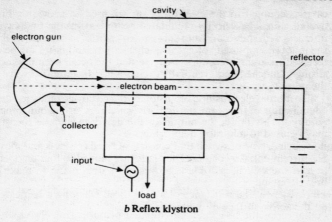

b **Reflex klystron**

which acts as both buncher and catcher (Fig. *b*). Velocity modulation of the electron beam is caused by the input radiofrequency wave in the cavity, and the modulated beam leaving the cavity is reflected back by a *reflector electrode*. Bunching occurs because the faster electrons travel further towards the reflector before reversing their direction of travel than do the slower ones. The bunches of electrons returning to the cavity that experience the maximum positive field give up the most energy since the direction of motion is now reversed.

As with the two-cavity klystron, optimum power transfer occurs when the transit time *t* of the electrons from and to the resonator is given by

$$\omega t = 2\pi(n + \tfrac{3}{4})$$

The klystron will only resonate around certain discrete values of the collector voltage, corresponding to the integers $n = 1, 2, 3$, etc. Oscillation is still possible for small excursions of collector voltage around these values, so that the reflex klystron is useful for providing automatic frequency control or in frequency-modulation transmission. This latter application requires a higher power output (up to about 10 watts) than for the more common low-power local-oscillator applications, where a typical power output of 10 milliwatts is needed.

Multicavity klystrons are used when either extremely high power pulses are required, as in the power source for a particle accelerator, or when continuous waves of moderate power are needed, as in ultrahigh-frequency television transmitters. Three or more resonant cavities coupled to the electron beam are used to provide a high overall gain. The velocity-modulated beam leaving the first cavity interacts with the second and subsequent cavities in such a way that the induced amplified voltage in each cavity remodulates the beam received from the preceding cavity so that the beam becomes more strongly bunched and eventually excites a highly amplified wave in the output circuit.

The mutual electrostatic repulsion of the electrons tends to cause *debunching of the beam, particularly when very strong bunching is required. This limits the output of the device. Magnetic focusing may be used to minimize debunching.

knife switch A type of switch that consists of fixed contacts and movable current-carrying blades. The blades are hinged so that on operation of the switch each blade moves in its own plane to make or break contact.

Koch resistance The resistance of a *photocell when light is incident on the active surface of the tube.

L

labyrinth loudspeaker *See* loudspeaker.

ladder network *See* filter.

lag (1) The amount, measured as a time interval or the angle in *electric degrees, by which one periodically varying wave is delayed in phase with respect to the similar phase in another wave. *Compare* lead[1].

(2) The time interval between the transmission of a signal and its detection by a receiver.

(3) The delay between a correcting signal of a control system and the response to it.

(4) The persistance of the electrical image in a television *camera tube; it may be several frames in duration.

lagging current An alternating current that has a *lag with respect to the applied electromotive force producing it. *Compare* leading current.

lagging load *Syn.* inductive load. A *reactive load in which the inductive *reactance exceeds the capacitive reactance and therefore carries a *lagging current with respect to the voltage across the terminals. A pure inductance introduces a lag of 90° or one quarter wavelength. *Compare* leading load.

laminated core *See* core.

lamination A thin stamping of iron or steel, oxidized or lightly varnished on the surface, that is employed in the assembly of a laminated *core for use in transformers, transductors, relays, chokes, or similar apparatus. The use of laminations reduces *eddy currents in alternating-current applications.

LAN *Abbrev. for* local area network.

land A metallic contact area in microminiature electronic devices or circuits.

Langevin ion An *ion that moves through a gas under the influence of an electric field.

language *Short for* programming language.

Laplace operator Symbol: ∇^2. The differential operator

$$\partial^2/\partial x^2 + \partial^2/\partial y^2 + \partial^2/\partial z^2$$

lapping A method of reducing the thickness of a *slice for applications where accuracy of the substrate thickness is critical to the operation of the device. Monolithic microwave *integrated circuits (MMICs) including microstrip transmission lines, where the impedance of the line depends on substrate thickness, have the most stringent requirements. After processing, a slurry of water and fine grit is used to wear down the back of the slice. The slurry is placed between a flat plate and the back of the slice, and the slice moved with respect to the plate in order to mechanically remove substrate material.

LARAM *Abbrev. for* line-addressable random-access memory.

large optical cavity laser (LOC laser) *See* injection laser.

large-scale integration (LSI) *See* integrated circuit.

laryngophone *Syn. for* throat microphone. *See* microphone.

laser Acronym from *l*ight *a*mplification by *s*timulated *e*mission of *ra*diation. A source of intense monochromatic coherent radiation in the visible, ultraviolet, or infrared regions of the electromagnetic spectrum. The narrow beam can be either pulsed or continuous. A laser operates on similar principles to the *maser but the transitions between the energy levels are in a higher energy range. Spontaneous transitions from the higher to the lower energy levels are a dominant mechanism in the laser and are used to stimulate further transitions in the material.

Laser action has been achieved in solid, liquid, and gaseous materials. The first laser was a *ruby laser,* in which transitions between energy levels of chromium ions in the crystal lattice were used. Population inversion (*see* maser) was achieved by light excitation.

A *gas laser* is excited by a continuous electrical discharge through the gas. The required population inversion is a result of inelastic collisions between ions of the gas, or between ions and the electrons in the discharge.

The *injection laser has a direct-gap semiconductor junction diode as the source of light, and differs from other types of laser in that the stimulated emission of light is initiated by injected minority carriers rather than by light.

A laser can be made to oscillate using mirrors that reflect the emitted radiation back and forth through the material; it therefore acts as its own cavity resonator. Further emission is stimulated by the reflected photons until oscillation starts.

laser Raman microprobe (LRM) *See* Raman spectroscopy.

laser scribing *See* scribing.

latching *See* locking.

lateral recording A method of recording a gramophone record or magnetic disk in which the groove modulations are parallel to the surface of the recording medium and perpendicular to the motion of the cutter.

lattice constant A parameter that describes the configuration of a crystal lattice. The lattice constant is given either as the lengths of the edges of a unit cell of a crystal or as the angle between the axes of the cell. The former description is also termed the *lattice parameter* or *lattice spacing.* The edge length of a cubic unit cell is usually given.

lattice filter *See* filter.

lattice network *See* network.

lattice parameter *Syn.* lattice spacing. *See* lattice constant.

Lauritzen's electroscope *See* electroscope.

lawnmower *Colloq.* A type of preamplifier used with radar receivers that reduces the level of *grass on the screen.

L band A band of microwave frequencies ranging from 0·39 to 1·55 gigahertz. *See* frequency band.

LCC *Abbrev. for* leadless chip carrier.

LCD *Abbrev. for* liquid-crystal display.

L-C network A tuned circuit that contains both inductance and capacitance. The product, *LC,* of inductance and capacitance is constant for any given frequency.

lead[1] (1) An electrical conductor, usually in the form of a wire or cable, that is used to make external connections between circuits or pieces of apparatus. *See* interconnection; intraconnection.

(2) The amount, measured as a time interval or as the angle in *electric degrees, by which one periodically varying wave is advanced in phase with respect to the similar phase in another wave. *Compare* lag.

lead[2] Symbol: Pb. A heavy metal, atomic number 82, that is mainly used either as an alloy, with other metals such as tin, to form solders or in storage batteries.

lead-acid cell A type of accumulator that contains sponge lead cathodes and lead dioxide anodes and employs dilute sulphuric acid as the electrolyte.

leader stroke *See* lightning stroke.

leadframe The complete interconnection pattern of leads inside an *integrated-circuit package. It is used to connect the integrated circuit to the outside world. The leadframe is formed from thin copper sheet (*see* wire bonding) or by plating the required pattern onto plastic tape (*see* tape automated bonding). The frame is relatively rigid and robust. *See also* dual in-line package; leadless chip carrier; pin grid array.

lead-in The cable that connects the active part of an aerial to the transmitter or receiver.

leading current An alternating current that has a *lead with respect to the applied electromotive force producing it. *Compare* lagging current.

leading edge (of a pulse) *See* pulse.

leading load *Syn.* capacitive load. A *reactive load usually containing resistance and capacitance that carries a *leading current with respect to the voltage across the terminals. A pure capacitance introduces a lead of 90° or one quarter wavelength. *Compare* lagging load.

leadless chip carrier (LCC) A common form of package used for *integrated circuits. It consists of a ceramic or plastic casing that contains a circuit using either *wire bonding or *tape automated bonding, and also forms the output contacts arranged around all four sides of the package. The metallic outer contacts are formed flush with the edges of the LCC package and can be flow-soldered to a printed circuit board or hybrid substrate. The number of contacts available varies from 28 with smaller packages up to 60 to 72 with larger more complex circuits. *Compare* dual in-line package; pin grid array.

leakage The passage of an electric current along a path other than that intended due to faulty insulation or isolation in a circuit, component, device, or other piece of apparatus.

leakage current A fault current occurring in any electronic device, circuit, etc., due to *leakage. It is small compared with a short-circuit current.

leakage flux Lost flux in any apparatus, such as a transformer, that contains a magnetic circuit. The leakage flux is flux that is outside the useful portion of the flux circuit.

leakage indicator An instrument that detects and measures the value of *leakage current to earth in any electrical system.

leakage reactance Unwanted reactance in a transformer or alternator caused by *leakage flux cutting one coil but not the other. A leakage inductance is produced by this effect and leads to losses in the system.

Leblanc connection *See* three-phase to two-phase transformer.

LEC *Abbrev. for* liquid-encapsulated Czochralski.

Lecher line A *transmission line that consists of a short section of two parallel open wires that may be used for the measurement of wavelength or for impedance matching.

Leclanché cell A primary *cell that contains a carbon-rod anode and a zinc cathode, which may be amalgamated. The electrolyte is 10–20 per cent ammonium chloride solution. *Polarization is minimized by means of a depolarizer consisting of manganese dioxide mixed with crushed carbon contained in either a fabric bag or a porous pot. The depolarizer is slow-acting and therefore the e.m.f. falls off fairly rapidly in closed-circuit use; the cell is particularly useful however for intermittent applications such as ringing electric bells or in telephony.

The *agglomerate cell* was developed from the Leclanché cell in an attempt to reduce the internal resistance. This form of cell has the depolarizer in the form of solid blocks held to a carbon plate by rubber bands, with the cathode usually as a large zinc cylinder surrounding the blocks. E.m.f. $\simeq 1\cdot 4$ volts.

LED *Abbrev. for* light-emitting diode.

Leduc effect *Syn.* Righi effect. If heat flows through a strip of conductor or semiconductor placed in an orthogonal magnetic flux density, a temperature gradient is set up across the strip at right angles to both the direction of heat flow and the magnetic flux density.

LEED *Abbrev. for* low-energy electron diffraction. *See* diffraction.

left-hand rule *See* Fleming's rules.

Lenard spiral A small noninductively wound spiral of bismuth wire, mounted between mica plates, that is used to measure magnetic fields. It depends for its operation on the significant increase in the resistance of bismuth under the influence of an orthogonal magnetic field. The change in resistance due to a magnetic field is measured by connecting the spiral to one arm of a *Wheatstone bridge.

Lenz's law *See* electromagnetic induction.

level compensator A device or circuit that automatically compensates for the effects of amplitude variations in a received signal.

Leyden jar An early type of capacitor.

LF *Abbrev. for* low frequency. *See* frequency band.

life test A test in which a sample or population of items is subjected to stated stress conditions for specified times with stated failure or success criteria, in order to determine its *reliability characteristics. The data from such tests will provide information giving *failure rate and *mean life of the item. The reliability of most semiconducor devices, etc., is so great that *accelerated life tests and *step stress life tests are employed to avoid unnecessarily long tests. *Truncated tests* are those terminated after a predetermined time or a predetermined number of failures or a combination of these. A *screening test* is a test designed to remove unsatisfactory items or those likely to exhibit early failures. This is sometimes called *burn-in* (*see* failure rate).

lifetime The mean time interval between generation and recombination of a charge *carrier in a *semiconductor.

lift-off A technique used in processing gallium arsenide slices to produce a required metallization pattern. The slice is covered with *resist, which is then exposed and developed to produce the desired pattern. Metal is then applied, usually by evaporation, and deposited on top of the resist as well as the substrate. The resist is dissolved using a suitable solvent and the metal on the resist is removed or 'lifted off' with it, leaving the desired pattern on the substrate. This technique contrasts with the usual etch process for silicon slices, where metal is applied to the slice first. The metal is covered with resist, which is exposed and developed to produce a protective layer for the desired pattern. The metal is then etched away before the remaining resist is dissolved. These two processes are shown in the diagram. For lift-off to be facilitated a small protruding lip at the top edge of the resist is desirable. *Assisted lift-off* is a multilevel resist technique used to produce such a protruding lip. Other meth-

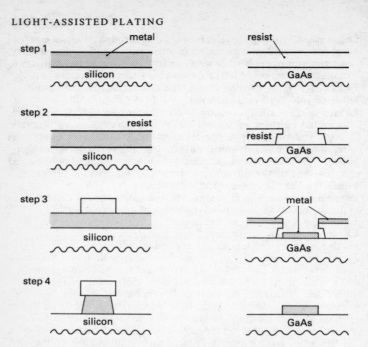

Comparison of lift-off (right) with the usual etch process (left)

ods of *edge profile modification* include baking the slice after development of the resist, which produces a crust on the resist with a suitable lip. The *chlorobenzene method* involves soaking the resist in chlorobenzene to remove residual solvents from the upper layers of the resist film. This retards development of the resist in the upper layer and produces a suitable undercut edge.

light-assisted plating A method of plating involving photogenerated electron-hole pairs in a semiconducting material. Light is used to generate electron-hole pairs in the semiconductor, below a junction. The built-in field causes the electrons and holes to drift in opposite directions. If the material is immersed in an appropriate solution the excess electrons at one surface reduce the solution and plating results. The technique was developed for plating *via holes in gallium arsenide slices, but could be used in other circumstances when it is difficult or impossible to apply an electric field directly.

light-emitting diode (LED) A p-n junction diode that emits light as a result of direct radiative recombination of excess electron-hole pairs

– the Lossev effect – which is one of the *recombination processes in a semiconductor. In *direct-gap semiconductors, such as gallium arsenide, it is a major part of recombination and a significant amount of light will be emitted following injection of excess minority carriers. The quantity of light produced in a forward-biased p-n junction diode formed from such material will be proportional to the numbers of excess minority carriers, i.e. to the bias current. The useful light obtained from the diode is dependent on the optical quality of the crystal surfaces. The frequency, i.e. colour, is a property of the material used, since the energy of the emitted photon is determined by the band-gap energy. Light-emitting diodes are useful for low-voltage display devices, such as calculators or digital watches.

light guide *See* fibre-optics system.

lightning arrester *See* surge diverter.

lightning conductor A lightning protective system that consists of a single conductor providing a path between an air terminal and earth along which a lightning stroke can pass.

lightning flash *See* lightning stroke.

lightning protective system *See* lightning stroke.

lightning stroke An electric discharge due to the discharge of one of the charged regions of a thunder cloud. The polarity of the lightning stroke is the polarity of the electric charge that comes to earth. A complete *lightning flash* is a complete discharge along a single path.

The path of the lightning flash is established by an initial discharge, the *leader stroke,* that can develop either downwards from the cloud to earth or upwards from earth to the cloud. A *dart leader stroke* is a leader stroke that develops continuously. One that develops in a series of relatively short steps is a *stepped leader stroke.* The *return stroke* consists of a high current discharge that flows upwards as soon as a downward leader stroke strikes the earth. If the flash is made up of more than one lightning stroke it is termed a *multiple-stroke lightning flash.*

A lightning stroke to any part of a power or communication system is described as a *direct stroke* and the *surge produced in the system by it or by a flashover from it is a *direct lightning surge.* An *indirect stroke* induces a voltage in such a system without actually striking it and the surge induced by it is an *indirect lightning surge.*

A *lightning protective system* is a complete system of conductors that is designed to protect a building or equipment from the effects of a lightning stroke. *See also* lightning conductor.

light-pen A pen-like device that produces data, in the form of a visible image on the screen of a cathode-ray tube, by 'writing' in a manner similar to that with a pen on paper. Light-pens are almost invariably used in conjunction with an on-line *visual display unit that inputs the data produced to a computer; they are thus extremely powerful design tools.

limb *See* transformer.

limited space charge accumulation mode (LSA mode). *See* Gunn diode.

limited stability The property of any system, circuit, device, etc., that is stable only for a specific range of values of input signal and unstable outside this range.

limiter Any device that automatically sets a boundary value or values upon a signal. The term is usually applied to a device which, for inputs below a specified instantaneous value, gives an output proportional to the input, but for inputs above that value gives a constant peak output. A limiter that sets one boundary value to the peak (either positive or negative) of a signal is a *clipper*. A *base limiter* is one whose output comprises that part of an input signal exceeding a predetermined value. A *slicer* is a limiter having two boundary values, the portion of the signal between these values being passed on. A limiter designed to give a constant current whatever the applied voltage is a *current limiter*.

linac *Short for* linear accelerator.

line (1) *Short for* transmission line. (2) *See* television.

line-addressable random-access memory (LARAM) *See* CCD memory.

linear (1) Denoting any system, device, or apparatus that has its essential physical parts arranged in a line, as in a linear accelerator.

(2) Describing any device that has an output directly proportional to the value of the input and varies continuously with it, as in a linear *amplifier.

linear accelerator A particle *accelerator in which electrons or protons are accelerated as they travel along a straight evacuated chamber. The acceleration is provided by the electric field vector associated with the radiofrequency (r.f.) output from a *klystron or *magnetron.

In the *standing-wave accelerator* standing waves from an r.f. supply are established between a series of cylindrical electrodes coaxial with the chamber. The standing waves are established with the electric vector aligned axially. The electrons are only accelerated in the gaps between the electrodes; inside the electrodes they drift towards the next electrode. The electrodes are therefore termed *drift tubes*. The lengths of the drift tubes and the frequency of the r.f. supply is arranged so that the phase of the electric field vector is always in accelerating mode when the electrons emerge from a drift tube. The length, l, is related to the energy, E, of a particle of mass m and the r.f. supply frequency, f, by

$$l = (1/f)(2E/m)^{1/2}$$

As the energy of the particles increases, the lengths of successive drift tubes must be correspondingly increased in order to maintain the required phase relationship.

Higher particle energies are achieved using a *travelling-wave accelerator,* in which the accelerating chamber is a long *waveguide. An r.f. *travelling wave is established from a high-power source and the waveguide is excited so that a large-amplitude travelling wave is produced, travelling with a phase velocity equal to the local velocity of the electrons to be accelerated. Power is transferred from the r.f. wave to the accelerated electrons and the r.f. power is boosted at regular intervals along the length using klystrons.

Very high energies are difficult to produce because of the extremely long lengths of accelerating chamber required: the 22 GeV linac at Stamford University, California, has a 3.5 km tube. Such tube lengths involve great engineering difficulties and expense. Extra high energies are produced using accelerators such as proton *synchrotrons.

linear amplifier *See* amplifier.

linear-beam microwave tube *Syn.* O-type microwave tube. *See* microwave tube.

linear circuit *Syn.* analog circuit. A circuit in which the output varies continuously as a given function of the input. *Compare* digital circuit.

linear detector *See* detector.

linear inverter *See* inverter.

linearly graded junction A junction between two different-polarity semiconductors (p-n, p-i, or n-i) in which the concentration of impurities varies linearly across the junction. *Compare* abrupt junction.

linearly polarized wave *See* plane-polarized wave.

linear network *See* network.

linear scan (1) A sweep of the electron beam in a cathode-ray tube in which the beam scans the screen with constant velocity, usually by application of a sawtooth waveform to the deflection plates or coils. *See also* time base.

(2) A scan using a *radar beam that moves with constant angular velocity.

linear time-base oscillator A *relaxation oscillator that is used to generate a sawtooth waveform for use as a *time base.

linear transducer *See* transducer.

line choking coil *Syn.* screening reactor. An inductive reactor that may be connected in series with electrical plant, such as a power transformer, in order to protect such plant from sudden steep-fronted or high-frequency surges in the power supply. Partial absorption of the surge occurs due to I^2R losses, core losses, and partial reflection of the surge.

303

line communication Communication, such as broadcasting or telephony, between two points by means of a physical path such as wire or waveguide.

line frequency *See* television.

line of flux An imaginary line drawn in a magnetic field whose direction at any point along its length is that of the magnetic flux density, *B*. The number of lines of flux through unit area perpendicular to the direction of *B* is equal to the magnetic flux density at that point. *See also* field.

line of force An imaginary line drawn in an electric field whose direction at any point along its length represents that of the field at that point. The number of lines of force through unit area perpendicular to the field is equal to the field strength at that point. *See also* field.

line printer An output device used with computers and data-processing systems that prints an entire line of characters at a time rather than a character at a time. Typical operating speeds vary from 200 to 3000 lines per minute.

line reflection *See* transmission line.

line-sequential colour television A *colour-television system in which each of the video signals (red, blue, and green) is transmitted in turn for the duration of one entire scanning line.

line voltage *See* voltage between lines.

link (1) A communication channel or circuit that is used to connect other channels or circuits.

(2) A path between two switches that form part of a central control system in automatic switching.

lin-log receiver A type of *radar receiver that has a linear output for small input signals and a logarithmic output for large ones. Such a receiver is useful over a large range of received signals.

lip microphone *See* microphone.

liquid crystal An organic liquid consisting of long-chain molecules that line up under the influence of an applied electric field to give a quasi-crystalline structure to the liquid. A change in the applied field causes a change in the reflectivity of the liquid making them extremely suitable for use as passive display devices.

Liquid crystals are also temperature-sensitive: there is an apparent change of colour with increasing temperature that makes them suitable for use in temperature indicators.

liquid-crystal display (LCD) A type of passive display that uses *liquid crystals, as in the seven-segment numerical display of digital watches and pocket calculators.

liquid-encapsulated Czochralski (LEC) A method of growing gallium arsenide crystals from molten gallium arsenide by slowly pulling the crystal vertically from the melt. A layer of liquid boric oxide (B_2O_3) is floated on the surface of the gallium arsenide to confine the melt: hence the description liquid-encapsulated (*see* diagram). The

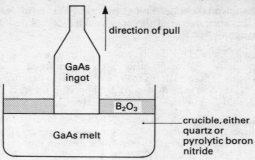

Schematic diagram of LEC growth method

growth is started using a seed crystal, which is introduced through the boric oxide when the appropriate temperature profile is achieved. The machines used to grow the crystals are referred to as *crystal pullers*. *High-pressure* LEC processes are performed under high external pressures – up to 50 atmospheres. *Low-pressure* LEC processes are carried out at pressures of about 1 atmosphere. The pullers required for each of these processes are different, and different heat-flow characteristics exist; the two types of growth machine are therefore not easily interchangeable. The liquid encapsulation and high pressures originally used were required to contain the rather violent reactions that occur as the gallium and arsenic react exothermically to produce the gallium arsenide melt. The low-pressure technique was developed by introducing the arsenic in a controlled manner into the molten gallium below the boric oxide surface.

liquid-phase epitaxy A method of growing an *epitaxial layer on a substrate from a molten material. The substrate crystal is placed in a slider and the material to be deposited is contained in molten form in a 'boat'. The melt is supercooled to just below the solidification temperature. As the slider containing the substrate material is moved slowly across the surface of the melt, atoms solidify onto the crystal substrate. This method of epitaxy is most useful for 3–5 or 2–6 compound semiconductors, particularly gallium arsenide substrates. It has limitations and is losing popularity, but is inexpensive and capable of growing many material compositions; it is therefore still used for some applications, such as light-emitting diodes, that do not require such thin uniform high-quality layers as are required for microwave devices.

liquid rheostat A variable resistor in which the resistance element is a liquid metal.

305

liquorice allsorts *Colloq.* Foil capacitors. The term arises from the physical appearance of the colour-coded component (*see* Table 2, backmatter).

Lissajous' figure The displacement pattern traced out when two sinusoidally varying quantities are superimposed at right angles to each other. These patterns may be obtained in practice on the screen of a *cathode-ray oscilloscope by applying the two signals to the horizontal and vertical deflection plates.

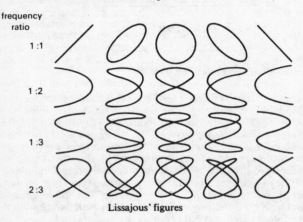

frequency ratio

1 : 1

1 : 2

1 . 3

2 : 3

Lissajous' figures

The simplest pattern is a straight line, which occurs when two signals of equal frequency and in phase with each other are superimposed. Much more complicated figures result when the ratio of the frequencies is not a simple one. Typical figures are shown in the diagram for various frequency ratios and phase angles. The patterns can be used for accurate frequency matching of two signals or for identifying the phase relationship of two signals of the same frequency.

lithography Patterning techniques used in the manufacture of *integrated circuits, *semiconductor components, *thin-film circuits, and *printed circuits. The techniques are used to create or transfer required patterns from masks to a substrate surface using energy-sensitive materials known as *resists.

The most common technique is *photolithography, but in VLSI applications where the dimensions of the pattern to be transferred are of the same order as the wavelengths of the light used other techniques have been developed. *See* electron-beam lithography; ion-beam lithography; X-ray lithography.

Litzendraht wire A multistranded wire formed from many fine conducting filaments. It is employed for high-frequency applications, as in coils used in radio, in order to reduce the high-frequency resistance. *See* skin effect.

live *Syn.* alive. *See* dead.

L network *Short for* inductance network. *See* network.

load (1) Any device or material that absorbs power from a source of electrical signals. Examples include loudspeakers, television and radio receivers, the material heated by dielectric or induction heating, logic circuits, and any driven circuit.

(2) The output power delivered by any electrical machine, generator, transducer, electronic circuit, or device. The machine, generator, etc., is *off-load* when it is operated under normal conditions but no absorbing load is connected at the output. It is *on-load* when connected to an absorbing load.

The maximum power that is absorbed by a load or delivered as a load in a predetermined time period is the *peak load.*

load characteristic A characteristic curve for an electronic device, such as a transistor, showing the dynamic relationship between the instantaneous values of two variables when all the supply voltages are maintained constant. Emitter current, I_e, plotted against emitter voltage, V_e, provides an example.

load circuit The output circuit of any electrical machine, generator, transducer, circuit, or device.

load-circuit efficiency The ratio of the useful power delivered to the load by the load circuit of any device, to the power input to the device.

load coil *See* induction heating.

load curve A graph in which the load power of a transmission or distribution system is plotted against time.

load factor The ratio of the average load power supplied over a specified time interval to the peak load power in that period. The load factor is usually expressed as a percentage.

The *plant load factor* relates specifically to an electrical generator or group of generators, and is the ratio of the actual number of electrical units (kilowatt-hours) supplied by the generators in a given time interval to the number that would have been supplied had the plant operated at its maximum continuous rating for that interval. This factor also is usually expressed as a percentage.

load impedance The *impedance presented by a load to the driver circuit that supplies power to it. The effect of variations in the load impedance on the performance of an oscillator is seen in a *load impedance diagram* in which the oscillator output is plotted against load impedance.

loading *See* transmission line.

loading coils *See* transmission line.

load leads The conductors or transmission lines that connect the power source for dielectric or induction heaters to the load, load coil, or applicator used.

load line A line drawn on the graph of a family of *characteristics of an electronic device, such as a transistor, that shows the relationship between current and voltage of the circuit under consideration for a given load.

load matching Adjustment of the output impedance of the load circuit of a dielectric or induction heater to optimize the energy transferred from the source to the load.

load regulator A circuit or device that maintains a load at a constant value or varies it in a predetermined manner.

load transfer switch A switch that is used to connect the power output from a generator or other power source to one or to other load circuits as required.

lobe *See* radiation pattern.

lobe switching *Syn.* beam switching. A form of scanning used in radar in which the direction of maximum radiation or reception is switched sequentially to each of a number of preferred directions, for example from one side to the other of the target area. Lobe switching is achieved by using a *steerable aerial and switching the output into the appropriate circuits, each of which has an optimum signal-to-noise response for the chosen direction.

local area network (LAN) A communication system that links several computers, usually microcomputers, together with printers, telex machines, FAX, etc., within a small and defined locality.

local exchange *See* telephony.

local feedback *Syn.* multiple-loop feedback. *See* feedback.

localizer beacon *See* beacon.

local oscillator *See* superheterodyne reception.

locator beacon *Syn. for* homing beacon. *See* beacon.

lock-in detector *Syn.* lock-in amplifier. A detector that responds only to an input signal whose frequency is synchronous with the frequency of a locally generated control signal. It may be used as a null-point detector in a bridge circuit.

locking (1) Controlling the frequency of an oscillator by means of an applied signal of constant frequency from an external source.
(2) *Syn.* latching. Holding a circuit in position or in a certain state until previous operating circuits are ready to change the circuit.

locking-in Synchronizing the frequencies of two coupled oscillators, as in a frequency doubler, so that the two frequencies have a desired ratio, usually of two integral numbers. One of the oscillators must be free-running and capable of being pulled to the desired frequency.

locking-on The following of a target, automatically, by a radar aerial.

locking relay *See* relay.

LOC laser *Abbrev. for* large optical cavity laser. *See* injection laser.

Locos *Tradename* Acronym from *loc*al *o*xidation of *s*ubstrate. *See* coplanar process.

logarithmic decrement *See* damped.

logarithmic resistor A form of variable resistor designed so that the fractional change of resistance is directly or indirectly proportional to the movement of the contact.

logical one The digit 1 used in *binary notation. It is equivalent to the value 'true' of a logical statement. *Logical zero* is the digit 0 in binary notation. It is equivalent to the value 'false'.

Popular (formerly BSI) symbol	Binary logic circuit	IEC approved symbol	Popular (formerly BSI) symbol	Binary logic circuit	IEC approved symbol
	AND gate	&		NOR gate, negated output	≥1
	NAND gate, negated output	&		NOR gate, negated inputs	≥1
	NAND gate, negated inputs	&		Exclusive-OR gate	=1
	OR gate	≥1		Inverter (NOT gate)	

logic circuit A circuit designed to perform a particular logical function based on the concepts of 'and', 'either-or', 'neither-nor', etc. Normally these circuits operate between two discrete voltage levels, i.e. *high* and *low logic levels,* and are described as *binary logic circuits.* Logic using three or more logic levels is possible but not common.

The basic digital *gates are:

AND gate. A circuit with two or more inputs and one output in which the output signal is high if and only if (sometimes written *iff*) all the inputs are high simultaneously.

*Inverter (*NOT gate*). A circuit with one input whose output is high if the input is low and vice versa.

NAND gate. A circuit with two or more inputs and one output, whose output is high if any one or more of the inputs is low, and low if all the inputs are high.

309

NOR gate. A circuit with two or more inputs and one output, whose output is high if and only if all the inputs are low.

OR gate. A circuit with two or more inputs and one output whose output is high if any one or more of the inputs are high.

Exclusive OR gate. A circuit with two or more inputs and one output whose output is high if one input is high.

The graphical symbols for the basic circuits are shown in the table. These circuits are for use with *positive logic:* that is, the high voltage level represents a *logical 1 and low a logical 0. *Negative logic* has the high level representing a logical 0 and low a logical 1. The same circuits may be used in negative logic but become the complements of the positive logic circuits, i.e. a positive OR circuit becomes a negative AND circuit.

Binary circuits are extensively used in computers to carry out instructions and arithmetical processes. Any logical procedure may be effected by using a suitable combination of the basic gates. *See also* truth table.

Binary circuits may be formed from discrete components or, more commonly, from *integrated circuits. Families of integrated logic circuits exist based on bipolar transistors (*see* diode-transistor logic (DTL), emitter-coupled logic (ECL), resistor-transistor logic (RTL), I^{2}L, nonthreshold logic (NTL), and transistor-transistor logic (TTL)). *MOS logic circuits are based on *field-effect transistors.

Bipolar logic circuits are capable of very high speed operation but have relatively complex structures compared to MOS logic circuits, and therefore a lower functional *packing density. MOS logic circuits have thus been widely used for large-scale integration (LSI) despite their lower speed of operation, and bipolar logic circuits have been used for circuits demanding high performance and high speeds. Recent improvements in bipolar technology, however, have improved the packing densities that can be achieved with bipolar circuits. For VLSI (very large scale integration) applications demanding high speeds of operation, bipolar circuits have great potential. I^{2}L circuits offer the highest density and lowest power dissipation, approaching that of MOS circuits. ECL circuits have the highest performance at present.

logic diagram A diagram that shows the logic elements of a computer or data-processing system, or a function thereof, and their interconnections but does not usually show any constructional or engineering details. A logic diagram is useful when designing such systems or smaller networks to perform a specific mathematical operation such as integration.

logic element The smallest circuit blocks in a computer or data-processing system that may be represented by mathematical operators in symbolic logic. *See also* logic circuit.

logic symbol A graphical symbol representing a logic element. *See* logic circuit.

long-line effect An effect sometimes observed when an oscillator is coupled to a load through a transmission line that is long compared with the wavelength of the oscillator output. The oscillator may jump from its desired frequency to a nearby unwanted frequency.

long-persistence screen A type of screen used in cathode-ray tubes that allows the image on the screen to persist for several seconds. This is usually achieved by mixing phosphorescent compounds with the usual fluorescent compounds of the screen.

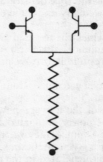

Long-tailed pair

long-tailed pair A pair of matched bipolar *transistors that have their emitters coupled together with a common emitter bias resistor acting as a constant current source. The name is derived from the bias resistor that physically resembles a tail (*see* diagram): the larger (therefore longer) the bias resistor the more nearly it resembles a constant source because of the relatively large voltage developed across it. The name was originally applied to matched thermionic valves with the cathodes connected to a common cathode bias resistor.

The long-tailed pair forms the basis of most *differential amplifiers and can be used for *push-pull operation to produce outputs in antiphase when antiphase inputs are applied to the bases of the transistors.

long wave A radiowave that has a wavelength in the range one to 10 km. *See* frequency band.

loop *See* feedback control loop.

loop aerial *Syn.* frame aerial; coil aerial. A type of *directive aerial with high directivity that consists essentially of a coil with one or more turns of wire of very small axial length compared to the diameter. The direction of maximum sensitivity or transmission is the direction coplanar with the coil so that this type of aerial is very useful for radio direction finders. It is also simple and light in construction enabling it to be used with portable radio receivers.

loop direction finding *Syn. for* frame direction finding. *See* direction finding.

loop signals *See* feedback control loop.

loose coupling *Syn. for* undercoupling. *See* coupling.

Lorentz force The force experienced by an electron or other charged particle moving in a region of magnetic flux density, **B**. The force acts in a direction that is perpendicular both to the direction of particle motion and of the flux density and is given by $q\mathbf{B} \times v$, where v is the particle velocity and q the charge.

loss *See* dissipation.

loss angle The angle by which the angle of *lead of the current is less than 90° when a capacitor or dielectric is subjected to sinusoidal alternating electric stress. It is mainly due to *electric hysteresis loss.

Lossev effect *Syn.* radiative recombination. *See* recombination processes.

loss factor (1) The ratio of the average power dissipation in a line, circuit, or device to the power dissipated at peak load.

(2) The product of the *power factor and the relative *permittivity of a *dielectric. For a given alternating field the power factor is proportional to the heat generated in the material.

lossless line A hypothetical transmission line in which no attenuation occurs.

lossy Denoting an insulator that dissipates more energy than is considered normal for that class of material.

lossy line A type of *transmission line that is designed to produce a high degree of attenuation.

loudspeaker An electroacoustic device that converts electrical energy into sound energy. It is the final unit of any sound reproducer or of the acoustic circuit of any broadcast receiver. Its action is the reverse of that of the *microphone but it is designed to handle far greater power, enabling the sound output to be audible over a large area.

Most types of loudspeaker use a coil and diaphragm arrangement in which a small coil is fixed at the centre of a diaphragm that is free to move in an annular gap. A strong magnetic field, produced by either a permanent magnet or an electromagnet, is applied across the gap. The audio signal is input to the coil (known as the *speech coil*) as alternating current, causing it to move in the magnetic field as a result of *electromagnetic induction. The diaphragm is thus caused to vibrate at the same frequency as the alternating current and sound waves are produced by it. Any loudspeaker using this arrangement can be described as a *magnetic loudspeaker*.

For high efficiency (up to 50 per cent of energy conversion) a *horn loudspeaker* is used. This type of speaker uses a small diaphragm at the mouth of a large exponential horn, and has marked directional properties. The horn speaker is impractical for most indoor uses as it is very large. A more convenient size is achieved using a large conical or ellipsoidal diaphragm, usually made of stiff paper, with the speech coil at its apex. The cone is supported round its edge by a metal frame, and the coil is maintained in position at the centre of the gap by thin flexible supports known as a *spider*. A large baffle, usually the cabinet housing the speaker, is used with this *cone loudspeaker* to prevent the direct passage of sound from front to back and thus improve the low-frequency response.

A simple cone loudspeaker has a uniform power output over a moderate range of frequencies but the output falls at the high- and low-frequency ends of the audible spectrum. For good reproduction at low frequencies a large cone is required in order to give a larger radiation resistance at these frequencies. At high frequencies the mass of the vibrating system sets an upper limit for good reproduction and a small cone is required to optimize the response at these frequencies. Two speakers are sometimes used together in order to overcome these requirements: a large cone for the low notes and a small one for the high notes. The large cone is designed to have a very low resonant frequency to improve its output and the strong resonant peaks, which would cause a boomy sound, are eliminated by using considerable damping (*see* damped). A frequency-selective network, known as the *loudspeaker dividing network,* is required with this type of speaker in order to divide the spectrum between the two sound radiators.

Some speakers use an arrangement of multiple coils and cones designed to reduce the effective mass of the speaker at high frequencies. A well-designed speaker gives a uniform response between about 80 hertz and 10 000 hertz but its conversion efficiency is only about five per cent. Cone loudspeakers are essentially omnidirectional but the use of a suitable cabinet can introduce directional properties if required.

Other methods of producing the sound vibrations are sometimes used. The *crystal loudspeaker* utilizes a *piezoelectric crystal as the vibrating part. *Magnetostriction is used in the *magnetostriction loudspeaker* and the vibrations of a magnetic *armature produce the sound of a *magnetic-armature loudspeaker*. The action of electrostatic fields produces the mechanical movement in an *electrostatic loudspeaker*. Acoustic standing waves can be reduced by placing a loudspeaker in a special housing containing air chambers. Such an arrangement is termed a *labyrinth loudspeaker*.

A *loudspeaker microphone* is a dynamic loudspeaker that may also be used as a microphone. This arrangement is often used in an intercommunication (intercom) system enabling a single unit to be used for both speaking and listening. A manually operated switch connects the device to the appropriate circuit for the desired function. The most convenient arrangement is a push-button that returns automatically to the loudspeaker condition when pressure is removed from it.

loudspeaker dividing network *Syn.* crossover network. *See* loudspeaker.

loudspeaker microphone *See* loudspeaker.

low-electron-velocity camera tube *Syn.* cathode-voltage-stabilized camera tube. *See* camera tube. *See also* image orthicon; vidicon.

lower sideband *See* carrier wave.

low frequency (LF) *See* frequency band.

low-frequency compensation Compensation applied to an amplifier when it is used with low-frequency signals. The compensation is designed to avoid distortion due to signal attenuation and phase shift caused by the reactance of coupling capacitors.

low-level injection *See* injection.

low-level modulation A method of modulation whereby the modulation is produced at a stage in a system where the power level is low compared with the level of power that is output by the system.

low-level programming language *See* programming language.

low logic level *See* logic circuit.

low-loss line A type of *transmission line that is designed to dissipate very little energy per unit length. The series resistance and shunt conductance of the line are therefore made low.

low-pass filter *See* filter.

low-pressure Czochralski *See* liquid-encapsulated Czochralski.

low-pressure puller *See* liquid-encapsulated Czochralski.

low-velocity scanning *See* scanning.

LRM *Abbrev. for* laser Raman microprobe. *See* Raman spectroscopy.

LSA mode *Short for* limited space-charge accumulation mode. *See* Gunn diode.

L-section *See* quadripole.

LSI *Abbrev. for* large-scale integration. *See* integrated circuit.

luminance flicker *See* flicker.

luminance signal *See* colour television.

luminescence The emission of electromagnetic radiation from a substance due to a nonthermal process. The term is also used to describe the radiation itself, particularly when it falls within the visible spectrum. Luminescence occurs when atoms of the material are excited and then decay to their *ground state with the emission of radiant energy. If the luminescence ceases as soon as the source of excitation is removed, i.e. the persistence is less than about 10^{-8} second, it is termed *fluorescence.* If it persists for longer than about

10^{-8} second it is termed *phosphorescence*. A luminescent material is known as a *phosphor*.

The most common source of energy that results in luminescence is other electromagnetic radiation or electrons or other charged particles. *Stokes' law* states that the radiation emitted is usually of longer wavelength than that of the exciting radiation (although it may sometimes be shorter). Ultraviolet radiation therefore can produce visible light from a phosphor. The light produced has a characteristic colour for a particular fluorescent material: fluorescene, yellow-green; quinine sulphate, blue; chlorophyll, red.

Fluorescence is used for examining the spectrum in the ultraviolet region, in fluorescent lighting, and for display purposes, such as with the screen of a cathode-ray tube. Phosphorescence is used when longer persistence is required, such as with a long-persistence screen.

Galvanoluminescence is a feeble glow emitted by the anode of an electrolytic cell or rectifier, and is caused by bombardment of the anode by the negative ions of the solution. *Thermoluminescence* is an indirect effect of bombardment by ionizing radiation and is seen when the material is heated after subjection to radiation. The radiation releases electrons within the material and these are trapped at defects within the solid. These electrons are released on heating and the energy thus produced is emitted as visible radiation. Other energy sources that excite luminescence include friction (*triboluminescence*) and chemical reaction (*chemiluminescence*).

luminescent lamp *See* electroluminescence.

luminophore *Syn. for* phosphor. *See* luminescence.

lumped parameter Any circuit parameter, such as inductance, capacitance, or resistance, that can be treated as a single parameter at a point in the circuit for the purposes of circuit analysis over a specified range of frequencies.

M

machine code *See* programming language.

machine equation *See* patching.

machine variable *See* patching.

magamp *Short for* magnetic amplifier.

magic eye (1) *Syns.* cathode-ray tuning indicator; electron indicator tube; electron-ray tube. A miniature *cathode-ray tube used in radio receivers and as an indicator for voltmeters and galvanometers in a.c. bridges. It exhibits a fluorescent pattern determined by the rectified output voltage of the receiver, and is an aid to tuning the receiver or finding the null point in a.c. bridges.

(2) A photocell used to control the automatic opening of doors in shops, lifts, etc. When the beam of light incident on the photocell is cut off, or in some cases made complete, the circuit is made to operate.

magic-T junction *Colloq. syn. for* T-junction. *See* hybrid junction.

magnesium Symbol: Mg. A light metal, atomic number 12, used extensively for the construction of electronic components, most commonly as an alloy with aluminium.

Magnestat *Tradename* A type of *magnetic amplifier.

magnet A body that possesses the property of *magnetism. The term is applied to those bodies that can produce an appreciable magnetic field external to themselves. Magnets are either temporary or permanent. *See also* permanent magnet; electromagnet; ferromagnetism; magnetite.

magnetic amplifier An *amplifier in which *transductors are used to produce power amplification of the input signal.

magnetic armature loudspeaker *See* loudspeaker.

magnetic balance A device that determines directly the force between two magnetic poles. A long magnet is balanced on a knife edge so that it takes up a horizontal position. One magnetic pole of a second long magnet is brought near to one end and the force (of attraction or repulsion) between the poles is balanced by the addition of weights or the action of a movable rider at the other end to restore the horizontal position. The magnets are made long in order to reduce interference by interaction between their second poles.

The magnetic balance can also be used to measure the strength of a magnetic field. In this type of balance the magnet that acts as the balance arm is replaced by a long conductor. A known current is passed through the conductor and the force exerted on one end by a pole of the magnet to be measured is balanced as above. The field due to the magnet may be calculated from the force between it and the known field due to the current. This type of balance can be calibrated to read magnetic field directly for a stated current and distance between the poles.

This method of measurement contains errors since the precise location of the magnetic poles is indeterminate. More accurate methods have been devised that measure the *magnetic moment or *magnetic flux density associated with permanent magnets and current-carrying conductors. *See* flip-coil; Cotton balance.

magnetic bias *Syn.* alternating-current bias. *See* magnetic recording.

magnetic blow-out *See* circuit-breaker.

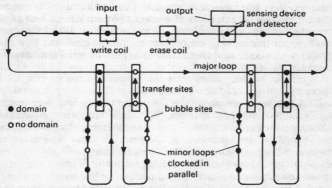

Major-minor loop organization of magnetic bubble memory

magnetic bubble memory A *serial magnetic memory that is fabricated as a solid-state device and in which information is stored as microscopic domains of magnetic polarization. The memory consists of a nonmagnetic garnet substrate – *G^3 – on which an epitaxial layer of magnetic garnet (usually calcium-germanium *YIG) is grown. In the presence of a sufficiently large steady magnetic flux density applied perpendicular to the surface, the direction of magnetic polarization tends to be all in one direction. Any small domains of opposite polarization, which can be produced (or destroyed) by local variation of the magnetic flux density, are stable and may be moved through the surface using suitable weaker magnetic flux densities applied parallel to the surface.

The small domains are known as *magnetic bubbles* and are formed by *electromagnetic induction using a single coil on the surface. They are caused to move through the surface by means of a periodic structure of suitably shaped magnetic electrodes formed from a different material, such as *permalloy. A rotating magnetic flux density parallel to the surface causes the electrodes alternately to attract and repel the bubbles.

The bubble memory is a digital device in which the main direction of polarization is usually vertically downwards; the presence of a domain polarized vertically upwards represents a logical 1; the absence of a domain represents a logical 0. The information may be sensed at the end of the electrode structure either by electromagnetic induction or *magnetoresistance.

Magnetic bubble memories are nonvolatile, i.e. the information is retained if the power supply is interrupted provided that the steady magnetic flux density is maintained, usually by using a permanent magnet. Very little power is consumed and, using suitably shaped electrodes and organization of storage cells, very large functional packing density can be achieved. Chips containing 100 kilobits have been produced using a major-minor loop arrangement (*see* diagram), which is the analogue of the serial parallel serial arrangement used for *CCD memory.

Magnetic bubble memory is the magnetic analogue of CCD memory, which it predates. Compared to CCD memory it is nonvolatile, requires fewer processing steps, and consumes less power. CCD memory however is faster, the silicon chips are considerably cheaper than the garnet chips, and sensing and control circuits can be formed on the same chip as the memory array. The magnetic bubble memory requires external control circuits and also requires extra circuits to generate the rotating magnetic fields.

magnetic circuit A completely closed path described by a given set of lines of *magnetic flux. The direction of the path at any point is that of the *magnetic flux density at that point.

magnetic constant *Syn. for* permeability of free space. *See* permeability.

magnetic contactor A *contactor that is operated by magnetic means, such as an alternating magnetic field.

magnetic controller *See* automatic control.

magnetic core *Syn. for* ferrite core. *See* core.

magnetic crack detection A method of detecting discontinuities at or near the surface of a ferromagnetic material (*see* ferromagnetism). A magnetizing field is applied to the material and leakage of flux or uneven magnetization will arise at the site of any discontinuity. This is detected when a *magnetic fluid is painted on the surface, the particles in the fluid concentrating above the discontinuities.

magnetic crossed-field modulator A type of *frequency doubler consisting of an annular coil surrounded by a hollow toroid of magnetic material. The toroid forms the core of an external toroidal winding (*see* diagram). The toroidal winding carries a direct voltage and the alternating current is input to the annular winding. The output volt-

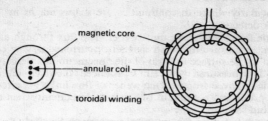

magnetic core

annular coil

toroidal winding

Magnetic crossed-field modulator

age developed in the toroidal winding has a frequency twice that of the input signal due to the interaction of the orthogonal magnetic fields in the coils.

magnetic cutter *See* recording of sound.

magnetic damping A method of damping (*see* damped) an indicating instrument in which the damper consists of a metal vane, connected to the pointer, that moves through a magnetic field. The induced currents in the vane are in such a direction as to oppose the motion (*see* electromagnetic induction).

magnetic deflection *See* electromagnetic deflection.

magnetic dipole moment *See* magnetic moment (both definitions).

magnetic disk *See* moving magnetic surface memory.

magnetic drum *See* moving magnetic surface memory.

magnetic field The space surrounding a magnet or a current-carrying conductor and containing *magnetic flux. It may be represented by *lines of force whose direction at any point is the direction of the force exerted on a small coil (a search coil) placed in the field at that point. The direction of the force is normal to the magnetic flux density at that point. It is assumed that the dimensions of the coil are sufficiently small so as not to disturb the magnetic conditions.

magnetic field strength *Syn.* magnetizing force. Symbol: H; unit: ampere/metre. The strength of a magnetic field at a point in the direction of the line of force at that point. It is defined in vacuo from the equation

$$B = \mu_0 H$$

where B is the magnetic flux density and μ_0 is a constant, the *permeability of free space. *See also* Ampere's law.

magnetic flip-flop A *flip-flop that uses one or more *magnetic amplifiers as the bistable element.

magnetic fluid A fluid that consists of particles of magnetic material dispersed in a suitable oil. The magnetic material used is usually very finely divided iron or an oxide of iron. A magnetic fluid is useful for demonstrating the magnetic flux distribution at a surface,

319

particularly where discontinuities are suspected, as in *magnetic crack detection.

magnetic flux Symbol: Φ; unit: weber. The flux through any area in the medium surrounding a magnet or current-carrying conductor, equal to the surface integral of the *magnetic flux density over the area. It is measured by the e.m.f. produced when a circuit linking the flux is removed from it. One weber of flux linking a circuit of one turn produces an e.m.f. of one volt in that circuit when the flux is reduced to zero.

magnetic flux density *Syn.* magnetic induction. Symbol: B; unit: tesla (weber/metre^2). The fundamental force vector in magnetism; the magnetic analogue of the electric field E. Both a magnet and a current-carrying coil exert forces on other coils or magnets. The magnetic flux density produced by such magnets or coils is a vector quantity and lines of flux can be drawn whose direction at any point is the direction of magnetic flux density. The value of B is given by the number of lines of flux per unit area and is expressed by the equation

$$\mathrm{d}F = I(\mathrm{d}s \times B)$$

where $\mathrm{d}F$ is the force exerted due to B on an element of length $\mathrm{d}s$ of wire carrying a current I. This defines the unit of magnetic flux density as that which exerts a force of one newton on a wire of length one metre carrying a current of one ampere.

magnetic focusing *Syn. for* electromagnetic focusing. *See* focusing.

magnetic head *See* magnetic recording.

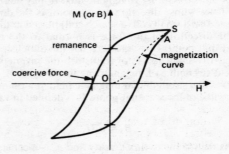

Magnetic hysteresis loop

magnetic hysteresis A phenomenon observed in ferromagnetic materials below the Curie point (*see* ferromagnetism) where the *magnetization of the material varies nonlinearly with the magnetic field strength and also lags behind it. The magnetic *susceptibility of such materials is large and positive, a large value of magnetization

(*M*) being produced for comparatively small fields. A characteristic plot of either magnetization, *M,* or magnetic flux density, *B,* against magnetic field strength, *H,* demonstrates the hysteresis effect and is termed a *hysteresis loop* (*see* diagram). If an initially unmagnetized sample of iron is subjected to an increasing magnetic field the magnetization follows the curve shown by the dotted line OAS. This is known as the *magnetization curve.* If the specimen is then subjected to a complete magnetizing cycle with magnetic field varying symmetrically between $+H$ and $-H$ the curve shown by the solid line is followed.

The value of *M* at zero field is termed the *remanence,* the corresponding value of *B* being known as the *residual induction.* The value of *H* at the point at which *M* (or *B*) falls to zero is known as the *coercive force,* and is the reverse field required to demagnetize the sample. The area enclosed by the $M-H$ curve represents the work done in taking the sample through a complete cycle and is therefore the energy dissipated in each complete cycle when ferromagnetic material is subjected to an alternating magnetic field. This is known as the *hysteresis loss.* The *incremental hysteresis loss* is the total energy dissipated in the sample when subjected to an alternating magnetic field and is the product of the hysteresis loss and the number of cycles.

The effect of hysteresis in a magnetic core is to cause an increase in the effective resistance of the coil surrounding the core; the *hysteresis factor* is the increase in the effective resistance of a coil carrying a current of one ampere at a specified frequency.

The general form of the hysteresis curve is shown in the diagram. The area enclosed by the curve depends on the nature of the ferromagnetic sample: minimum area (and minimum coercive force) occur with soft iron, rising to a value some twenty times greater with tungsten steel. Any complete magnetizing cycle (say between the values of $H + h$ and $H - h$) will give rise to a hysteresis loop. *See also* ferromagnetism.

magnetic induction *See* magnetic flux density.

magnetic intensity *Obsolete syn. for* magnetic field strength.

magnetic leakage The loss of magnetic flux from a magnetic circuit due to a portion of the total magnetic flux following a path that renders it ineffective for the desired function of the circuit. Magnetic leakage reduces the overall efficiency of the operation. In a transformer, for example, magnetic leakage occurs when some of the flux from the primary circuit does not link with the secondary circuit.

The *magnetic leakage coefficient,* σ, is defined by the ratio of the total magnetic flux to the effective (or useful) magnetic flux; i.e.

$$\sigma = 1 + \text{leakage flux/useful flux}$$

A typical value of σ is 1·2 in electrical machines.

magnetic lens *See* electromagnetic lens.

magnetic link *See* surge-current indicator.

magnetic loudspeaker *See* loudspeaker.

magnetic memory A *memory in which digital information is represented by the direction of magnetization of magnetic material. *See* moving magnetic surface memory; magnetic bubble memory.

magnetic microphone *See* microphone.

magnetic mirror A region of high magnetic field strength that reflects charged particles.

magnetic modulation *See* transductor.

magnetic moment (1) *Syns.* magnetic dipole moment. Symbol: m; unit: ampere metre^2. A measure of the strength of a *magnet. When a magnet is placed in a homogeneous magnetic flux of magnetic flux density B it experiences a torque T such that

$$T = m \times B$$

where m is the magnetic moment. If the magnet is a small coil of area dA carrying a current I, then the magnetic moment m is equal to IdA.

(2) *Syn.* (*obsolete*) magnetic dipole moment. Symbol: p_m; unit: weber metre. The product of magnetic moment m and the *permeability of free space, μ_0:

$$p_m = \mu_0 m$$

Historically magnets and small circulating currents were considered to be *dipoles consisting of two equal and opposite magnetic poles, analogous to electric charges, separated by a distance r (the magnetic length). The magnetic dipole moment p_m was the product of the magnetic pole strength and the magnetic length; it was determined by the torque, T, experienced when the magnet was placed in a magnetic field H. Thus p_m was determined by the relation:

$$T = p_m \times H$$

Modern usage strongly favours the first definition, the second definition being very rarely used. Small magnets, however, are frequently referred to as dipoles, particularly on an atomic scale, and the magnetic moment m termed the magnetic dipole moment. In modern texts these terms are defined by the first definition above. The term 'magnetic moment' has been used wherever possible throughout this dictionary as defined in (1) above.

magnetic monopole A hypothetical magnetic particle with a single magnetic charge of either north or south. Such a particle would be analogous to the electrical particles the electron and proton. *Maxwell's equations would prove completely symmetrical if the existence of such particles were to be proved. Magnetic monopoles were postulated on conservation and symmetry principles and are thought to be more massive than nucleons. Their existence is not barred by quantum theory nor classical electromagnetic theory and it has been suggested that they occur in extremely high energy cosmic rays. No proof of their existence or otherwise has yet been found despite intensive searches for them.

magnetic pick-up *See* pick-up.

magnetic pole A region of concentrated magnetism in a magnet analogous to an electrostatic point charge. Historically a magnet was considered as being formed from two magnetic poles of opposite types (north and south) located near its ends. Lines of magnetic force converge on or diverge from the magnetic poles. Use of the concept of magnetic poles allowed the theory of magnetostatics to be developed along similar lines to that of electrostatics, by applying the inverse square law of forces to these imaginary poles. This approach to magnetostatics however requires the use of the magnetic monopole for its development; also the precise location of the magnetic poles is indeterminate. Modern practice favours the use of *magnetic moment instead.

magnetic potential *Obsolete syn. for* magnetomotive force.

magnetic printing (1) *Syn.* magnetic transfer. Copying of a recorded signal on one section of a magnetic recording medium by making intimate contact between the section containing the recorded signal and the magnetic material on to which it is to be copied.

(2) A method of printing using ink containing magnetic material so that the print may be read by a suitable sensing device as well as visually.

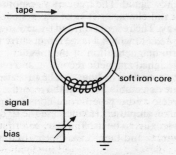

Magnetic tape recording

magnetic recording A method of recording electrical signals on a magnetic medium to produce a reasonably permanent easily reproducible record that is flexible in use and produces an accurate reproduction of the original signal.

323

The best known application of magnetic recording is sound recording on *magnetic tape* (*tape recording*). In magnetic sound recording the magnetic tape is moved at a uniform speed past the poles of an electromagnet and is longitudinally magnetized. Variations in the audiofrequency current supplying the electromagnet produce corresponding variations in the magnetization. During reproduction the process is reversed: the tape is fed past an electromagnet (the *replay head*) and the variations of magnetization induce currents in the coils corresponding to the original magnetizing currents.

The recording medium is usually made from finely divided ferrous oxide particles deposited on a plastic backing known as the base. The base is cellulose acetate tape 0·25 inches wide. Heat-treated steel tape or wire has also been used as the recording medium.

The electromagnet used to record, reproduce, or erase the signal on the tape is called a *head* (or *magnetic head*). It is possible to use a single *read-write head* to perform each separate function but commercial tape recorders usually use separate heads for recording, reproducing, and erasing. A typical head consists of soft iron pole pieces wound with wire coils. The distance between the pole pieces is the *gap length* and a good recorder may have a gap length as small as 0·6 mm allowing a sharper record and thus more faithful reproduction. The magnetic tape completes the magnetic circuit (*see* diagram) and the magnetization produced represents the flux pattern in the gap at the moment the tape leaves the gap. During recording *magnetic bias* is applied to the *recording head*. This is an alternating current of frequency between 60 and 100 kilohertz superimposed on the audiofrequency signal. The frequency response, distortion, and signal-to-noise ratio characteristics of the system are improved by biasing in this way. The record is erased by causing the tape to move past the *erasing head* to which a large direct current is applied. This produces uniform magnetization of the magnetic material.

With a well-designed magnetic recording and reproducing system very high quality sound recordings can be produced with very little difference in tone detectable between the recording and the original source. A complete audio tape device consists of microphone, or other signal source, amplifiers (to increase the signal amplitude and allow small gap lengths to be used), erase, record, and replay heads, tape-carrying system and tape-drive motor, output amplifiers, and loudspeakers.

Multitrack tapes are available containing one, two, or four separate recording tracks. The ratio of maximum to minimum sound levels that may be recorded and reproduced successfully is about 60 decibels; this figure is better than that for other sound recording systems. Crosstalk between adjacent tracks on a tape is small. Audio tape devices for broadcast purposes are required to have a signal-to-crosstalk ratio not less than 60 decibels in the range 200 hertz to 10 kilohertz.

Commercial tape recorders are usually either reel-to-reel or cassette types. The former type can usually take large reels of tape allowing very much longer playing time than the cassette type, which are limited by the minimum thickness of tape that can be used without breakage. The cassette recorder however is easier to use and the cassettes, which are permanently encased, are easy to store without damage from dust or mishandling. They are however less versatile.

Magnetic recording is also used for magnetic memories in computers and data-processing systems. Data can be stored on and retrieved from magnetic tape or from a disk or drum (*see* moving magnetic surface memory).

magnetic residual loss *See* ferromagnetism.

magnetic resistance *See* reluctance.

magnetic saturation *See* ferromagnetism.

magnetic screening The use of a screen of high *permeability magnetic material in order to protect electric circuits, devices, or other apparatus from the effects of magnetic fields. Magnetic screening is often used to surround sensitive alternating-current measuring instruments and thus shield them from stray external magnetic fields.

magnetic shell A thin sheet of ferromagnetic material, such as iron, magnetized across its thickness so that the two sides have opposite polarity. It can be considered as an infinite number of small bar magnets and the strength is expressed as *magnetic moment per unit area.

magnetic shift register A *shift register consisting of an array of magnetic cores. As each new pulse is received at the input the magnetization pattern of each core shifts along one position.

magnetic shunt A means of varying the useful *magnetic flux of a magnet in an electrical measuring instrument. It consists of a piece of magnetic material mounted near the magnet and adjustable in position relative to the magnet. It is often used to extend the range of the measuring instrument.

magnetic susceptibility *See* susceptibility.

magnetic tape (1) *See* magnetic recording. (2) *See* moving magnetic surface memory.

magnetic-tape unit (MTU) *See* moving magnetic surface memory.

magnetic transfer *See* magnetic printing.

magnetic transition temperature *Syn. for* Curie point. *See* ferromagnetism.

magnetic tuning Tuning of a very high frequency (microwave) oscillator by means of a ferrite rod in the cavity resonator of the oscillator. The magnetization of the rod is determined by an external steady magnetic flux density, and is varied to provide a different frequency in the cavity by altering the flux density.

magnetism The phenomena associated with regions containing magnetic flux. Magnetic properties were first noticed in the naturally occurring oxide of iron, magnetite. Ampère discovered that a small coil carrying a current behaves like a magnet; he suggested that the origin of all magnetism lay in small circulating currents associated with each atom. Ampère's theory, which gave a natural explanation of the fact that no isolated magnetic pole had ever been observed, is essentially similar to modern atomic theory: his elementary current circuits (Amperean currents) are the motions of the negatively charged electrons in closed orbits around the positively charged atomic nucleus.

All materials exhibit magnetic properties, the nature of those properties depending on the distribution of electrons in the outer orbits of the atoms. *Diamagnetism is a weak effect, common to all materials and resulting from the orbital motion of the atomic electrons. *Paramagnetism occurs in certain materials that have a permanent molecular *magnetic moment due to electron *spin. It is a stronger effect than diamagnetism, opposed to it, and masks it in paramagnetic materials. Some paramagnetic materials, such as iron, also display *ferromagnetism (see also ferrimagnetism, antiferromagnetism) at temperatures below the Curie point. Ferromagnetic materials can produce a substantial magnetic flux density and some are suitable for use as *permanent magnets. A magnetic field can also be produced by an electric current (see electromagnet).

magnetite *Syn.* lodestone. A naturally occurring oxide of iron ($FeO.Fe_2O_3$) that exhibits *ferrimagnetism.

magnetization *Syn.* intensity of magnetization. Symbol: M; unit: ampere per metre. A measure of the magnetic polarization that occurs when a material is placed in a magnetic flux. It is defined as the magnetic moment per unit volume and is the product of magnetic field strength, H, and magnetic *susceptibility, χ_m:

$$M = H\chi_m$$
$$M = B/\mu_0 - H$$

where B is the magnetic flux density and μ_0 is the *permeability of free space.

magnetization curve *See* magnetic hysteresis; ferromagnetism.

magnetize To induce a magnetic flux density in a material; to cause a material to exhibit magnetic properties.

magnetizing force *See* magnetic field strength.

magneto An alternating-current electrical *generator, particularly of the *synchronous a.c. generator type coupled to an induction coil, that uses a *permanent magnet to provide the magnetic flux rather than an electromagnet.

magnetocaloric effect *Syn.* thermomagnetic effect. A fall in temperature occurring in certain paramagnetic salts when demagnetized adiabatically. *Adiabatic demagnetization* is achieved by magnetizing the salt between the poles of an electromagnet, cooling to remove the heat produced by the magnetization, and then isolating the system thermally before switching off the electromagnet. The resultant demagnetization occurs adiabatically and a fall in temperature of the specimen is produced. The effect increases as the initial temperature of the salt is decreased and by a process of repeated magnetization and adiabatic demagnetization, temperatures close to absolute zero may be achieved.

magnetomotive force (m.m.f.) Symbol: F_m; unit: ampere; ampere-turn. The line integral of *magnetic flux density, B, for a closed path, divided by the magnetic constant μ_0:

$$F_m = \mu_0^{-1} \oint B.dl$$

It is equal to the total conduction current linked, during one traverse of the closed path. If the path encloses a current I the value of F_m increases by I for every traverse, i.e. F_m is multivalued unlike electromotive force, which is single-valued.

magneto-optical effect *See* Kerr effects.

magnetoresistance A change in the resistance of a ferromagnetic material when it is subjected to a magnetic flux. It is closely allied to *elastoresistance and is caused by the stress in the material due to *magnetostriction. The magnetoresistive effect is an inverse function of the magnetostriction, i.e. positive magnetostriction causes a decrease in the resistance and negative magnetostriction causes an increase in the resistance.

magnetostriction Mechanical deformation of a ferromagnetic material when subjected to a magnetic field. The effect arises because of internal stresses in the material due to the anisotropy energy required to magnetize it in certain directions relative to the crystal axes (*see* ferromagnetism). Conversely, when subjected to mechanical stress a change in the magnetization of the material is observed.

Joule magnetostriction (or *positive magnetostriction*) is an increase in the length of a rod or tube of ferromagnetic material when an axial magnetic flux is applied to it. *Negative magnetostriction* occurs in materials, such as nickel, that suffer a decrease in length with an increase in magnetic flux density. Magnetostriction also affects deformed ferromagnetic samples. A bent bar tends to straighten under the influence of a magnetic flux applied along its length (*Guillemin effect*); a twisted bar also tends to straighten under similar conditions (*Wiedemann effect*).

High-frequency alternating fields set up longitudinal vibrations in the sample, which thus acts as a source of sound waves; if the frequency of these vibrations corresponds to the natural frequency of the sample large values of amplitude occur. Magnetostrictive vibrations have many applications. At ultrasonic frequencies they form a source of ultrasonic energy that can be used commercially, for example for ultrasonic cleaning or for breaking the oxide film on aluminium for soldering. *Magnetostriction oscillators utilize the magnetostrictive effect of alternating currents to produce frequency-controlled oscillations at frequencies in the range 25 000 hertz downwards.

Magnetostriction is also used to provide the vibrating part of a *loudspeaker; conversely audio vibrations applied to a magnetostrictive rod produce corresponding flux changes, as in a magnetostriction *microphone.

magnetostriction loudspeaker *See* loudspeaker.

magnetostriction microphone *See* microphone.

magnetostriction oscillator An *oscillator that uses the phenomenon of *magnetostriction to produce oscillations or to control the frequency of an oscillator. In a simple magnetostriction oscillator a bar of magnetostrictive material is magnetized by a direct magnetic field and a coil or coils wound coaxially around it. A *tuned circuit is connected in series with the coil. Direct voltage applied to the tuned circuit causes oscillations that are normally quickly damped (*see* free oscillations). If the frequency of the tuned circuit corresponds to the natural frequency of vibration of the magnetostrictive bar, the oscillations are maintained because of the interaction of the stresses in the bar and the associated changes of permeability.

A magnetostrictive rod may be used to control the frequency of an oscillator in a similar manner to a piezoelectric crystal (*see* piezoelectric oscillator). The frequency of the oscillator is adjusted to be almost that of the natural frequency of the magnetostrictive rod; the induced vibrations in the bar are used to pull the frequency of the oscillator and maintain it at a substantially constant value.

A magnetostriction oscillator may also be used as a source of sound energy. The induced longitudinal vibrations of the bar are used to provide sound waves. Such oscillators are very efficient when used under water.

magnetostriction strain gauge *See* strain gauge.

magnetostriction transducer *See* magnetostrictor.

magnetostrictor *Syn.* magnetostriction transducer. A transducer that utilizes *magnetostriction to convert electrical energy into mechanical energy or vice versa. The most common method is to clamp a rod or tube of ferromagnetic material at its centre and pass an alternating current through a coil wound around the length of the material.

magnetron *Syn.* magnetron oscillator. A crossed-field *microwave tube that produces radiofrequency (r.f.) oscillations in the microwave region. An early magnetron was used as a rectifier – the *recti-*

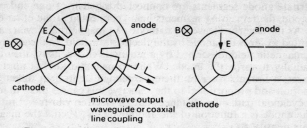

Resonant-cavity magnetron   **Split-anode magnetron**

a Basic types of magnetron

fier magnetron – but all modern magnetrons are designed as oscillators.

The basic magnetron consists of a central cylindrical cathode surrounded by a cylindrical anode divided into two or more segments or containing several *cavity resonators (Fig. *a*). A steady electrostatic field is applied between the anode and cathode. A steady magnetic flux density is applied parallel to the cylindrical axis and orthogonal to the electrostatic field. The magnetic flux is usually provided by a permanent magnet or sometimes by an electromagnet. The symbol for the permanent magnet type is shown in Fig. *b*. Electrons, emitted from the cathode, move under the influence of these two fields. The interaction of the electrons with the gaps or the resonant cavities of the anode produces radiofrequency oscillations. These oscillations are output through a coupled waveguide or coaxial line.

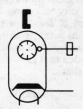

b Symbol for magnetron with permanent magnet

Most magnetrons are of the *multicavity* type in which the anode contains several resonant cavities coupled together, although the *split-anode* type is sometimes used (Fig. *a*). The most common form of split-anode magnetron is the *interdigital magnetron,* in which al-

329

ternate anode segments are connected together at one end of the anode, the remaining segments being connected at the other end.

As the electrons leave the cathode of a magnetron they are accelerated towards the anode by the electrostatic field. In the absence of a magnetic field, described by a magnetic flux density, they travel radially towards the anode. When a magnetic field is applied it exerts a Lorentz force on them perpendicular to their direction of motion and proportional to the velocity; this causes them to follow a cycloidal path. The distance that an electron can travel towards the anode is a function of the anode voltage, V, and the magnetic flux density B.

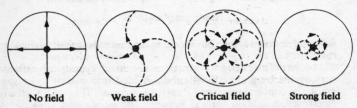

| No field | Weak field | Critical field | Strong field |

c **Effect of magnetic field on electrons in magnetron**

For a given value of anode voltage, the *critical field* is that value of B at which an electron just fails to reach the anode and returns to the cathode with zero kinetic energy. The *critical voltage* is the maximum anode voltage, in the presence of a fixed magnetic flux density, at which an electron just fails to reach the anode. The effect of an increasing magnetic field with fixed anode voltage is shown in Fig. c. Under the strong field condition, when the value of B exceeds the critical field, an electron gains kinetic energy; it returns to the cathode with a nonzero velocity following a cycloidal path of relatively small radius. The effect of the strong field is thus to produce a narrow *sheath* of electrons rotating about the cathode with an angular velocity ω. If the radii of the anode and cathode are b and a respectively, then provided that $(b - a)$ is small compared to both b and a it can be shown that

$$\omega = 2V/[B(b^2 - a^2)]$$

The rotating sheath of electrons interacts with the resonant cavities or gaps in the anode structure to produce radiofrequency oscillations in them; the radiofrequency oscillations in turn interact with the electrons in a complex manner: the electrons are either accelerated by the r.f. field and turned back to the cathode or are decelerated by it and travel to the anode giving up energy to the r.f. field as they do so (Fig. d). On average the net power gained by the r.f. fields

330

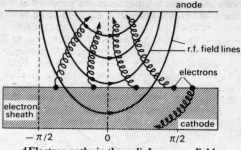

d Electron paths in the radiofrequency field

when an electron loses kinetic energy is greater than that required to return one to the cathode; r.f. oscillations are therefore sustained by the system. The closed nature of the circuit effectively supplies the positive feedback required for the oscillations to occur.

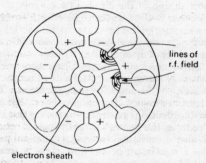

e Electron configuration in a π-mode magnetron

There are various possible modes of operation depending on the geometrical structure of the anode, the magnitudes of the fields, and the phase differences of the r.f. fields between successive cavities. For a particular design of anode the magnitudes of V and B must be adjusted so that the angular velocity of the electrons is synchronized with the alternation of the r.f. fields in the cavities so as to produce the optimum transfer of energy. When properly adjusted, an efficiency of about 70% is possible with *π-mode operation;* this is the simplest and most efficient mode of operation in which the phase difference between successive cavities is π. The electrons are retarded by several successive cavities when the proper phase relationship is maintained and travel towards the anode in 'spokes' (Fig. *e*).

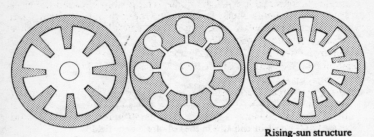

Rising-sun structure

f Typical anode structures

Typical anode structures are shown in Fig. *f*. The *rising-sun magnetron* contains alternate cavities of two different resonant frequencies, allowing different modes of operation to be selected at the output. Suitable design of the anode of a magnetron allows either standing waves or travelling waves to be output from the device. A *travelling-wave magnetron* has several possible modes of oscillation; the relationship between these different modes and the anode voltage for various values of magnetic flux density is represented graphically on a *Hartree diagram*. The performance of a *standing-wave magnetron* is usually shown on a *Rieke diagram.

Excessive heating of the anode by incident electrons is avoided by constructing it from a material, such as copper, that has good thermal conductivity. Electrons returning to the cathode produce *back heating;* this reduces the heater current required when the tube is running and also stimulates *secondary emission of electrons, which provides a significant proportion of the total cathode emission.

Large power outputs are possible by running the tube in short pulses (*pulsed magnetrons*) rather than continuously; this gives an improvement of output power of up to 1000. A typical medium high power magnetron, pulsed for one microsecond at a repetition rate of 1000 hertz, can produce 0·1 mm wavelength waves with a power output during the pulse of 750 kilowatts. It requires an anode voltage, V, of 31 kilovolts and magnetic flux density, B, of 0·28 tesla.

Most magnetrons are *fixed-frequency magnetrons* but *tunable magnetrons* have been produced in which a variation of 10–20% in the frequency is achieved by means of plungers moved into the resonators from one end.

Magnetron arcing is a fault condition that happens most frequently in high-power devices: internal breakdown occurs in the form of an arc between the cathode and anode. Sudden changes in frequency sometimes occur due to *mode jumping,* when the operational mode changes suddenly if the phase relationship between the electrons and r.f. fields alters. Faults in the input or output circuits can cause small changes in the output frequency known as *frequency pushing* or *pulling,* respectively.

magnetron effect *See* thermionic valve.

main anode *See* mercury-arc rectifier.

main gap *See* starter electrode.

mains The source of domestic electrical power distributed nationally throughout the U.K. The *mains frequency* is the frequency at which the electrical power is supplied. This is 50 hertz in the U.K. and usually 60 hertz in the U.S.

mains hum *See* radio noise.

main store *See* memory.

maintaining voltage *See* gas-discharge tube.

major cycle *See* serial memory.

majority carrier The type of charge *carrier in an extrinsic *semiconductor that constitutes more than half of the total charge carrier concentration.

major lobe *See* radiation pattern.

make (1) To close a circuit by means of a switch, *circuit-breaker, or similar device.

(2) The maximum gap between the contacts of such a device for the circuit to be closed. *Compare* break.

make-and-break A type of switch that is automatically activated by the operation of the circuit in which it is incorporated, and that repetitively makes and breaks the circuit. The best known application of a make-and-break is in an electric bell circuit.

Malter effect An effect observed when a *semiconductor of high *secondary-emission ratio, such as caesium oxide, is subjected to electron bombardment. The semiconductor can become strongly positively charged and if a layer of such a material is separated by a thin insulating layer from a metal plate it can be used to develop a potential difference. With an insulating layer of about 0·1 micrometre thickness a potential of up to 100 volts can be produced.

manganese Symbol: Mn. A metal, atomic number 25, that is used in alloys such as *manganin and *Heusler alloys. The element is also used in some forms of primary *battery.

manganin An alloy that consists of 70–86% copper, 15–25% manganese, and 2–5% nickel and has a high coefficient of electrical *resistivity, low *temperature coefficient of resistance, and low contact potential (*see* contact). It is used in the manufacture of precise wire-wound *resistors.

man-made noise *See* noise.

Mansbridge capacitor *See* self-sealing capacitor.

Marconi aerial Originally the simple vertical wire *aerial, earthed at its lower end, used by Marconi. The term is now applied to vertical aerials one quarter wavelength long. Usually the aerial is connected to earth by a series reactance, which may be either an inductance or a capacitor, of such a value that together with the aerial it is electrically equivalent to one quarter wavelength. Usually the series reactance is variable, allowing the aerial to be tuned to any resonant frequency within a designated range.

marker pulse *See* time-division multiplexing.

mark-space ratio The ratio in a pulse waveform of the pulse duration to the time between successive pulses. In a perfect square wave the mark-space ratio is unity.

Marx effect A decrease in the stopping potential required to prevent the escape of photoelectrons from the surface of illuminated sodium or potassium. It occurs when the material is also illuminated by light of lower frequency than that causing the photoelectric release.

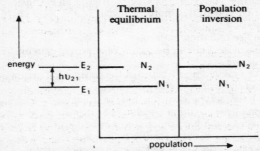

a Energy levels and populations in a maser

maser Acronym from *mi*crowave *a*mplification by *s*timulated *e*mission of *r*adiation. A source of intense coherent monochromatic radiation in the *microwave region of the electromagnetic spectrum. The maser can be used as a microwave amplifier or oscillator.

Planck's law states that an atom can jump from one allowed energy level, E_2, to another, E_1, accompanied either by the emission or absorption of a *photon of electromagnetic radiation:

$$|E_2 - E_1| = h\nu$$

where h is the *Planck constant and ν the frequency of the radiation. If E_2 is greater that E_1 emission occurs; if E_2 is less than E_1 absorption occurs. A *spontaneous transition* from a higher energy level to a lower energy level occurs when the atom changes energy spontaneously without any external stimulus. A *stimulated transition* occurs when the atom changes energy in response to externally applied electromagnetic radiation. Energy is absorbed from the external source (i.e. attenuation occurs) when the stimulated transition is from a lower to a higher level; energy is given to the source (i.e. amplification occurs) in the opposite direction.

The overall power absorption, P_{abs}, is given by

$$P_{abs} = h\nu_{21}(W_{12}N_1 = W_{21}N_2)$$

where N_1 is the number of atoms of energy E_1, N_2 is the number of atoms of energy E_2 ($E_2 > E_1$), and W_{12}, which is equal to W_{21}, is the probability of a transition occurring per unit time and is proportional to the intensity of the applied radiation.

The basic maser process occurs when the external signal is amplified, i.e. the net power flow is from the atoms to the signal and $P_{abs} < 0$. This condition is satisfied when $N_2 > N_1$, i.e. when more electrons are in the higher energy state than the lower energy state, and is known as *population inversion*. Population inversion is a non-equilibrium state and various relaxation processes tend to restore the population to the thermal equilibrium state (Fig. *a*). Power must therefore be supplied to the system in order to maintain an inverted population.

Many different types of maser exist. They differ in the material used and the method of producing the population inversion. The first was a *gas maser* in which the maser material was ammonia. A gas maser has a low density of atoms and therefore the total power available is low and the bandwidth is very narrow. Gas masers are unsuitable as microwave amplifiers but when used as oscillators have high spectral purity and may be employed as frequency standards. There are two main types of gas maser – the beam maser and the optically pumped gas-cell maser (Fig. *b*).

In the *beam maser* a beam of gas molecules is focused by an inhomogeneous electric field so that excited molecules are concen-

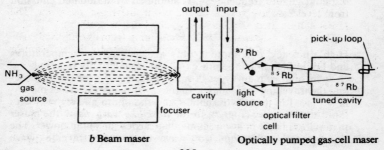

b Beam maser   **Optically pumped gas-cell maser**

335

trated in the centre of the beam. The beam then passes through a small hole into a *cavity resonator. The focusing effect is such that an inverted population enters the cavity resonator, where oscillations are built up by stimulated emission in the cavity. The *optically pumped gas-cell maser* consists of a cavity resonator containing the gas, such as rubidium (Rb). The gas is exposed to a light source of the appropriate frequency to produce population inversion. With sufficient light intensity oscillations build up in the cavity.

Solid-state masers use paramagnetic materials, such as ruby, which have multiple energy levels and transitions involving photons of microwave energies. In the *three-level maser* three energy levels are involved (Fig. *c*). Population inversion is produced by applying radiation of energy $h\nu_{31}$, where

$$hv_{31} = E_3 - E_1$$

The frequency ν_{31} is termed the *pump frequency*. The population of level 3 is increased and that of level 1 decreased, until when N_3 is

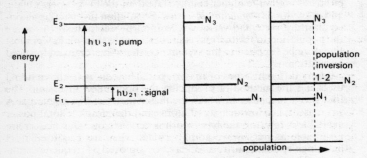

c Energy levels and populations in a 3-level maser

approximately equal to N_1 the 1-3 transition is said to be saturated. The relative populations of level 2 and level 1 are then inverted and a small signal of frequency ν_{21} is amplified by stimulated emission from level 2:

$$hv_{21} = E_2 - E_1$$

High-efficiency operation is only achieved at low temperatures and liquid helium temperature (4·2 kelvin) is usually required.

There are two main types of solid-state maser – the cavity maser and the travelling-wave maser. In the *cavity maser* the maser material is placed in a cavity resonator and the pump and signal waves input to the cavity (Fig. *d*). In the *travelling-wave maser* the maser material is placed in a waveguide that carries the pump power. The signal wave is carried on a slow-wave structure (*see* travelling-wave

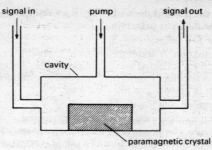

signal in   pump   signal out

cavity

paramagnetic crystal

d Solid-state cavity maser

tube) surrounding the maser material and interacts with it along its length so that radiofrequency power amplification results.

Microwave maser amplifiers are used in radio and radar astronomy, radiometry, terrestrial radar, communications, telemetry, satellite tracking and communications, and space communication.

mask (1) A device used to shield selected areas of a semiconductor *chip during the manufacture of semiconductor components and *integrated circuits. During the *photolithography process, a set of photographic masks is required to define the openings in the oxide layer through which the various *diffusions are made, the windows through which the metal contacts are formed, and the pattern in which the desired metal interconnections are formed. The photographic masks used are either emulsion on glass or an etched thin film of chromium or iron oxide on glass; they are produced by photographic reduction from large-scale layouts.

(2) A device made from metal foil and used during the manufacture of *thin-film circuits in order to define the pattern of material deposited as a thin film by vacuum evaporation on to the substrate.

Massey formula *See* secondary emission.

mass resistivity The product of the mass and resistance of a conductor divided by the square of its length. It equals the product of resistivity and density of the material.

master *See* recording of sound.

master frequency meter *See* integrating frequency meter.

master oscillator An *oscillator of extremely high inherent frequency stability that is used to drive a power *amplifier in order to supply a substantial power output to the load. This arrangement is used where substantial output power is required with a high degree of frequency stability, as in the generation of a *carrier wave, since the frequency of an oscillator is partially dependent on the load supplied by it. The master oscillator 'sees' a constant load from the

*buffer between it and the multistage power amplifier and hence the frequency stability is maintained.

master trigger *See* radar.

matched termination A termination to a *network or *transmission line at which no reflected waves are produced. A load that absorbs all the power incident from a transmission line and forms a matched termination is known as a *matched load*.

matched waveguide *See* waveguide.

matching *See* impedance matching; load matching.

Mathiessen's rule The product of the electrical *resistivity and *temperature coefficient of resistance of a metal is constant and independent of the number of impurities or defects present in the crystal. The presence of crystalline defects causes a marked increase in the resistance of a metal but this is offset by a corresponding decrease in the temperature coefficient of resistance.

matrix parameters *See* network; transistor parameters.

maximum power theorem If a variable load is to be matched to a given power source in order to achieve the maximum possible power dissipation in the load, the resistance of the load must be made equal to the internal resistance of the source. The *available power* then obtainable is $V_o^2/4R_i$, where V_o is the open-circuit electromotive force of the source and R_i the internal resistance.

The converse problem, that of matching a power source to a given load in order to achieve maximum power dissipated in the load, is not solved by this theorem. In this case the power source with the lowest internal resistance gives the maximum power.

This theorem can be modified to apply to alternating-current linear networks. The impedances Z_S and Z_L of the voltage generator and load, respectively, contain an imaginary term:

$$Z_S = R_S + iX_S$$
$$Z_L = R_L + iX_L$$

For maximum power dissipation in the load it is necessary to satisfy the conditions given by

$$X_S + X_L = 0$$
$$R_S = R_L$$

maxwell Symbol: Mx. The unit of magnetic flux in the obsolete CGS electromagnetic system of units. One maxwell equals 10^{-8}weber.

Maxwell bridge A four-arm *bridge for measuring inductance in terms of a capacitance and resistances (*see* diagram). At balance, as indicated by a null response on the instrument, I,

$$R_s R_x = R_b R_a$$
$$L_x = R_b R_a C_s$$

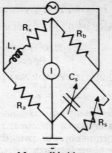

Maxwell bridge

Maxwell's equations A set of classical equations relating the vector quantities applying at any point in a varying electric or magnetic field. The four basic equations are:

$$\text{curl } \boldsymbol{H} = \partial \boldsymbol{D}/\partial t + \boldsymbol{j}$$
$$\text{div } \boldsymbol{B} = 0$$
$$\text{curl } \boldsymbol{E} = -\partial \boldsymbol{B}/\partial t$$
$$\text{div } \boldsymbol{D} = \rho$$

$\boldsymbol{H}$ is the magnetic field strength, $\boldsymbol{D}$ the electric displacement, t is time, $\boldsymbol{j}$ is the current density, $\boldsymbol{B}$ the magnetic flux density, $\boldsymbol{E}$ the electric field strength, and ρ the volume charge density.

From these equations Maxwell deduced that each field vector obeys a wave equation; he also showed that in free space, where $\boldsymbol{j} = 0$ and $\rho = 0$, the solutions represent a transverse wave travelling through space with the velocity of light. These waves are known as electromagnetic waves (*see* electromagnetic radiation). Further work showed that certain properties of these waves, i.e. reflection, refraction, and diffraction, are identical to the properties of light waves and that light waves are a form of electromagnetic radiation.

Maxwell's theory deals only with macroscopic phenomena and does not offer an explanation of phenomena arising from interactions on an atomic scale, such as dispersion and the photoelectric effect. On an atomic scale it has been found necessary to introduce the quantum mechanical theory of electromagnetic radiation.

From Maxwell's equations it is possible to deduce the wave velocity in a medium as

$$v = 1/\sqrt{(\mu\varepsilon)}$$

where ε is the *permittivity of the medium and μ is its *permeability, $\mu = \boldsymbol{B}/\boldsymbol{H}$. In vacuo the wave velocity is given by

339

$$c = 1/\sqrt{(\mu_0\varepsilon_0)}$$

Thus in a nondispersive medium of refractive index n, where

$$n = c/v,$$
$$n^2 = \mu_r\varepsilon_r$$

or in a nonferromagnetic material where $\mu_r \simeq 1$,

$$n^2 = \varepsilon_r$$

where ε_r is the relative permittivity of the medium and μ_r the relative permeability. This is known as *Maxwell's formula*. In a dispersive medium the above formula applies provided that all measurements are carried out at the same frequency.

Maxwell's formula *See* Maxwell's equations.

Maxwell's rule In an electric circuit linked by a magnetic flux each part of the circuit experiences a force causing it to tend to move in such a direction as to enclose the maximum possible magnetic flux.

MBE *Abbrev. for* molecular beam epitaxy.

mean current density *See* current density.

mean life The mean time to *failure of a device, component, or any part or subsystem that can be separately tested. For any particular item the mean life will be based on the results of *life tests and one or more of the following will be quoted.

The *observed mean life* is the mean value of the observed times to failure of all specimens in a sample of items under stated stress conditions. The criteria for what constitutes failure should be stated.

The *assessed mean life* is the mean life of an item determined as a limiting value of the confidence interval with a stated probability level based on the same data as the observed mean life of nominally identical items. The following conditions apply: the source of the data should be stated; results may be combined only when all conditions are similar; it should be stated whether one- or two-sided intervals are being used; the lower limiting value is usually used for mean life statements and the assumed underlying distribution should be stated.

The *extrapolated mean life* is the extension of the observed or assessed mean life by a defined extrapolation or interpolation for stress conditions different from those applying to the conditions of the assessed mean life.

The elapsed time at which a stated proportion (q per cent) of a sample or population of items has failed is the q-*percentile life*. This is quoted as the observed percentile life, assessed percentile life, and extrapolated percentile life under similar conditions to those quoted for mean life.

measurand *See* transducer.

medium frequency (MF) *See* frequency band.

medium-scale integration (MSI) *See* integrated circuit.

medium voltage A voltage in the range 250 to 650 volts: used in electrical power transmission and distribution systems.

medium wave A radiowave that has a wavelength in the range 0.1 to 1 km, i.e. of frequency between 3000 and 300 kilohertz. *See* frequency band.

mega- Symbol: M. (1) A prefix to a unit, denoting a multiple of 10^6 (i.e. 1 000 000) of that unit: one megavolt equals 10^6 volts.

(2) A prefix used in computing to denote a multiple of 2^{20} (i.e. 1 048 576): one megabyte equals 2^{20} bytes.

megaphone A portable device used to amplify and direct sound. It contains a microphone, amplifier, and loudspeaker in a conical horn together with a battery power supply and acts as a directional loudspeaker.

megger *Tradename* A portable insulation tester calibrated directly in megohms.

Meissner effects *See* superconductivity.

mekometer *See* Kerr effects.

meltback transistor A type of bipolar junction *transistor in which the junctions are formed by melting the semiconductor, doping to the required polarity, and then allowing it to solidify.

memory *Syn.* store. Any device associated with a *computer that is used to store information, such as *programs or data, in digital form. The minimum region of any memory device required to store one unit of information is a *storage element*. The *memory capacity* is the total number of *bits that can be stored in any given memory device or computer system. A large computer system contains several different types of memory organized into a distinct hierarchy according to the availability of information stored in it. The *internal memory* comprises various memory circuits that form part of the main computer and are under the direct control of the *central processing unit. The *external memory* consists of various storage devices that are separate from the main computer but available to it through appropriate *peripheral devices.

The internal memory of a computer includes the *main store* in which the main program and data are stored; part of the main store is used as a *temporary memory* in which intermediate or partial results can be stored during the performance of a program. Various standard programs and sets of commonly used data are also stored in the internal memory, in various fixed *read-only memories (ROMs), and can be accessed by the main program.

The performance of a computer depends on the speed of operation. Various memory devices have very different *access times.* The access time is the time interval between a read pulse being received by the memory and the output of the required *bits from the memory. The *write time* is the time interval needed for a bit of information to be entered into a memory location. The method of retrieving the information depends on the nature of the memory device: *random access memories (RAMs) are inherently faster than *serial memories. The various types of memory can also be classified according to the physical nature of the device: *volatile memory* is a device in which the stored information is lost when the power supply is switched off; *permanent memory* retains the information when the power ceases and is thus *nonvolatile*. Volatile memory is either *static or *dynamic memory. Nonvolatile memory can be either *read-only memory or *read-write memory; read-only volatile memory is not possible.

The memory hierarchy is organized so that the fastest and therefore most expensive memory is that portion of the memory that interacts directly with the central processing unit (CPU). A large computer system contains a fast *buffer memory* between the CPU and the main store. The control unit of the CPU organizes the transfer of data and program routines so that only those portions required for immediate use are contained in the buffer, other portions being retained in the slightly slower main store and transferred as required into the buffer. The slowest memory devices are those that are peripheral to the main computer and contain information that is required only infrequently.

Memory devices can be roughly classified into two categories: *moving magnetic surface memories, such as disk or magnetic tape units, usually have the largest storage capacity, are nonvolatile, and are the cheapest devices; *electronic memories, such as *solid-state memories, have no moving parts and today have thousands of memory locations and high operating speeds. *See also* magnetic bubble memory; Josephson memory.

memory capacity *Syn.* storage capacity. *See* memory.

memory location *Syn.* storage location. A storage element that has a unique *address in a *memory.

mercury Symbol: Hg. A metal, atomic number 80, that is the only metal that is a liquid at room temperatures. Its main applications in electronics are as the cathode in *mercury-arc rectifiers, and as an amalgam in some electrolytic cells.

mercury-arc rectifier *Syn.* pool rectifier. A cold-cathode arc-discharge tube that has a mercury-pool cathode. When a sufficiently high voltage is applied to the anode an *arc is struck across the tube. The tube is used as shown in the diagram. As the alternating-current input is applied the arc oscillates between anode 1 and anode 2 as each becomes positive in turn. Direct current is output to the load.

Many types of mercury-arc rectifier contain an extra anode known as the *holding anode*. This anode is supplied with a direct voltage and is used to maintain the arc when the *main anode(s)*, i.e.

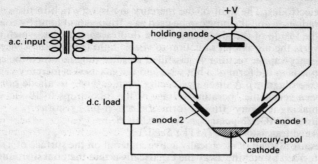

Mercury-arc rectifier

the anode(s) supplying the load, draw no current. Use of such an electrode to maintain the arc prevents *misfire,* i.e. a failure to produce the arc during an intended conduction interval.

The main anode is often surrounded by a metal shield – the *anode shield* – to protect it from massive ionization or radiation. One type of mercury-arc rectifier has an anode in the form of a *steel tank*. This type of anode is very robust. An *arc baffle* is often used to prevent mercury splashing from the cathode on to the anode. One version of the mercury-arc rectifier uses a small auxiliary electrode dipped into the cathode to start the arc (*see* ignitor).

After the arc has been struck a typical voltage drop across the tube is 20 to 25 volts; such tubes are thus more efficient than high-vacuum rectifier tubes.

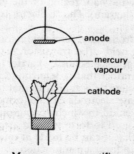

Mercury-vapour rectifier

mercury-vapour lamp An arc-discharge tube used as a lamp. The gas in the tube is mercury vapour and the arc is struck between mercury

343

electrodes. The light of the mercury arc is very rich in ultraviolet radiation and the tube can be used as a *fluorescent lamp by coating the inside of the glass tube with a fluorescent powder, which converts the ultraviolet radiation to visible light (*see* luminescence).

mercury-vapour rectifier A gas-filled rectifier tube in which the cathode is in the form of a hot wire and the gas used is mercury vapour (*see* diagram). A discharge occurs from cathode to anode but not vice versa; the operating voltage of the tube drops to 10–15 volts making it a very efficient form of electron-tube rectifier.

merged CMOS/bipolar *Syn. for* bi-CMOS.

merged transistor logic (MTL) *See* I²L.

mesa A plateau of electrically active material on the surface of a slice, formed by etching away the electrically active material surrounding the location where a mesa is to be formed. Formation of mesas is a simple way of providing isolation between portions of a slice.

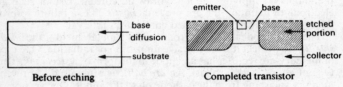

base diffusion	emitter / base
substrate	etched portion
	collector
Before etching	**Completed transistor**

Mesa transistor

mesa transistor A type of bipolar junction *transistor in which the *base is diffused nonselectively into a suitable substrate and the portions around the base then etched away to leave a plateau above the substrate. A double *diffusion technique may be used to form the *emitter inside the base region or the emitter may be alloyed into the base (*see* diagram). The *collector is the region of the substrate below the base.

MESFET Acronym from *metal semiconductor field-effect transistor*. *Syn.* Schottky-gate field-effect transistor. A type of junction *field-effect transistor that has a Schottky barrier (*see* Schottky effect) as the gate electrode rather than a semiconductor junction. The current-voltage characteristic is similar to that of a junction FET. The Schottky barrier gate electrode can be formed on semiconductors, such as gallium arsenide, in which doping cannot be easily effected, and is most often used to form gallium arsenide FETs. It has the added advantage that it can be made at much lower temperatures than are required to form a p-n junction.

mesh *See* network.

mesh contour *See* network.

mesh current *Syn.* cyclic current. *See* network.

mesh voltage *Syn. for* hexagon voltage. *See* voltage between lines.

metal-ceramic *See* cermet.

metal film resistor A *film resistor that uses a metal as the resistive element.

metallic bond *See* metallic crystal.

metallic circuit A circuit composed entirely of metallic conductors that connects two points in a *telecommunication system and that does not use an earth return (*see* earth-return circuit).

metallic crystal A crystal that consists of a regular array of positive ions together with a 'cloud' of free electrons. This type of structure arises when an element, usually metallic, forms a bond with adjacent atoms in which the valence electrons are not bound to a particular atomic site but are free to move throughout the crystal lattice. Elements that form a *metallic bond* of this nature are good electrical conductors.

metallization pattern *See* metallizing.

metallized paper capacitor *See* paper capacitor.

metallizing *Syn.* silvering. Depositing thin films of metal (originally silver) on a glass, semiconductor, or other substrate in order to render it electrically conducting. The metal is then etched into the required *metallization pattern* using a specially designed *mask. The technique is widely used in solid-state electronics for the formation of interconnections on *integrated circuits or *thin-film circuits and to form *bonding pads on integrated circuits or discrete components.

There are several processes used for metallizing. The chosen method depends on the substrate and the metal to be deposited. Chemical reduction is more often used to deposit silver. *Electroplating, *vacuum evaporation, and *cathode sputtering are commonly used methods. The last is not suitable for depositing aluminium on a surface. *Vapour plating is less often employed, as is *burning-on*, in which an oily solution of a salt of the metal is heated on the surface to be coated. This method is most suited to the deposition of platinum, gold, silver, or iridium.

metal migration *See* critical current density.

metal rectifier A *rectifier that consists of a metal in contact with a suitable solid, such as a semiconductor or an oxide of the metal. It depends for its action on the asymmetrical conductance of the junction, the resistance being very much less in one direction of current flow (usually from the compound into the metal) than the other. Typical materials used are cuprous oxide on copper and selenium on copper. Metal rectifiers only operate at low voltages of the order of a few volts and for high operating voltages a series of rectifiers is required. The magnitude of the operating current flowing depends on the area of contact (the *active area*). *Compare* semiconductor diode.

meter (1) Any measuring instrument, such as a voltmeter or ammeter.

345

(2) US spelling of *metre.

meter-protection circuit A circuit designed to protect a meter, such as a voltmeter, from an overload current. A common type of circuit is a device, such as a neon *glow lamp in parallel with the meter, that short circuits at a predetermined voltage and shunts the current away from the meter.

meter resistance The internal resistance of a meter, such as a voltmeter, measured at its terminals at a given temperature. The resistance of particular types of meter, such as rectifier instruments, varies with the frequency, magnitude, and waveform of the input signal. In such cases families of curves can be drawn at specified temperatures relating the resistance to these quantities.

metre Symbol: m. The *SI unit of length defined (since 1983) as the length of the path travelled by light in vacuum during a time interval of 1/299 792 458 of a second. It was previously defined as the length equal to 1 650 763.73 wavelengths in a vacuum of the radiation associated with the transition between the levels $2p_{10}$ and $5d_5$ of the krypton-86 atom. The wavelength of this radiation is about 605.8 nanometres. The metre can be measured very accurately and, unlike the original *international prototype metre*, which was a bar of platinum, does not change with time.

metre bridge A form of *Wheatstone bridge with the ratio arms in the form of a tappable uniform wire of length one metre.

MF *Abbrev. for* medium frequency. *See* frequency band.

mho Symbol: Ω^{-1}. The reciprocal *ohm, formerly used to measure *conductance but now replaced by (and equal to) the *siemens.

mica A naturally occurring mineral that consists of complex aluminium-potassium silicates and has a monoclinic structure; it can therefore be readily cleaved into thin plates. It has a large dielectric constant virtually independent of frequency and retains its properties as an insulator even at very high temperatures. It also has low loss and high dielectric strength and is used as the dielectric material in *mica capacitors. It is widely used for electrical insulation because of its unequalled combination of physical and electrical characteristics.

mica capacitor A *capacitor that uses *mica as the dielectric material. Mica capacitors have characteristics of low loss, low temperature coefficient of capacitance, and good frequency stability. There are several types.

Clamp-type mica capacitors (Fig. *a*) have the mica clamped between tin-foil electrodes. Alternate layers of foil are brought out on opposite sides. Metal lugs are soldered to the electrodes. This type of construction is used in the manufacture of standard capacitors but has been largely superseded for other applications.

Eyelet-construction mica capacitors consist of silvered mica plates fixed together with metal eyelets to form stacks. The possibility of

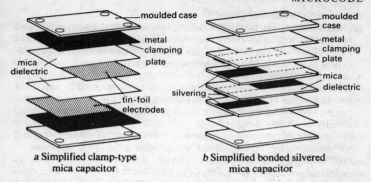

a Simplified clamp-type mica capacitor

b Simplified bonded silvered mica capacitor

relative movement and bowing of the plates in this construction leads to poor stability of capacitance and of temperature coefficient.

Bonded silvered mica capacitors (Fig. *b*) have mica plates that, except for the outer plates, are silvered on both sides, the silvering forming the appropriate electrode areas. The stacked plates are bonded together by firing. This arrangement gives a dimensionally stable stack with good stability of capacitance.

Button mica capacitors are circular capacitors with terminals in the form of a metal band round the perimeter and a metal eyelet in the centre. This type of capacitor also has good dimensional stability and is particularly suitable for high-frequency operation when mounted in a true coaxial arrangement.

The extremely good stability of most forms of mica capacitor both with temperature and frequency makes them highly suitable for use in *filters.

micro- Symbol: μ. A prefix to a unit, denoting a submultiple of 10^{-6} of that unit: one microsecond equals 10^{-6} seconds.

microalloy transistor A type of bipolar junction *transistor in which the semiconductor substrate forms the base; the emitter and collector are formed by recrystallization from a suitable impurity-metal/semiconductor alloy in small pits etched in the surface of the base. *Microalloy diffused transistors* are formed in a similar manner but a gaseous diffusion is carried out first to produce a nonuniform base before the emitter and collector are formed. These transistors are now obsolete.

microchip *Colloq.* A *chip containing complex *microcircuits, as used in computers, electronic control circuits, etc. The word is a misnomer because such a chip is considerably larger than those containing discrete components or simple circuits, and the term usually refers to the circuitry contained within it rather than the chip itself.

microcode *See* microprogram.

347

microcomputer *See* computer.

microelectronics The branch of *electronics concerned with or applied to the realization of electronic circuits or systems from extremely small electronic parts. It includes the design, production, and application of any *microminiaturization* technique to reduce the cost, size and weight of electronic parts, subassemblies, and assemblies and to replace vacuum-tube circuits with solid-state compatible parts. Increased miniaturization is particularly desirable in the field of computers and therefore *integrated circuits of ever increasing packing density are being designed.

microfaceting Small-scale roughness on an otherwise uniformly etched surface, causing the etched surface to have a cloudy appearance.

microfilm plotter *See* incremental plotter.

microminiaturization *See* microelectronics.

micron Symbol: μ. A unit of length equal to 10^{-6} metre, now renamed as the micrometre (μm).

microphone A device that converts sound energy into electrical energy. It forms the first element in a telephone, a broadcast transmitter, and all forms of electrical sound recording. It is the converse of the *loudspeaker and one type of device, the loudspeaker-microphone (*see* loudspeaker), may be used for both purposes.

There are many different types of microphone: the most common types are the *carbon, *capacitor, *crystal, moving-coil, and *ribbon microphones. Most types of microphone operate by converting the sound waves into mechanical vibrations that in turn produce electrical energy. The most common method is to use a thin diaphragm mechanically coupled to a suitable device. The force exerted is usually proportional to the sound pressure but in the case of the ribbon microphone it is proportional to the particle velocity.

In the *moving-coil microphone* a small coil is attached to the centre of the diaphragm; when the diaphgragm is caused to move, by sound waves, in a steady magnetic flux an e.m.f. is produced in the coil by electromagnetic induction. The e.m.f. is a function of the incident sound pressure. The *moving-iron microphone* operates in a similar manner but a small piece of iron is moved by the diaphragm. This induces an e.m.f. in a current-carrying coil surrounding it (*compare* induction microphone). Microphones in which the electrical energy is produced by the motion of a coil or conductor in a magnetic flux density are described as *magnetic microphones*.

In the *magnetostriction* and crystal microphones the sound pressure is converted into electrical pressure by direct deformation of suitable magnetostrictive or piezoelectric crystals. In the *hot-wire microphone no mechanical vibrations are involved: the sound pressure produces a change in the resistance of a hot wire. In the *glow-discharge microphone* the current of the glow discharge is modulated by the sound waves.

Specially shaped microphones have been designed. The *ear microphone* is specially shaped to fit into the human ear; the *lip microphone* is designed to be held close to the lips and thus cut down on extraneous external sounds. The *throat microphone* is shaped to be worn on the throat and responds directly to vibrations of the larynx, thus cutting out background noise.

Most microphones have strong directional properties that often vary with frequency. Their usual lack of sensitivity is not a great disadvantage since the output from the microphone is usually amplified and the directional properties tend to minimize background noise. Good quality sound reproduction is usually achieved but resonance in the mechanical system must be avoided. This is usually done by making the resonant frequency of the moving parts either much higher or much lower than the sound frequency to be reproduced.

microprocessor The physical realization of the *central processing unit of a given computer system on either a single *chip of semiconductor or on a small number of chips.

microprogram A *program that is used to define the *instruction set of a computer. Computers that can have their instruction sets entered in this way are *microprogrammable.* The languages in which the microprograms are written are called *microcodes.*

microstrip A strip of conducting material in close proximity to an earthed conducting plane; it is used as a *transmission line in ultra-high frequency applications. A *parallel-wire line* is a microstrip in which the earth plane is replaced by a wire that is parallel to but insulated from the conductor, which is also in the form of a wire.

microstrip line A transmission line formed on a monolithic microwave *integrated circuit (MMIC). A microstrip line functions essentially as a distributed inductance in the microcircuit, but treatment as a transmission line takes account of associated capacitance, mutual coupling, and discontinuities. The impedance of the microstrip line is determined by the ratio of conductor width to substrate thickness, dielectric constant of the substrate, and, to a lesser extent, the thickness of the conductor. The physics is complex and exact analytical analyses are not available. Good analytical approximations have been obtained using computer-aided design techniques that closely match the behaviour of the line in reality.

microwave An electromagnetic wave with a wavelength in the range 3 millimetres to 1·3 metres, i.e. between infrared radiation and radiowaves on the frequency spectrum. Microwaves are used in *radar and telecommunications and are also used commercially for extremely rapid cooking. *See also* maser; frequency band.

microwave choke *See* choke.

microwave frequency *See* frequency band.

microwave generator A device, such as a *klystron or *magnetron, used to generate microwaves.

microwave integrated circuit (MIC) An *integrated circuit that operates at microwave frequencies. Strictly, the term refers to a hybrid circuit; a monolithic microwave circuit is referred to as an MMIC.

microwave tube An *electron tube that is suitable for use as an amplifier or oscillator at microwave frequencies (*see* frequency band). These tubes usually employ *velocity modulation of the electron beam rather than density modulation as in the valves used in *audio-frequency valve amplifiers or oscillators.

Microwave tubes may be classified into two main types: *linear-beam tubes,* in which the electron beam travels in an essentially linear direction, and *crossed-field tubes*, in which the electron beam follows a curved path under the influence of orthogonal electric and magnetic fields.

The *klystron and most forms of *travelling-wave tubes are linear-beam tubes; the *magnetron and carcinotron (*see* travelling-wave tube) are examples of crossed-field tubes.

mike *Colloq.* A microphone.

mil *US syn. for* thou.

milking generator A low-voltage d.c. generator that is used to charge one or more cells of an accumulator independently of the others.

Miller effect The phenomenon by which an effective feedback path between the input and output of an electronic device is provided by the *interelectrode capacitance of the device. This can affect the total input admittance of the device, which results in the total dynamic input capacitance of the device being always equal to or greater than the sum of the static electrode capacitances.

Miller indices Triplets of numbers corresponding to the three spatial directions and used to specify particular directions and planes within a crystal structure orientated with respect to the unit cube of the crystal.

Miller integrator An *integrator that contains an active device, such as a *transistor, in order to improve the linearity of the output from a pulse generator. Miller integrators are used particularly with sawtooth pulse generators, such as those used to generate a *time base.

Miller sweep generator *See* time base.

milli- Symbol: m. A prefix to a unit, denoting a submultiple of 10^{-3} of that unit: one millimetre equals 10^{-3} metre.

Millman travelling-wave tube *See* travelling-wave tube.

MIM capacitor *Short for* metal-insulator-metal capacitor. A thin-film capacitor consisting of two metal plates separated by a dielectric material and used in integrated circuits. MIM capacitors are the most commonly used type of *monolithic capacitor in MMICs (monolithic microwave *integrated circuits).

minimum discernible signal (mds) *Syn.* threshold signal. The smallest value of input signal to any circuit or device that just produces a discernible change in the output. For example, in a *radio receiver it is the smallest value of received signal that just produces an audible output at the loudspeaker. *See also* radar.

minimum sampling frequency *See* pulse modulation.

minority carrier The type of charge *carrier in an extrinsic *semiconductor that constitutes less than half of the total charge carrier concentration.

mirror galvanometer *See* galvanometer.

MISFET *See* field-effect transistor.

misfire *See* gas-discharge tube; mercury-arc rectifier.

mismatch A condition that occurs when the impedance of a *load does not equal the output impedance of the source to which it is connected.

MIST *See* field-effect transistor.

misuse failure *See* failure.

mixed coupling *See* coupling.

mixer *Syn.* frequency-changer. A device that is used in *superheterodyne reception to produce an output signal of different frequency from the input signal. The received amplitude-modulated carrier wave is mixed with a locally generated signal from a beat-frequency oscillator (*see* beats) to produce an intermediate-frequency signal that retains the amplitude-modulation characteristics of the original signal. The amplitude of the output signal has a fixed relationship to the input-signal amplitude and is usually a linear function of it.

The *conversion conductance* of the device is defined as the current output at frequency f_2 divided by the voltage input at frequency f_1 under short-circuit output-load conditions. It is usually expressed in milliamperes per volt.

MKS system A system of *absolute units in which the fundamental *units of mass, length, and time are the kilogram, metre, and second and the fourth fundamental quantity necessary to define completely electric and magnetic quantities is the *permeability of free space, μ_0. The permeability was orginally defined as having the value 10^{-7} henry/metre (*compare* CGS system).

It was later shown that if μ_0 were given the value $4\pi \times 10^{-7}$ henry/metre then electric and magnetic equations would appear more rational: equations concerned with spherical systems would contain the factor 4π, which characterizes any concept of sphericity; those concerned with circular or cylindrical systems, such as a coil of wire, would contain the characteristic factor 2π; those concerned with a linear system, such as a straight wire, would not contain the factor π. Adoption of the value $4\pi \times 10^{-7}$ H/m for μ_0 leads to the *rationalized MKS system*, which is the basis of the system of *SI units.

351

MLR *Abbrev. for* multilevel resist.

MMIC *Abbrev. for* monolithic microwave integrated circuit. *See* integrated circuit.

MNOS *Abbrev. for* metal-nitride-oxide semiconductor. *See* read-only memory.

MOCVD *Abbrev. for* metal-organic chemical vapour deposition. *Syn.* OMCVD. A form of *vapour phase epitaxy developed for 3–5 compound semiconductors in which the material to be deposited is transported to the substrate using organic molecules such as trimethyl-gallium, -arsine, and trimethyl-aluminium. This method is particularly useful for growing AlGaAs layers as aluminium is particularly difficult to transport in nonorganic form.

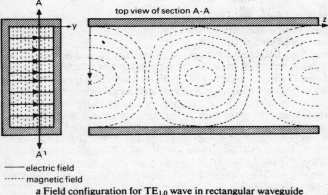

——— electric field

------ magnetic field

a Field configuration for $TE_{1,0}$ wave in rectangular waveguide

mode *Syn.* transmission mode. Any one of the several different states of oscillation of an electromagnetic wave of given frequency. The mode of an electromagnetic wave depends upon the configurations of the vectors describing the wave; three main types of wave exist:

TE waves (or *H waves*) are transverse electric waves in which the electric field vector (*E*-vector) is always perpendicular to the direction of propagation, *z*, i.e.

$$E_z = 0$$

TM waves (or *E waves*) are transverse magnetic waves in which the magnetic vector (*H*-vector) is always perpendicular to the direction of propagation, *z*, i.e.

$$H_z = 0$$

352

TEM waves are transverse electromagnetic waves in which both the *E*-vector and the *H*-vector are perpendicular to the direction of propagation, i.e.

$$E_z = H_z = 0$$

The TEM mode is the one most commonly excited in coaxial lines. It cannot be propagated in a waveguide.

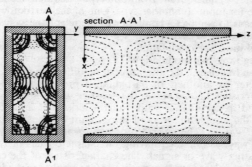

b Field configuration for TE$_{2,1}$ wave
in rectangular waveguide

Solving Maxwell's equations for the particular conditions existing, it is commonly found that the solutions are characterized by the presence of one or more integers (m, n) that can take values from zero to infinity; this allows several possible modes for each type of wave. The physical constraints and the frequency of the radiation usually limit the number of permitted values of m and n and therefore the number of possible modes.

A particular example may be considered. In a *waveguide each component of the wave contains a factor given by

$$\exp(i\omega t - \gamma_{m,n}z)$$

where ω is 2π times the frequency and i is $\sqrt{-1}$; $\gamma_{m,n}$ is the *propagation coefficient and determines the phase and amplitude of the wave components. For each mode a lower limit exists where the complex quantity $\gamma_{m,n}$ is purely real and is equal to the *attenuation constant, $\alpha_{m,n}$. The amplitude of the component then decreases exponentially. The frequency is below cut-off and the wave does not propagate. The *dominant mode* of the oscillation is the mode that has the minimum cut-off frequency. When $\gamma_{m,n}$ is imaginary the phase of the wave varies with distance z and the wave is propagated with no attenuation. In practice $\gamma_{m,n}$ is never purely imaginary and

353

some attenuation always occurs due to energy losses in the transmission line. The effects of different m,n modes on the configuration of a wave travelling along a waveguide can be seen from the diagrams. A waveguide that is designed to separate different modes of oscillation of a wave is termed a *mode filter*.

mode filter *See* mode.

modem Acronym from *mo*dulator-*dem*odulator. A device that is used to convert signals output from one type of equipment into a form suitable for input to another type. One of the best-known uses for modems is to connect a computer to the telephone system.

mode number *See* klystron.

modulated wave *See* modulation.

modulating wave *See* modulation.

modulation In general, the alteration or modification of any electronic parameter by another. In particular, a process by which certain characteristics of one wave, the *carrier wave, are modulated or modified in accordance with a characteristic of another wave or signal, the *modulating wave*. The resultant composite signal is the *modulated wave*. The reverse process is *demodulation,* by which an output wave is obtained having the characteristics of the original modulating wave or signal. The characteristics of the carrier that may be modulated are the amplitude (*see* amplitude modulation) or phase angle (*angle modulation*), particular forms being *phase and *frequency modulation. Modulation by an undesired signal is *cross modulation. Multiple modulation* is a succession of processes of modulation in which the whole or part of the modulated wave from one process becomes the modulating wave for the next. *Intermodulation* is the modulation of the components of a complex wave by each other in a nonlinear system, producing waves having frequencies, among others, equal to the sums and differences of those of the components of the original wave. *See also* velocity modulation; pulse modulation.

modulation factor *See* amplitude modulation.

modulation index *See* frequency modulation.

modulator (1) Any device that effects *modulation.

(2) *Syn.* master trigger. A device, usually a *multivibrator, that is used in *radar to generate a train of short pulses, each of which acts as a trigger for the oscillator.

modulator electrode An electrode that is used to modulate the flow of current in an electrode device. In a *field-effect transistor it is the gate electrode(s), used to modulate the channel conductivity. In a *cathode-ray tube it is the electrode used to control the electron-beam intensity.

molecular beam epitaxy (MBE) A method of *epitaxy that is carried out in ultrahigh vacuum. It is a sophisticated evaporation technique in which the elements to be deposited are evaporated from ovens and impinge on the substrate where they are deposited in crystalline order. The substrate slice is rotated to ensure a uniform epitaxial layer. Under suitable conditions the process can be controlled to produce almost any required epitaxial layer composition, thickness, and doping level with a resolution of virtually one atomic layer, to a high degree of accuracy and uniformity across the slice. Disadvantages are the high vacuum requirements, complex and costly equipment, and the slow growth rate of the epitaxial layer.

monochromatic radiation Electromagnetic radiation of a single frequency. In practice radiation of a single frequency is never achieved and the term is applied to a narrow range of frequencies. The term is also applied to particulate radiation when the particles are all of the same type and energy but in this case the description *homogeneous* or *monoenergetic* is usually preferred. *Compare* polychromatic radiation.

monochrome television *Syn.* black-and-white television. *See* television.

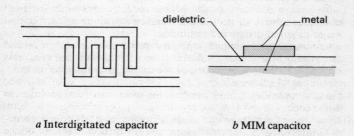

a Interdigitated capacitor   *b* MIM capacitor

monolithic capacitor A capacitor that is formed as part of a monolithic microwave *integrated circuit (MMIC). It can be used as a *blocking capacitor, a *bypass capacitor, or a tuning capacitor as part of a *tuned circuit. The four main types of monolithic capacitor are a *stub in a *microstrip transmission line; *interdigitated capacitors* (Fig. *a*), in which the capacitance is produced by capacitive coupling between adjacent conductors separated by the substrate dielectric material; *MIM (metal-insulator-metal) capacitors (Fig. *b*), which are parallel plate capacitors and the most common type used; and *Schottky diodes used as *varactors in voltage controlled oscillator circuits.

monolithic integrated circuit *See* integrated circuit.

monophonic sound reproduction *See* reproduction of sound.

monoscope A type of *electron tube that is used to produce single images, such as a test pattern, suitable for television broadcasting. It contains an aluminium plate electrode with the desired image printed on it in carbon. The plate is scanned by an electron beam and video signals are produced because of the difference in *secondary emission of electrons of aluminium and carbon. *Compare* camera tube.

monostable *Syn.* one-shot; univibrator. A type of circuit that has only one stable state but on the application of a *trigger pulse it can take up a second quasi-stable state. A common form of monostable consists of a *multivibrator with resistive-capacitive coupling. Monostables can be used to provide pulses of fixed duration, for pulse stretching or shortening, or as a delay element.

Moog synthesizer *See* electrophonic instrument.

Morse code An internationally agreed code for the transmission of signals, particularly in *telegraphy. It is a two-condition code in which each character to be transmitted consists of a number of dots and dashes, each group being separated by spaces. The dots and dashes are represented electrically by pulses of different duration. *Cable code,* a variation of the Morse code, is commonly used in submarine cables: the dots, dashes, and spaces all have equal duration in time but different amplitudes.

A *Morse key* is a manually operated device used to convert information into Morse code. A *Morse printer* is a device that converts received information, recorded on a suitable medium, into the corresponding characters.

Morse telegraphy Telegraphy in which the signals are transmitted as a Morse code.

MOS *Abbrev. for* metal-oxide semiconductor. *See* MOS capacitor; MOS integrated circuit; MOS logic circuit; field-effect transistor.

mosaic *Short for* photomosaic. *See* iconoscope.

mosaic crystal An imperfect crystal composed of a number of smaller crystals that have been grown together so that the corresponding crystal planes are nearly or exactly parallel but have discontinuities at their mutual surfaces. Such crystals are effectively single crystals for most practical purposes but the discontinuities form crystal defects that can affect their properties, for example by providing extra energy levels in *semiconductors. *Compare* perfect crystal.

mosaic electrode The light-sensitive electrode of a television *camera tube on which the image is formed.

MOS capacitor The basic metal-insulator-semiconductor structure that consists of an upper *gate electrode* separated from the surface of a semiconductor substrate by a thin layer of insulating material (Fig. *a*). The gate electrode is either a metal or highly doped *polysilicon; the semiconductor substrate is usually silicon and the insulator silicon dioxide (oxide). The structure is the fundamental

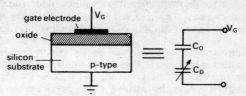

a MOS capacitor and equivalent circuit

part of all MOS devices and integrated circuits including insulated-gate *field-effect transistors, *charge-coupled devices, and MOS random-access memory (*see* solid-state memory). The behaviour of all MOS devices depends on the conditions present at the surface of the semiconductor substrate: the MOS capacitor is an extremely useful tool for studying such surfaces since the total capacitance of the device depends on the surface conditions.

The total capacitance of the device is a function of the voltage, V_G, applied to the gate electrode and is also frequency dependent. An applied potential difference, V_G, across the device appears partly across the oxide layer and partly across the semiconductor:

$$V_G = V_O + \phi_S$$

where V_O is the potential difference across the oxide and ϕ_S is the potential difference across the silicon. By convention the bulk potential of the silicon is zero and ϕ_S is therefore referred to as the *surface potential*.

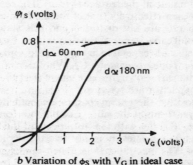

b Variation of ϕ_S with V_G in ideal case

The capacitance C_O of the oxide layer is constant and is given by

$$C_O = \varepsilon_{ox}/d$$

357

where ε_{ox} is the relative *permittivity of the oxide and d is the thickness. The capacitance C_D of the silicon is a function of the surface potential ϕ_S. The variation of ϕ_S with applied voltage V_G is shown in Fig. b.

The effect of surface potential on the energy bands of a p-type semiconductor is shown in Figs. c and d under conditions of thermal equilibrium, i.e. the *Fermi level, E_F, is constant throughout the semiconductor. E_i is the intrinsic Fermi level, i.e.

$$E_i = E_v + E_g/2$$

In the ideal case the *work functions, ϕ_m and $\phi_{silicon}$, of the electrode material and the substrate are identical and $\phi_S = 0$ when $V_G = 0$.

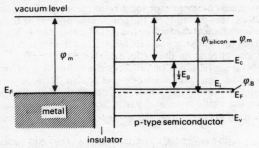

c Flat-band condition in ideal case when ϕ_{ms} is zero

If ϕ_S is negative the energy bands are bent upwards and majority carriers accumulate at the surface (Fig. d). The total capacitance of the device is then effectively that of the oxide layer. When $\phi_S = 0$, the energy bands are not distorted and this state is termed 'flat-band'. The total capacitance of the device is the oxide capacitance in series with a small capacitance due to the bulk material and is termed the *flat-band capacitance*, C_{FB}. Variation of total capacitance with gate voltage is shown in Fig. e where the full C-V curves for an ideal MOS capacitor are given.

If ϕ_S is positive the energy bands are bent downwards and a depletion layer forms at the surface of the semiconductor. This acts as a dielectric in series with the insulator and the total capacitance decreases. As ϕ_S becomes more positive the depletion-layer width increases and the capacitance falls. When $\phi_S = \phi_B$, the semiconductor surface becomes inverted, where ϕ_B is given by $(E_F - E_i)$ in the bulk material. Minority carriers begin to accumulate at the surface of the semiconductor and can contribute to the capacitance. As V_G is further increased *strong inversion* occurs at a value of ϕ_S corresponding approximately to $2\phi_B$. At this point the semiconductor is

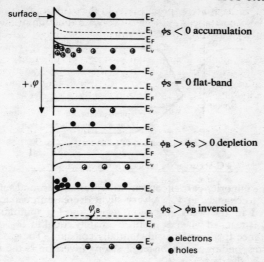

d **Effect of ϕ_S on simplified energy bands of a p-type semiconductor**

effectively shielded by the presence of the inversion layer and further penetration by the electric field due to V_G is prevented. The depletion-layer width, W_m, is therefore effectively a maximum and ϕ_S remains substantially constant. The inversion capacitance is frequency dependent since it depends on the ability of the minority-carrier concentration in the inversion layer to respond to changes in V_G by thermal generation or recombination. Similar curves are however obtained at all frequencies in the accumulation and depletion regions, i.e. when $\phi_S < \phi_B$.

The low-frequency capacitance is usually measured using a *quasi-static* method. A linearly varying ramp voltage dV_G/dt is applied to the gate electrode and at any given instant the output current, *i*, is directly proportional to the capacitance since

$$i = dQ/dt = CdV_G/dt$$

In the low-frequency case the minority-carrier concentration responds to the changes in applied voltage: after inversion the total capacitance rises from a minimum value, $C_{min\ LF}$, until with strong inversion the capacitance approaches the value of the oxide capacitance.

With a high-frequency small-signal variation superimposed on the ramp component the inversion capacitance does not increase

359

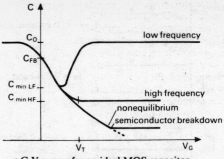

e C-V curves for an ideal MOS capacitor

since the minority-carrier concentration cannot respond sufficiently rapidly to changes in V_G. A very slight decrease in capacitance is observed in the weak inversion region until a minimum value $C_{\min HF}$ is reached when ϕ_S is approximately equal to $2\phi_B$. The total capacitance then remains substantially constant. The gate voltage V_T corresponding to the onset of strong inversion is the *threshold voltage* of the device.

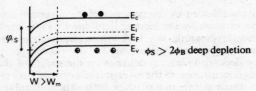

f Nonthermal equilibrium case

If a sufficiently rapid large-signal variation of V_G is applied, thermal equilibrium is not maintained and the inversion layer does not form. The depletion region therefore continues to increase in width (Fig. *f*); this is known as *deep depletion*. The total high-frequency capacitance thus continues to decrease below the value $C_{\min HF}$ until the value of ϕ_S becomes sufficiently large for *breakdown of the semiconductor substrate to occur (Fig. *e*). The *relaxation time* of the capacitor is the time interval required for the inversion layer to form following a large-signal variation – usually a voltage pulse – and is measured by the time required for the capacitance to recover to $C_{\min HF}$.

· In the case of an n-type substrate a similar effect occurs but the polarity of the applied voltage is reversed.

360

The value of ϕ_S in the depletion state is determined by the number of fixed charges in the depletion region, i.e.

$$\phi_S = eN_A W$$

where e is the electron charge, N_A is the doping level of the substrate, and W the width of the depletion region. Since

$$\phi_S = V_G - V_O$$

and

$$V_O = Q_S / C_O = Q_S d / \varepsilon_{ox}$$

where Q_S is the number of charges per unit area in the semiconductor, then for a given value of V_G the surface potential ϕ_S is a function of the oxide thickness and the doping level, N_A.

In practice a work function difference, ϕ_{ms}, exists between the electrode and the semiconductor and there is a fixed charge concentration, Q_{FC}, at the oxide-semiconductor interface. The flat-band condition ($\phi_S = 0$) does not therefore occur at $V_G = 0$ but at a definite applied voltage, V_{FB}, where

$$V_{FB} = -Q_{FC} / C_O + \phi_{ms}$$

A value of about -1 volt is obtained with a typical device geometry, and the flat-band voltage, V_{FB}, may be used to calculate Q_{FC}.

The shape of the C-V curve is modified by various surface conditions in the oxide and the substrate and it can be used for quality control during MOS processing. In particular, the migration of mobile impurity ions, especially sodium, through the oxide as a result of voltage or thermal stress causes a shift, ΔV_{FB}, in the flat-band voltage.

MOS capacitors may be used as fixed or variable capacitors in MOS integrated circuits. They may be used as fixed capacitors either in the accumulation region or more commonly in the strong inversion region. In the latter case an n-type diffusion is made into the substrate to act as a source of minority carriers and therefore maintain the inversion capacitance.

MOS capacitors in the nonequilibrium state are used as dynamic storage capacitors, as in the MOSRAM memory cell (*see* solid-state memory), or to form *charge-coupled devices. In this mode of operation the operating speed must be fast compared to the relaxation time.

MOSFET *See* field-effect transistor.

MOS integrated circuit A type of *integrated circuit based on insulated-gate *field-effect transistors. MOS circuits have several advantages compared to *bipolar integrated circuits and account for a substantial proportion of all semiconductor devices produced.

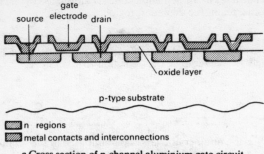

a Cross section of n-channel aluminium gate circuit

MOS transistors are self-isolating and no area-consuming isolation diffusions are required; this enables a very high functional *packing density to be obtained. MOS transistors may be used as active load devices (*see* field-effect transistors) and a separate process is not therefore required to form resistors. When used as load devices *pulse operation of the circuit is easily achieved using the gate electrodes to activate the device: power dissipation is greatly reduced, involving less complicated heating problems. A characteristic of MOS transistors is their exceptionally high input impedance. This allows the gate electrodes to be used as temporary storage capacitors thus keeping the circuits comparatively simple. This is called *dynamic operation.* Usually these circuits operate above a minimum specified frequency.

The relatively few processing steps required in manufacture, compared to bipolar integrated circuits, enables large chips to be made thus further increasing the functional compactness and reducing the costs. MOS circuits tend however to be slower than their bipolar counterparts due to their inherently lower *mutual conductance and to the fact that their speed is extremely dependent on the load capacitance.

Two main types are used, having slightly different structures: *aluminium gate circuits* and *self-aligned gate circuits.* A typical section of an aluminium gate circuit is shown in Fig. *a.* *Source/*drain diffusions are made through holes etched in a thick oxide layer into the substrate and a new oxide layer grown over the diffused regions. These layers may be additionally thickened if required using a low temperature vapour phase reaction. Openings are then made in the oxide layer between source and drain regions and a thin oxide layer grown to form the *gate insulators. Contact windows to the source and drain regions are opened, and aluminium deposited to form the gate electrodes and the circuit interconnection pattern. Difficulties

362

can arise in these circuits if the alignment of the successive process-ing steps is not absolutely precise.

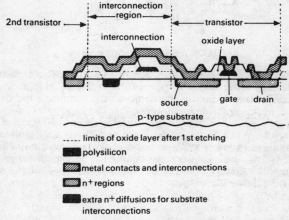

b Cross section of n-channel self-aligned circuit

In the self-aligned gate circuit the gate electrodes are formed before the source/drain diffusions are made; the most widely used method is known as the *silicon gate technology* (Fig. *b*). Openings are made in the initial thick oxide layer and a thin layer of oxide grown to form the *gate insulators. This is immediately covered with a layer of polycrystalline silicon (polysilicon) using a vapour-phase reaction. The polysilicon is then etched to form the gate electrodes together with some interconnections. The gate oxide is then re-moved from the regions not covered with polysilicon leaving open-ings through which the source/drain diffusions are made. The edges of these diffused regions are defined by the previously etched gate regions thus providing the required precision of alignment. The dif-fusing material also enters the polysilicon regions and dopes them, which has the desirable effect of reducing their resistivity. The whole slice is covered with a further oxide layer, contact windows etched, and the final metal layer is deposited and etched to form the interconnections. Two layers of interconnections are possible using this technique, plus some interconnections in the substrate itself by diffusing extra regions when the source/drain diffusions are made.

MOS logic circuit A *logic circuit constructed in an *MOS integrated circuit. It consists of a combination of MOS *field-effect transistors in series or in parallel that perform the logic functions, i.e. act as

363

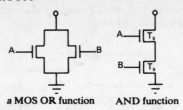

a MOS OR function   AND function

AND or OR gates, etc. (Fig. *a*), coupled to other MOS transistors
that determine the output voltages of the circuit. MOS logic circuits
are classified according to the method of determining the output
voltage, i.e. into *ratio* or *ratioless circuits*. The logic gates effectively
act as switches when used with a suitable choice of high and low
logic levels. When the required input conditions are fulfilled the
combination switch is 'on' and provides a conducting path. If the
switch is 'off' the gate does not conduct. The high logic level is
chosen to be greater than the threshold voltage, V_T, of the MOS
transistor; the low level is lower.

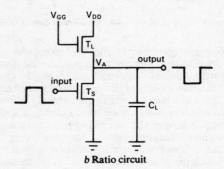

b **Ratio circuit**

In a *ratio circuit* the logic gate, represented as a single switch
transistor, T_S, is connected in series with a load transistor T_L. The
drain of the load transistor is connected to the power supply and the
source of the switch transistor to earth (Fig. *b*). The output is taken
from the *node, A, between the transistors. The circuit will usually
be driving similar MOS logic gates. These have a very high input
impedance and are represented by a capacitor C_L. A voltage of
magnitude greater than or equal to the drain voltage, V_{DD}, is applied to the gate of T_L. In static operation the voltage is applied
continuously; in dynamic operation the voltage is applied on the
application of a *clock pulse in order to reduce dissipation.

364

A low logic level input to the gate of the switch transistor T_S results in T_S being 'off'; C_L is then charged by T_L until the output voltage, V_A, reaches a value sufficient to cause T_L to turn off, i.e. until V_A reaches $(V_{GG} - V_T)$ or V_{DD}, whichever is lower. Application of a high logic level to the gate of T_S causes T_S to be 'on' and C_L discharges through T_S. The output voltage V_A falls to a level determined by the relative impedances of the two transistors.

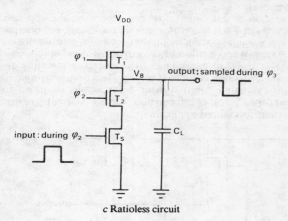

c Ratioless circuit

It can be shown that the voltage V_A at the node depends on the ratio of the *aspect ratios of the devices, and these are manufactured to ensure an output voltage suitable for a low logic level, i.e. less than the threshold voltage of the following gate. The circuit provides inversion of the logic function; thus an AND function in T_S provides a NAND output, etc.

If the dynamic version of the circuit (with the gate voltage of the load transistor clocked) is used a minimum rate of clocking must be specified to prevent loss of information at the output due to leakage paths causing the charge on the load capacitor to decay.

In a *ratioless circuit* (Fig. *c*) a second load transistor, T_2, is connected in series between the first load transistor T_1 and the logic gate, represented by a single switch transistor T_S; the output voltage V_B is taken from the node, B, between T_1 and T_2. A clocking system is employed, usually a four-phase system, to apply a bias to the gates of the load transistors T_1 and T_2 in turn. During phase one (ϕ_1) bias is applied to the gate of T_1, T_1 is turned on, and the load capacitor C_L (usually the gate of the switch transistor of the following stage) is charged to $(V_{GG} - V_T)$. During phase two (ϕ_2) bias is applied not to T_1 but to the gate of T_2. If a high logic level is applied

365

to T_S at this time, both T_2 and T_S will be turned on, C_L will discharge through them, and the output voltage V_B will fall to the low logic level. If T_S is not turned on, i.e. a low logic level is applied to the gate of T_S during ϕ_2, C_L will not discharge since no conducting path to earth exists and V_B will remain at the high logic level. The output of the circuit is sampled by the following circuit during phases ϕ_3 and ϕ_4; information may thus only be supplied to T_S once in every four clock phases.

Operation of this circuit does not depend on the impedances of the devices and it is therefore termed ratioless. Power dissipation is very low since no conducting path ever exists directly between the power supply and earth and the circuits depend solely on charge storage in the load capacitance. The circuit is inverting and two gates are frequently combined to provide a noninverting circuit. If used in a dynamic *shift register, for example, six transistors are needed for each *bit of information. Ratioless circuits are used only for certain specialized applications.

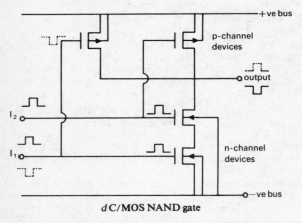

d C/MOS NAND gate

A *C/MOS logic circuit* uses *complementary MOS transistors to provide the basic logic functions. The basic NAND gate is shown in Fig. *d*. C/MOS have the advantage that the power required is extremely low and they are suitable for applications where very little power consumption is a condition, as in digital watches. They have a lower packing density than ratio circuits since every transistor requires its complement and therefore isolation of p-channel devices from n-channel devices is required. For convenience, groups of n-channel devices (and p-channel devices) are formed in the same area of the chip. The speed of operation is relatively slow, compared

366

to *transistor-transistor logic, because of the relatively large bulk capacitance of the substrate. C/MOS circuits are however very resistant to stray noise pulses. Faster versions of C/MOS circuits have been designed, the fastest version being the silicon-on-sapphire type of circuit.

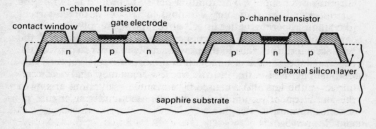

oxide layer
----- etched region of epitaxial layer for isolation

e Silicon-on-sapphire construction

In *silicon-on-sapphire C/MOS* the substrate is made from sapphire and an epitaxial layer (*see* epitaxy) of silicon is grown on the surface. Selective diffusions and etchings are used to fabricate the individual transistors. The speed of this type of circuit is extremely fast since the sapphire is an insulator and the bulk capacitance is therefore very small. Individual transistors are easily isolated by etching the epitaxial layer completely away (Fig. *e*). Unfortunately this type of circuit is still very expensive to manufacture due to difficulties in growing the epitaxial layer satisfactorily.

MOSRAM *Abbrev. for* MOS random-access memory. *See* solid-state memory.

MOS random-access memory (MOSRAM) *See* solid-state memory.

MOST *See* field-effect transistor.

MOS transistor (MOST; MOSFET) An insulated-gate *field-effect transistor in which the gate electrode is metal (either aluminium or very highly doped polysilicon), the insulator is silicon dioxide (usually shortened to *oxide), and the substrate is silicon.

mother *See* recording of sound.

motional impedance *Syns.* driving impedance; driving-point impedance. A component of the impedance of an electromechanical or acoustic *transducer that results from the counter e.m.f. generated by the motion of the transducer. It is equal to the vector difference between the input impedance measured under specified load conditions and the *blocked impedance.

motor A machine that converts electrical energy into mechanical motion. The motion is produced by the torque due to the magnetic fields associated with the currents in the windings (*see also* electromagnetic induction).

An alternating-current (a.c.) motor operates with an alternating-current power supply. The rotating parts are termed the *armature* (or *rotor*) and the stationary windings the *stator*. A direct-current (d.c.) motor operates with a direct-current power supply and usually contains a *commutator* that connects each of the sections of the primary winding in turn to the power supply in order to provide the necessary torque. *See also* induction motor; universal motor.

motorboating *Noise that occurs in low-frequency and audiofrequency amplifiers and is caused by unwanted oscillations arising in the amplifiers. It results in a sound from the loudspeaker that resembles a motorboat engine.

mount *See* waveguide.

mountain effect *See* direction finding.

moving-coil galvanometer *See* galvanometer.

moving-coil instrument A measuring instrument that depends for its operation on the interaction between a steady magnetic field and the magnetic field induced in a movable coil carrying the current to be measured. The instrument is designed so that the coil rotates in the magnetic flux density and the angular deflection is directly proportional to the current to be measured. Moving-coil instruments have a relatively low power consumption and a uniformly divided scale. The magnetic flux density is usually provided by the suitably shaped pole pieces of a permanent magnet although an electromagnet may be used. The instrument is only suitable for use with direct current but may be adapted for alternating-current measurements by means of a suitable *rectifier. *See* rectifier instrument; galvanometer.

moving-coil microphone *See* microphone.

moving-iron instrument A measuring instrument that depends for its operation on the interaction between current-carrying fixed coils and soft-iron pieces in the moving system. The attraction-type of instrument depends on the attraction of the coil for a soft-iron armature when the magnetic flux due to the current in the coil changes (Fig. *a*). The repulsion-type of instrument depends on the mutual repulsion between two soft-iron rods located inside the coil (Fig. *b*). Since both are contained within the coil similar poles are induced in adjacent ends.

moving-iron microphone *See* microphone.

moving-magnet galvanometer *See* galvanometer.

moving magnetic surface memory A *memory device in which digital information is stored in magnetic material, such as iron oxide, formed as a thin magnetic film on a nonmagnetic substrate. The

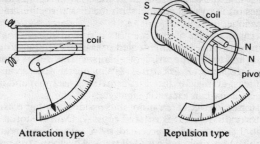

Attraction type   Repulsion type

Moving-iron instruments

direction of magnetization of small localized areas of the magnetic material represents the stored information. Information is stored and accessed to and from the material by means of small electromagnets – read and write heads (*see* magnetic recording). The same electromagnet can be used for both reading and writing – a read/write head – using time-division multiplexing. The portion of the magnetic medium available to a head is the *track. The nature of the device and the speed of operation depend on the substrate material, which can be either flexible or rigid, and the physical geometry used.

Magnetic-tape units (MTU) use magnetic tape (*see* magnetic recording) as the storage medium. A flexible tape is wound on reels and moves past the read/write head. Laminar-flow air bearing is used to reduce friction; the read/write head is held in a position so that there is no physical contact between the head and the tape, allowing very high tape velocities to be used. MTUs are *serial memories.

Disk storage devices are *random-access memories in which the magnetic material is coated on to circular substrates. Data is stored on a series of concentric tracks on the surface of the disk and a read/write head moves radially over the surface to select a particular track. The disk itself is rotated in a disk drive to bring a particular location to the read/write head.

Large disk memories employ a stack of rigid metal plates (*hard disks*) coated on both sides with the magnetic material and rotating together on a common spindle. A set of read/write heads is used: two are required for each plate. A typical large disk memory can store 10^9 to 10^{10} *bits with an access time (*see* memory) of about 20 milliseconds. Small disk memories are formed on one surface of a flexible plastic disk, known as a *floppy disk*. The disk is permanently encased in a stiff plastic or cardboard envelope. The envelope is held rigidly and the disk is rotated inside it. A movable read/write head comes into physical contact with the disk through a radial slit in the envelope.

An intermediate size of disk memory is provided by a single hard disk mounted in a hermetically sealed container together with read/write heads and their supporting mechanism. This unit is commonly known as a *Winchester disk drive*. A Winchester disk is capable of storing at least 20 megabytes of information (*see* byte) and in some cases the capacity can exceed a gigabyte.

Drum storage devices are random-access devices that consist of a rigid coated cylinder that is rotated rapidly. Data is stored on circumferential tracks and is accessed by a set of fixed read/write heads: one head is provided for each track. The access time is usually shorter than for disk memory since the heads do not move, but the storage capacity is lower. Drums have therefore been displaced by disks. Disks and drums are both block-access devices (*see* random-access memory). Laminar-flow air bearing is used in both disk and drum systems.

Moving surface memories are nonvolatile and are used as *peripheral devices; it is usually possible to remove them physically from the reading systems. Interchange of disks or reels of tape greatly increases the volume of information available to a given computer system.

moving-magnet instrument A measuring instrument that depends for its operation on the interaction between the magnetic flux density induced in a fixed coil by the current to be measured and a small permanent magnet suspended in the plane of the coil. It is essentially the converse of the *moving-coil instrument and has the disadvantage that it must be accurately aligned relative to the magnetic meridian. *See also* galvanometer.

MTL *Abbrev. for* merged transistor logic. *See* I^2L.

M-type microwave tube *Syn. for* crossed-field microwave tube. *See* microwave tube.

mu circuit *See* feedback.

multicavity microwave tube *See* klystron; magnetron.

multichannel analyser An instrument that assigns an input waveform to a number of channels according to a specific parameter of the input. A device that sorts a number of pulses into selected ranges of amplitude is known as a *pulse-height analyser*. A circuit that selects the different frequency components of the input is known as a *spectrum analyser*. Multichannel analysers usually have facilities for performing both these operations.

multielectrode valve (1) A *thermionic valve that contains more than three electrodes.

370

(2) *Syn.* multiple-unit valve. A *thermionic valve that contains within one envelope two or more sets of electrodes each having its own independent stream of electrons.

multilevel resist (MLR) *Syn.* portable conformable mask (PCM). A technique in *photolithography that uses two or more layers of resist to produce a desired pattern. The bottom layer of resist is relatively thick and produces a very planar topography. The topmost layer is very thin and is used for the optical exposure. The exposed pattern in the top layer is then replicated in the lower layer or layers, i.e. the topmost layer acts as a *mask for the lower layers. *Bilevel resist* consists of only two layers, one thick and one thin. Intermixing between the two layers of resist can be a problem, and to prevent this *trilevel resists* can be used. In this case a very thin *transfer layer* of metal or dielectric film is used to completely separate the two layers of resist. Multilevel resist techniques are rather complex but offer several advantages, particularly the planar surface produced and initial exposure in a thin planar resist which provides the optimum conditions for photolithography.

multiple access *See* communications satellite.

multiple folded dipole *See* dipole aerial.

multiple-loop feedback *Syn. for* local feedback. *See* feedback.

multiple modulation *See* modulation.

multiple-unit valve *See* multielectrode valve.

multiplexer *See* multiplex operation.

multiplex operation The simultaneous transmission of several signals along a single path without any loss of identity of an individual signal. The various signals are input to a *multiplexer* that is used to allocate a transmission path to the input according to a particular parameter of the signal (*see* frequency-division multiplexing; time-division multiplexing). The original signals are reconstructed at the receiver by a *demultiplexer* that is operated so as to synchronize with the multiplexer. The transmission path may use any suitable medium, such as wire, waveguide, glass fibres (*see* fibre optics system), or radiowaves. The communication channel chosen is termed a *multiplex channel*.

multiplier (1) *See* electron multiplier; photomultiplier.

(2) A device that has two or more inputs and that produces an output of magnitude equal to the product of the magnitudes of the input signals.

(3) A device or circuit, such as a *frequency multiplier, that produces an output equal to a specified multiple of the input signal.

multistable A circuit or device that has more than one stable state.

multistage feedback *Syn.* single-loop feedback. *See* feedback.

multivibrator An oscillator that contains two linear *inverters coupled in such a way that the output of one provides the input for the other.

371

There are several types of multivibrator, the action of which depends on the type of coupling used.

Capacitive coupling produces an *astable multivibrator* that has two quasi-stable states; once the oscillations are established the device is free-running, i.e. a continuous waveform is generated without the application of a *trigger.

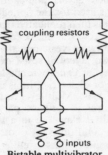

coupling resistors

inputs

Bistable multivibrator

Resistive-capacitive coupling produces a *monostable multivibrator.

Resistive coupling (*see* diagram) produces a bistable circuit that has two stable states and can change state on the application of a trigger pulse. *See* flip-flop.

musa *See* steerable aerial.

mush area *See* service area.

music centre *See* reproduction of sound.

muting switch A switch that automatically activates a noise-suppression circuit in a particular piece of equipment when required, as when the *noise in the system exceeds a predetermined level.

mutual branch *Syn.* common branch. *See* network.

mutual capacitance An indication of the extent to which two capacitors can interact. It is expressed as the ratio of the electric charge transferred to one to the corresponding potential difference of the other.

mutual conductance *Syn.* transconductance. Symbol: g_m. The ratio of the incremental change in the output current, I_{out}, of any amplifying circuit or device, to the incremental change of input voltage, V_{in}, causing it, when the output voltage, V_{out}, is held constant.

In the case of alternating current the mutual conductance ought strictly to refer to the in-phase component of the output current, but common usage refers to the total complex output current, i.e. the parameter is actually the *transadmittance* of the device. *See also* transfer parameters.

mutual impedance *Syn.* transfer impedance. *See* network.

mutual inductance *See* inductance; electromagnetic induction.

mutual-inductance coupling *See* coupling.

mutual parameter *See* transfer parameter.

Mylar *Tradename* A polyester, usually produced in sheets of various thicknesses, that is used for a variety of applications such as insulation, the base material of a magnetic tape, or as the dielectric in certain types of capacitor.

myoelectrical control Control of machines by means of the potential differences developed by muscle contractions. It is widely used in power-assisted devices for the physically handicapped and for controlling artificial limbs.

N

NAA *Abbrev. for* neutron activation analysis.

NAND circuit (or gate) *See* logic circuit.

nano- Symbol: n. A prefix to a unit, denoting a submultiple of 10^{-9} of that unit: one nanometre equals 10^{-9} metre.

NASA *Abbrev. for* National Aeronautical and Space Administration. *See* standardization.

National Bureau of Standards (NBS) *See* standardization.

natural frequency The frequency at which *free oscillations occur in an electrical or mechanical system. It is the frequency at which *resonance occurs in such a system in response to a periodic driving force.

NBS *Abbrev. for* National Bureau of Standards. *See* standardization.

n-channel Denoting an MOS circuit or device, such as a *charge-coupled device, *MOS integrated circuit, or *MOS transistor, in which the conducting channel is formed as n-type semiconductor. The term also describes junction field-effect transistors that have an n-type channel. *Compare* p-channel.

NDR effect *Short for* negative differential resistance effect. *See* Gunn effect.

near-end crosstalk *See* crosstalk.

needle gap *See* spark.

Néel temperature *See* antiferromagnetism; ferrimagnetism.

negative bias A voltage that is applied to an electrode of an electronic device and is negative with respect to some fixed reference potential, usually earth potential.

negative booster *See* booster.

negative feedback *Syns.* inverse feedback; degeneration. *See* feedback.

negative glow *See* gas-discharge tube.

negative ion *See* ion.

negative logic *See* logic circuit.

negative magnetostriction *See* magnetostriction.

negative phase sequence *See* phase sequence.

negative photoresist *See* photoresist.

negative resistance A property of certain devices whereby a portion of the current-voltage characteristic has a negative slope, i.e. the current decreases with increasing applied voltage. Devices that exhibit negative resistance include the *thyristor, the *tunnel diode, and the *magnetron.

negative-resistance oscillator *See* oscillator.

374

negative sequence *See* phase sequence.

negative transmission *See* television.

neon Symbol: Ne. An inert gas, atomic number 10, that exhibits a characteristic red glow when ionized and is extensively used as the gas in *glow lamps. Neon lamps can be made extremely large for use in illuminated signs or extremely small for use as indicators and voltage regulators.

neon lamp *See* glow lamp.

neper Symbol: Np. A dimensionless unit used in telecommunications to express the ratio of two powers. It is the natural logarithm of the square root of the power ratio. Thus if two values of power, P_1 and P_2, differ by n nepers then

$$n = \log_e[\sqrt{(P_2/P_1)}]$$
$$= \frac{1}{2} \log_e(P_2/P_1)$$
$$P_2/P_1 = e^{2n}$$

In a single transmission line or other transmission network the neper can be used to compare two currents, usually the input and output currents, the current being proportional to the square root of the power. Thus if two currents, I_1 and I_2, differ by N nepers then

$$N = \log_e(I_2/I_1)$$

One neper equals 8·686 *decibels.

Nernst effect If a conductor or semiconductor is placed in a magnetic flux density and a temperature gradient maintained across it at right angles to the magnetic flux density, an e.m.f. is developed in the material orthogonal to both the magnetic flux density and the temperature gradient. This is the converse of the *Ettinghausen effect and is related to the *Leduc effect.

nerve current A naturally occurring extemely small current that flows through a nerve path in a human or other live animal.

net loss The difference between the attenuation and the gain in any circuit, device, network, or transmission line.

network (1) In electronics, a number of impedances connected together to form a system that consists of a set of inter-related circuits and that performs specific functions. The behaviour of the network depends on the values of the components, such as the resistances, capacitances, and inductances, from which it is formed and the manner in which they are interconnected. The values of the components are termed the *network constants*. The nomenclature of networks describes either the type of component, the method of interconnection, or the expected behaviour of the network.

Networks are described as *resistive, resistance-capacitance (R-C), inductance-capacitance (L-C), inductance (L) networks,* etc., depending on their components.

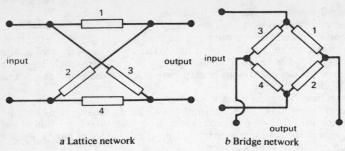

a Lattice network   b Bridge network

Lattice networks have the input and output terminals at a junction between two or more conductors (Fig. *a*); a *bridge network* is a particular type of lattice network (Fig. *b*).

Series networks and *parallel networks* have their elements connected in series and in parallel, respectively.

Linear networks have a linear relationship between the voltages and currents; otherwise they are *nonlinear*.

Bilateral networks conduct in both directions whereas those that conduct in only one direction are *unilateral*.

Passive networks contain no energy source or sink other than normal ohmic losses; those that do contain an energy source or sink are *active*.

All-pass networks attenuate all frequencies equally; other networks are described according to their frequency response (*see* filter).

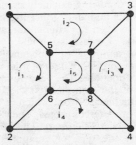

c Conducting paths of a network

A point within a network at which three or more of the elements are joined is termed a *branch point* (or *node*); points 1–8 in Fig. *c* are branch points. A conducting path between two such points is termed a *branch* (1–2, 3–4, etc., in Fig. *c*). A voltage at a point in the network measured relative to the voltage at a designated branch point is termed a *node voltage*. A closed conducting loop in the network (e.g. 1, 3, 5, 7, 1) forms a *mesh contour*, and the portion of the network bounded by it is termed a *mesh*. Any branch that is common to two or more meshes is a *mutual branch* (e.g. 5–6). Two branches of a network are said to be *conjugate* if an e.m.f. in one of them does not produce a current in the other. The currents circulating in the meshes are known as *mesh currents*.

The behaviour of a network may be analysed by applying Kirchhoff's laws to each mesh in the network in turn; both the real and imaginary parts of the complex impedances involved must be satisfied simultaneously. For a large network containing many meshes, as with many types of *filter network, this method is very cumbersome. An alternative is to apply *Thévenin's theorem or Norton's theorem to a linear network; these theorems however cannot be applied to nonlinear networks.

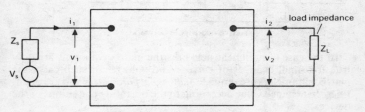

d Passive quadripole network

Analysis of linear networks can most usefully be done by considering the network as a *quadripole and deriving sets of equations relating the currents, voltages, and impedances at the input and output. Fig. *d* shows a passive quadripole network with an input source consisting of a voltage generator V_s of internal impedance Z_s. Fig. *e* shows an active quadripole network presenting an input impedance z_1 to the input source V_s of internal impedance Z_s, and appearing as a current generator $g_m v_1$ shunted by a resistance r_0 and producing a voltage v_2 in the output circuit.

Three different sets of equations can be written down: the impedance equations, the admittance equations, and derived from these the hybrid equations. The impedance equations can be written:

$$\begin{vmatrix} v_1 \\ \\ v_2 \end{vmatrix} = \begin{vmatrix} z_{11} & z_{12} \\ \\ z_{21} & z_{22} \end{vmatrix} \begin{vmatrix} i_1 \\ \\ i_2 \end{vmatrix}$$

This is known as the Z-matrix. Equivalent matrices can be written for the hybrid and admittance equations. These are the H- and Y-matrices, respectively. The constants in these equations are known as the *matrix parameters* (*see also* transistor parameters). Three-terminal devices, such as transistors, can be represented as quadri-poles that have two terminals joined together.

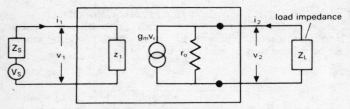

e Active quadripole network

In the case of nonlinear networks the matrix equations are only true for small changes of current and voltage. In such cases the matrix parameters are termed *small-signal parameters* and are quan-tities that change value according to the operating conditions of the device.

The input and output impedances of a network, v_1/i_1 and v_2/i_2, can be calculated from the matrix equations; it can be shown that v_1/i_1 depends on the load impedance Z_L connected to the output and conversely that v_2/i_2 depends on the impedance Z_s of the source connected to the input.

The *driving-point impedance is the impedance presented at a pair of terminals of a network of four or more terminals, under designated conditions at the other pair(s) of terminals. In the limit-ing case, for a quadripole, if the input (or output) is open circuit, the output (or input) impedance is the *open-circuit impedance*. The other limiting case is when the input (or output) is a short circuit in which case the output (or input) impedance is the *short-circuit impedance*. The quantities v_2/i_1 and v_1/i_2 are termed the *mutual impedances* of the network under open-circuit conditions, i.e. when $i_2 = 0$ and $i_1 = 0$, respectively.

See also quadripole; filter; transistor parameters.

378

(2) In computing systems, the linking together either of several microcomputers with each other and with peripheral devices such as printers so that several users can share access to the peripheral units, or of a mainframe computer with remote input/output terminals (outstations) so that several users can have access to the computer simultaneously.

(3) In information transmission and reception, the linking together of various electronic transmit/receive systems, such as computers, telex, or FAX units, using the telephone cable network to allow rapid communications between remote users.

See also local area network; wide area network; videotext.

network constants (or **parameters**) *See* network.

Neumann's law *Syn. for* Faraday-Neumann law. *See* electromagnetic induction.

neuroelectricity Electricity generated in the nervous systems of humans and other live animals.

neutral (1) Having no net positive or negative electric charge; at *earth potential.

(2) Denoting the line that completes the domestic *mains supply and is connected to earth at the power station.

neutralization The provision in an amplifier of negative *feedback of such amplitude and phase that it counteracts any inherent positive feedback. If the positive feedback is not neutralized, unwanted oscillations may occur.

Unwanted positive feedback in an amplifier can arise from the *Miller effect and is counteracted by *Rice neutralization*. In this type of neutralization a circuit is used to provide a 180° phase shift in the voltage fed back to the input (base) circuit. *Parasitic oscillations produced during *push-pull operation can be counteracted by *cross neutralization,* in which a portion of the output voltage of each device is fed back by means of a neutralizing capacitor to the input (base) circuit of the other device.

The voltage fed back to the input is termed the *neutralizing voltage*. An amplifier that derives the neutralizing voltage from a capacitor in the output circuit is known as a *neutrodyne*. The degree of neutralization in an amplifier may be observed using an indicating device termed a *neutralizing indicator*.

neutral temperature *See* thermocouple.

neutrodyne *See* neutralization.

neutron An elementary particle that has zero charge, spin $\frac{1}{2}$, and a rest mass very slightly greater than that of the *proton. Neutrons are present with protons in the nuclei of atoms and when bound in a nucleus are stable. Free neutrons are unstable. Their absence of electric charge makes them difficult to detect but does allow them to penetrate the open structure of an atom.

neutron activation analysis (NAA) *Syn.* nuclear reaction analysis (NRA). A method of detecting impurities in a sample of semiconductor material. The sample is exposed to a high neutron flux by being placed in or near to a nuclear reactor. The neutrons generate radioactive isotopes of the stable elements present in the sample, and the radiation emitted as the isotope subsequently decays is detected and measured using a multichannel analyser. The identity and concentration of many elements can be determined using this method, to a high degree of accuracy.

newton Symbol: N. The *SI unit of force defined as the force that, when applied to a mass of one kilogram, gives it an acceleration of one metre per second per second.

Nichrome *Tradename* An alloy of approximately 62% nickel, 15% chromium, and 23% iron. It has a very high resistivity and can operate at extremely high temperatures; it is therefore suitable for use in making thin-film resistors, wire-wound resistors, and heating elements.

nickel Symbol: Ni. A metal, atomic number 28, that is widely used in electronics. It exhibits strong *ferromagnetism and is used in magnetic alloys. It is also used as a conductor and in electrolytic cells.

nickel-cadmium cell A secondary *cell that has a nickel hydroxide/nickel oxide mixture as the anode, a cadmium cathode, and potassium hydroxide as the electrolyte.

nickel-iron cell *See* Edison cell.

Ni-Fe cell *See* Edison cell.

Nixie tube *See* digitron.

noctovision A television system that views an infrared image rather than a visible light image and is sometimes used at night. A specially designed infrared-sensitive *camera tube produces a video signal that can be detected by a standard receiver.

node (1) Any point, line, or surface in a distributed field at which some specified variable of a standing wave, such as voltage or current, attains a minimum value, usually zero. A *partial node* has a nonzero minimum. A point at which maximum magnitude is attained is an *antinode*.

(2) *Syn. for* branch point. *See* network.

node voltage *See* network.

nodon rectifier An *electrolytic rectifier containing an aluminium cathode, a lead anode, and ammonium phosphate solution as the electrolyte.

noise Unwanted electrical signals that occur within electronic or electrical devices, circuits, or apparatus and that result in spurious signals occurring at the output. In telecommunications, such as *telephony, and in the *reproduction of sound the presence of noise can result in the production of unwanted audible signals at the loudspeaker that can mask the required sounds. In television systems, noise can result in unwanted signals appearing on the picture.

*Interference may produce noise but does not always do so. *Man-made noise* is any noise that arises from human sources, such as car ignition systems.

In general, noise may be classified into two main types: *white noise,* which has a relatively wide flat frequency spectrum, and *impulse noise,* which is due to a single momentary disturbance or a series of such disturbances.

Thermal noise is a form of white noise that is caused by the random thermal agitation of electrons in resistive components. *Random noise* is noise that arises from any randomly occurring transient disturbance. If the rate of occurrence is sufficiently high it results in white noise, similar to thermal noise. If the rate of occurrence is low, random noise contributes to the impulse noise. All electronic circuits and devices suffer from thermal noise and random noise. Additional sources of noise arise in telecommunications from many different sources. *Atmospheric* and *galactic noise* both contribute to *radio noise. *Carrier noise* arises from unwanted variations in the frequency or amplitude of a *carrier wave that are not a result of the modulating wave.

The relative magnitudes of each different form of noise is a function of frequency, location on the earth's surface, the type of aerial and receiver used, and prevailing weather conditions. At any point in a telecommunication system the *basic noise* is the total noise present at that point.

The amount of noise that occurs in a receiver is given by the *noise factor, F,* where

$$F = P_o/N_o$$

P_o is the available carrier-wave power and N_o is available noise output power measured in a given resistance. The values of P_o and N_o are obtained by replacing the received signal by an unmodulated radiofrequency wave from a signal generator. The *noise figure, F_{dB},* of the receiver is the noise factor expressed in decibels, i.e.

$$F_{dB} = 10 \log_{10} F$$

See also acoustic feedback; frying; gas noise; grass; motor boating; partition noise; radio noise; Schottky noise; snow.

noise factor *See* noise.

noise figure *See* noise.

no-load Operation of any electronic or electrical machine, apparatus, circuit, or device under rated operating conditions but in the absence of a *load.

nondestructive read operation *See* read.

noninductive Denoting an electric circuit, device, or winding that, for its desired function, has negligible *inductance. *Bifilar winding is often employed in the manufacture of coils in order to minimize the inductance of the coil. A circuit that contains zero inductance is extremely difficult to realize. *Compare* inductive.

nonlinear amplifier *See* amplifier.

nonlinear distortion *See* distortion.

nonlinear network *See* network.

nonreactive Denoting any electric circuit, device, or winding that for its desired function, has negligible *reactance.

nonreactive load A *load in which the alternating current is in phase with the alternating voltage at the terminals. *Compare* reactive load.

nonresonant line *See* transmission line.

nonsaturated mode *See* saturated mode.

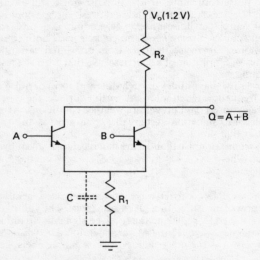

Nonthreshold logic–NAND gate

nonthreshold logic (NTL) A family of integrated *logic circuits that operates at a high speed at relatively low power. The basic NTL gate is shown in the diagram. It may be considered as a mixture of *resistor-transistor logic (RTL) and a simplified form of *emitter-coupled logic (ECL). The input transistors A and B form a *long-tailed pair with resistor R_1. The voltage gain is determined by the ratio R_2/R_1 and is adjusted to be slightly more than unity. Unlike other logic circuits the transistors conduct a relatively large current in the 'off' state, i.e. a logical 0 at the input. If a logical 1 is applied to the base of either transistor, that transistor saturates and the collector voltage falls. If both inputs are high the output is low, but if either input is low (or if both inputs are low) then the output voltage is high. The voltage swing is low, thus avoiding deep saturation and optimizing the switching time. A capacitor C may also be used, in parallel with resistor R_1, to further improve the switching speed. The relative simplicity of the basic gate gives a good potential for VLSI applications, but the low voltage gain employed makes the circuits susceptible to stray noise.

nonvolatile memory *Syn. for* permanent memory. *See* memory.

NOR circuit (or **gate**) *See* logic circuit.

Norton's theorem *See* Thévenin's theorem.

NOT circuit (or **gate**) *Syn. for* inverter. *See* logic circuit.

note frequency *Syn. for* beat frequency. *See* beats.

n-p junction A *p-n junction.

n-p-n transistor *See* transistor.

NRA *Abbrev. for* nuclear reaction analysis. *See* neutron activation analysis.

NTL *Abbrev. for* nonthreshold logic.

NTSC *Abbrev. for* National Television System Committee. *See* standardization.

n-type conductivity Conduction in a semiconductor in which current flow is caused by the movement of electrons through the semiconductor. *Compare* p-type conductivity. *See also* semiconductor.

n-type semiconductor An extrinsic semiconductor that contains a higher density of conduction *electrons than of mobile *holes, i.e. electrons are the majority carriers. *Compare* p-type semiconductor. *See also* semiconductor.

nuclear reaction analysis (NRA) *Syn. for* neutron activation analysis.

nucleon *See* nucleus.

nucleus The central and most massive part of an atom. It carries a positive charge Ze, where Z is the atomic number of the atom and e the electronic charge. A nucleus consists of tightly bound protons and neutrons collectively termed *nucleons;* the total number of nucleons is called the mass number, A. The number of neutrons associated with a given number of protons can vary within limits, giving rise to various isotopes of an element.

Most naturally occurring isotopes have stable nuclei although some, together with all artificially produced isotopes, have unstable nuclei; these are radioactive and undergo chemical or nuclear transformation as they decay.

nuclide An atom that is completely characterized by its atomic number, Z, mass number, A, and nuclear energy state.

null method *Syn.* balance method. An accurate method of measurement in which the quantity being measured, such as resistance or capacitance, is balanced by another of a similar kind: voltages in different circuits are adjusted so that the response of an indicating instrument falls to zero. The best known example of the null method is in the *Wheatstone bridge although it is used in many other bridge circuits.

null-point detector An instrument that is used with *bridge circuits to give a zero response when the balance point is achieved. Direct-current bridge circuits employ a sensitive galvanometer: balance is achieved when a zero indication is obtained. Alternating-current bridge circuits frequently employ a sensitive loudspeaker, usually as a headset: the sound becomes inaudible at the balance point.

number of poles The number of different conducting paths that may be simultaneously opened or closed by a switch, circuit-breaker, or other similar apparatus. The device is described as single pole, double pole, triple pole, or multi-pole depending on the number of poles that it can operate.

numerical control A type of automatic control in which a number generated by the controlling device, such as a digital computer, is used to control another device, particularly automatic machines such as machine tools.

Nyquist diagram A diagram that can be used as a criterion of stability in an *amplifier. It shows the relation, in rectangular coordinates, between the amplification and the feedback of the device.

Nyquist noise theorem The law relating the power P dissipated in a resistor due to thermal noise with the frequency f of the signal. At ordinary temperatures the law is expressed as:

$$\mathrm{d}P = kT\mathrm{d}f$$

where k is the Boltzmann constant and T the thermodynamic temperature.

O

observed failure rate *See* failure rate.

observed mean life *See* mean life.

OCR *Abbrev. for* optical character reader.

octave An interval that has the frequency ratio 2:1.

octode A *thermionic valve that contains a total of eight electrodes arranged as five grid electrodes between the cathode and the main anode and an additional anode placed between the two innermost grids. Its most common application has been as a combined *mixer and local oscillator.

oersted Symbol: Oe. The unit of magnetic field strength in the obsolete CGS electromagnetic system of units. One oersted equals 79·58 (i.e. $1000/4\pi$) amperes per metre.

off-line (1) Denoting a *peripheral device, possibly an *on-line device, that is switched off, broken, or disconnected from a computer.

(2) Denoting a *peripheral device that is controlled by instructions generated earlier and held on a computer storage device. *See also* batch processing. *Compare* on-line.

off-line plotter *See* incremental plotter.

ohm Symbol: Ω. The *SI unit of electrical *resistance, *reactance, and *impedance. It is defined as the resistance between two points on a conductor when a constant current of one ampere flows as a result of an external potential difference of one volt applied between the points. This unit was formerly called the *absolute ohm* and replaced the *international ohm* (Ω_{int}) as the standard of resistance: one Ω_{int} equals 1·000 49 Ω.

ohmic *See* Ohm's law.

ohmic contact An electrical contact in which the potential difference across it is linearly proportional to the current flowing through it. Ohmic contacts between a metal and a semiconductor can be formed, but it is necessary to have a highly doped surface layer on the semiconductor to ensure that the dominant method of transfer of carriers (electrons in n-type semiconductor) is due to the *tunnel effect. Otherwise a *Schottky diode is formed and a nonlinear current-voltage relationship occurs.

ohmic loss Power dissipation in an electrical circuit, network, or device that is due to the resistance present rather than to other causes such as eddy currents or back e.m.f.

ohmmeter An instrument that measures the electrical *resistance of conductors or insulators. The indicating scale is calibrated in ohms or suitable multiples or submultiples of ohms.

ohm metre Symbol: Ω m. The *SI unit of electric *resistivity.

Ohm's law The electric current, *I,* flowing in a conductor or resistor is linearly proportional to the applied potential difference, *V,* across it. From the definition of *resistance, *R,* Ohm's law can be written:

$$V = IR$$

Any electrical component, circuit, or device that maintains such a linear relationship between current and voltage can be described as *ohmic.*

oil-break (of a switch, *circuit-breaker, etc.) Having contacts that separate in oil. *Compare* air-break.

OMCVD *Abbrev. for* organo-metallic chemical vapour deposition. *Syn. for* MOCVD.

omni-aerial *Short for* *omnidirectional aerial.

omnidirectional aerial An *aerial whose radiative or receiving properties at any instant are the same on all bearings. *Compare* directive aerial.

ondograph An instrument that produces a graph of an alternating voltage. It has a digital response and hence the graph of the waveform appears as a series of small steps rather than as a smooth curve.

ondoscope A glow-discharge tube that detects the presence of high-frequency radiation.

one-shot multivibrator *Syn.* single-shot multivibrator. *See* monostable.

one-sided abrupt junction *See* abrupt junction.

O-network *See* quadripole.

on-line Denoting a *peripheral device of a computer that is directly controlled by the computer and executes instructions as they are generated. *See also* interactive. *Compare* off-line.

on-line plotter *See* incremental plotter.

opamp *Short for* operational amplifier.

open arc An *arc struck between carbon electrodes with the air being permitted free access to it. The products of the combustion tend to dissipate throughout the atmosphere rather than remaining in the immediate vicinity of the arc. Shielding from stray draughts is often provided. *Compare* enclosed arc.

open circuit *See* circuit.

open-circuit impedance *See* network.

open-circuit voltage The voltage across the output terminals of any electrical or electronic network, device, machine, or other apparatus at *no-load when designated operating conditions are carried out.

open-ended Denoting any circuit, system, or process that can be readily augmented by the addition of further stages, parts, or steps.

opening time *See* time to trip.

open wire *Syn.* overhead line. A conductor that is not surrounded by insulation and is supported separately above ground.

operating point The point on the family of *characteristic curves of an active electronic device, such as a transistor, that represents the magnitudes of voltage and current when designated operating conditions are applied to the device.

operational amplifier An *amplifier that is constructed from preassembled units. The term is usually reserved for an amplifier of two or more stages that is designed for insertion into other equipment.

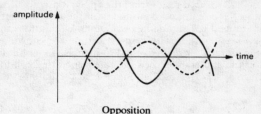

Opposition

opposition Two periodic quantities of the same frequency are said to be *in opposition* (or *antiphase*) if the phase difference between them is one half-period, i.e. 180° (*see* diagram).

optical ammeter An instrument that measures the current flowing through the filament of an *incandescent lamp by comparing, photometrically, the illumination produced with that produced by a current of known magnitude in the same filament.

optical character reader (OCR) A device that is used to produce coded signals suitable for a given digital *computer from information in the form of characters, numbers, or other symbols printed on paper. Maximum efficiency normally requires a specially designed typeface.

optical image *See* camera tube.

optically erasable memory *Syn. for* floating-gate PROM. *See* read-only memory.

optically pumped gas-cell maser *See* maser.

optical stepper *See* photolithography.

optimum bunching *See* velocity modulation.

Oracle *Tradename. See* teletext.

OR circuit (or **gate**) *See* logic circuit.

origin distortion *See* distortion.

orthicon *See* iconoscope.

orthogonal Mutually perpendicular; at right angles.

oscillating current (or **voltage**) A current (or voltage) waveform that periodically increases and decreases in amplitude with respect to time, according to a particular mathematical function. Oscillating waveforms can be sinusoidal, sawtooth, square, or many other shapes.

oscillation (1) A periodic variation of an electrical quantity, such as current or voltage.

(2) A phenomenon that occurs in an electrical circuit if the values of self-inductance and capacitance in the circuit are such that an oscillating current results from a disturbance of the electrical equilibrium of the circuit.

A circuit that produces oscillations freely is termed an *oscillatory circuit*. Oscillations that result from the application of a direct-voltage input to the circuit and continue until the direct voltage is removed are termed *self-sustaining oscillations*. Oscillations that tend to decrease in amplitude with respect to time are known as *stable oscillations; unstable oscillations* tend to increase in amplitude with respect to time and soon exceed the rated operating conditions of the circuit. *See also* free oscillations; forced oscillations; damped; parasitic oscillations.

resonant elements

Tunnel-diode oscillator   **Equivalent circuit**

a **Negative-resistance** oscillator

oscillator A circuit that converts direct-current power into alternating-current power at a frequency that is usually greater than can be achieved by rotating electromechanical alternating-current generators. Application of the direct-voltage supply to the circuit is usually sufficient to cause it to oscillate and for the oscillations to be maintained until the direct voltage is switched off.

There are two broad categories of oscillator: *harmonic oscillators* generate essentially sinusoidal waveforms and contain one or more active circuit elements continuously supplying power to the passive components; *relaxation oscillators are characterized by non-sinusoidal waveforms, such as sawtooth waveforms, and the switched exchange of electrical energy between the active and passive circuit elements.

A simple harmonic oscillator consists essentially of a frequency-determining device, such as a resonant circuit, and an active element that supplies direct power to the resonant circuit and also compensates for damping due to resistive losses. In the case of a simple L-C circuit, application of a direct voltage causes *free oscillations in the circuit that decay because of the inevitable resistance in the circuit (*see* damped). In the absence of the resistance no damping would occur and the free oscillations would continue at a constant amplitude until the direct voltage was removed. The active element in an oscillator can be considered as supplying a *negative resistance of sufficient value to compensate for the positive resistance; consequently the complete oscillator contains effectively zero resistance and when shocked will oscillate continuously.

The effective negative resistance is provided either by a device, such as a unijunction transistor, that exhibits a negative-resistance portion of its characteristic or by employing positive *feedback of power in order to overcome the damping. Any particular oscillator may be studied from the negative-resistance approach or from a feedback approach. In the latter case internal positive feedback is considered to be present in the negative-resistance device. Usually *negative-resistance oscillators* are those that contain a device such as a unijunction transistor or tunnel diode (Fig. *a*), operated in the negative-resistance portion of the characteristic determined by the applied voltage, V_A, and external source resistance, R_s. *Feedback oscillators* are those that employ external positive feedback. An inherent phase shift of 180° occurs between the base and collector of the common-emitter connection shown in Fig. *b*. Various types of feedback circuit are used in order to provide the necessary counterbalancing phase-shift. Transformer coupling is shown in the diagram; the resonant circuit is formed by the transformer primary (L) and the capacitor C.

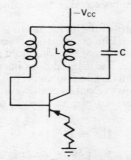

b Common-emitter oscillator with transformer feedback

The frequency-determining device may consist of a component, such as a piezoelectric or magnetostrictive crystal, that converts mechanical stress into electrical impulses; alternatively such a de-

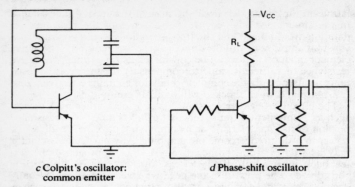

c Colpitt's oscillator:
common emitter

d Phase-shift oscillator

vice may be coupled to the resonant circuit to prevent frequency drift.

Colpitt's oscillator and the *phase-shift oscillator* are shown in Figs. *c* and *d*.

oscillatory circuit *See* oscillation.

oscilloscope An instrument that produces a visible image of one or more rapidly varying electrical quantities. An image can be produced showing the variation of a signal with respect to time or with respect to another electrical quantity. The most usual type of oscilloscope is the *cathode-ray oscilloscope. A three-dimensional oscilloscope contains a screen consisting of a cube of luminescent material that is suitably excited by three signals to produce a three-dimensional image.

An oscilloscope provided with the means to produce a permanent record of the signal on film or magnetically sensitive paper is known as an *oscillograph;* the record thus produced is an *oscillogram.*

Ostwald's dilution law A law that indicates the degree of dissociation of an electrolyte when dissolved in water. It is obtained by applying the law of mass action to electrolytic dissociation. If unit amount of an acid, HA, is dissolved in a volume, V, of water it dissociates according to

$$HA \rightleftharpoons H^+ + A^-$$

At equilibrium, if the fraction of acid dissociated is α, then the concentrations of acid and dissociated ions are related by:

$$[H^+][A^-]/[HA] = \alpha^2/(1 - \alpha)V = K$$

where K is a constant termed the *dissociation constant* of the reaction. For weak electrolytes α is very much less than unity and the law reduces to $\alpha^2 = KV$. From this it can be seen that if the dilution of the electrolyte is increased the degree of dissociation also increases.

O-type microwave tube *Syn. for* linear-beam microwave tube. *See* microwave tube.

outgassing (1) The removal by heating of some of the air adsorbed on the interior surfaces of a vacuum system or tube.

(2) The slow deterioration of a vacuum due to the release of residual adsorbed gases from the interior surfaces of a vacuum system or tube.

out of phase *See* phase.

output (1) The power, voltage, or current delivered by any circuit, device, or apparatus.

(2) The part of any circuit, device, or apparatus, usually in the form of terminals, at which the power, voltage, or current is delivered. *See also* input/output.

(3) To deliver as an output signal.

output gap An *interaction space in a *microwave tube where power is extracted from the electron beam; it hence constitutes the output section of the tube.

output impedance The *impedance presented at the output of an electronic circuit or device.

output meter An instrument that is coupled to the output of an electronic circuit or device and that measures the power output by the device.

output transformer A transformer that is used for coupling an output circuit, particularly that of an amplifier, to the *load.

outside broadcast A television broadcast that originates from a source other than a television studio. An outside broadcast uses mobile cameras and transmitting or recording apparatus. A live outside broadcast uses a portable transmitter to relay the broadcast signal to the main control centre where it is broadcast from the main transmitters. If the broadcast is not live it is usually recorded on film; *videotape recorders are now also used. *See also* electronic news gathering.

overall efficiency The ratio of the power absorbed by the *load of a device, such as a power amplifier or induction heater, to the power supplied by the source, either as mains supply or batteries.

overbunching *See* velocity modulation.

overcoupling *See* coupling.

overcurrent *See* overcurrent release.

overcurrent release *Syn.* overload release. A switch, circuit-breaker, or other tripping device that operates when the current in a circuit exceeds a predetermined value. A current that causes the release to operate is an *overcurrent*.

The device is often designed so that a delay occurs after the overcurrent is sensed and before the device trips. Several different delay conditions can be used: *definite time lag* overcurrent release has a predetermined delay independent of the magnitude of the overcurrent; *inverse time lag* overcurrent release has a delay that is an inverse function of the magnitude of the overcurrent; *inverse and definite minimum time lag* overcurrent release occurs when the delay is an inverse function of the magnitude of the overcurrent until a minimum value of the delay is reached. *Compare* undercurrent release.

overdamping *Syn.* periodic damping. *See* damped.

overdriven amplifier An *amplifier that is operated with an input voltage greater than that for which the circuit was designed. Overdriving results in *distortion being introduced into the output waveform.

overhead line *See* open wire.

overlapping gate CCD *See* charge-coupled device.

overlay structure *See* program.

overload Any *load delivered at the output of an electrical device, circuit, machine, or other apparatus that exceeds the rated output of the equipment. It is expressed numerically as the difference between the overload value and the rated value; it can also be quoted as a percentage of the rated output. The magnitude of the load at which operation of the equipment becomes unsatisfactory, due to overheating or distortion, is the *overload level;* the value of the maximum load that can be tolerated without permanent physical damage to the equipment is the *overload capacity*.

overload capacity *See* overload.

overload level *See* dynamic range; overload.

overload release *See* overcurrent release.

overscanning *See* scanning.

overshoot *See* pulse.

overvoltage *Syn.* excess voltage. (1) A voltage that exceeds the normal voltage applied between two conductors or between a conductor and earth.

(2) (at a given electrode in an electrolytic cell) The amount by which the applied e.m.f. necessary to release hydrogen on that electrode from a particular electrolyte exceeds the e.m.f. required to liberate hydrogen from the same electrolyte on a standard platinum electrode. *See also* oxygen overvoltage.

(3) The difference between the operating voltage of the tube of a *Geiger counter and the threshold voltage.

overvoltage release A switch, circuit-breaker, or other tripping device that operates when the voltage in a circuit exceeds a predetermined value. A voltage that causes the release to operate is an overvoltage. *Compare* undervoltage release.

Owen bridge A four-arm *bridge used for the measurement of inductance in terms of known resistance and capacitance (*see* diagram).

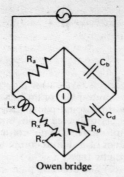

Owen bridge

At balance, as shown by the null response of the indicating instrument, I,

$$L_x = C_b R_a R_d$$
$$R_x = (C_b R_a / C_d) - R_c$$

oxidation A chemical reaction used in electronics in which a thin portion of the surface of a silicon chip is converted to silicon dioxide.

oxide In general, a chemical compound of an element with oxygen. In particular, *short for* silicon dioxide (*silica). It is the most widely used insulating and/or passivating material in the construction of devices, components, and integrated circuits fabricated in silicon. Silicon dioxide can be readily grown on the surface of silicon and layers of varying thicknesses are grown during the fabrication of *silicon devices and circuits.

Oxide acts as a barrier to the diffusion of impurities into the substrate material and is used as an 'on-chip' *mask during the *planar process. A thin oxide layer is used to form the insulator in the manufacture of MOS devices, such as insulated-gate *field-effect transistors and *charge-coupled devices. An extra thick oxide layer can be formed by means of the *coplanar process in order to prevent spurious MOST formation. A layer of oxide is formed over both bipolar and MOS integrated ciruits and devices for *passivation of the surface.

oxide masking The use of the silicon dioxide layer (*see* oxidation) on the surface of a silicon chip to provide a mask for selective processing of the chip. The oxide formed by oxidation of the surface of the silicon is selectively etched to expose the underlying silicon prior to *diffusion or *metallizing.

oxygen overvoltage (at an anode of an electrolytic cell) The difference between the actual e.m.f. necessary to release oxygen at the anode and the theoretical value for release of the gas.

393

P

PA *Abbrev. for* public address (system).

PABX *Abbrev. for* private automatic branch exchange. *See* telephony.

packing density (1) *Syn.* functional packing density. The number of devices or logic gates per unit area of an *integrated circuit.

(2) The amount of information that is contained in a given dimension of a storage system of a digital computer, e.g. the number of *bits per inch of magnetic tape.

pad (1) *See* attenuator. (2) *See* bonding pad.

pair A *transmission line consisting of two similar conductors that are insulated from each other but are associated to form part of a communication channel or channels. If the two conductors are twisted around each other they form a *twisted pair*. A *coaxial pair* is a pair of cylindrical conductors that are coaxial and may be used to form a *coaxial cable.

paired cable *Syn* twin cable. A type of cable composed of several *pairs of conductors: each pair is twisted together but no two sets of pairs are twisted.

pairing A fault occurring in television picture tubes that employ interlaced scanning (*see* television). Lines of alternate fields tend to coincide instead of interlacing with each other and the vertical resolution is halved.

PAL *Abbrev. for* phase alternation line. A colour-television system developed in Germany. The chrominance signal (*see* colour television) is resolved into two components in *quadrature that are used for *amplitude modulation of the chrominance subcarriers. In the PAL system the relative phase of the quadrature components is reversed on alternate lines in order to minimize phase errors. This system has been generally adopted in Europe. *Compare* SECAM.

pam *Abbrev. for* pulse-amplitude modulation *See* pulse modulation.

pancake coil A flat coil in which the common method of winding is to have the turns arranged as a flat spiral.

panoramic radar indicator *See* radar indicator.

panoramic receiver A *radio receiver that is automatically tuned so as to receive certain frequency bands, each for a preselected frequency. The period of the tuning variation is preselected. In this way a range of radiofrequencies can be regularly monitored, for example, when listening for distress calls.

paper capacitor A capacitor of medium loss and medium capacitance-stability that is used in high-voltage a.c. and d.c. applications. These capacitors are manufactured by winding together aluminium foils interleaved with layers of tissue paper. The moisture content of the paper is removed by impregnation with a suitable oil or wax. *Metallized paper capacitors* use an evaporated metal film as the electrode instead of aluminium.

paper tape *Syn.* punched tape. A strip of paper on which information can be carried in the form of holes punched across the width. Each combination of between five to eight holes represents a specific character, number, or symbol. Paper tape has been used for the input and output of data or programs to digital *computers but has been largely superseded.

A *tape punch* is a device that prepares the paper tape. The signals to be punched may be provided manually from a keyboard, from a source of experimental data, or automatically as the output of a computer. A *tape reader* is a device that decodes the information on a paper tape into electrical signals and displays, stores, or transmits the information to a computer. *Compare* punched card.

PAR *Abbrev. for* precision approach radar.

parabolic reflector *Syn.* paraboloid reflector. A radiofrequency or microwave-frequency reflector (*see* directive aerial) that has a hollow concave paraboloid shape so that all waves passing through the focus will be reflected parallel to the axis of rotation.

parallel Circuit elements are said to be *in parallel* if they are connected so that the current divides between them and later reunites (*see* diagram).

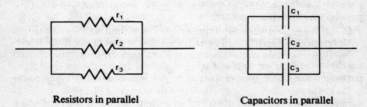

Resistors in parallel   **Capacitors in parallel**

For n resistors in parallel the total resistance, R, is given by:

$$1/R = 1/r_1 + 1/r_2 + \ldots 1/r_n$$

where $r_1, r_2, \ldots r_n$ are the values of the individual resistors.

For n capacitors in parallel the total capacitance, C, is given by:

$$C = c_1 + c_2 + c_3 + \ldots c_n$$

where $c_1, c_2, \ldots c_n$ are the individual capacitances. The capacitors thus behave collectively as a large capacitor having the total plate area of the component capacitors.

Machines, transformers, and cells are said to be in parallel when terminals of the same polarity are connected together. Several cells connected together in parallel have a lower total internal resistance than a single cell and can therefore supply a larger maximum current. *Compare* series. *See also* shunt.

parallel circuit A circuit containing two or more elements connected in parallel across a pair of lines or terminals (*see* diagram).

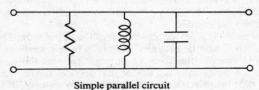

Simple parallel circuit

parallel memory *See* random-access memory.

parallel network *Syn.* shunt network. *See* network.

parallel-plate capacitor A *capacitor formed from two parallel metal electrodes separated by dielectric material.

parallel resistance *See* resonant frequency.

parallel resonant circuit *Syns.* rejector; antiresonant circuit. *See* resonant circuit; resonant frequency.

parallel supply *See* series supply.

parallel-T network *Syn. for* twin-T network. *See* quadripole.

parallel transmission A communication system in which the characters of a word are transmitted in parallel along separate lines. Transmission of the characters is usually simultaneous. *Compare* serial transmission.

parallel-wire line *See* microstrip.

paramagnetic Curie temperature *Syn. for* Weiss constant. *See* paramagnetism.

paramagnetism An effect observed in certain materials that possess a permanent atomic or molecular *magnetic moment. Each orbital electron in an atom constitutes an individual current and hence has a magnetic moment; however the possible *energy levels available are such that only unfilled shells of electrons contribute to the magnetic moment of the atom as a whole. The *spin of the atomic electrons also has a magnetic moment associated with it but in an atom only unpaired spins contribute to the magnetic moment of the atom as a whole. Most free atoms would have a magnetic moment due to orbital electrons in unfilled outer shells but practical substances combine in general so as to complete the outer shells; most gaseous molecules and ionic or homopolar liquids or solids have no overall magnetic moment and are diamagnetic (*see* diamagnetism).

Permanent magnetic moments are only possessed by molecules or ions containing unpaired electron spins (oxygen, O_2, for example contains two unpaired spins) or by particular ions of multivalent transition elements in which there is an unfilled inner shell. Paramagnetism is most commonly associated with electron spin but a few compounds also have an orbital contribution.

In the absence of an applied magnetic flux density, thermal motion causes the individual magnetic moments to be randomly orientated throughout the sample and the net magnetization is zero. In the presence of a magnetic flux the magnetic moments tend to align themselves in the direction of the field; this tendency however is opposed by the thermal agitations and a paramagnetic substance has a small positive *susceptibility, χ, which is temperature dependent.

The behaviour of a paramagnetic gas can be approximately described by the Langevin function, which shows that at ordinary magnetic fields and temperatures the gas obeys *Curie's law:*

$$\chi = C/T$$

where C is a constant and T the absolute temperature. However, at sufficiently high flux density and low temperature a saturation point is achieved with all the molecules aligned along the field with neglibible thermal effects. Very dilute paramagnetic liquids also obey Curie's law.

The magnetic properties of paramagnetic solids and liquids depend on the complex intra-atomic and interatomic forces operating within them and the behaviour cannot always be described by a simple equation. Nonhydrated liquids and many paramagnetic solids at ordinary temperatures and fields obey the *Curie–Weiss law*:

$$\chi = C/(T - \theta),$$

where θ, the *Weiss constant,* can be either positive or negative. The Curie–Weiss law is only obeyed at temperatures $T > |\theta|$ and is a modification of Curie's law arising from the mutual interactions of the ions or molecules.

Certain metals, such as sodium or potassium, exhibit *free-electron paramagnetism* or *Pauli paramagnetism* in which a small positive susceptibility with only slight temperature dependence is observed. Both effects are due to the conduction electrons in the metals. The individual atoms in the solid are left as diamagnetic ions and the conduction electrons exhibit both diamagnetism and paramagnetism. In most metals these effects are of the same order of magnitude but Pauli paramagnetism occurs when the paramagnetism effect is greater than the diamagnetism.

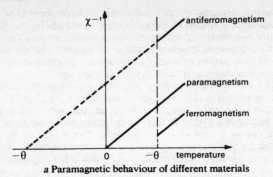

a Paramagnetic behaviour of different materials

Another effect arising in certain molecules is *Van Vleck paramagnetism*. It occurs in molecular systems where the splitting, Δ, between the ground state and its associated excited state is very great, i.e. $\Delta \gg kT$ where k is the Boltzmann constant. There is then an additional small positive temperature-independent contribution to the susceptibility of the material.

At temperatures below certain critical temperatures and approximately equal to the modulus of the Weiss constant, the intermolecular forces of many paramagnetic solids become much greater than the thermal agitations. The magnetic moments are no longer randomly orientated but take up an appropriate ordered state and the materials become either ferromagnetic, antiferromagnetic, or ferrimagnetic. The paramagnetic behaviour of different materials can be compared by plotting the inverse susceptibility χ^{-1} against absolute temperature in the paramagnetic region (Fig. *a*).

Paramagnetism causes an increase of magnetic flux density within a sample; this is represented schematically by a concentration of

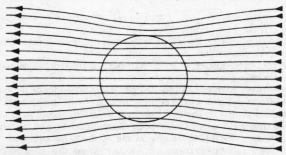

b Change in magnetic flux density in a paramagnetic substance

398

the lines of magnetic flux density passing through it (Fig. *b*). If a paramagnetic substance is placed in a nonuniform magnetic field it tends to move from the weaker to the stronger region of the field; a bar of paramagnetic material placed in a uniform magnetic flux tends to orientate itself with the longer axis parallel to the flux.

parameter A quantity that is constant in a given case but has a particular value for each different case considered. Examples include the values of the resistances, capacitances, etc., that form an electrical *network or the constants appearing in the equations connecting currents and voltages at the terminals of a network. *See also* transistor parameters.

parametric amplifier A microwave *amplifier in which the *reactance of the device is varied with respect to time in a regular manner, i.e. it is pumped. An alternating voltage, the pump voltage, is the most common form of *pump*. Provided that a suitable relationship is maintained between the pump frequency and the signal frequency, energy is transferred from the pump and amplification of the signal is achieved.

paraphase amplifier An amplifier that converts a single input into an input suitable for *push-pull operation.

parasitic aerial *Syn. for* passive aerial. *See* directive aerial.

parasitic oscillations Unwanted oscillations that can occur in *amplifier or *oscillator circuits. The frequency of such oscillations is usually very much higher than the frequencies for which the circuit has been designed since it is mainly determined by stray inductances and capacitances, such as in connecting leads, and by interelectrode capacitances.

Parasitic stoppers are devices incorporated into a circuit to prevent parasitic oscillations being generated. They are commonly resistors used in the input and output circuits of the device. They take their name from the particular part of the circuit in which they are incorporated. For example, an *anode stopper* is connected to an *anode, a *grid stopper* to a *grid, and a *base stopper* to a *base.

parasitic radiator *Syn. for* passive aerial. *See* directive aerial.

parasitic stopper *See* parasitic oscillations.

partial node *See* node.

particle-induced X-ray emission (PIXE) *Syn. for* soft X-ray appearance spectroscopy.

partition noise *Noise that occurs in active electronic devices due to random fluctuations in the distribution of the currents between the electrodes.

pascal Symbol: Pa. The *SI unit of pressure, defined as the pressure that results when a force of one newton acts uniformly over an area of one square metre.

Pascal *See* programming language.

Paschen's law The *breakdown voltage that initiates a discharge between electrodes in a gas is a function of the product of pressure and distance. For example, if the distance between the electrodes is doubled breakdown will only occur at the same potential difference if the gas pressure is halved. *See also* Hittorf's principle.

pass band *See* filter.

passivation Protection of the junctions and surfaces of solid-state electronic components and integrated circuits from harmful environments. Passivation is most commonly achieved by forming a layer of silicon dioxide on the surface of the silicon chip. An alternative method is *glassivation,* in which the passivating layer is a melt of vitreous material that is deposited on the surface of the semiconductor and allowed to harden.

passive Denoting any device, component, or circuit that does not introduce *gain or does not have a directional function. In practice only pure resistance, capacitance, inductance, or a combination of these three is passive. *Compare* active.

passive aerial *Syns.* secondary radiator; parasitic aerial; parasitic radiator. *See* directive aerial.

passive filter *See* filter.

passive network *See* network.

passive satellite *See* communications satellite.

passive substrate A substrate, such as glass or ceramic, that is used in microelectronics for its lack of transistance (*see* transfer parameter).

passive transducer *See* transducer.

patch board *Syn.* plug board. *See* patching.

patch cord *See* patching.

patching (1) Forming temporary connections between circuits of an analog *computer in order to simulate a particular physical process. Patching is carried out in a particular location that forms part of the computer and in which the circuits available for patching are terminated. There is usually a board or panel – the *patch board* – that contains an array of jack sockets, each of which is connected to a circuit. Connections are formed between the sockets by means of flexible insulated conductors – *patch cords.*

Suitable patching of chosen circuits in an analog computer results in the establishment of a desired relationship between the voltages in the computer. The mathematical equations that express the relationships are the *machine equations.* A given voltage that represents one of the mathematical variables in the machine equation is a *machine variable.*

(2) Altering a computer *program by the direct modification of the machine code (*see* programming language).

Pauli exclusion principle No two electrons in an atom can exist in the same quantum energy state, i.e. they cannot be described by the same four quantum numbers, l, m, n, and s. A particular atomic orbital of given values of *n*, *l*, and *m* can contain only two electrons: these electrons must have *spins (*s*) equal to $+\frac{1}{2}$ and $-\frac{1}{2}$. An atomic orbital that contains two electrons of opposing spins is said to contain spin-paired, or paired, electrons. An orbital containing only one electron is said to contain an unpaired electron, or unpaired spin.

Pauli paramagnetism *See* paramagnetism.

P band A band of microwave frequencies ranging from 0·225 to 0·390 gigahertz. *See* frequency band.

PBT *Abbrev. for* permeable base transistor. *See* vertical FET.

PBX *Abbrev. for* private branch exchange. *See* telephony.

p-channel Denoting an MOS circuit or device, such as a *charge-coupled device, *MOS integrated circuit, or *MOS transistor, in which the conducting channel is formed as p-type semiconductor. N-channel devices are however preferred (*see* field-effect transistor). The term also describes junction field-effect transistors that have a p-type channel. *Compare* n-channel.

PCM *Abbrev. for* portable conformable mask. *See* multilevel resist.

pcm *Abbrev. for* pulse code modulation. *See* pulse modulation.

p.d. (or **pd**) *Abbrev. for* potential difference.

pdm *Abbrev. for* pulse-duration modulation. *See* pulse modulation.

peak factor *Syn.* crest factor. The ratio of the *peak value of a periodically varying quantity to the *root-mean-square value. If the quantity varies sinusoidally the peak factor is $\sqrt{2}$.

peak forward voltage The maximum instantaneous voltage applied to a device in the forward direction, i.e. in the direction in which the device is designed to pass current with the minimum resistance.

peak inverse voltage The maximum instantaneous voltage applied to a device in the *reverse direction, i.e. in the direction of maximum resistance. If the peak inverse voltage becomes too great, *breakdown of the device will occur. The breakdown is avalanche breakdown in a semiconductor device and arc formation in a valve. A rated value of peak inverse voltage for a device specifies the maximum inverse voltage that the device can tolerate without breakdown.

peak limiter *Syn. for* clipper. *See* limiter.

peak load *See* load.

peak point (1) *See* unijunction transistor. (2) *See* tunnel diode.

peak pulse amplitude *See* peak value.

peak-riding clipper A *limiter that automatically adjusts the value of voltage at which it operates according to the peak value of the pulse train input to the circuit.

peak-to-peak amplitude *See* amplitude.

peak value *Syn.* amplitude; crest value. (1) The maximum positive or negative value of any alternating quantity, such as current or voltage, during a given time interval. The positive and negative values are not necessarily equal in magnitude.

(2) *Syn.* peak pulse amplitude. The maximum value of an *impulse voltage or current.

pedestal A *pulse waveform that is 'flat-topped', i.e. the amplitude is made constant for a pulse interval large compared to the rise and fall times (*see* pulse). A pedestal is combined with a second waveform in order to increase the magnitude of the second waveform by a constant amount.

Peltier effect *See* thermoelectric effects.

Peltier element *Syn.* thermoelectric module. *See* thermoelectric effects.

pentagrid *See* heptode.

pentatron A *thermionic valve that contains five electrodes and is effectively two triodes with a single common cathode, two anodes, and two control grids.

pentode A *thermionic valve containing five electrodes. It is equivalent to a tetrode containing an additional electrode, the *suppressor grid,* between the screen and the anode. The suppressor grid is at a negative potential relative to both anode and screen and is used to prevent secondary electrons from the anode reaching the screen. The suppressor grid must be of an open mesh design otherwise the passage of the primary electron beam would be impeded by it.

percentage modulation *See* amplitude modulation.

percentile life *See* mean life.

perfect crystal A single crystal that has a uniform arrangement of atoms throughout and contains no impurities or dislocations: an *ideal single crystal. *Compare* mosaic crystal.

perfect dielectric *See* dielectric.

perfect transformer *Syn. for* ideal transformer. *See* transformer.

period *Syn.* periodic time. Symbol: T. The time required to complete a single cycle of regularly recurring events. The period of an oscillation is related to the frequency, ν, and the angular frequency, ω, by

$$T = 1/\nu = 2\pi/\omega$$

periodic Denoting any variable quantity that has regularly recurring values with respect to equal increments of some independent variable, such as time. The interval between two successive repetitions is the *period.

periodic damping *Syn. for* overdamping. *See* damped.

periodic duty *See* duty.

periodic table The classification of chemical elements, first introduced by Mendeléev, that demonstrates a periodicity of chemical properties when the elements are arranged in order of their atomic numbers. One form of periodic table is given in Table 11, backmatter. Elements of similar chemical properties occur in the same vertical group in the periodic table. The periodic table was used to predict the existence of undetected elements.

peripheral devices Devices that are connected to a *computer, form part of the computer system, and whose operation is controlled by the *central processing unit of the computer. *Terminals, *visual display units, *printers, *magnetic tape units, and disks (*see* moving magnetic surface memory) are examples of peripherals.

permalloy Originally an alloy composed of 78·5% nickel and 21·5% iron but now any of a variety of alloys made by the addition of copper, cobalt, manganese, etc., to the original. Permalloys are characterized by a high magnetic *permeability at low values of magnetic flux density and by low *hysteresis loss.

permanent magnet A magnetized sample of a ferromagnetic material, such as steel, that possesses high retentivity and is stable against reasonable handling. It requires a definite demagnetizing flux in order to destroy the residual magnetism. A *simple magnet* consists of a single bar, which can be horseshoe shaped, of the material. A *compound magnet* has several suitably shaped bars or laminations fastened together. *See also* ferromagnetism; magnetic hysteresis.

permanent memory *Syn.* nonvolatile memory. *See* memory.

permeability Symbol: μ; unit: henry per metre. The ratio of the magnetic flux density, B, in a medium to the external magnetic field strength, H, i.e.

$$\mu = B/H$$

The *permeability of free space* is designated μ_0 and is termed the *magnetic constant*. In the system of *SI units it has the value $4\pi \times 10^{-7}$ henry/metre. In other systems, such as the *CGS and *MKS systems, it has been given different values. Using *Maxwell's equations it can be shown that

$$\mu_0\varepsilon_0 = 1/c^2$$

where ε_0 is the *permittivity of free space and c is the velocity of light.

The *relative permeability*, μ_r, is the ratio of the magnetic flux density in a medium to the magnetic flux density in free space for the same value of external magnetic field strength, i.e.

$$\mu_r = \mu/\mu_0$$

For most materials μ_r is a constant. Diamagnetic materials have a value of μ_r less than unity. Paramagnetic materials have a value just greater than unity (*see* diamagnetism; paramagnetism). Ferromagnetic materials have values of μ_r that are very much greater than unity and depend on the magnetic flux density (*see* ferromagnetism; magnetic hysteresis).

The *incremental permeability* is the permeability measured when a small alternating magnetic field is superimposed on a large steady one.

permeability of free space *Syn.* magnetic constant. *See* permeability.

permeability tuning *Syn. for* slug tuning. *See* tuned circuit.

permeable base transistor (PBT) *See* vertical FET.

permeameter An instrument that measures the magnetic properties of a ferromagnetic material, particularly its permeability.

permeance Symbol: Λ; unit: henry. The reciprocal of *reluctance.

permittivity Symbol: ε; unit: farad per metre. The ratio of the electric *displacement, D, in a dielectric medium to the applied electric field strength, E, i.e.

$$\varepsilon = D/E$$

It indicates the degree to which the medium can resist the flow of electric charge and is always greater than unity.

The *permittivity of free space* is designated ε_0 and is termed the *electric constant*. It is related to the *permeability of free space, μ_0, by the equation

$$\varepsilon_0\mu_0 = 1/c^2$$

where c is the velocity of light. Thus since μ_0 has the value $4\pi \times 10^{-7}$ henry per metre in the system of *SI units, then

$$\varepsilon_0 = (1/4\pi c^2) \times 10^7 \text{ farad/metre}$$
$$= 8.854 \times 10^{-12} \text{ farad/metre}$$

The *relative permittivity, ε_r, is the ratio of the electric displacement in a medium to the electric displacement in free space for the same value of applied electric field strength, i.e.

$$\varepsilon_r = \varepsilon/\varepsilon_0$$

The dimensionless quantity ε_r is also termed the *dielectric constant* when it is independent of electric field strength and refers to the dielectric medium of a capacitor. It can then be defined as the ratio of the capacitance of the capacitor containing the dielectric medium to the capacitance it would have were the dielectric removed.

permittivity of free space *Syn.* electric constant. *See* permittivity.

persistence *Syn.* afterglow. (1) The time interval after excitation during which a phosphor continues to emit light (*see* luminescence), particularly the phosphor on the screen of a *cathode-ray tube. A curve that shows the magnitude of the luminance with respect to time after excitation of a luminescent screen is a *persistence characteristic*. The persistence depends on the nature of the phosphor and for luminescent screens is commonly chosen to be less than the persistence of the image on the human retina (about 0.1 seconds); it can however vary from fractions of a second to several years.

(2) A faint luminosity observed after the passage of an electric discharge through certain gases; it can last for several seconds.

persistor A device, commonly in the form of a miniature bimetallic printed circuit, that depends for its operation on the sharp changes of resistance in a metal as it passes from a state of superconductivity to its normal resistive state. It can be used as a low-temperature storage element or very fast switch.

perveance The space-charge-limited characteristic between the electrodes in an *electron tube. It is a function of the current density, j, and the collector voltage, V, being equal to $j/V^{3/2}$.

pfm *Abbrev. for* pulse-frequency modulation. *See* pulse modulation.

PGA *Abbrev. for* pin grid array.

phantom circuit A circuit that is superimposed on two other circuits, which consist of two pairs of transmission lines, such as the pairs of wires of a telephone circuit. The two pairs of wire, termed the *physical* or *side circuits,* are effectively in parallel from the point of view of the phantom circuit. The individual circuits must be carefully balanced to avoid *crosstalk.

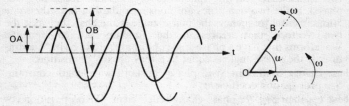

Phase angle between two quantities of the same frequency

phase (1) The stage or state of development of a regularly recurring quantity; it is the fraction of the *period that has elapsed with respect to a fixed datum point.

The amplitude variations of a sinusoidally varying quantity are similar to simple harmonic motion; such a quantity may be represented as a rotating vector *OA* (*see* diagram) of length equal to the maximum amplitude and rotating through an angle 2π during the *period, T, of the waveform. The vector has an angular velocity, ω, equal to $2\pi/T$ and related to the frequency, ν, of the waveform by

$$\nu = \omega/2\pi$$

The phase of such a quantity with respect to another quantity, represented by the vector *OB,* is given by the angle, α, between them (*see* diagram). This is the *phase angle (*see also* phase difference), which is constant if the two quantities have the same frequency.

405

Particles in a travelling wavefront move in the same direction with the same relative displacement and are said to be in the same phase of vibration. The wavelength is equal to the distance travelled between two points, in the direction of propagation of a wavefront, at which the same phase recurs.

Periodic quantities that have the same frequency and waveform and that reach corresponding values simultaneously are said to be *in phase;* otherwise they are *out of phase.*

Quantities that are nonsinusoidal or that have different waveforms can be considered as vectors that represent their fundamental properties; the same terms can then be applied to them: the frequency of a nonsinusoidal waveform is the frequency of a sinusoidal waveform having the same period as the waveform under consideration (*see* fundamental frequency).

(2) One of the separate circuits or windings of a *polyphase system or apparatus.

(3) One of the lines or terminals of a *polyphase system or apparatus.

phase angle Symbol: α. The angle between two vectors that represent two sinusoidally varying quantities of the same frequency (*see* phase). If the two quantities are nonsinusoidal but have the same *fundamental frequency, the phase angle is the angle between the two vectors that represent the fundamental components. Waveforms that have a phase angle of $\pi/2$ are said to be in *quadrature. If the phase angle is equal to π they are in *opposition.

phase-change coefficient *Syns.* phase constant; wavelength constant. *See* propagation coefficient.

phase constant *Syn. for* phase-change coefficient. *See* propagation coefficient.

phase control *See* silicon-controlled rectifier.

phase corrector A network that restores the original phase of a waveform that has suffered phase *distortion.

phase delay The ratio of the inserted *phase shift undergone by a periodic quantity to the frequency.

phase deviation *See* phase modulation.

phase difference (1) Symbol: ϕ. The difference in phase between two sinusoidally varying quantities of the same frequency. It may be expressed as an angle – the *phase angle – or as a time.

(2) The angle between the reversed secondary vector of an *instrument transformer and the corresponding primary vector. The vectors represent current in a current transformer and voltage in a voltage transformer. The phase difference is positive if the reversed secondary vector *leads the primary and negative if it *lags. The term *phase error* has been used in this application but this is deprecated.

phase discriminator A *detector circuit that produces an output wave in which the amplitude is a function of the *phase of the input.

phase distortion *See* distortion.

phase error *See* phase difference.

phase inverter A circuit that changes the phase of an input signal by π. A common application is for driving one side of a *push-pull amplifier.

phase lag *See* lag.

phase modulation (p.m.) A type of *modulation in which the phase of the *carrier wave is varied about its unmodulated value by an amount proportional to the amplitude of the modulating signal and at the frequency of the signal wave, the amplitude of the carrier wave remaining constant. If the modulating signal is sinusoidal the instantaneous amplitude, e, of the phase-modulated wave may be written:

$$e = E_m \sin(2\pi Ft + \beta \sin 2\pi ft)$$

where E_m is the amplitude of the carrier wave, F the unmodulated carrier wave frequency, β the peak variation in the phase of the carrier wave due to modulation, and f is the modulating signal frequency. The peak difference between the instantaneous phase angle of the modulated wave and the phase angle of the carrier is the *phase deviation. Compare* frequency modulation.

phase sequence The order in which the three *phases of a three-phase system (*see* polyphase system) reach a maximum potential of given polarity. The normal order in a particular system is termed the *positive sequence;* the reverse order is the *negative sequence.* A *phase-sequence indicator* is an instrument used to indicate the phase-sequence of such a system.

phase shift Any change that occurs or is introduced into the phase of a periodic quantity or in the phase difference between two or more such quantities. Phase shift can occur as a result of errors introduced by a particular device or circuit or can be deliberately inserted by a phase-shifting network.

phase-shift oscillator *See* oscillator.

phase splitter A circuit that has a single input signal and produces two separate outputs with a predetermined phase difference. An example is the driver for a *push-pull amplifier.

phase velocity The velocity at which an equiphase surface of a travelling wave is propagated through a medium, i.e. the velocity at which the crests and troughs travel. It is equal to λ/T where λ is the wavelength and T the period of the wave. If the frequency is ν and the number of wavelengths per unit distance is σ (the wavenumber) then

$$\lambda/T = \nu/\sigma$$

phase voltage The voltage in one *phase of a polyphase system. The term is sometimes used to indicate the voltage relative to neutral but such usage can give rise to ambiguities and is deprecated.

phasing Adjusting the position of the picture transmitted in television or facsimile transmission along the scanning line.

phonon One quantum of thermal energy associated with the vibrations of the atoms in a crystal lattice. If the frequency of vibration is ν and h the Planck constant, the phonon is equal to $h\nu$.

phonon-assisted photoconductivity *Syn. for* indirect photoconductivity. *See* photoconductivity.

phosphor *Syn.* luminophore. A luminescent material. *See* luminescence.

phosphor-bronze Bronze that contains at least 0·18% of added phosphorus. The addition of the phosphorus enhances the tensile strength, ductility, and shock resistance of the alloy. Phosphor-bronze in strip form has been widely used for galvanometer suspensions and other similar applications.

phosphorescence *See* luminescence.

photoactor A device that provides the light source to activate a photoconductive *photocell when the latter is used as a switch.

photocathode A cathode that emits electrons as a result of the *photoelectric effect.

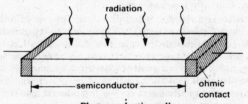

Photoconductive cell

photocell A light-electric transducer. The term was originally short for *photoelectric cell,* which is a vacuum *diode containing a *photocathode and an anode. When the photocathode is illuminated electrons are emitted. The electrons are driven to the anode by a positive potential, the *driving potential,* applied to the anode and a photocurrent flows in the external circuit. The small current flowing when the device is not illuminated, but the driving potential is on, is the dark current.

The term is now most commonly used to designate a *photoconductive cell,* which consists of *semiconductor material sandwiched between two *ohmic contacts (*see* diagram). The semiconductor may be either bulk material in the form of a rod or bar or a thin polycrystalline film on a glass substrate. The conductivity of the sample increases markedly when it is subjected to light or other radiation of suitable wavelength (*see* photoconductivity) and a photocurrent, superimposed on the small dark current, flows in an external circuit. The gain of such a device is defined as:

$$\text{gain} = \Delta I / e G_{\text{pair}}$$

where ΔI is the incremental current due to photoconductivity (the photocurrent), e is the electron charge, and G_{pair} the number of electron-hole pairs created per second.

The term 'photocell' is also sometimes applied to a *photodiode or a *photovoltaic cell. The resistance of photocells drops markedly upon illumination and the *dark resistance* – i.e. the resistance when the devices are not illuminated – is much higher than the *dynamic resistance of the devices.

photoconductive camera tube *See* camera tube. *See also* vidicon.

photoconductive cell *See* photocell.

photoconductive storage tube *See* storage tube.

photoconductivity An enhancement of the conductivity of certain semiconductors due to the absorption of *electromagnetic radiation. Photoconductivity can result from the action of radiation in the visible portion of the spectrum in some materials.

Radiation, of frequency v, can be considered as a stream of *photons of energy hv, where h is the Planck constant. If a semiconductor is exposed to radiation an electron in the *valence band can be excited by the radiation. If the photon energy is sufficiently great the energy absorbed by the electron causes it to be excited across the forbidden band into the *conduction band and a hole remains in the valence band. Thus when the photon energy, hv, exceeds the energy gap, E_{g}, there is a sudden marked change in the conductivity of the material due to the creation of excess charge carriers. For small intensities of illumination the increase in conductivity is approximately proportional to the intensity. If the photon energy exceeds the *work function, Φ, of the material, the electron is liberated from the solid by the *photoelectric effect. Photoconductivity resulting under the condition

$$\Phi > hv > E_{\text{g}}$$

is sometimes termed the *internal photoelectric effect*. Band-to-band transitions as described above result in *intrinsic photoconductivity*.

Measurement of the absorption spectra of semiconductors can yield valuable information on the magnitude of E_{g}. *Direct-gap semiconductors have an absorption edge corresponding exactly to E_{g} (Fig. *a*). Indirect-gap semiconductors yield a value larger than E_{g} (Fig. *b*). Even if the photon energy is only equal to E_{g} so that a direct transition is not possible, *indirect photoconductivity* can occur in indirect-gap semiconductors provided that a *phonon is simultane-

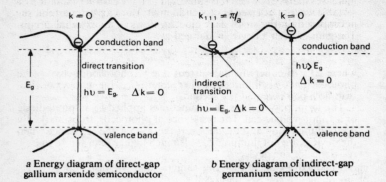

a Energy diagram of direct-gap gallium arsenide semiconductor

b Energy diagram of indirect-gap germanium semiconductor

ously created or destroyed. If k is the wave vector and thus represents the momentum of the electron in the crystal lattice, transitions of energy E_g can only take place if there is a change in momentum so that $\Delta k \neq 0$. This results in a small absorption edge in the spectrum corresponding to E_g. The momentum change required for the occurrence of indirect transitions has to be absorbed or released by the assisting phonon in order to conserve momentum.

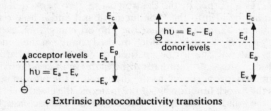

c Extrinsic photoconductivity transitions

Extrinsic photoconductivity is an effect observed in some semiconductors when the photon energy, $h\nu$, of the incident radiation is not sufficiently large to cause band-to-band transitions; instead the energy corresponds to the energy gap required to excite an electron from the valence band, energy E_v, into an acceptor level or from a donor level into the conduction band, energy E_c (Fig. *c*). E_a and E_d are the acceptor and donor energy levels. In this case electron-hole pairs are not created: an increase in the p- or n-type carriers results, respectively.

Photoconductive materials have a short response time to the incident radiation because the excess carriers generated disappear very quickly due to recombination. They are very useful as switches and photodetectors and as *photocells can produce an a.c. signal if the incoming radiation is suitably modulated, as by a mechanical chopping device.

photoconverter A photoelectric *transducer that produces digital electrical signals from an optical pattern. It consists of an array of photocells each of which produces a current proportional to the intensity of light falling on it. The illumination may be due to an optical image produced on a transparent screen or a pattern resulting from an illuminated photographic *mask or cut-out of opaque material.

photocurrent An electric current produced in a device by the effect of incident electromagnetic radiation. *See also* photocell; photoconductivity; photoelectric effect; photoionization.

photodetachment The interaction of electromagnetic radiation with a negative ion to detach an electron from the ion and leave a neutral atom or molecule, i.e.

$$M^- + h\nu \rightarrow M + e^-$$

where $h\nu$ represents a photon of frequency ν and h is the Planck constant. The mechanism in this process is exactly the same as for *photoionization of a neutral species. The *ionization potential of the negative ion is equal to the *electron affinity of the atom or molecule.

photodetector Any electronic device that detects (*see* detector) or responds to light energy. *See* photocell; photodiode; phototransistor.

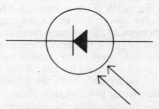

a General symbol for photodiode

photodiode A semiconductor *diode that produces a significant *photocurrent when illuminated. There are two main classes of photodiode: *depletion-layer photodiodes* and *avalanche photodiodes*. The general symbol is shown in Fig. *a*.

A common form of depletion-layer photodiode consists of a reverse-biased *p-n junction operated below the *breakdown voltage. When exposed to electromagnetic radiation of suitable frequency excess charge carriers are produced as a result of *photoconductivi-

411

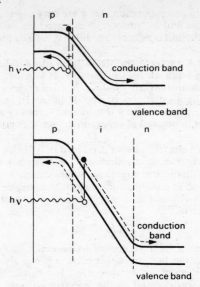

b Energy bands of reverse-biased p-n
junction (top) and p-i-n photodiodes

ty; these carriers are in the form of electron-hole pairs. They normally recombine very quickly but those generated in or near to the *depletion layer present at the junction cross the junction and produce a photocurrent (Fig. *b*). The photocurrent is superimposed on the normally very small reverse saturation current, or dark current. The p-n junction can be formed by any of the usual methods and illumination can be either normal to the plane of the junction or parallel to it. Examples of p-n junction photodiodes are shown in Fig. *c*.

The *p-i-n photodiode* (Figs. *b, d*) contains a layer of intrinsic (i-type) semiconductor material sandwiched between the p- and n-regions. The depletion layer is wholly contained within the i-region. The thickness of the intrinsic region can be adjusted to produce devices with optimum sensitivity and frequency response. The p-i-n photodiode is the most common type of depletion-layer photodiode.

Depletion-layer photodiodes may also be realized using a metal-semiconductor junction, a heterojunction, or a point-contact diode. The *Schottky photodiode* (Fig. *e*) uses a metal-semiconductor junc-

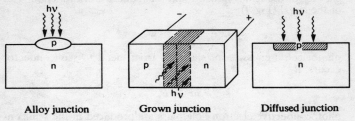

Alloy junction　　　**Grown junction**　　　**Diffused junction**

c Types of p-n junction photodiode

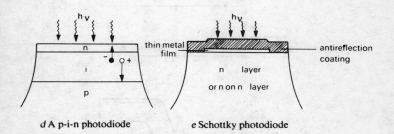

d A p-i-n photodiode　　　*e* Schottky photodiode

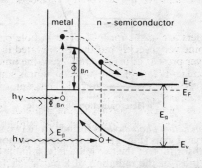

f Simplified energy diagram for Schottky photodiode

tion. For optimal operation, in order to avoid losses due to absorption and reflection the metal film must be very thin (about 10 nanometres) and an antireflection coating must be used. The photocurrent can be caused by two mechanisms, which depend on the value of the incident photon energy, *hv*, relative to the energy

413

gap, E_g, of the semiconductor and the reduced work function, Φ_{Bn}, of the metal (Fig. *f*): if

$$E_g > h\nu > \Phi_{Bn}$$

photoelectric emission of electrons from metal to semiconductor occurs; if

$$h\nu > E_g$$

photoconductive electron-hole pairs are produced in the semiconductor.

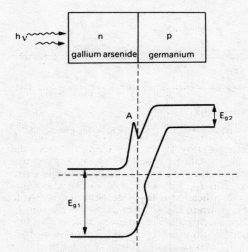

g Heterojunction photodiode

The *heterojunction photodiode* is formed from two semiconductors of different band gaps (Fig. *g*). The frequency response depends on the relative absorption of the two materials, suitable choice of material causing most of the radiation of a particular frequency to be absorbed close to the junction. Such diodes have a high speed of response and are highly frequency-selective. The small barrier, A, in the energy diagram is due to the discontinuity in the conduction bands and can be crossed by tunnelling (*see* tunnel effect) or surmounted by electrons with sufficiently high energy.

Point-contact photodiodes are constructed with a *point-contact junction. The active volume of such devices is extremely small and as a result it has a high speed of operation. It is limited to applications in which the radiation can be focused on to an extremely small spot.

Modulation of the intensity of illumination falling on a photodiode produces a modulated photocurrent that is dependent on the incident illumination.

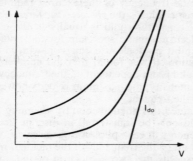

h I-V characteristics of an avalanche photodiode

The other class of photodiodes, avalanche photodiodes, are reverse-biased p-n junction diodes that are operated at voltages above the breakdown voltage. Current multiplication of electron-hole pairs, generated by the incident electromagnetic radiation, occurs due to the *avalanche process. The photomultiplication factor, M_{ph}, is defined as the ratio of the multiplied photocurrent, I_{ph}, to the photocurrent, I_{pho}, at voltages below breakdown where no avalanche multiplication takes place. Static current-voltage characteristics are shown in Fig. *h*. The current I is the sum $I_{ph} + I_{do}$, where I_{do} is the dark current. The variation of I_{do} is also shown. The maximum photomultiplication depends inversely on the square root of I_{do}, and in order to achieve optimum operation I_{do} must be kept as small as possible. The device is thus operated with the voltage, V, approximately equal to the breakdown voltage, V_B. Avalanche photodiodes provide a substantial gain at microwave frequencies. Schottky photodiodes may also be operated in the avalanche region.

photoelectric alarm An electronic alarm system that employs a *photocell used as a switch. The most common form of operation is to arrange for the photocell to be subjected to a constant beam of light. If the light beam is interrupted the photocurrent ceases, causing the alarm to be automatically activated.

photoelectric cell *See* photocell.

photoelectric constant The ratio of the Planck constant, *h*, to the electron charge, *e*. The photoelectric constant can be determined by photoemission experiments and has been used to calculate the Planck constant.

photoelectric effect An effect, first noticed by Heinrich Hertz, whereby electrons are liberated from matter when it is exposed to electromagnetic radiation of certain energies. In solids electrons are only liberated when the frequency of the exciting radiation is greater than a characteristic value – the *photoelectric threshold* of the material. The photoelectric threshold is usually in the mid-ultraviolet region of the electromagnetic spectrum for most solids, although some metals exhibit photoelectric emission with visible or nearultraviolet radiation. It was found that the numbers of electrons ejected from a solid is not dependent on the frequency of the radiation but on the intensity; the maximum velocity however is directly proportional to the frequency.

Einstein explained this phenomenon by assuming that the radiant energy could only be transferred in discrete amounts, i.e. as *photons. The energy of each photon is given by $h\nu$, where h is the Planck constant and ν the frequency of the incident radiation. Provided that $h\nu$ exceeds the *work function of the material, Φ, an electron in the material absorbing a photon is ejected from the surface. The maximum kinetic energy, E, of the electrons is given by the *Einstein photoelectric equation:*

$$E = h\nu - \Phi$$

Only the least strongly bound electrons in the solid possess this maximum energy. More strongly bound electrons have an effectively greater value of Φ and are ejected with energies less than E.

Gases and liquids can also emit electrons under the influence of electromagnetic radiation. In a gas each electron is emitted by a single atom or molecule and the photoelectric threshold is determined by the *ionization potential, which replaces the work function in Einstein's equation (*see* photoionization).

Photoelectric emission from a material may be prevented by applying a negative potential to the surface. The minimum potential required to prevent photoelectric emission is the *stopping potential.* The *inverse photoelectric effect* has also been observed: certain materials emit radiation from their surface when subjected to an electron beam.

The photoelectric effect is utilized in some types of *photocell and in *photomultipliers. *See also* photoconductivity.

photoelectric galvanometer *See* galvanometer.

photoelectric threshold *See* photoelectric effect.

photoelectron spectroscopy A method of analysis of the composition of a semiconductor by detecting photoelectrons generated by photons impinging on the surface of the material (*see* photoelectric effect). The method is nondestructive and only photoelectrons generated at or very near to the surface are detected. The photoelectrons have an energy typical of the parent atom, but also characterized by neighbouring atoms: chemical bonding between the parent atom and its neighbours causes perturbations of the energies detected.

X-ray photoelectron spectroscopy (XPS) uses photons in the X-ray range of energy and *ultraviolet photoelectron spectroscopy* (UPS) uses photons in the ultraviolet energy range. This technique contrasts with *Auger electron spectroscopy in that the electrons result from transitions between the outer atomic orbitals (XPS) or the valence and conduction bands (UPS) as opposed to electrons in the inner orbitals (Auger ES). XPS yields sharper more easily identifiable energy peaks than UPS but since focusing is not easy the lateral resolution is poor.

photoemission Emission of electrons from a material as a result of exposure to electromagnetic radiation. *See* photoionization; photoelectric effect.

photoemissive camera tube *See* camera tube. *See also* iconoscope.

photoglow tube *See* phototube.

photoionization Ionization of an atom or molecule that results from exposure to electromagnetic radiation. The mechanism of photoionization is the same as that operating in the *photoelectric effect, but in photoionization an electron is ejected from a single atom or molecule, such as occur in the gaseous state.

The incident radiation, of frequency v, can be considered to consist of discrete *photons of energy hv, where h is the *Planck constant. If the energy hv exceeds the first *ionization potential, I_1, of the atom or molecule an electron absorbing the photon will be removed from the atom or molecule. The excess energy ($hv - I_1$) is distributed as kinetic energy between the ejected electron and the positive ion, according to the conservation of energy principle. Since the relative mass of the ion is very much greater than that of the electron, the excess ionic energy is negligible and the kinetic energy, E, of the electron is given by

$$E = hv - I_1$$

This is similar to Einstein's photoelectric equation with the work function, Φ, replaced by the ionization potential of the single atom (or molecule).

Photoionization can only occur when the photon energy, hv, exceeds the ionization potential, I_1; there is thus a *photoionization threshold* of frequency below which the effect does not occur. The value of the threshold frequency usually lies in the ultraviolet region of the electromagnetic spectrum. If the energy, hv, of the incident radiation is sufficiently great more strongly bound electrons can be ejected from the neutral species. If I_2 is the second ionization potential, then electrons with energy E_2 ($< E_1$) can be produced where

$$E_2 = h\nu - I_2$$

The resulting positive ion is in an excited state rather than the *ground state.

In the case of photoionization of molecules, part of the excess energy may be used to produce ions in a vibrationally or rotationally excited state, in which case the energy, E, of the emitted electrons will be less than the maximum predicted by Einstein's equation.

photoionization threshold *See* photoionization.

photolithography *Syn.* (*colloq.*) printing. A technique used during the manufacture of *integrated circuits, *semiconductor components, *thin-film circuits, and *printed circuits. Photolithography is used in order to produce a desired pattern from a photographic *mask on a substrate material preparatory to a particular processing step.

The clean substrate is covered with a solution of *photoresist by spincoating, spraying, or immersion. The solution is allowed to dry and is then exposed to light or near ultraviolet radiation through the mask. *Deep ultraviolet exposure* can be used where greater resolution is required because of the shorter wavelength. Quartz masks rather than glass ones must then be used together with different resist material. The depolymerized portions of the photoresist are removed using a suitable solvent, such as trichloroethylene, and the polymerized portion remains and acts as a barrier to etching substances or as a mask for deposition processes. When the processing step is completed the remaining photoresist is removed using another suitable solvent.

Different methods are used to expose the resist through the mask. *Proximity printing* places the mask close to the slice but not in actual contact with it (Fig. *a*). The diffraction that occurs at the edges of the patterns on the mask causes divergence of the light and this method is only suitable for applications not requiring a high lateral resolution. *Contact photolithography* has the mask in contact with the surface of the slice. After alignment it is vacuum clamped to the slice for exposure (Fig. *b*). The resolution and uniformity of the technique depend critically on the mask being undamaged and the degree of contact that can be achieved. Any slight curvature of the slice or the mask causes *runout*, i.e. errors in the exposed pattern. The actual contact tends to damage the masks and they therefore have a limited life. Contact lithography however is a cheap and rapid technique and suitable for use with smaller slices and small-scale integrated circuits.

For large slices and VLSI circuits other methods are used. *Projection photolithography* uses an optical system to produce an image of the mask on the slice. The mask and slice are moved in synchronism

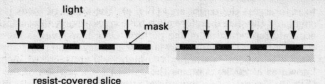

a Proximity printing   *b* Contact photolithography

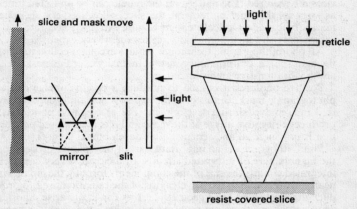

c Projection photolithography   *d* Optical stepping

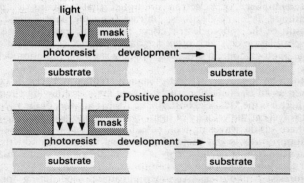

e Positive photoresist

f Negative photoresist

419

to scan across the entire area (Fig. *c*). The depth of focus is very small and the slice therefore must have an extremely flat surface for accurate reproduction of the pattern. *Optical stepping* is the most commonly used technique for large-scale integration (LSI) applications, where good resolution and high yield are required. A mask, known as a reticle, contains the pattern for a portion of the slice (Fig. *d*). The pattern is imaged onto the slice, then the slice is moved and the exposure repeated. The step and repeat process continues until the entire slice has been exposed. The technique is particularly suitable for digital circuits where tens of thousands of identical devices are required. The pattern on the reticle can be up to ten times as large as the final pattern on the slice, allowing very accurate masks to be made. Optical steppers require extremely accurate optical systems and their complexity makes them very expensive.

All photolithographic techniques require excellent collimation of the light source, uniform and constant intensity over the mask area, and a vibration-free environment.

Positive photoresist is used to produce a positive image of the photographic mask. In this case the exposed portion is depolymerized and removed during development (Fig. *e*). Negative photoresist produces a negative image of the mask and it is the exposed portion that is polymerized and remains after development (Fig. *f*).

Photolithography is an important part of the *planar process for the manufacture of integrated circuits. In the case of very complex integrated circuits the size of the components forming the circuits is nearly comparable to the wavelengths of the radiation used to produce the masks and large geometric errors can easily arise. Shorter wavelengths are therefore required and other lithography systems have been developed. *See* electron-beam lithography; ion-beam lithography; X-ray lithography.

photomultiplier An *electron multiplier that contains a *photocathode. Primary electrons, emitted from the photocathode as a result of the *photoelectric effect, initiate the cascade. A suitable *scintillation crystal is often used as the source of illumination for the photocathode in order to provide a sensitive radiation detector or *counter.

photon The *quantum of electromagnetic radiation. It can be considered as an elementary particle of zero mass and having energy $h\nu$, where h is the *Planck constant and ν is the frequency of the radiation. It travels at the velocity of light, c, and has momentum $h\nu/c$ or $h\lambda$, where λ is the wavelength of the radiation. Photons can cause *excitation of atoms and molecules resulting in *photoconductivity in semiconductors; if the energy is sufficiently great *photoionization or the *photoelectric effect can result.

photoresist A photosensitive organic material used during *photolithography. *Negative photoresists* are materials that form polymers on exposure to light. *Positive photoresists* are polymers that are de-polymerized by the action of light. The polymerized material acts as a barrier during processing steps in the manufacture of solid-state devices, etc.

photosensitive recording *See* recording of sound.

photosensitivity The property of responding to electromagnetic radiation, particularly in the ultraviolet, visible, or infrared portions of the electromagnetic spectrum. Various responses are observed, which can be either physical or chemical. *See* photoconductivity; photoelectric effect; photoionization; photoresist; photovoltaic effect.

phototelegraphy *See* facsimile transmission.

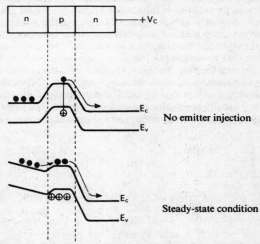

Simplified energy bands and carrier transport in an n-p-n phototransistor

phototransistor A *photodetector that consists of a bipolar junction *transistor operated with the base region *floating. The potential of the base region is determined by the number of charge carriers stored in it. The electromagnetic (usually ultraviolet) radiation to be detected is applied to the base of the transistor and produces the base current; the transistor is operated essentially in *common-emitter connection.

The action of the electromagnetic radiation produces excess electron-hole pairs in the base region. Minority carriers produced in or near to the *depletion layers associated with the reverse-biased collector junction are swept into the collector region (*see* diagram);

421

majority carriers remain in the base region and are stored as excess majority carriers. The emitter junction therefore becomes more forward biased.

Once the base potential rises sufficiently, due to majority-carrier storage, injection of minority carriers from the emitter occurs and a steady-state condition is reached (*see* diagram). The transistor can then behave like a normal transistor with the base signal resulting from the illumination and being a function of the intensity of illumination.

The collector current is essentially equal to the current generated in a p-n junction photodiode multiplied by β_{FE}, the *beta-current gain factor. β_{FE} is measured when the same structure is used as a simple transistor with a base contact. A typical structure can have a very large value of β (about 100) and therefore a greatly increased sensitivity is possible compared to the p-n junction photodiode. The speed of operation of the phototransistor however is comparatively less due to the time required to charge the base region to a sufficient potential to realize the transistor action.

During the response time the emitter current rises from zero to a steady-state value determined by the rate of generation of excess minority carriers in the base. If the intensity of illumination reaches a sufficiently high value the collector current becomes limited by the external circuit components and the transistor saturates.

phototube An *electron tube that contains a photosensitive electrode, usually the cathode. A *vacuum phototube* is evacuated to a sufficiently low pressure that ionization of the residual gas in the tube does not affect the characteristics. A *gas phototube* is one that contains a gas, such as argon, at very low pressure in order to minimize the space charge effect of the photoelectrons in the tube. If the gas pressure is such that a glow discharge (*see* gas-discharge tube) occurs across the tube, the tube is termed a *photoglow tube*. The sensitivity of such a tube is increased by the presence of the glow discharge.

The *dynamic sensitivity* of a phototube used as a *photodetector is defined as the ratio of the alternating component of the anode current to the alternating component of the incident radiant flux. Such tubes respond only to the proportion of the incident radiant flux falling on the photocathode; the *acceptance angle* is the solid angle at the photocathode within which all the exciting flux reaches the cathode. This is determined by the geometry of the tube. Like other types of photodetector, phototubes exhibit a definite *dark current on which the photocurrent is superimposed and which contributes to the total current flowing through the device. *See also* photocell; photomultiplier.

photovaristor *See* varistor.

photovoltaic cell A cell that utilizes the *photovoltaic effect in order to produce an e.m.f. An example is the *solar cell, the basis of which is an unbiased p-n junction. Metal-semiconductor junctions are a common form of photovoltaic cell. They depend for their action on the formation of a potential barrier across the unbiased junction (*see* Schottky effect). They are often termed *barrier-layer photocells* (or *rectifier photocells*).

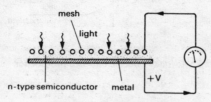

Front-wall photovoltaic cell

A typical structure is shown in the diagram. The contact to the n-type semiconductor is in the form of a mesh to minimize reflection of the incident radiation. A cell of this type is termed a *front-wall* type since the metal-semiconductor junction is the region exposed to the radiation. If the arrangement is reversed, with the Schottky barrier at the back, the cell is termed a *back-wall* photovoltaic cell. Front-wall cells are more blue-sensitive and have a higher output than the back-wall type since most of the radiation is absorbed in or near the potential barrier and the absorption losses are small. Back-wall cells are usually red-sensitive since other components of the light are absorbed in the semiconductor before reaching the potential barrier layer.

photovoltaic effect An effect arising when a junction between two dissimilar materials, such as a metal and a semiconductor or two opposite polarity semiconductors, is exposed to electromagnetic radiation, usually in the range near-ultraviolet to infrared. A forward voltage appears across the illuminated junction and power can be delivered from it to an external circuit. The effect results from the depletion region and resulting potential difference invariably associated with an unbiased junction (*see also* p-n junction).

The energy bands are shown for a p-n junction (Fig. *a*) and a metal-semiconductor junction (Fig. *b*). The incident radiation imparts energy to electrons in the valence band and electron-hole pairs are produced in the depletion region around the p-n junction or in the barrier layer at the metal-semiconductor junction. As the electron-hole pairs are produced they cross the junction due to the inherent field (Figs. *a, b*) and produce the forward bias: an excess of holes migrating into the p-type semiconductor or the metal produc-

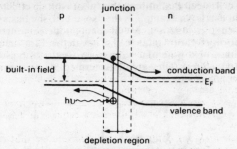

a Photovoltaic effect in unbiased p-n junction

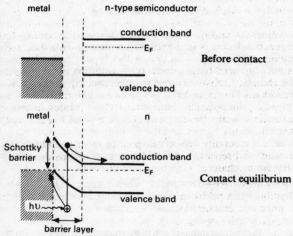

b Photovoltaic effect in metal-semiconductor contact

es a positive bias; electrons migrating into the n-type semiconductor produces a negative bias.

The photovoltaic effect is utilized in *photovoltaic cells, such as *solar cells.

photronic cell *Syn. for* barrier-layer photocell. *See* photovoltaic cell.

pick-up A *transducer that converts information, usually recorded, into electrical signals. The term is particularly applied to the electro-mechanical transducers used to reproduce the signals recorded in the grooves of a gramophone record. There are several types of pick-up in common use.

Ceramic pick-ups are constructed from a suitable ceramic material, such as barium titanate, that exhibits the *piezoelectric effect. The mechanical vibrations produced by the grooves of the rotating record stress the ceramic material and result in a corresponding e.m.f. Ceramic materials are mechanically reliable and relatively stable under ambient conditions.

Crystal pick-ups are piezoelectric crystals that behave in a similar manner to ceramic pick-ups but are less reliable and stable than the ceramic type.

Magnetic pick-ups contain a small inductance that can move in the field of a *magnet. Mechanical vibrations in the groove of the record cause movement of the coil and hence a current is induced in it due to the changing *magnetic flux density linking it.

Other less common types of pick-up include the *capacitor pick-up,* in which the mechanical vibrations cause a variation in the capacitance, the *variable-resistance pick-up,* in which variations in the resistance result, and the *variable-reluctance pick-up,* in which there is a magnetic circuit whose *reluctance varies with the mechanical motion.

A *pick-up cartridge* is a removable assembly containing the electromechanical transducer and the reproducing stylus that couples the mechanical variations of the groove to the transducer. In a stereophonic gramophone system the stylus vibrates in two dimensions and two transducers are used that respond to each component of the vibrations.

In a *compact disc system, the pick-up is an assembly of injection laser and light sensor. It differs from the pick-ups described above in that no mechanical vibrations are involved.

pick-up cartridge *See* pick-up.

pico- Symbol: p. A prefix to a unit, denoting a submultiple of 10^{-12} of that unit: one picofarad is 10^{-12} farads.

picture carrier *See* television.

picture element The smallest portion of a picture area in a *television system that is resolved by the scanning processes. In transmission it effectively corresponds to the smallest area of the optical image that produces a discernible video signal. In reception it effectively corresponds to the area of smallest detail that can be resolved on the screen of the picture tube.

picture frequency *Syn. for* frame frequency. *See* television.

picture inversion *See* facsimile transmission.

picture noise *See* grass.

picture signal *Syn. for* video signal. *See* television.

picture telegraphy *See* facsimile transmission.

picture tube A *cathode-ray tube used in a *television receiver to reproduce the transmitted picture. The electron beam is intensity-modulated by the transmitted video signal in order to reproduce the transmitted luminance; it is then caused to traverse the screen by *sawtooth waveforms applied to deflection coils around the tube. Automatic focusing is applied to convergence coils to maintain a clear image at the edges of the screen. The scanning process (*see* television) must be performed in synchronism with the transmitted information in order to maintain a satisfactory picture. *See also* colour picture tube.

picture white *Syn. for* white peak. *See* television.

Pierce crystal oscillator *See* piezoelectric oscillator (Figs. *b, c*).

piezoelectric crystal A crystal that exhibits the *piezoelectric effect. All *ferroelectric crystals are piezoelectric as well as certain nonferroelectric crystals and some ceramics. The best-known examples of piezoelectric crystals include quartz crystal, Rochelle salt, and barium titanate. *See also* piezoelectric oscillator.

piezoelectric cutter *Syn.* crystal cutter. *See* recording of sound.

piezoelectric effect An effect that occurs when certain materials are subjected to mechanical stress. An electrical polarization is set up in the crystal and the faces of the crystal became electrically charged. The polarity of the charges reverses if the compression is changed to tension. Conversely an electric field applied across the material causes it to contract or expand according to the sign of the electric field.

The piezoelectric effect is reversible with an approximately linear relation between deformation and electric field strength. The *piezoelectric strain constant, d,* is defined as

$$d_{i,k} = \delta e_k / \delta E_i$$

where δe_k is the incremental stress and δE_i the change in electric field strength along defined axes in the crystal ($i \equiv x, y, z$ and $k \equiv xx, yy, zz, yz, zx, xy$).

The piezoelectric effect is observed in all *ferroelectric crystals and in nonferroelectric crystals that are asymmetric and have one or more polar axes. The magnitude of the piezoelectric effect depends on the direction of the stress relative to the crystal axes. The maximum effect is obtained when the electrical and mechanical stresses are applied along the *X*-axis (the electric axis) and the *Y*-axis (the mechanical axis), respectively. The third major axis of a piezoelectric crystal is the *Z*-axis (the optical axis).

The piezoelectric effect is important because it couples electrical and mechanical energy and thus has many applications for electromechanical *transducers. Piezoelectric crystals are used to provide frequency standards and in *piezoelectric oscillators.

piezoelectricity Electrical signals generated as a result of the *piezoelectric effect.

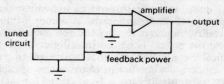

a Basic piezoelectric oscillator

piezoelectric oscillator An oscillator that utilizes a piezoelectric crystal in order to determine the frequency. Such oscillators are very stable. If an alternating electric field is applied across a suitable direction of a piezoelectric crystal, mechanical vibrations result (*see* piezoelectric effect). If the frequency corresponds to a natural frequency of vibration of the crystal, substantial mechanical vibrations result. These in turn produce an alternating electric field across the crystal. The mechanical vibrations suffer little from damping and have a sharp resonance peak; piezoelectric crystals are therefore suitable for use as frequency standards.

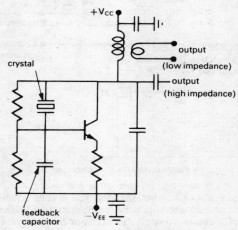

b Pierce crystal oscillator with crystal between base and collector

A suitably cut piezoelectric crystal is mounted between the plates of a capacitor in order to apply the alternating voltage. The capacitor is usually formed by *sputtering a metallic film on the large faces of the crystal in order to minimize the mechanical loading. The crystal is supported by lightweight supports that touch it at a mechanical node. In order to produce extremely high frequency stability the crystal can be supported in vacuo; for the highest frequency stability, required for the control of powerful frequency transmitters, the crystal is placed in an electrically heated oven, thermostatically controlled to within 0·1 kelvin. In the latter case, where the temperature coefficient of frequency of oscillation is required to be substantially zero, a T-cut crystal is usually used. This is cut as a thin plate whose faces contain the X-axis and a line in the YZ-plane inclined at an angle to the Z-axis. This cut exhibits lower piezoelectric activity than X- or Y-cuts.

Piezoelectric crystals can be connected in various ways to an oscillator circuit. The circuits used may be classified into two main types.

In the *crystal oscillator* the crystal replaces the *tuned circuit in the oscillator and thus provides the resonant frequency (Figs. *a, b, c*).

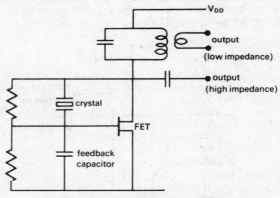

c Pierce crystal oscillator using an FET

In the *crystal-controlled oscillator* the crystal is coupled to the oscillator circuit, which is tuned approximately to the crystal frequency. The crystal controls the oscillator frequency by *pulling the frequency to its own natural frequency and thus preventing frequency drift (Fig. *d*).

piezoelectric strain constant *See* piezoelectric effect.

piezoelectric strain gauge *See* strain gauge.

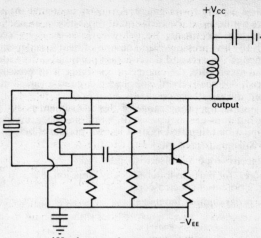

d **Hartley crystal-controlled oscillator**

π-mode operation *See* magnetron.

pinch-off voltage *See* field-effect transistor.

pincushion distortion *See* distortion.

p-i-n diode A semiconductor *diode that contains a region of almost *intrinsic (i-type) semiconductor between the p-type and n-type regions. The *depletion layer associated with a p-n junction is entirely contained within the i-type region in the p-i-n structure. At low signal frequencies the diode behaves similarly to a normal *p-n junction, but at high frequencies it exhibits a variable resistance. It is suitable for IMPATT mode operation (*see* IMPATT diode) and for use as a depletion-layer *photodiode.

pin grid array (PGA) A form of package used for *integrated circuits that is capable of providing up to several hundred connections to one chip and is used for LSI and VLSI applications. It consists of a ceramic or moulded plastic casing that contains a *leadframe. The leadframe is connected to the *bonding pads of the chip using either *wire bonding or *tape automated bonding, and is connected to an array of output pins around the edges of the PGA package. The grid array of pins may be formed as several parallel rows of pins at two opposite sides of the package or around all four sides of the package, depending on the size and complexity of the integrated circuit. *See also* dual in-line package; leadless chip carrier.

p-i-n photodiode *See* photodiode.

429

Pirani gauge *Syn.* hot-wire gauge. An instrument that measures low pressure by means of the variation of resistance of a conductor with pressure. An electrically heated wire loses heat by conduction through the low-pressure gas. The rate of heat loss depends on the resistance of the wire and if a constant potential difference is maintained across a wire, the change in resistance with pressure can be measured. Alternatively, the applied p.d. can be varied in order to maintain the resistance constant.

The normal operating range of this instrument is 10^{-2} to 10^{-4} mmHg but pressure as low as 10^{-6} mm has been measured. The gauge must be calibrated against known pressures before use.

π-section *See* quadripole; filter.

pith-ball electroscope *See* electroscope.

PIXE *Abbrev. for* particle-induced X-ray emission. *See* soft X-ray appearance potential spectroscopy.

PL/1 *See* programming language.

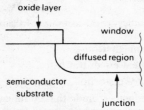

a Planar process

planar process The most commonly used method of producing junctions during the manufacture of *semiconductor devices. A layer of silicon dioxide is thermally grown on the surface of a silicon substrate of the desired conductivity type. *Photolithography is used to etch holes in the oxide layer, which then acts as a *mask for the *diffusion of suitable impurities into the substrate in order to produce a region of opposite polarity. The junction between the two semiconductor types actually meets the surface of the substrate below the oxide since the diffusion occurs in directions both normal to and parallel to the surface of the silicon (Fig. *a*).

Several diffusions can be carried out serially. Usually a final layer of oxide is grown to cover the entire chip (except for the contacts) in order to provide a stable surface for the silicon and to minimize surface-leakage effects. The characteristics of early junction transistors tended to be dominated by surface-leakage effects and the planar process proved to be one of the most important single advances in semiconductor technology. A planar transistor is shown in Fig. *b*.

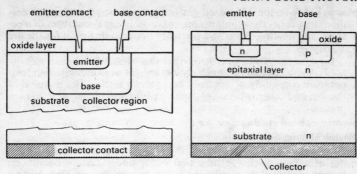

b Planar transistor   *c* Planar epitaxial transistor

The n-p-n planar *epitaxial transistor* shown in Fig. *c* is a transistor formed by a combination of *epitaxy and diffusion. The lightly doped epitaxial layer is grown on to the highly doped substrate and the junctions are formed by diffusion into the epitaxial layer. In this technique the highly doped substrate forms the bulk of the collector and the collector series resistance is therefore reduced while the lightly doped epitaxial layer maintains the collector-base breakdown characteristics.

Planck constant Symbol: *h*. A universal constant arising from *Planck's law,* which states that the energy of *electromagnetic radiation is confined to discrete packets or *photons. The energy of each photon is given by $h\nu$, where ν is the frequency of the radiation. Planck's law is fundamental to quantum theory. The value of *h* is $6.626\ 196 \times 10^{-34}$ joule seconds.

Planck's law *See* Planck constant.

plane-polarized wave *Syn.* linearly polarized wave. An electromagnetic wave in which the vibrations are rectilinear and parallel to a plane that is transverse to the direction of propagation of the wave.

planetary electron An *electron that orbits around the nucleus of an atom.

Planox *Tradename. See* coplanar process.

plan position indicator (PPI) A presentation used in a *radar receiver in which the signal from the target appears as a bright spot on the screen. The distance and bearing of the target are given by the polar coordinates of the spot with respect to the centre of the screen.

Planté cell The first secondary cell to be constructed. It consisted of rolled lead sheets that were dipped into dilute sulphuric acid.

plant load factor *See* load factor.

431

plasma (1) A region of ionized gas in an arc-discharge tube (*see* gas-discharge tube) that contains approximately equal numbers of electrons and positive *ions and provides a conducting path for the arc discharge.

(2) *Syn. for* positive glow. *See* gas-discharge tube.

(3) A substantially completely ionized gas at extremely high temperature consisting of electrons and atomic nuclei. The material of the sun and other substances capable of thermonuclear reaction are plasmas.

plasma assisted etching *Syn. for* dry etching. *See* etching.

plasma etching *See* etching.

plasma oscillations Oscillations of the ions and electrons in the *plasma of a *gas-discharge tube that occur under certain conditions independently of the conditions in the external circuit. They cause a larger scattering of the electron stream than is explicable by ordinary gas collisions.

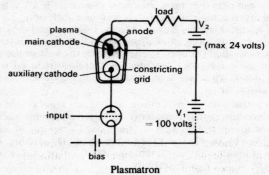

Plasmatron

plasmatron A gas-filled electron tube that contains a thermionic cathode and can pass large currents at only a few volts anode potential (*see* diagram). An auxiliary circuit is used to generate a *plasma of electrons and positive ions in the tube before the main discharge occurs in order to provide a conducting path for it. Unlike the *thyratron the plasmatron can operate continuously.

plastic-film capacitor A capacitor the dielectric of which is a plastic film. The electrical properties depend on the molecular structure. Material made from polar (asymmetrical) molecules will have an increased dielectric constant that is frequency dependent; nonpolar (symmetrical) molecules give a material the properties of which are independent of frequency.

Two main types of plastic-film capacitor exist: *polystyrene film* and *polyester film capacitors.* The former contains a nonpolar plastic with excellent properties that is used with metal-foil electrodes to produce a low-loss capacitor with good capacitance stability. Polyester-film capacitors are slightly polar plastic-film capacitors with either foil or metallized electrodes and are useful for d.c. applications and for operation up to 125°C.

plastics A group of organic materials that are plastic at some stage during their manufacture although stable in use at normal temperatures; they can be shaped by the application of heat and pressure. Plastics have a wide variety of uses in electronics, particularly in the packaging of solid-state circuits, components, or devices, as the substrate material for magnetic tape or disk and gramophone records, and as insulating and dielectric material.

plate (1) *US syn. for* anode.

(2) An electrode in an electrolytic *cell or in a *capacitor.

plateau A region of the current-voltage characteristic of an electronic device that exhibits a substantially constant value of current for a significant voltage range.

plated bridge *Syn. for* air bridge.

plated heat sink A *heat sink formed on a power FET (*field-effect transistor) by plating a thick layer of metal on the back of the slice. Plated heat sinks are often used in conjunction with *via holes in *integrated circuits containing power FETs.

plated magnetic wire A magnetic wire (*see* magnetic recording) that has a surface of ferromagnetic material plated on to a nonmagnetic core.

plating *See* electroplating.

platinum Symbol: Pt. A metal, atomic number 78, that is extremely stable and noncorroding and is used as an electrical contact or conductor at high temperatures or when chemical attack is likely.

platinum resistance thermometer *See* resistance thermometer.

plug and socket A device that enables electrical apparatus to be connected or disconnected from a source of supply or other equipment. It consists of two separable portions, the 'male' plug and the 'female' socket, with metal contacts that engage each other when connection is effected. A *polarized plug* is constructed so that engagement with the socket is only possible in one position.

plug bars *See* electron beam lithography.

plug board *Syn. for* patch board. *See* patching.

plug-in Denoting a device that may be rapidly connected or disconnected from a complex electrical or electronic system or apparatus. The device usually consists of a standardized *printed-circuit board or *integrated-circuit package that may be simply inserted in or removed from the main equipment and thus facilitates repair or maintenance of the equipment.

plumbicon *See* vidicon.

plumbing *Colloq.* The metal parts and connections of *waveguides and coaxial lines.

p.m. *Abbrev. for* phase modulation.

PMBX *Abbrev. for* private manual branch exchange. *See* telephony.

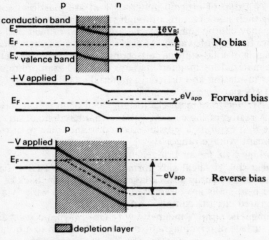

a Energy diagrams of a p-n junction

p-n junction The region at which two *semiconductors of opposite polarity meet, i.e. at which a p-type and n-type semiconductor meet. A simple p-n junction is formed from the same material in which approximately equal doping levels lead to two different conductivity types; this is known as a *homojunction*. The properties of p-n junctions are used in many semiconductor devices, such as *diodes and *transistors.

The simplified energy diagrams of an unbiased junction, a forward-biased junction (positive voltage applied to the p-region), and a reverse-biased junction are shown in Fig. *a*. In the unbiased state equilibrium considerations demand that the Fermi level, E_F (*see* energy bands), is constant throughout the bulk of the material. This causes distortion of the energy bands at the junction and results in an electric field across the junction. This field is known as the *built-in field*. At equilibrium there is a small *depletion layer containing fixed ionized atoms and substantially no mobile charge carriers.

Under reverse-bias conditions the depletion layer is increased as is the electric field across the junction, which thus acts as a barrier to current flow. Thermally generated charge carriers diffusing into the depletion layer are swept across the junction by the electric field and thus produce a small *reverse saturation current*, I_o.

Under forward-bias conditions the built-in field virtually disappears and charge carriers are attracted across the junction into the opposite polarity type (where they become *minority carriers) and cause a current to flow in an external circuit.

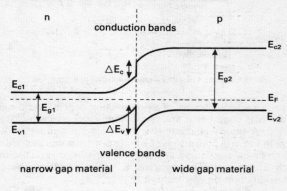

b Ideal n-p heterojunction at equilibrium (zero bias)

A *heterojunction* is formed from two dissimilar materials, such as silicon and germanium. The simplified energy diagram of an unbiased n-p heterojunction is shown in Fig. *b*. Under equilibrium conditions the distortion of the energy levels at the junction is such that discontinuities in both valence and conduction bands exist at the junction.

The current-voltage relationship of an ideal p-n homojunction under forward-bias conditions obeys an exponential relationship:

$$I = I_o(e^{eV/kT} - 1)$$

where e is the electronic charge, V the applied voltage, k the Boltzmann constant, and T the absolute temperature. This is the *Schockley equation* (or *ideal diode equation*) and assumes that the current is due to the diffusion of charge carriers across the junction, that the injected minority carrier density is small compared to the majority carrier density, and that there is no generation or recombination of charge carriers in the depletion region. In practice these conditions are not always fulfilled, particularly in silicon where the intrinsic carrier concentration is low, and it is found that the Schockley equation may be modified so that the term eV/kT is replaced by eV/nkT, where n has a value between one and two.

435

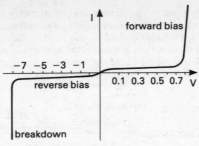

c Current-voltage characteristic of p-n junction

The characteristic curve of a p-n junction depends on the geometry, the bias conditions, and the doping level on each side of the junction. A typical characteristic curve for a p-n junction diode is shown in Fig. *c*. Applications of simple p-n junctions include rectification, voltage regulation, and use as a varistor or varactor and as a switch. Heterojunctions are used in *injection lasers and HJBTs (*heterojunction bipolar transistors).

See also diode; catching diode; charge-storage diode; Gunn diode; IMPATT diode; fast-recovery diode; light-emitting diode; photodiode; tunnel diode; Zener diode.

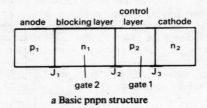

a **Basic pnpn structure**

pnpn (or **p-n-p-n**) **device** *Syn.* thyristor. A semiconductor device that is almost invariably fabricated in silicon and consists of alternating p-type and n-type layers and contains at least three p-n junctions (Fig. *a*). The device has a bistable current-voltage characteristic (Fig. *b*) and is used for power-switching purposes. A pnpn device with no gate electrodes (Fig. *c*) is termed a *four-layer diode* (or *Shockley diode*). Its switching function is controlled by the anode voltage, V_A. If V_A exceeds the *breakover voltage*, V_{Bo}, junction J_2 becomes forward-biased and the device conducts. The energy diagrams are shown in Fig. *d*. If the anode voltage is then reduced so that the anode current falls below the value I_H, the *holding current*, conduc-

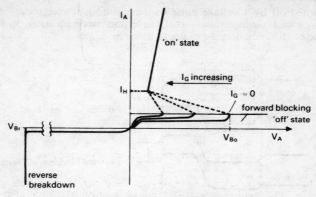

b Current-voltage characteristics of pnpn devices

tion ceases and the device is 'off'. This state is known as the *forward blocking state*. If reverse bias is applied across the device, i.e. if negative bias is applied to the p-type anode and positive bias to the cathode, both the anode and cathode junctions (J_1 and J_3) are re-verse-biased and a small reverse saturation current occurs until *breakdown is reached.

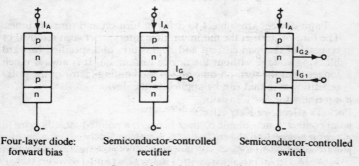

Four-layer diode: forward bias

Semiconductor-controlled rectifier

Semiconductor-controlled switch

c Contacts for various pnpn devices

If the device has a gate terminal to the second p-layer, known as the control layer (Fig. *c*), the device is a *silicon-controlled rectifier (SCR). It can be turned on by a voltage pulse to the gate at anode voltages less than V_{Bo}. A pnpn device that has connections to all four layers (Fig. *c*) is termed a *silicon-controlled switch* (SCS). This device can be turned on by a voltage pulse to the control gate and

437

turned off by a voltage pulse to the second gate electrode – the blocking layer electrode – that causes the anode junction, J_1, and the control junction, J_2, to become reverse-biased.

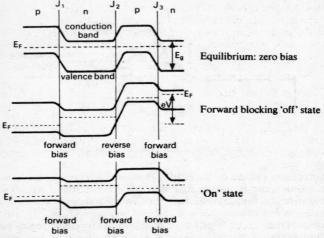

d Simplified energy bands for pnpn structure

Pnpn devices are subject to delay in turn-on and turn-off times. The turn-off time is the minimum time interval between removal or reversal of the load current and application of a specified forward blocking voltage without the device turning on. It is always much longer than the turn-on time and is the limiting factor in the pulse repetition rate that can be applied to the device.

p-n-p transistor *See* transistor.

Pockel's effect *See* Kerr effects.

point contact A non-ohmic contact between a pointed metallic wire (a *cat's whisker*) and the surface of a *semiconductor.

point-contact diode A diode rectifier formed by making a point contact between a small metal wire with a sharp point and a semiconductor. The contact may be a simple mechanical contact or a small alloyed contact formed by electrical discharge, as in the *gold-bonded diode.

The point-contact diode has a very small area, which results in a very small capacitance, and it is therefore suitable for microwave applications. It has however a large spreading resistance, a large leakage current, and poor reverse breakdown characteristics. Its characteristics are difficult to predict since the device is subject to wide variations in its physical properties, including the pressure of the wire whisker, the contact area, and the crystal structure.

point-contact photodiode *See* photodiode.

point-contact transistor *See* transistor.

poison An impurity that is present in a material and inhibits electron *emission from the material.

Poisson's equation If the electric potential is V at a point (x, y, z) and the charge density is ρ, then Poisson's equation is given as

$$\partial^2 V/\partial x^2 + \partial^2 V/\partial y^2 + \partial^2 V/\partial z^2$$
$$= -4\pi\rho$$

This can be rewritten as

$$\nabla^2 V = -4\pi\rho$$

polar Denoting a component, such as an *electrolytic capacitor, that can only operate normally in one direction of applied voltage or any system, such as a molecule, or any physical property that exhibits asymmetry.

polar axis A crystal axis of rotation that does not contain a centre of symmetry and is not normal to a reflection plane of the crystal. Certain properties of the crystal are dissimilar at opposite ends of a polar axis.

polarity (1) (magnetic) The manifestation of two types of regions in a magnet at which the inherent magnetism appears to be concentrated (poles). There are two types of magnetic pole: north-seeking (N) and south-seeking (S).

(2) (electrical) The manifestation of positive or negative parameters in an electrical circuit or device. The parameters include voltage, charge, current, and majority-carrier type.

polarization (1) A phenomenon occurring in a simple electrolytic *cell containing dissimilar electrodes. The current obtained from the cell decreases substantially soon after commencing the operation of the cell. This is due to the accumulation of bubbles of hydrogen gas, released from the electrolyte, around one of the electrodes. The bubbles partially cover the plate and thus increase the internal resistance of the cell and also create an e.m.f. of opposite polarity to the cell e.m.f. Cells such as the *Daniell cell and the *Leclanché cell are designed to minimize polarization.

(2) *Short for* dielectric polarization.

polarization diversity *See* diversity system.

polarized plug *See* plug and socket.

polarized relay *Syn.* directional relay. *See* relay.

pole (1) Each of the terminals or lines to a piece of electrical apparatus, a circuit, or network between which the circuit voltages are applied or produced. *See also* number of poles; quadripole.

(2) *See* magnetic pole.

(3) An electrode of an electrolytic *cell.

polychromatic radiation Electromagnetic radiation that contains more than one frequency. The term is also applied to particulate radiation when the particles are all the same type but of different energies. The description *inhomogeneous* is usually preferred in this latter case. *Compare* monochromatic radiation.

polyester-film capacitor *See* plastic-film capacitor.

polyphase system An electrical system or apparatus that has two or more alternating supply voltages displaced in phase relative to each other. In a symmetrical polyphase system each voltage is of the same magnitude and frequency and is displaced by an equal amount. If there are n sinusoidal voltages the mutual phase displacement is $2\pi/n$ radians and the system requires n lines at least. Thus in the *three-phase system* there is a phase difference of $2\pi/3$ between three voltage lines. An exception is the *two-phase system* that has a phase difference of $\pi/2$ between the two voltages.

polyphase transformer A *transformer that is used with a *polyphase system. The magnetic circuits required for each of the phase windings usually have portions in common with each other in order to retain the correct voltages.

polysilicon *Colloq.* Polycrystalline silicon. Silicon in polycrystalline form is most often used to form the gate electrodes in silicon-gate *MOS integrated circuits and *charge-coupled devices. In this application the silicon is doped with a sufficiently high doping concentration so that it becomes degenerate and exhibits metallic properties.

polystyrene-film capacitor. *See* plastic-film capacitor.

pool rectifier *See* mercury-arc rectifier.

population inversion *See* maser.

port An access point in an electronic circuit, device, network, or other apparatus where signals can be input or output or where the variables of the system may be observed or measured.

portable conformable mask (PCM) *Syn. for* multilevel resist.

positive booster *See* booster.

positive column *Syn. for* positive glow. *See* gas-discharge tube.

positive electron *Syn. for* positron. *See* electron.

positive feedback *Syns.* direct feedback; regeneration. *See* feedback.

positive glow *Syns.* positive column; plasma. *See* gas-discharge tube.

positive ion *See* ion.

positive logic *See* logic circuit.

positive magnetostriction *Syn. for* Joule magnetostriction. *See* magnetostriction.

positive phase sequence. *See* phase sequence.

positive photoresist *See* photoresist.

positive sequence *See* phase sequence.

positive transmission *See* television.

positron *Syn.* positive electron. *See* electron.

post-office box A box containing a number of resistance coils connected to form three arms of a *Wheatstone bridge, the resistance to be measured forming the fourth arm. Each resistance coil is connected between adjacent metal blocks so that it may be shorted out of the circuit by the insertion of a metal plug between the blocks. Resistances from 0.1 to 10^6 ohms can be measured.

pot *Colloq.* A potentiometer.

potential *Short for* electric potential.

potential barrier A region in a field of force in which the potential is of such polarity as to oppose the motion of a particle subject to the field.

potential difference (p.d.) Symbol: ΔV, U. The difference in electric potential between two points, equal to the line integral of the electric field strength between the points. If a charge is moved from one to the other of the points by any path, the work done is equal to the product of the potential difference and the charge. The *potential gradient* at a point is the potential difference per unit length. *See also* electromotive force.

potential divider *Syn.* voltage divider. A chain of resistors, inductors, or capacitors arranged in series. It is tapped at one or more points along the chain in order to obtain one or more predetermined fractions of the total voltage across the chain.

potential equilibration injection *Syn. for* fill-and-spill injection. *See* charge-coupled device.

potential gradient *See* potential difference.

potential transformer *Syn. for* voltage transformer. *See* transformer.

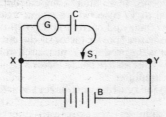

Potentiometer

potentiometer (1) A form of *potential divider that uses a uniform wire as the resistive chain. A movable sliding contact is used to tap off any potential difference less than that between the ends of the wire.

A typical use is for the measurement of potential difference or e.m.f. by balancing the unknown e.m.f. with that of a standard. For example, to measure the e.m.f. of a cell, C, the cell is arranged as shown in the diagram. The slider, S, is moved along XY until the null position is found on the galvanometer, G. The e.m.f. due to C just balances the potential across XS_1 (of length l_1). The cell is replaced by a standard cell, C_S, and the new balance point found at XS_2 (length l_2). Then

$$E_C/E_S = l_1/l_2$$

where E_C and E_S are the e.m.fs. of the unknown and standard cells, respectively.

More elaborate forms of potentiometer are available for precision applications, such as the *Kelvin-Varley slide.

(2) Any variable resistor, usually wire-wound, used in electronic circuits that has a third movable contact. The geometry of the device can be arranged so that the output voltage is a particular function of the applied voltage. The uniform wire can be arranged as a single coil or a spiral with the movable contact rotating about the axis through the centre of the coil. A *sine potentiometer* or a *cosine potentiometer* produce an output proportional to the sine or cosine of the angular displacement of the shaft, respectively.

Poulsen arc converter An *arc converter that can generate alternating-current signals of frequencies up to 100 kilohertz.

powdered-iron core A magnetic *core that is constructed of finely divided particles of iron embedded in a plastic or ceramic binding material. The low dissipation of such a core makes it very useful for high-frequency applications.

power Symbol: $P;$ unit: watt. The rate at which energy is expended or work is done. In a direct-current circuit or device the power developed is equal to VI, where V is the potential difference in volts and I the current in amperes. In an alternating-current circuit the power developed is equal to $VI\cos\phi$, where V and I are the root-mean-square values of voltage and current and ϕ is the *phase angle between them. $\cos\phi$ is the *power factor of the circuit or device and the *apparent power* is the product VI, measured in volt-amperes. The product $VI\sin\phi$ is the *reactive power. See* var.

power amplifier *See* amplifier.

power component (1) (of a current or voltage) *See* active current; active voltage.

(2) (of the volt-amperes) *See* active volt-amperes.

power detection A form of detection in which the *detector supplies a substantial power output directly to the *load without using an intermediate amplifying stage.

power efficiency The ratio of the energy output by a specified operating conditions. The term is especially applied to an electroacoustic transducer, such as a loudspeaker. The inverse of the power efficiency is the *power loss* of the device.

power factor The ratio of the actual power in watts developed by an a.c. system, as measured by a *wattmeter, to the apparent power in volt-amperes, indicated by *voltmeter and *ammeter readings. If the voltage and current are sinusoidal the power factor, P/VI, is equal to the cosine of the *phase angle between them. The power factor is also equal to the ratio of the *resistance, R, to the *impedance, Z, and thus indicates the dissipation in an *insulator, *inductor, or *capacitor.

power frequency The frequency at which domestic and industrial mains electricity is supplied and distributed. In the U.K. the standard value is 50 hertz; in the U.S. it is 60 hertz.

power line *See* transmission line.

power loss *See* power efficiency.

power pack A device that converts power from an a.c. or d.c. supply, usually the *mains, into a form that is suitable for operating electronic devices.

power relay *See* relay.

power station *Syn.* generating station. A complete assembly consisting of all necessary plant, equipment, and buildings at a suitable site for the conversion of energy of one type, such as nuclear energy, into electrical power.

power supply Any source of electrical power in a form suitable for operating electronic circuits. Alternating-current power may be derived from the mains either directly or by means of a suitable *transformer. Direct-current power may be supplied from batteries, suitable rectifier/filter circuits, or from a converter.

A *bus is frequently employed to supply power to several circuits or to several different points in one circuit. Suitable values of voltage are derived from the common supply by *coupling through dropping resistors or by capacitive coupling.

power transformer *See* transformer.

power transistor A *transistor designed to operate at relatively high values of power or to produce a relatively high power gain. Power transistors are used for switching and amplification. They usually require some form of temperature control since the power dissipation in them ranges from 1 watt to 100 watts.

power winding *See* transductor.

PPI *Abbrev. for* plan position indicator.

ppm *Abbrev. for* pulse-position modulation. *See* pulse modulation.

practical units *See* CGS system.

preamp *Short for* preamplifier.

preamplifier An amplifier that is used in a sound-reproduction system, such as a *radio receiver, in order to amplify the received signals before they are input to the main part of the system.

precision approach radar (PAR) A *radar system that is used at an airport to present accurate information about the location of incoming aircraft in the vicinity of the airport and is used as an aid for air traffic control. An associated *airport surveillance radar* system (ASR) is usually employed separately to scan the surrounding area and presents continuous information to the air traffic controller about the distance and bearing of all aircraft within a given radius of the airport.

preconducting current *See* gas-discharge tube.

pre-emphasis A technique used to improve the *signal-to-noise ratio in a radio-communication system that employs *frequency modulation or *phase modulation. A network that increases the modulation index of the higher modulation frequencies relative to the lower ones is inserted at the transmitter. De-emphasis is used at the receiver in order to restore the relative strengths of the audiofrequency signals, i.e. a network is inserted that reduces the relative strength of the higher frequencies. Amplitude-modulation systems rarely use pre-emphasis and de-emphasis as the resulting improvement in signal-to-noise ratio is only slight.

The technique is also used for magnetic tape recordings and gramophone records.

preferred values *See* standardization.

preselector *See* superheterodyne reception.

pressure-type capacitor A *capacitor used at high voltage that contains an inert gas under pressure as the dielectric.

Prestel *Tradename. See* teletext.

prf *Abbrev. for* pulse repetition frequency. *See* pulse.

Price's guard wire A wire conductor used to prevent surface leakage current associated with an insulator from reaching the measuring instrument during determination of insulation resistance, particularly of cables.

primary cell *See* cell.

primary electrons Electrons that impinge on a surface and cause *secondary emission of electrons from the surface. The term is also sometimes used to describe electrons released from atoms by one of the processes of *electron emission other than secondary emission. Such electrons however are more commonly described in terms of the relevant process of emission, as with thermionic electrons.

primary emission *Electron emission other than *secondary emission.

primary failure *See* failure.

primary radiator *Syn. for* active aerial. *See* directive aerial.

primary service area *See* service area.

primary standard A standard that is used nationally or internationally as the basis for a given unit. *Compare* secondary standard.

primary voltage (1) The voltage across the primary (input) winding of a *transformer.

(2) The voltage developed by a primary *cell.

primary winding *See* transformer.

principal axis The longest crystallographic axis.

printed circuit An electronic circuit, or part of a circuit, in which the conducting interconnection pattern is formed on a board. A thin board of insulating material is coated with a conducting film, usually copper. *Photolithography is then used to coat part of the film with protective material. The unprotected metal is removed by etching, leaving the desired pattern of interconnections. Discrete components or packaged *integrated circuits may then be added to complete the circuit.

Double-sided printed circuits are commonly produced in which both sides of the board have a circuit formed on them, with *feed-throughs to connect the two sides as required. Printed circuits have been produced with several alternating layers of metal film and thin insulating film mounted on a single board. Boards with up to 12 layers of interconnections have been produced, although this is not common.

The use of plug-in printed circuits in electric or electronic equipment facilitates maintenance and repair. Printed circuits are reasonably robust when subjected to careful handling.

printer A device that produces printed characters, numbers, or symbols on paper from a source of information, such as a source of experimental data, the output of a *computer, information transmitted by a *telecommunication system, or a data-bearing medium such as punched cards. *See also* line printer; teleprinter.

printing *Colloq.* Photolithography.

printing receiver *See* teleprinter.

print-through A form of *distortion arising during magnetic tape recording. It is caused by a region of strongly magnetized tape affecting adjacent layers.

private branch exchange (PBX) *See* telephony.

private exchange *See* telephony.

probe (1) An electric lead that connects to a measuring or monitoring circuit, or contains such a circuit at its end or along its length, and that is used for testing purposes. The measuring or monitoring circuit may be formed from either active or passive components.

(2) A resonant conductor that is inserted into a *waveguide or *cavity resonator in order to inject or extract energy.

program A complete set of instructions written in a particular *programming language that causes a computer to perform a set of defined operations. A given program includes all the necessary instructions that cause the computer to input or output data or results, to perform mathematical operations, to store data in designated locations, to transfer data or instructions from one part of the system to another, and to perform any necessary operations required for the successful completion of the program.

A *subroutine* is a section of a program to which control may be transferred from a number of points throughout the program. When the instructions in the subroutine have been obeyed, control is returned to the point from which the transfer was made. This saves the repetition of identical sections of code in different places in the program. The word 'SUBROUTINE' is part of the FORTRAN language. A *diagnostic routine* is used to check automatically the operation of a program or part of a program and to detect errors (*see* bug).

A large computer program is often arranged in an *overlay structure* so that the portions of the program in current use are copied from disk storage into the main memory as required.

programmable read-only memory (PROM) *See* read-only memory.

programme signal The complex wave containing the information corresponding to the sound information (audio signal) and vision information (video signal) during a specific radio or television broadcast.

programming language A language that is designed to be mutually comprehensible to computers and humans. A particular digital computer operates in *binary notation and all instructions must ultimately be produced in the appropiate binary code, known as *machine code*.

A *high-level programming language* is a language that resembles natural language or mathematical notation more closely than machine code. A *compiler* is a program that is used to convert a high-level language to machine code. There are many different high-level languages suitable for different applications. Examples include *Pascal, FORTRAN, ALGOL-60, BASIC,* and *PL/1*.

A *low-level programming language* is one that resembles machine code more closely that natural language. Low-level languages are usually known as assembly languages and require an *assembler* program to convert them to machine code.

projection lithography *See* photolithography.

PROM *Abbrev. for* programmable read-only memory. *See* read-only memory.

propagation coefficient *Syn.* propagation constant. Symbol: γ, P. A complex quantity that expresses the effect of a *transmission line on a sinusoidal progressive wave. The propagation coefficient is defined for a uniform transmission line of infinite length supplied with a sinusoidal current of specified frequency at its sending end.

Under steady-state conditions, if the currents at two points along the line, separated by unit length, are I_1 and I_2, where I_1 is nearer the sending end of the line, then

$$\gamma = \log_e(I_1/I_2)$$

at the specified frequency; I_1/I_2 is the vector ratio of the currents. γ is a complex quantity and may be written

$$\gamma = \alpha + i\beta$$

where $i = \sqrt{-1}$. The real part, α, is the *attenuation constant and is measured in nepers per unit length of line. It measures the transmission losses in the line. The imaginary part, β, is the *phase-change coefficient* and is measured in radians per unit length of line. It is the *phase difference between I_1 and I_2 introduced by the transmission line. Thus

$$I_1/I_2 = \exp(\alpha + \beta i) = \exp(\alpha).\exp(i\beta)$$

If the displacement of the vibration is a maximum at a given point and equal to p_1, then at the same instant the displacement, p_2, at a distance x along the transmission line is given by

$$p_2 = p_1 \exp(-\alpha - i\beta)x$$

An infinite transmission line is not physically possible but conditions simulating those in an infinite line are realized when a transmission line of finite length is terminated by its characteristic impedance. *Compare* image transfer constant.

propagation constant *See* propagation coefficient.

propagation loss Energy loss from a beam of *electromagnetic radiation as a result of absorption, scattering, and spreading of the beam.

proportional control A control system that operates by first determining the difference between the actual value of the quantity to be controlled and the desired value, and then applying a correction proportional to this difference.

proportional counter A *gas-filled radiation detection tube operated in the region of proportionality on the characteristic curve of the detector. In this region gas multiplication occurs so that the output size is proportional to the number of ion-pairs produced by the initial ionizing event and therefore to the energy of the ionizing radiation. *Compare* Geiger counter; ionization chamber.

protective gap A *surge diverter that is used to protect an electrical system or apparatus from excessively high voltage *surges. It consists of a spark gap connected between a conductor of the system and earth. If a surge of voltage occurs a spark passes across the gap thus diverting part of the surge to earth. The term does not apply to any series resistor that may be present in the earth connection.

protective horn *See* arcing horn.

protective relay A relay that is used to protect electrical apparatus against overload and fault conditions. If an abnormal condition occurs the relay causes a circuit-breaker to open and thus disconnect the faulty apparatus from the power supply and from other associated equipment.

proton An elementary particle having a positive charge equal to that of the electron, mass 1.6725×10^{-27} kg, which is about 1836 times the electron mass, and spin $\frac{1}{2}$. It also has an intrinsic *magnetic moment. The proton forms the nucleus of the hydrogen atom and is a constituent part of all atomic nuclei. The number of protons in an atomic nucleus is equal to the number of orbital electrons and is the atomic number of the atom.

proton synchrotron *See* synchrotron.

proximity effect (1) An effect observed when two or more conductors carrying alternating current are placed close to one another. The distribution of current across the cross section of any one conductor is changed under the influence of magnetic fields due to the other(s). The *effective resistance of the conductor is modified by this effect and it is particularly significant in coils used at high (radio) frequencies.

(2) In *electron-beam lithography, if two lines or patterns are exposed very close to each other, unwanted exposure of the resist between them.

proximity printing *See* photolithography.

psophometer An instrument that measures the amount of *noise in a transmission system, particularly a telephone system. It includes a device for measuring noise power through a weighting network in order to produce objective results that approximately parallel subjective results with human observers. Standard psophometric weighting characteristics have been produced by the *CCITT.

Crosstalk and other noise may be measured by passing a suitable noise signal down one telephone channel and measuring the amount of this noise that appears on an adjacent line with a psophometer with standard weighting characteristics.

PTFE *Abbrev. for* polytetrafluoroethylene. *See* Teflon.

ptm *Abbrev. for* pulse-time modulation. *See* pulse modulation.

p-type conductivity Conduction in a semiconductor in which current flow is caused by the effective movement of mobile *holes through the semiconductor. *See also* semiconductor. *Compare* n-type conductivity.

p-type semiconductor An extrinsic semiconductor that contains a higher density of mobile *holes than of conduction *electrons, i.e. holes are the majority carriers. *See also* semiconductor. *Compare* n-type semiconductor.

public address system (PA) A complete reproducing system that is designed to render sounds audible to a large gathering of people. The system usually consists of one or more microphones, power amplifiers, and several loudspeakers suitably placed to minimize

448

interference and echo. PA systems are extensively used for large open-air meetings, in railway stations, and in large halls.

puff *Colloq.* One picofarad.

pulling A change of frequency observed in an electronic oscillator when it is coupled to a circuit containing another independent oscillation. The frequency of the oscillator tends to change towards that of the independent oscillation; the tendency is particularly marked if the difference in frequency is small. Complete synchronization can sometimes be achieved. Pulling is used to control the frequency of an oscillator, as in crystal-controlled oscillators (*see* piezoelectric oscillator).

pulsating current A current that exhibits regularly recurring variations of magnitude. The term implies that the current is unidirectional. A pulsating current or other pulsating quantity can be considered as the sum of a steady component and a superimposed alternating component whose average value cannot be zero.

pulse A single transient disturbance manifest as an isolated wave, one of a series of transient disturbances recurring at regular intervals, or a short train of high-frequency waves, such as are used in echo sounding or *radar. A pulse consists of a voltage or current that increases from zero (or a constant value) to a maximum and then decreases to zero (or the constant value), both in a comparatively short time. The zero or constant value in the absence of the pulse is termed the *base level*. A pulse is described according to its geometrical shape when the instantaneous value is plotted as a function of time: it can be rectangular, square, triangular, etc. Unless otherwise specified a pulse is assumed to be *rectangular*.

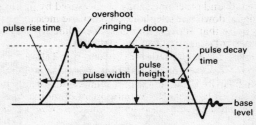

a Practical rectangular pulse

In practice a perfect geometrical shape is never achieved and a practical rectangular pulse is shown in Fig. *a*. The portion of the pulse that first increases in amplitude is the *leading edge*. The time interval during which the leading edge increases between specified limits, usually between 10% and 90% of the pulse height, is termed the *rise time*. The pulse decays back to the base level with a finite *decay time,* usually taken between the same limits as the rise time. The major portion of the decay time is termed the *trailing edge* of the pulse.

The time interval between the rise time and decay time is the *pulse width.* The magnitude of the pulse taken over the pulse width is normally substantially constant, ignoring any spikes and ripples, and is the *pulse height.* The pulse height can be quoted either as the maximum value, the average value, or the root-mean-square value, all measured over the pulse width and ignoring spikes or ripples. The height of a nonrectangular pulse, e.g. a triangular one, is normally the maximum amplitude.

A practical rectangular pulse can suffer from *droop,* which occurs when the pulse height falls slightly below the nominal value. The degree of droop is given by the *pulse flatness deviation;* this is the ratio of the difference between the maximum and minimum values of the pulse amplitude to the maximum amplitude, during the pulse width. A *valley* occurs when the droop exists only over a portion of the pulse width and the pulse height then recovers. The pulse *crest factor* is the ratio of the peak amplitude of the pulse to the root-mean-square amplitude.

A *spike* is an unwanted pulse of relatively short duration superimposed on the main pulse; *ripple* is unwanted small periodic variations in amplitude. A practical pulse frequently rises to a value above the pulse height and then decays to it with damped oscillations. These phenomena are known as *overshoot* and *ringing.* A similar effect occurs as the pulse decays to the base level. Pulse circuits may frequently incorporate a *smearer,* which is a circuit designed to minimize overshoot.

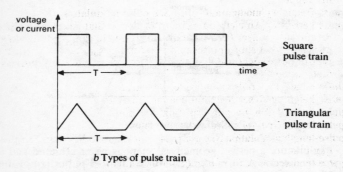

b Types of pulse train

A group of regularly recurring pulses of similar characteristics is called a *pulse train;* it is usually identified by the type of pulses in the train, e.g. square waves or sawtooth waves (Fig. *b*). The time interval

450

between corresponding portions of the pulses in the train, e.g. the pulse rise times, is the *pulse spacing* (or pulse-repetition *period*), *T*. The *pulse-repetition frequency* (or *pulse rate*) is the reciprocal of the period and is the rate at which pulses are transmitted in the pulse train; it is measured in hertz. Minor variations in the pulse spacing in a pulse train are known as pulse *jitter*. The *duty factor* of a pulse train is the ratio of the average pulse width to the average pulse spacing of pulses in the train.

pulse amplitude *Syn. for* pulse height. *See* pulse.

pulse-amplitude modulation (pam) *See* pulse modulation.

pulse carrier *See* pulse modulation.

pulse code modulation (pcm) *See* pulse modulation.

pulse coder *Syn. for* coder. *See* pulse modulation.

pulse communications Telecommunications involving the transmission of information by means of *pulse modulation. Pulse communication is used in systems that operate by means of *time-division multiplexing.

pulse detector *Syn. for* decoder. *See* pulse modulation.

pulse discriminator A *discriminator that selects and responds only to a *pulse with a particular characteristic of amplitude, period, etc., and that is used in *pulse operation.

pulse duration *Syn. for* pulse width. *See* pulse.

pulse-duration modulation (pdm) *Syns.* pulse-width modulation; pulse-length modulation. *See* pulse modulation.

pulse-flatness deviation *Syn.* pulse tilt. *See* pulse.

pulse-forming line An *artificial line that contains inductances and capacitors in series and is used to produce fast high-voltage pulses in *radar.

pulse-frequency modulation (pfm) *See* pulse modulation.

pulse generator An electronic circuit or device that generates voltage or current *pulses of a desired waveform. It can be designed to produce either single pulses or a pulse train. A *pulser* is a particular type of pulse generator that produces fast (short-duration) pulses of high voltage.

pulse height *Syn.* pulse amplitude. *See* pulse.

pulse-height analyser *See* multichannel analyser.

pulse interval *Syn. for* pulse spacing. *See* pulse.

pulse length *Syn. for* pulse width. *See* pulse.

pulse-length modulation *Syn. for* pulse-duration modulation. *See* pulse modulation.

pulse modulation A form of modulation in which *pulses are used to modulate the *carrier wave or, more commonly, in which a pulse train is used as the carrier (the *pulse carrier*). Information is conveyed by modulating some parameter of the pulses with a set of discrete instantaneous samples of the message signal. The *minimum*

451

sampling frequency is the minimum frequency at which the modulating waveform can be sampled to provide the set of discrete values without a significant loss of information.

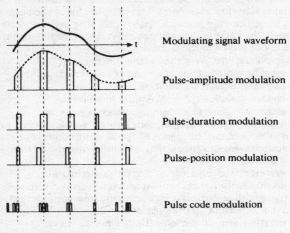

Forms of pulse modulation

Different forms of pulse modulation are shown in the diagram. In *pulse-amplitude modulation* (pam), the amplitude of the pulses is modulated by the corresponding samples of the modulating wave. In *pulse-time modulation* (ptm), the samples are used to vary the time of occurrence of some parameter of the pulses. Particular forms of pulse-time modulation are *pulse-duration modulation* (pdm), in which the time of occurrence of the leading edge or trailing edge is varied from its unmodulated position, *pulse-frequency modulation* (pfm), in which the repetition frequency (*see* pulse) of the carrier pulses is varied from its unmodulated value, and *pulse-position modulation* (ppm), in which the time of occurrence of a pulse is modulated from its unmodulated time of occurrence, i.e. the pulse repetition period is varied. All these types of pulse modulation are examples of uncoded modulation.

In *pulse code modulation* (pcm) only certain discrete values are allowed for the modulating signals. The modulating signal is sampled, as in other forms of pulse modulation, but any sample falling within a specified range of values is assigned a discrete value. Each value is assigned a pattern of pulses and the signal transmitted by means of this code. The electronic circuit or device that produces the coded pulse train from the modulating waveform is termed a *coder* (or pulse coder). A suitable *decoder* must be used at the receiver in order to extract the original information from the transmitted pulse train. *Morse code is a very well known example of a pulse code.

Pulse modulation is commonly used for *time-division multiplexing.

pulse operation Any method of operation of an electronic circuit or device that transfers electrical energy in the form of pulses.

pulse-position modulation (ppm) *Syn.* pulse-phase modulation. *See* pulse modulation.

pulser *See* pulse generator.

pulse radar *See* radar.

pulse rate *Syn. for* pulse repetition frequency. *See* pulse.

pulse regeneration The process of restoring a *pulse or pulse train to the original form, timing, and magnitude. Pulse regeneration is required in most forms of pulse operation since the circuits or circuit elements used can introduce *distortion.

pulse repetition frequency (prf) *Syn.* pulse rate. *See* pulse.

pulse repetition period *Syn. for* pulse spacing. *See* pulse.

pulse separation *Syn. for* pulse spacing. *See* pulse.

pulse shaper Any circuit or device that is used to alter any of the characteristics of a *pulse or pulse train. *Pulse regeneration is a special case of pulse shaping.

pulse spacing *Syns.* pulse separation; pulse interval; pulse repetition period. *See* pulse.

pulse tilt *Syn. for* pulse flatness deviation. *See* pulse.

pulse-time modulation (ptm) *See* pulse modulation.

pulse train *See* pulse.

pulse width *Syns.* pulse duration; pulse length. *See* pulse.

pulse-width modulation *Syn. for* pulse-duration modulation. *See* pulse modulation.

pump *See* parametric amplifier.

pump frequency *See* maser.

punched card A card on which information can be carried in the form of holes punched in columns. There are usually 80 columns per card. Each column has a combination of holes punched in it that represents a specific character, number, or symbol. Punched cards were the original storage medium for digital computers and other devices. The use of cards for the input of information to computers has largely ceased despite the ease with which small decks of cards may be hand-edited.

A *card punch* is a device that prepares the punched cards. The signals to be punched can be provided manually from a keyboard – a *key punch* – or can be provided directly from a source of experimental data or from the output of a computer. A *card reader* is a

453

device that decodes the patterns punched in the card, usually by means of *photocells, into electrical signals and displays, stores, or transmits the information to a computer. *Compare* paper tape.

punched tape *See* paper tape.

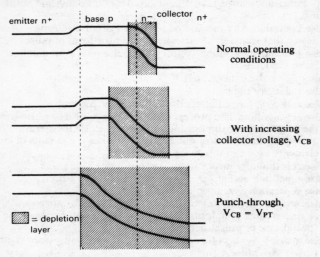

a Energy diagrams for an n-p-n transistor at different collector voltages

punch-through A type of *breakdown that can occur in both bipolar junction *transistors and unipolar *field-effect transistors. If the collector-base voltage, V_{CB}, applied to a bipolar junction transistor is increased, the depletion layer associated with the collector-base junction spreads across the base region. At a sufficiently high collector voltage, known as the punch-through voltage, V_{PT}, the depletion layer spreads through the entire base region and reaches the emitter junction. A direct conducting path is therefore formed from emitter to collector and charge carriers from the emitter 'punch through' to the collector. The associated energy diagram is shown in Fig. *a.*

In a unipolar transistor the effect operates in a similar manner. When the drain voltage, V_D, reaches a sufficiently large value (V_{PT}) the depletion layer associated with the drain spreads across the substrate and reaches the source. Charge carriers then punch through the substrate.

Punch-through can be a problem in a short-channel device used as a switch. Due to the relatively high doping level of the drain of an

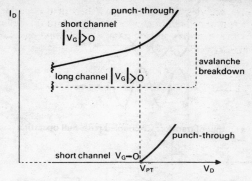

b **FET characteristics above pinch-off**

FET the drain-substrate depletion layer spreads readily across the substrate. Punch-through therefore must be avoided in the 'off' state of the switch, when the gate voltage, V_G, is zero. Punch through can also occur in the 'on' state ($|V_G| > 0$) when the device is operated in the saturated region above pinch-off (*see* field-effect transistor); charge carriers punch through the substrate from source to drain at sufficiently large values of V_D. In unipolar transistors that have relatively long channel lengths and high doping levels, punch-through does not occur; *avalanche breakdown occurs first (Fig. *b*).

puncture voltage The value of voltage that causes an insulator to be punctured when it is subjected to a gradually increasing voltage. *See also* impulse voltage.

purple plague The formation of purple-coloured areas on the bonds of a silicon *integrated circuit. The purple colour is caused by the formation of an unwanted aluminium-gold eutectic mixture ($AuAl_2$) at the bond between the gold connecting wire and the aluminium *bonding pads. The resulting bond is mechanically weak, since the eutectic is very brittle, and is therefore susceptible to failure.

push-pull amplifier *Syn.* balanced amplifier. *See* push-pull operation.

push-pull operation The use of two matched devices in such a way that they operate with a 180° phase difference. The output circuits combine the separate outputs in phase (Fig. *a*). One common means of achieving the desired 180° phase shift in the inputs is a transformer-coupled input circuit (Fig. *b*). *Complementary transistors may also be used (Fig. *c*), in which case no phase shift is required in the inputs.

Push-pull circuits are frequently used for *class A and *class B amplification and are then termed *push-pull amplifiers*. A push-pull

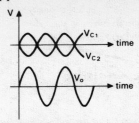

a Output of transformer-coupled push-pull operation

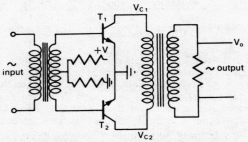

b Transformer-coupled class A push-pull operation

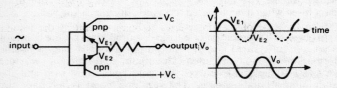

c Complementary transistor push-pull operation

amplifier that is suitably biased to give negligible output current when no input signal is present is termed a *quiescent push-pull amplifier*.

pyroelectricity The development of opposite electric charges at the ends of the polar axes in certain crystals when subjected to a temperature gradient. Tourmaline and lithium sulphate are two of the crystals that exhibit pyroelectricity.

Q

Q band A band of microwave frequencies ranging from 36·0 to 46·0 gigahertz. *See* frequency band.

Q-factor *Syn.* quality factor. Symbol: Q. A factor that is associated with a *resonant circuit and describes both the ability of the circuit to produce a large output at the resonant frequency and the selectivity of the circuit. The Q-factor is given by

$$Q = (1/R)\sqrt{(L/C)}$$

where R is the resistance of the circuit, L the inductance, and C the capacitance.

In a circuit capable of free oscillations

$$Q = \pi/\Lambda$$

where Λ is the logarithmic decrement (*see* damped) of the circuit. In the case of forced oscillations

$$Q = \omega_0 E_s/\eta$$

where ω_0 is 2π times the resonant frequency, E_s is the stored energy and η is the rate of energy dissipation. This relation is used to define Q in complicated resonant circuits or in resonant systems, such as cavity resonators, where the values of L, C, and R cannot be specified.

In the case of a simple series resonant circuit at the resonant frequency

$$\omega_0 L = 1/\omega_0 C$$

and therefore

$$Q = \omega_0 L/R$$

or

$$Q = 1/\omega_0 CR$$

This is effectively the ratio of the total inductive or total capacitive *reactance to the total series resistance at resonance. It can be shown that

$$|V_c/V_{app}| = Q$$

where V_c is the voltage across the capacitor and V_{app} the applied voltage. The series resonant circuit therefore can act as a tuned voltage *transformer.

In the case of a parallel resonant circuit the value of the resonant frequency is given by

$$\omega_0 = \sqrt{(1/LC)}\sqrt{(1 - 1/Q^2)}$$

This is slightly different from the series case where

$$\omega_0 = \sqrt{(1/LC)}$$

If Q is large however the difference may be neglected. It can be shown that at resonance, if Q is large,

$$|I_c/I| = Q$$

where I_c is the current in the capacitor and I the current drawn from the external source. The parallel resonant circuit may therefore be used as a tuned current *transformer. The parallel resistance, R_p, of a parallel resonant circuit is defined as $1/Y$ where Y is the admittance at resonance and is real (see resonant frequency). R_p is given by

$$R_p = Q^2 R_s$$

where R_s is the series resistance of the same components arranged as a series resonant circuit.

The *selectivity of a resonant circuit is defined as $\nu_0/2\delta\nu$ where ν_0 is the resonant frequency and $2\delta\nu$ is the difference in frequency between the two half-power points. It can be shown that

$$\nu_0/2\delta\nu = 1/Q$$

A single reactive component may be capable of resonance without any other components, i.e. if the self-capacitance of an inductance coil or the self-inductance of a capacitor is sufficiently large. The Q-factor of a single component is defined as the ratio of the reactance to the effective series resistance of the component, i.e. for an inductance

$$Q = \omega L/R$$

and for a capacitor

$$Q = 1/\omega CR$$

Q-point See quiescent point.

quadrant electrometer An *electrometer that consists of a flat cylindrical metal box formed from individually isolated quadrantal segments and containing a light foil-covered vane that is supported by a quartz fibre and is free to move (see diagram). Opposite quadrants are connected together. The instrument is usually arranged so that the vane hangs symmetrically within the quadrants when it and both pairs of quadrants are at zero potential. In addition the potential, V_c, applied to the vane must be large compared to the potentials

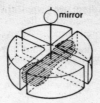

Quadrant electrometer

V_A and V_B applied to the quadrants. Under these circumstances the deflection, θ, of the vane is given by

$$\theta = k_1(V_A - V_B)$$

or if one pair of quadrants is at earth potential by

$$\theta = k_2 V$$

where k_1 and k_2 are constants and characteristic of the instrument. The deflection of the vane is observed by using a small mirror that reflects a spot of light attached to the torsion thread.

quadraphonic broadcasting *See* quadraphony.

quadraphony A sound reproduction system that is an extension of stereophonic sound reproduction (*see* reproduction of sound) and uses four channels, feeding four separate loudspeakers. The sound may be recorded either using four coincident directional microphones at right angles to each other and placed at the position of the listener, or using separated microphones whose outputs are divided between the channels and combined in the correct proportions to achieve the desired effect.

Various methods exist for recording and reproducing quadraphonic signals, all of which combine the signals to produce two channels that may be recorded by stereophonic techniques. *Compatible discrete four-channel recording* has four separate channels. Each stereo channel contains signals from two of the quadraphonic channels, multiplexed together. The sum signal of the two channels occupies one frequency band and the difference signal another frequency band. With a suitable *pick-up both the sum and difference signals in each of the stereo channels are detected and the original information extracted. *Matrix recording* systems combine the outputs of the four channels together in fixed proportions and phase relationships in order to produce two channels of information. The two channels are then decoded during reproduction in order to restore the original four channels.

Quadraphonic broadcasting adopts a similar approach. A second subchannel may be added to a stereo broadcast signal in quadrature with the subcarrier channel carrying the stereo difference signal. A matrix system may be used and the information transmitted in a coded form. International standards have yet to be decided.

quadrature Two periodic quantities that have the same frequency and waveform are *in quadrature* when the phase difference between them is $\pi/2$ (90°). They therefore differ by one quarter of a period, one wave reaching its *peak value when the other passes through zero.

quadrature component (1) (of current) *See* reactive current.
  (2) (of voltage) *See* reactive voltage.
  (3) (of volt-amperes) *See* reactive volt-amperes.

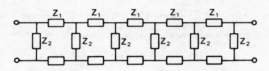

a Generalized ladder filter

quadripole *Syns.* four-terminal network; two-port network. A *network that has only four terminals, i.e. a pair of input terminals and a pair of output terminals. The behaviour of a quadripole is usually described by the impedances presented at its terminals at specified frequencies. If the electrical properties are unchanged when the input and output terminals are reversed the quadripole is *symmetrical;* otherwise it is *dissymmetric.* A quadripole is described as *balanced* if its electrical properties are unchanged when both the input and output pairs of terminals are interchanged simultaneously; otherwise it is unbalanced.

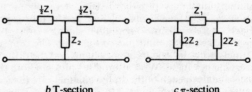

b T-section   *c* π-section

A very common arrangement, used particulary for *attenuators and *filters, is a ladder network consisting of a number of series and shunt impedances (Fig. *a*). This arrangement may be broken down for analysis into identical sections each with the same characteristic impedance. In order to avoid power dissipation in the load by re-

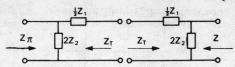

d L-sections

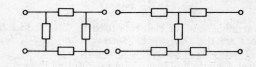

e Q-network   *f* H-network

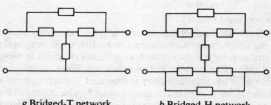

g Bridged-T network   *h* Bridged-H network

flections, a ladder network must be terminated by an impedance equal in value to the *iterative impedance of the sections. The ladder filter shown in Fig. *a* may be analysed as a series of *T-sections* (Fig. *b*) terminated in the iterative impedance Z_T. The same network may be considered as a series of *π-sections* (Fig. *c*) that must be terminated in the iterative impedance Z_π. Comparison of the two shows that the half-section or *L-section* (Fig. *d*) acts as an impedance transformer from Z_π to Z_T. Such half-sections may be used to match a network to a load; they are especially important when composite networks are being designed. It can be shown that if the impedances used in the ladder are Z_1 and Z_2, as shown in Fig. *b*, then

$$Z_T Z_\pi = Z_1 Z_2$$

In general Z_1 and Z_2 and hence Z_T and Z_π will be dependent on frequency. If the product $Z_T Z_\pi$ is substantially independent of frequency the network is a *constant-R network*.

461

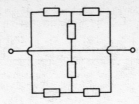

i Twin-T network

Other arrangements of elements used to form ladder filters or attenuators are the *O-network, H-network, bridged-T, bridged-H,* and *twin-T networks* (Figs. *e, f, g, h,* and *i*).

quality control A method of inspection used during the mass production of electronic components, circuits, devices, or other apparatus in order to ensure that the finished product conforms to specifications. It usually involves exhaustive checking of random samples at every stage during manufacture.

quality factor *See* Q-factor.

quantization A method of producing a set of discrete or quantized values that represents a continuous waveform. The waveform is divided into a finite number of subranges each of which is represented by an assigned value within the subrange. This technique is used whenever a set of discrete values is required, for example to produce data suitable for a digital computer or in *pulse modulation. Some loss of information of the original waveform is inherent in this process, although suitable choice of the subranges can keep the loss to a minimum; this leads to a certain amount of noise, or *quantization distortion.*

quantum *See* quantum theory.

quantum efficiency *See* quantum yield.

quantum mechanics A mathematical physical theory that grew out of the *quantum theory proposed by Planck. The subject developed in several different equivalent forms that involve the wave-particle duality of electromagnetic radiation and Heisenberg's uncertainty principle; it has been extended to satisfy the principles of relativity and other branches of physics and chemistry.

Quantum mechanics in the form of *wave mechanics* is used to solve the energy states of atoms and molecules and hence explains the electrical behaviour of *metals, *insulators, and *semiconductors. *See also* energy bands.

quantum numbers A set of numbers that are used to label the various possible values of certain physical properties. The property is restricted to certain discrete values (quantized) as a result of applying the principles of quantum mechanics to a physical system. A set of quantum numbers is usually in the form of a set of integers and half-integers.

quantum theory A theory introduced by Max Planck in 1900 that departed radically from classical Newtonian mechanics. It was found that certain physical phenomena could not be satisfactorily explained by classical mechanics. Planck postulated that radiant energy could be radiated, transported, or absorbed in the form of discrete indivisible packets known as *quanta*. The quantum of electromagnetic radiation is the *photon. The photon energy, *E,* is related to the radiation frequency, ν, by Plank's law:

$$E = h\nu$$

h is the *Planck constant and has dimensions of energy times time.

Physical quantities that are restricted to a number of discrete values by quantum theory are said to be *quantized*. The changes of energy that correspond to a quantum, or an integral number of quanta, are small and only significant on an atomic scale. Large-scale systems are adequately described by classical mechanics (*see* Maxwell's equations).

Quantum theory successfully described the *photoelectric effect, atomic structure, and the Compton effect. Despite its success in the above applications it gave misleading results in many other problems and has been further refined and developed into the system known as *quantum mechanics.

quantum yield *Syn.* quantum efficiency. The number of reactions of a particular type induced per photon of absorbed electromagnetic radiation. It can, for example, be the average number of photoelectrons produced per photon at a specified frequency in a photoelectric *photocell or *phototube.

quarter-phase system *Syn. for* two-phase system. *See* polyphase system.

quarter-wavelength line *Syn.* quarter-wavelength transformer. A *transmission line of length equal to one quarter of the wavelength of the *fundamental frequency. It is used for *impedance matching, particularly in systems designed to operate at high radiofrequencies, for the suppression of even-order harmonics in filter networks, and for coupling and feeders used with aerials.

quarter-wavelength transformer *See* quarter-wavelength line.

quartz Naturally occurring crystalline silicon dioxide (SiO_2). It exhibits marked piezoelectric properties and is frequently used as the piezoelectric crystal in *piezoelectric oscillators. It also exhibits marked *dielectric strength.

Quartz may be readily drawn into extremely fine uniform filaments that are very strong, elastic, and physically and chemically stable. Such *quartz fibres* are frequently used as torsion threads in delicate measuring instruments, such as *electrometers.

quartz-crystal oscillator *See* piezoelectric oscillator.

quasi-bistable circuit An astable *multivibrator that is caused to operate by a *trigger (i.e. it is *clocked) rather than being allowed to run freely. Provided that the frequency of application of the trigger is high compared to the natural frequency of vibration of the free-running circuit it operates as a bistable circuit, i.e. a *flip-flop.

quasi-static measurement *See* MOS capacitor.

quench A capacitor, resistor, or combination of the two, that is used in parallel with a contact to an inductive circuit and that inhibits *spark discharge across the contact when the current ceases. A quench is commonly employed across the make-and-break contacts of an induction coil.

quench frequency *See* super-regenerative reception.

quick-break switch A manually operated switch that is designed to operate at a speed independent of that at which the operating handle is moved. The speed of separation of the contacts is usually controlled by a spring. *Compare* slow-break switch.

quiescent-carrier telephony *See* telephony.

quiescent component A component of an electronic circuit that, at a specific instant, is not in operation but at a short time following the specified instant will become operative.

quiescent current The current that flows in any circuit under specified normal operating conditions but in the absence of an applied signal.

quiescent period The period between transmissions in a pulse transmission system.

quiescent point *Syn.* Q-point. The region on the characteristic curve of an active device, such as a transistor, during which the device is not operating.

quiescent push-pull amplifier *See* push-pull operation.

quiet automatic gain control *See* automatic gain control.

quieting sensitivity *See* radio receiver.

R

radar Acronym from *r*adio *d*irection *a*nd *r*anging. A system that locates distant objects using reflected radiowaves of microwave frequencies. Modern radar systems are highly sophisticated and can produce detailed information about both stationary and moving objects and can be used for navigation and guidance of ships, aircraft, and other vehicles and systems.

A complete *radar system* contains a source of microwave power, such as a *magnetron, a modulator to produce pulses of microwave energy where necessary, transmitting and receiving aerials, a receiver that detects the echo, and a cathode-ray tube (CRT) that displays the output in a suitable form. Several types of radar system are in common use.

Pulse radar systems transmit short bursts of high-frequency radiowaves and the reflected pulse is received during the time interval between the transmitted pulses.

Continuous-wave systems transmit energy continuously and a small proportion is reflected by the target and returned to the transmitter.

Doppler radar utilizes the *Doppler effect in order to distinguish between stationary and moving objects. The change in frequency between the transmitted and received waves is measured and hence the velocity of moving targets deduced.

Frequency-modulated radar is a system that transmits a frequency-modulated radar wave. The reflected echo *beats with the transmitted wave and the range of the target is deduced from the beat frequency produced.

Volumetric radar systems can produce three-dimensional positional information about one or more targets. Two transmitters used simultaneously are commonly employed. *V-beam radar* is a volumetric system using two fan-shaped beams.

In any of the above systems the direction and distance of the target is given by the direction of the receiving aerial and the time interval between transmission of the radar signal and reception of the echo.

The direction of the transmitting and receiving aerials can be periodically varied in order to scan a given area. *Coarse scanning* is often used to obtain an approximate target location before repeating the scan more accurately. A common arrangement is to rotate the aerials in a horizontal plane, and produce a synchronous circular scan on the CRT, in order to display any targets within the vicinity of the transmitter. Such a presentation is termed a *plan position indicator (PPI). Scanning in a vertical plane can also be used.

Pulsed radar systems frequently use the same aerial as both a transmitting and receiving aerial. The appropriate transmitting and receiver circuits are connected to it using a *transmit-receive (TR) switch. The pulse repetition frequency of the transmitted signals is determined by a *multivibrator known as the *master trigger*.

The *radar range* is the maximum distance at which a particular radar system is effective in detecting a target. It is usually defined as the distance at which a designated target is distinguished for at least 50% of the transmitted pulses. The range is dependent on the *minimum discernible signal that the radar receiver can accept, i.e. the minimum power input to the receiver that produces a discernible signal on the radar indicator. The power in the return echo is dependent on the peak power of the transmitted pulse. In general the larger the output power of the transmitter, the greater the range of the system. A given radar system is characterized by the performance figure; this is the ratio of the peak power of the transmitted pulse to the minimum discernible signal of the receiver. The ability of a radar system to differentiate objects along the same bearing is usually defined as the minimum radial distance separating targets at which they can be separately resolved.

Radar systems are used for the detection and control of aircraft (*see* precision approach radar), guiding of ships in fog, and for locating distant storm centres when an echo is produced by the associated heavy rainfall. Radar is used in astronomy and also has an extremely wide range of military applications.

radar beacon *See* beacon.

radar indicator *Syns.* radar screen; radarscope. A *cathode-ray tube that provides a visual display of the return echoes in a *radar system. A *panoramic radar indicator* simultaneously displays all the received echoes of different frequencies.

radar range *See* radar.

radarscope *Syn. for* radar indicator.

radar screen *Syn. for* radar indicator.

radial-beam tube An *electron tube in which the electron beam travels in a radial path from a central cathode to one of a set of circumferential anodes. The beam is rotated between the anodes by means of a rotating magnetic field. The tube can be used as a high-speed switch or a commutation switch.

radiation Any form of energy that is propagated as waves or streams of charged particles. *See* electromagnetic radiation.

radiation counter *See* counter.

radiation diagram *See* radiation pattern.

radiation effect *See* solar cell.

radiation efficiency *See* aerial efficiency.

radiation pattern *Syn.* radiation diagram. A diagram that represents the distribution of radiation in space from any source, such as a transmitting aerial, or conversely shows the effectiveness of reception of a receiving aerial. The pattern can be produced in either Cartesian or polar coordinates.

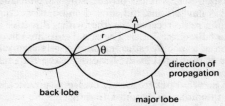

Figure-of-eight radiation pattern

A typical radiation pattern consists of one or more *lobes* that represent regions of enhanced response of an aerial in a horizontal or vertical plane. The *major lobe* is the lobe that contains the region of maximum radiation intensity or maximum sensitivity of detection of an aerial. It usually points forwards along the direction of propagation of the radiation. A *back lobe* is a lobe that points in the reverse direction to the direction of propagation. A typical radiation pattern is a polar figure of eight (*see* diagram). Point A (r, θ) on the diagram indicates the sensitivity of the aerial represented by r at an angle θ to the direction of propagation of the radiation.

radiation potential *See* ionization potential.

radiation resistance *See* aerial radiation resistance.

radiative recombination *Syn. for* Lossev effect. *See* recombination processes.

radio (1) The use of *electromagnetic radiation of frequency within the radiofrequency portion of the electromagnetic spectrum (*see* frequency band) for the transmission and reception of electrical impulses or signals without connecting wires or waveguides. It is also the process of transmitting or receiving such signals.

The unqualified term usually denotes the *telecommunication system that transmits audio information (*wireless*). Any communication channel, circuit, or link in which information is transmitted by radio is described as a radiocommunication or radio channel, radio circuit, or radio link.

(2) *Short for* radiofrequency. Denoting electromagnetic radiation in the *radiofrequency range or any device, component, or other apparatus used to transmit or receive information at frequencies

467

within this range, as in radio telephony, radio telegraphy, or radio facsimile.

(3) *Colloq.* A *radio receiver.

radioastronomy The study of astronomical bodies and events by means of the radio signals associated with them.

radio beacon *See* beacon.

radio broadcasting *Syn.* wireless broadcasting. *See* broadcasting.

radio compass A navigational aid carried on board aircraft and ships. It consists of a radio receiver together with a directive aerial and is essentially a *direction finder. The aerial is rotated in order to find the direction of a specific radio transmitter relative to the craft; the information is presented as the heading of the craft relative to the transmitter.

radio device *Short for* radiofrequency device. *See* radiofrequency.

radio direction finding *See* direction finding.

radio effect *Short for* radiofrequency effect. *See* radiofrequency.

radiofrequency (r.f.) Any frequency of *electromagnetic radiation or alternating currents in the range three kilohertz to 300 gigahertz (*see* frequency band). An electronic device, such as an *amplifier, *choke, or *transformer, that operates in this range is known as a *radiofrequency device* or *radio device;* similarly any associated effect such as *distortion is termed a *radiofrequency effect* or *radio effect.*

radiofrequency heating *Dielectric heating or *induction heating that is carried out using an alternating field of frequency greater than about 25 kilohertz.

radiogoniometer *See* goniometer.

radiography *See* X-rays.

radio interferometer *See* radiotelescope.

radiolocation *Obsolete name for* *radar.

radio noise Any unwanted sound or distortion appearing at the loudspeaker of a *radio receiver. A portion of the noise is due to *interference from various sources: disturbances in the atmosphere or discharges in the ionosphere (*atmospheric noise*), extraterrestrial sources such as the sun (*galactic noise*), signals from other radio transmitters (*radio-station interference*), and other man-made sources. The remaining portion of the noise is an inherent property of the electronic circuits and devices and is caused by the random motions of electrons in the circuits.

A very common form of noise is *mains hum,* which is caused by harmonics of the mains frequency being detected and amplified by the radio receiver, and results in a humming noise. It tends to worsen as the components in the receiver age. *See also* interference; noise.

radio pill A small capsule that contains a miniature radio transmitter and may be swallowed. Miniature *transducers respond to conditions within the body and the signals produced are transmitted to a

468

suitable receiver outside the body. The pill may be recovered after passing through the alimentary tract.

radio receiver *Syns.* radio; wireless; radio set. A device that converts transmitted radiowaves into audible signals. A simple receiver contains an *aerial, a *tuner that can be adjusted to the desired *carrier frequency, a preamplifier, detector, audiofrequency amplifier, and a loudspeaker. Modern radio receivers commonly employ *superheterodyne reception in order to improve the signal-to-noise ratio of the output.

Radio receivers can detect either *amplitude-modulated signals or *frequency-modulated signals. They are described as *AM receivers* or *FM receivers,* respectively. A receiver that has the facility to detect both types of signal is an *AM/FM receiver.* High-fidelity (hi-fi) radio receivers usually contain extra circuits that are associated with the audiofrequency amplifier and are used to restore the *bass response* and *treble response* of the output to that of the original audible source. The *bass boost* circuit restores the lower audiofrequency signals; the *treble compensation* acts on the higher audiofrequencies.

Stereophonic radio receivers contain suitable detecting circuits that demodulate stereophonic radio transmissions and produce two outputs, each of which is separately amplified and output to a loudspeaker.

The sensitivity of a radio receiver is determined by the *minimum discernible signal at the input that produces a discernible audio output. The maximum sensitivity is usually defined as the minimum input signal that produces a specified output power. In an FM receiver the sensitivity is commonly quoted as the *quieting sensitivity:* this is the minimum input signal under specified operating conditions that produces a specified output signal-to-noise ratio. *See also* reproduction of sound.

radio set (1) *See* radio receiver.

(2) A combined radio transmitter and receiver such as is used by amateur radio operators, in aircraft, or in ships.

radiosonde A small radio transmitter together with suitable transducers that is carried by a balloon or kite into the upper atmosphere and transmits meteorological and other scientific data to the ground.

radiospectroscope A device that analyses and displays the total radio-frequency signals received at an aerial. The signals are usually displayed on the screen of a cathode-ray tube and some indication of the modulation and field strength at the transmitted carrier frequency can be obtained from the height and spread of the trace.

radio telegraphy *See* telegraphy.

radio telephone *See* telephony.

469

radiotelescope A telescope used in *radioastronomy that detects extra-terrestrial radio signals. There are two basic forms of radio-telescope. The first consists of a large steerable parabolic reflector, or dish, that can be directed at a desired region of the sky. Radio-waves occurring within a pencil-beam emanating from the portion of sky being studied are received and amplified. The 250 ft diameter Jodrell Bank telescope is a well-known example of this type of tele-scope. In some cases, as with the 1000 ft instrument at Arecibo, Puerto Rico, the reflector is formed out of a natural hollow in the ground and cannot be steered.

The second basic type is the *radio interferometer*. This type con-sists of an array of small fixed or steerable aerials at a known dis-tance apart. Radio signals received at the aerials are allowed to interfere with each other at the radio receiver used and the source of the signals is deduced from the interference pattern produced. The interferometer has a greater resolution than the reflector type.

radiotherapy *See* X-rays.

radiowave An electromagnetic wave that has a frequency lying in the *radiofrequency range.

radio window The range of *radiofrequencies that are not reflected by the *ionosphere but pass straight through it. The radio window extends from about 50 gigahertz to about 15 megahertz (approxi-mate wavelength range: 6 mm to 20 m). The effect of the ionosphere is still noticeable up to 100 MHz but decreases as the radiofre-quency is increased. At frequencies above 10 GHz heavy rain can severely affect transmission.

High-frequency *television broadcasts fall within the radio win-dow and long-distance television communication therefore requires the use of *communications satellites as reflectors. *Radioastrono-my is also restricted to this frequency range.

RAM *Abbrev. for* random-access memory.

Raman spectroscopy A method of analysing the surface of a material using the light scattered from a small volume near the surface as a result of inelastic scattering between the incident light and molecu-lar and lattice vibrations in the material. The *laser Raman microprobe* (LRM) uses visible laser light as the incident light. Par-ticular molecules in the material can be identified by measuring the difference in wave number (the reciprocal of wavelength) between the incident laser light and the reflected Raman light. This differ-ence is due to molecular vibrations. Lattice vibrations within the crystal structure also contribute to the inelastic scattering, and Raman spectroscopy may be used in addition to detect stress within the crystal lattice or other nonstoichiometric effects.

random-access memory (RAM) *Syns.* direct-access memory. A *mem-ory designed so that the location of the data stored in it is indepen-dent of the content and any location in the memory may be directly accessed without having to work through from the beginning. The access time (*see* memory) to a given *memory location of such a memory is substantially constant for all memory locations.

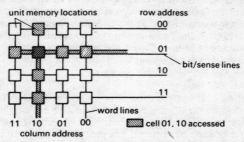

unit memory locations   row address

00

01

bit/sense lines

10

11

word lines

11   10   01   00   ▨ cell 01, 10 accessed
column address

Array of 4 × 4 storage cells

RAMs are formed either as *moving magnetic surface devices, such as a disk, or as *electronic memories, such as solid-state memory. In the latter case, the memory is arranged as a rectangular array of memory cells forming rows and columns. Each memory cell in the array forms an intersection between the rows and columns. An array of 16 storage cells arranged as a 4 × 4 matrix is shown in the diagram. Any individual cell in the array, such as the cell indicated, is defined by the address of one row and one column, as shown, since each row and column intersect once only.

In order to retrieve information from a particular location, the address codes of the row and column are specified. The output is sampled by suitable sensing devices, known as *sense amplifiers,* that are attached to each row and the rows are therefore termed *bit lines.* The columns are known as *word lines.* In practice a solid-state RAM consisting of an array of 128 × 128 (16 K) memory cells can be produced on one chip together with the associated address decoders, read, write, and refresh circuits.

Random-access memories are either *block-access memory* or *parallel memory.* The block-access memories, such as disk, are organized so that a set of *bits are stored together in adjoining locations and are output sequentially once located. Parallel memory, such as solid-state memory, is organized so that a set of bits is stored one in each of several separately accessed memory devices, and is output synchronously so that the bits are available in parallel. *Compare* serial memory.

random logic A *logic circuit that contains an arrangement of different interconnected logic gates rather than linked arrays of similar gates.

random noise *Syns.* background noise; ground noise. *See* noise.

471

range (1) The maximum distance from a *radio or *television transmitter at which reception of the signal is possible.

   (2) *See* radar.

range tracking Operation of a *radar system when viewing a moving target so that the *transmit-receive switch is automatically adjusted to switch to the receive mode at the correct instant for reception of the return echo; i.e. the *gate is adjusted to account for the alteration in distance of the target from the system.

raster *Syn. for* field. *See* television.

raster scanning *See* electron-beam lithography.

rated conditions *See* rating.

rating Stipulating or the stipulation of operating conditions for a machine, transformer, or other device or circuit and stating the performance limitations of such equipment. Rating is carried out by the manufacturer of such equipment. The designated limits to the operating conditions within which the device or equipment functions satisfactorily are the *rated conditions* (current, load, voltage, etc.). If the rated conditions are not adhered to the device is likely not to produce its rated performance.

ratio adjuster *See* tap changer.

ratio circuit *See* MOS logic circuit.

ratioless circuit *See* MOS logic circuit.

rationalized MKS system *See* MKS system.

rat-race *Colloq. syn. for* ring junction. *See* hybrid junction.

RBS *Abbrev. for* Rutherford back scattering.

R-C *Abbrev. for* resistance-capacitance. It is used as a prefix to describe circuits or devices that depend critically for their operation on their resistance and capacitance or that employ resistance-capacitance *coupling.

R-C network *Short for* resistance-capacitance network. *See* network.

reactance Symbol: X; unit: ohm. The part of the total *impedance of a circuit not due to pure *resistance. It is the imaginary part of the complex impedance, Z, i.e.

$$Z = R + \mathrm{i}X$$

where R is the resistance and i is equal to $\sqrt{-1}$. Reactance is due to the presence of *capacitance or *inductance in a circuit. The effect of reactance is to cause the voltage and current to become out of *phase.

   If an alternating voltage, given by

$$V = V_0 \cos\omega t$$

where ω is the angular frequency, is applied to a circuit containing capacitance the impedance of the circuit is given by

$$Z = R - \mathrm{i}/\omega C$$

472

where $1/\omega C$ is the *capacitive reactance,* X_C, which decreases with frequency. The current *leads the voltage: the phase angle is 90° in a purely capacitive circuit.

In a circuit containing inductance the impedance is given by

$$Z = R + i\omega L$$

where ωL is the *inductive reactance,* X_L, which increases with frequency. The current *lags the voltage: the phase angle is 90° in a purely inductive circuit.

reactance chart A chart that is presented in a form that enables the user to read directly the *reactance of any given *capacitor or *inductor at a specified frequency, and conversely to deduce the capacitance or inductance of a given reactance at a particular frequency.

reactance coil *See* inductor.

reactance drop *See* voltage drop.

reactance transformer A device that consists of pure reactances arranged in a suitable circuit and commonly used for impedance matching at radiofrequencies.

reactivation A process used with a thoriated tungsten filament cathode in order to improve the emission of electrons from the surface. An abnormally high voltage is applied to the filament and this causes a layer of thorium atoms to migrate to the surface.

reactive current *Syns.* reactive component, wattless component, idle component, quadrature component of the current. The component of an alternating current vector that is in *quadrature with the voltage vector.

reactive factor The ratio of the *reactive volt-amperes of any *load, circuit, or device to the total *volt-amperes.

reactive ion beam etching *See* etching.

reactive ion etching *See* etching.

reactive load A *load in which the current and voltage at the terminals are out of phase with each other. *Compare* nonreactive load.

reactive power *See* power.

reactive voltage *Syns.* reactive component, wattless component, idle component, quadrature component of the voltage. The component of an alternating voltage vector that is in *quadrature with the current vector.

reactive volt-amperes *Syns.* reactive component, wattless component, idle component, quadrature component of the volt-amperes. The product of the current and the *reactive voltage or the product of the voltage and the *reactive current. *Compare* active volt-amperes.

reactor A device or apparatus, particularly a capacitor or inductor, that possesses *reactance and is used because of that property.

read (1) To generate an output signal from a charge *storage tube that corresponds to the information stored in the tube.

(2) To remove information from a computer storage device or memory. *Destructive read operation* (DRO) is a read operation that leaves no information in the storage device. If the information is to be preserved DRO must be immediately followed by a write operation to restore it. *Nondestructive read operation* does not destroy the information. The type of read operation depends on the nature of the memory.

read-around number The number of times a particular location of a charge *storage tube may be consulted before information is lost from surrounding locations due to *spill.

Read diode *See* IMPATT diode.

reading beam *See* storage tube.

read-mostly memory (RMM) *See* read-only memory.

read-only memory (ROM) *Syn.* fixed store. A *memory that retains information permanently and in which the stored information cannot be altered by a program or normal operation of a computer. ROMs are therefore used to store the control programs in pocket calculators and commonly used *microprograms in large computer systems.

Information may be placed in the storage array during manufacture of the memory – so called '*hardwiring*' – and the two possible binary states of each memory location are determined by the physical construction of the device. ROM is commonly formed by replacing the storage capacitors of a *solid-state (random-access) memory by either open circuits or connections to earth. A form of ROM in which the information may be placed in the array by the user rather than during the initial manufacture is a *fusible-link ROM*. Each memory cell is provided with a fusible link to earth; information is placed in the array by applying a particular pattern of electrical impulses to the array that is strong enough to blow the fuses at locations where open circuits are required. Once the pattern has been formed it is retained permanently. The fusible-link memory is a form of *programmable read-only memory* (PROM) but can be programmed only once.

For some applications *read-mostly memories* (RMMs) are required. In this case the stored information is altered very infrequently and the write time can be very much greater than the access time (*see* memory). RMMs are used when long-term nonvolatile storage of information is required. A form of PROM is used to form a read-mostly memory. The most common type is the *floating-gate PROM*. In this type of PROM the basic memory cell is an *MOS transistor that has two gate electrodes, one above the other, separated by a layer of silicon dioxide (*see* diagram). The lower gate is entirely surrounded by oxide and is therefore *floating. The threshold voltage that must be applied to the upper gate electrode in order to cause the MOS transistor to operate is determined by the amount

474

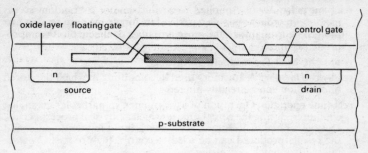

Floating-gate memory cell

of charge stored on the floating gate. Charge can be placed on the floating gate by applying a relatively high voltage, of the order of 25 volts, between the gate electrode and the drain while holding the substrate material at a low voltage. Some electrons gain sufficient energy – about four electronvolts – to cross the potential barrier of the insulating silicon dioxide and hence charge the floating gate. Silicon dioxide is an excellent insulator and the charges remain on the floating gate for up to several years. The charges can be removed from the gates by optical erasure: exposure to ultraviolet radiation causes the silicon dioxide to become sufficiently conductive to allow the stored charges to leak away. The entire memory chip is exposed and then recharged with a complete charge pattern. The small size of individual cells makes individual erasure impractical.

The *electrically alterable read-only memory* (EAROM) is a form of RMM that does not require exposure to ultraviolet light. The most usual type of EAROM contains *MNOS devices:* each individual memory cell consists of an MOS transistor with an overlying signal gate electrode of similar structure to the gate of the optically erasable ROM. The insulating layer between the substrate and central portion of the signal gate is formed from two different insulating materials, usually silicon dioxide and silicon nitride, and charges are stored at the interface between these materials. Charge patterns are placed in the cells in a similar manner to the floating-gate memory and may also be discharged by signals applied to the gate. Selective erasure of cells can be achieved.

An alternative form of EAROM is formed from an amorphous semiconductor or semiconducting glass that retains a pattern of local charges. Erasure is achieved by a strong current pulse that removes the entire charge pattern, ready for a new pattern to be produced electrically.

read operation *See* read.

475

read-out pulse A pulse applied to a word line of a *random-access memory in order to enable a particular storage location on that line to become available to the sense amplifier connected to the appropriate bit line.

read-write head *See* magnetic recording.

read-write memory A *memory used in computing in which the stored information can be readily altered.

real-time operation Operation of a *computer, in particular an analog computer, during the actual time in which a physical process occurs. Data generated by the physical process is input to the computer and the results produced can be used to control the process. *Compare* interactive; batch processing.

receiver The part of a telecommunication system that converts transmitted waves into a desired form of output. The range of frequencies over which a receiver operates with a selected perfomance, i.e. a known sensitivity, is the *bandwidth* of the receiver. The process of limiting reception to a desired portion of a cycle of operation is termed *gating*. The *minimum discernible signal is the smallest value of input power that produces an output. *See* radio receiver; television; superheterodyne reception.

receiving aerial *See* aerial.

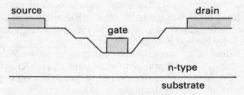

Recessed gate FET

recessed gate FET A power FET (*field-effect transistor) that has the gate electrode formed in a slot etched in the substrate between the source and drain electrodes (*see* diagram). The use of the same mask for etching the slot and depositing the metal to form the gate electrode results in the gate metal being placed in the centre of the slot. This is known as a *self-aligned* recessed gate. Use of a recessed gate structure has the advantage that the extra channel thickness on each side of the gate reduces parasitic resistances between the gate and the source and drain; the position of the gate below the substrate surface does not restrict the ability of the gate to modulate the source-drain current under positive gate bias despite the shrinkage of the depletion region that occurs under positive bias. The disad-

476

vantage is that the gate-to-drain feedback is increased, which reduces the gain of the FET.

reciprocal lattice *See* energy bands.

reciprocal theorem If an electromotive force, E, applied at a point in a *network produces a current, I, at a second point, then if the same e.m.f. is applied at the second point it will produce the same current at the original point.

recombination processes Various processes by which excess electrons and holes in a semiconductor recombine and tend to restore the system to the thermal equilibrium condition given by

$$pn = n_i^2$$

p is the number of holes, n the number of electrons, and n_i the number of holes or electrons in the intrinsic semiconductor at the same temperature.

The basic recombination processes are *band-to-band recombination,* when an electron in the conduction band recombines with a hole in the valence band, and *trapping recombination,* when *electron* or *hole capture* by a suitable acceptor or donor impurity occurs in the semiconductor (*see* diagram).

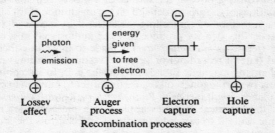

Recombination processes

The energy lost by the conduction electron involved in band-to-band recombination may be emitted as a photon of radiation (*Lossev effect*) or may be transferred as kinetic energy to a free electron or hole (*Auger process*). The Lossev effect is the inverse process to *photoconductivity and forms a significant proportion of the total recombination in *direct-gap semiconductors. The Auger process is the inverse process to *impact ionization.

recombination rate The rate at which recombination of electrons and holes in a *semiconductor occurs.

recombination velocity *See* semiconductor.

record Any permanent or semipermanent presentation of electrical data, in particular a gramophone record.

recorder *See* graphic instrument.

477

recording The process of making a permanent or semipermanent record of electrical data of various types, particularly *recording of sound.

recording channel (1) One of the independent recorders in a recording system that uses two or more recorders, as in stereophonic recording.

(2) An independent track on a recording medium, such as magnetic tape, that can accomodate two or more tracks.

recording head *See* magnetic recording.

recording instrument *See* graphic instrument.

recording of sound The process of producing a permanent or semipermanent record of sounds that may be used in suitable replay or reproducing apparatus (*see* reproduction of sound) in order to reproduce the original sounds. The three main methods of sound recording are *electromechanical sound recording, photosensitive recording* (or *sound-on-film*), and *magnetic recording.

Electromechanical sound recording is used to produce gramophone records. The sound to be recorded is detected with one or more suitable microphones and the signals are amplified. The amplified signals are then used to drive a suitable *cutter* that produces an undulating groove in the surface of a wax or cellulose disc. The undulations of the groove depend on the magnitude of the signals. A common method of controlling the cutter is by means of a varying magnetic flux density produced by the audiosignal in a coil. The cutter (termed a *magnetic cutter*) is made to move in either a horizontal (lateral recording) or vertical (vertical recording) plane by the changing magnetic flux density of the coil. In stereophonic recording two orthogonal coils cause the cutter to move in two orthogonal planes simultaneously. *Piezoelectric* (or *crystal*) *cutters* are also used: the electrical impulses applied to a piezoelectric crystal or ceramic cause it to expand or contract and controls the movement of the cutter. In *compact disc systems the sound to be recorded is coded using *pulse code modulation, and the resulting coded signals are used to cut a series of tiny bumps on the wax or cellulose disc.

Gramophone records are mainly used in commercial mass-production for home reproduction or for broadcast entertainment. Many copies of the original recording can be made comparatively easily. Once the recording has been made on to a wax or cellulose disc, the disc is plated with copper to form a negative *master*. A copper plated positive *mother* is usually produced from the master before mass production so that further copies of the master may be made as required. Commercial production is easily effected by pressing suitable plastic pieces – biscuits – with the master to produce any number of positive discs. Mass production of compact discs is effected by pressing in a similar manner.

Photosensitive or sound-on-film recording is used mainly in the cinema or for broadcast entertainment when film is used. It may also be used for the analysis of sounds when an expanded time-scale is employed. This method allows recordings to be made of much longer duration than on disc. The amplified audiosignals are used to modulate a light source, to which the recording film is exposed. The film is then developed and the record appears as a strip of varying density known as the *soundtrack*. The frequency of the light source is usually such that for convenience a *carrier wave is commonly used that is either amplitude-modulated or frequency-modulated by the audiosignal. *White recording* is a recording system in which the minimum density of the developed film corresponds to the maximum received power of an amplitude-modulated signal or the lowest received frequency of a frequency-modulated system. In *black recording* the maximum density corresponds to the above parameters.

Magnetic recording is widely used for a variety of modern applications. The overall performance is as good as, or better than, the above two systems and it is very convenient in use. It is replacing gramophone recording for many functions.

A sound recording and reproduction system is required to produce sounds at one place that are as nearly as possible a faithful reproduction of the original, at a sufficient volume to give a sound field to the hearer approximating in amplitude to that which the hearer would have obtained from the original source. The acoustics of the reproducing room and the limitations of the reproduction equipment generally militate against these conditions and they are rarely satisfied in practice. Systems that most nearly satisfy these requirements are described as *high-fidelity* (hi-fi).

recording wattmeter *See* watt-hour meter.

record player *See* gramophone.

rectangular pulse *See* pulse.

rectification efficiency *See* rectifier.

rectifier A device that passes current only in the forward direction and can therefore be used as an a.c. to d.c. converter. A single device usually suppresses or attenuates alternate half-cycles of the alternating-current input (*see* half-wave rectifier circuit). *Full-wave rectifiers usually contain two devices in a back-to-back arrangement, although with one type of *mercury-arc rectifier the single device can be used. The most common type of rectifier is the semiconductor *diode; other types include *mercury-vapour rectifiers, *metal rectifiers, and mechanical devices such as a commutator. The *rectification efficiency* of any rectifier is the ratio of the direct-current output power to the alternating-current input power.

The output of a rectifier consists of a unidirectional current that rises to a maximum value periodically; this value corresponds to the peak value of the alternating-current input. The output is usually smoothed by a *smoothing circuit before being applied to the load in order to reduce the amount of *ripple. The fluctuating output can be considered as a steady d.c. component with an a.c. component superimposed on it. A portion of the ripple is sometimes due to the *rectifier leakage current*, i.e. alternating current that flows through the rectifier without being rectified.

rectifier filter *See* smoothing circuit.

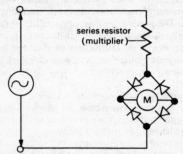

series resistor
(multiplier)

M

Rectifier instrument used as an a.c. voltmeter

rectifier instrument A d.c. instrument that can be made suitable for a.c. measurements by using a *rectifier to convert the alternating current to be measured into a unidirectional current. A common arrangement is a *bridge circuit formed from four metal or semiconductor *diodes and a moving-coil instrument, M (*see* diagram). The indicated value on the instrument is usually the *root-mean-square value of a sinusoidal a.c. input waveform; rectifier instruments are subject to waveform error if used with non-sinusoidal input waveforms.

rectifier leakage current *See* rectifier.

rectifier photocell *Syns.* barrier-layer photocell; blocking-layer photocell; photronic cell. *See* photovoltaic cell.

rectifier voltmeter A voltmeter containing a *rectifier circuit in the input that converts alternating voltage into an essentially unidirectional voltage, which is then measured.

rectilinear scanner *See* scanner.

rectilinear scanning *See* scanning.

recurrent-surge oscilloscope An instrument used to investigate electrical *surges. A *surge generator is used in conjunction with a *cathode-ray oscilloscope (CRO). The repetition rate of the surges produced by the surge generator is made to synchronize with the time base of the CRO so that a steady picture is obtained on the screen, suitable for visual or photographic inspection.

red gun *See* colour picture tube.

reduction factor *See* tangent galvanometer.

redundancy (1) The fraction of the information in a transmission system that may be eliminated without loss of essential information. The excess information is often deliberately included to allow for loss in the transmission system.

(2) The extra components, devices, or circuits included in an electronic circuit or apparatus to increase the reliability of the system. If therefore a fault develops in one portion of the system the redundant circuits or components provided can take over the function of the faulty part. Redundancy is very important in systems, such as *communications satellites, in which high reliability is essential.

(3) *See* compact disc system.

reed relay *See* relay.

Reed-Solomon code *See* compact disc system.

reflected current *Syn. for* return current. *See* reflection coefficient.

reflected impedance *See* transformer.

reflected power Power that is returned from the *load back to a generator.

reflected wave (1) *See* travelling wave. (2) *See* indirect wave.

reflection coefficient *Syn.* return-current coefficient. The vector ratio of the *return current*, I_R, to the *incident current*, I_0, when a *transmission line is incorrectly terminated with an impedance, Z_R, not equal to the characteristic impedance, Z_0, of the line. The incident current is that current flowing in the line at termination when Z_R is made equal to Z_0; the return current is the portion of the current flowing back along the line due to incorrect matching. The actual current in the line at termination is the vector sum of I_R and I_0. In this case the reflection coefficient may be expressed in terms of the impedances Z_0 and Z_R, i.e. the reflection coefficient can be given by

$$(Z_0 - Z_R)/(Z_0 + Z_R)$$

Compare reflection factor.

reflection error An error arising in radionavigation systems, *direction finding, and *radar systems due to undesired reflections of the transmitted energy.

reflection factor The vector ratio of the current, I, delivered to a load of impedance Z_B by a source of impedance Z_A when the impedances are not matched, to the current, I_0, delivered to the load when the impedances are matched by an impedance matching network of *image transfer constant, θ, equal to zero (*see* diagram). The reflection factor is thus the ratio of the current delivered to an unmatched load to the current that would be delivered to a perfectly matched load. It is given by

$$I/I_0 = \sqrt{(4Z_A Z_B)/(Z_A + Z_B)}$$

The ratio of the electrical powers in Z_B for these matched and unmatched cases is also given by I/I_0. The *reflection loss* between Z_A and Z_B is the ratio of the powers in Z_B expressed in decibels, i.e. it is given by

$$10 \log_{10}(I/I_0)$$

If the reflection loss is negative it represents a reflection gain in the system. *Compare* reflection coefficient.

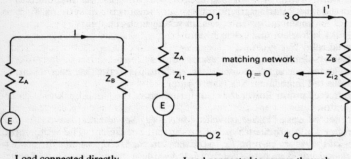

Load connected directly to source

Load connected to source through matching network

reflection loss *See* reflection factor.

reflector *See* directive aerial.

reflector electrode *See* klystron.

reflex bunching *See* velocity modulation.

reflex circuit A circuit that is used to amplify a signal at one frequency and also after it has been converted to a second frequency.

reflex klystron *See* klystron.

refresh *Syns.* rewrite; regenerate. To restore the condition of a memory cell in a computer to its original state in order to maintain the integrity of the information stored in it. A refresh is required following destructive read operation (*see* read). In dynamic memories periodic refresh must be carried out to prevent loss of information during standby intervals.

regenerate *See* refresh.

regeneration *Syn. for* positive feedback. *See* feedback.

regenerative receiver A *radio receiver used with amplitude-modulated radiowaves in which positive *feedback is used in order to increase the sensitivity and *selectivity of the receiver by reducing the damping. *See also* super-regenerative reception.

register *Syn.* accumulator. One of a number of word-sized locations (*see* word) in the *central processing unit of a computer in which the arithmetic and logical operations required by a *program are performed on data obtained from *memory, *input/output devices, or other registers.

regulator An electronic device that is used to maintain the voltage (*voltage regulator*) or current (*current regulator*) constant at a given point in a circuit or to vary it in a controlled manner. Regulators are used to control the output of a load or to regulate the voltage or current of an electronic device, despite fluctuations in the circuit conditions, particularly variations in the supply voltage.

Reicz microphone *See* carbon microphone.

reignition voltage The voltage required to re-establish the discharge in a *gas-discharge tube after the tube has ceased conducting. The reignition voltage is often applied during the period of deionization of the tube, i.e. the interval during which the ions in the tube recombine to form gas molecules, and can therefore be less than the voltage required to initiate the first discharge.

rejection band A band of frequencies that are attenuated or suppressed by a *transducer or *filter. In the latter case the term *attenuation band* is more usually used.

rejector *Syn. for* parallel resonant circuit. *See* resonant circuit; resonant frequency.

relative permeability *See* permeability.

relative permittivity *See* permittivity.

relaxation oscillator An oscillator in which one or more voltages or currents change suddenly at least once during each cycle. The circuit is arranged so that during each cycle energy is stored in and then discharged from a reactive element (e.g. a capacitor or induc-

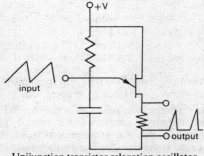

Unijunction transistor relaxation oscillator

483

tance), the two processes occupying very different time intervals. An oscillator of this type has an asymmetrical output waveform that is far from being sinusoidal.

A commonly produced output waveform is a *sawtooth waveform; square or triangular waveforms can be easily produced when required by means of a suitable circuit. Sawtooth waveforms are particularly useful as the internal *time base of a *cathode-ray tube.

The output waveform is very rich in harmonics and for some purposes this is particularly useful. Common types of relaxation oscillator include the *multivibrator and *unijunction transistor (*see* diagram) but many other circuit arrangements are possible.

relaxation time (1) Symbol: τ. The time interval during which the *dielectric polarization of a point in a dielectric falls to $1/e$ of its original value due to the electric conductivity of the dielectric.

(2) *See* MOS capacitor.

relay An electrical device in which one electrical phenomenon (current, voltage, etc.) controls the switching on or off of an independent electrical phenomenon. There are many types of relay, most of which are either *electromagnetic* or *solid-state relays*.

The *armature relay* is an electromagnetic relay in which a coil wound on a soft-iron core attracts a pivoted armature that operates contacts or tilts a mercury switch (Fig. *a*). There are several different designs for the armature: *differential relays* have two coils and only operate when the currents in the coils are additive not subtractive. The armature may be split so that a small section of the metal operates with small currents independently of the main contacts, which require large currents to move the whole armature; *polarized relays* have a central permanently magnetized core and operate differently with currents in different directions.

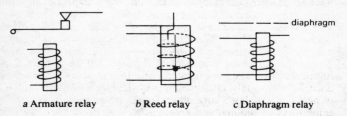

a **Armature relay**   *b* **Reed relay**   *c* **Diaphragm relay**

The electromagnetic *reed relay* has a coil wound around a glass envelope containing fixed contacts and a centrally placed reed contact in the form of a thin flat metal strip (Fig. *b*). When the coil is energized the reed is deflected and can either make or break contact with the fixed leads.

The electromagnetic *diaphragm relay* has a coil wound around a central core with a thin metal diaphragm plate mounted close to its end (Fig. *c*). When the coil is energized the central portion of the diaphragm moves towards the core and makes contact with it.

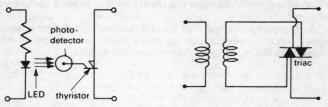

d LED-coupled solid-state relay   *e* Transformer-coupled solid-state relay

A true solid-state relay has all its components made from solid-state devices and involves no mechanical movement. Isolation between input and output terminals is provided using a *light-emitting diode (LED) in conjunction with a *photodetector. The switching is achieved using a *silicon-controlled rectifier (SCR) or more commonly two SCRs (a triac). This type of relay is compatible with *digital circuitry and has a wide variety of uses with such circuits. The relay cannot normally be formed on a single chip since the LED is usually formed in gallium arsenide and the photodetector in silicon. Isolation may also be achieved by transformer-coupling on the input. Again a single chip may not be used. Examples of solid-state relays are shown in Figs. *d* and *e*.

Solid-state relays have advantages over electromechanical relays because of increased lifetime, particularly at a high rate of switching, decreased electrical noise, compatibility with digital circuitry, and ability to be used in explosive environments since there are no contacts across which *arcs can form; the lack of physical contacts and moving elements also gives increased resistance to corrosion. No mechanical noise is associated with them. This is particularly important for certain applications where noise could be an annoyance, as in hospitals. Disadvantages include the substantial amount of heat generated at a current above several amperes, necessitating some form of cooling, and greatly increased production costs for multipole devices compared to single pole devices; in certain applications a physical disconnection may be required for safety purposes and this is not available in solid-state relays.

Other types of relay include thermionically operated relays in which the heating effect of a current is used to operate contacts or the effect of a heating coil on a bimetallic strip is employed. Gas-

485

filled relays, such as the *thyratron, have also been used but these are being superseded by solid-state relays using SCRs.

Relays are frequently described by the electrical parameter that causes them to operate or according to their function. *Current relays, voltage relays,* and *power relays* operate when a predetermined value of current, voltage, or electrical power is applied to the input circuit.

A *frequency relay* operates when a predetermined change in the input signal frequency occurs; *frequency-selective relays* operate only when the input signal has a preselected value.

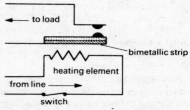

f **Relay incorporating a thermal time delay**

A *locking relay* is used to render a circuit, device, or other apparatus inoperative under particular conditions, especially fault conditions. A *self-locking relay* is one that, having been operated, remains in the operated state even when the energizing input has ceased. Such a relay requires a reset signal to restore the original state of the contacts.

A *slow-operating relay* has an intentional delay between the energizing input and operation of the contacts; it includes some form of time-delay mechanism to achieve this. An example is shown in Fig. *f*.

Relay *See* communications satellite.

reliability The ability of any device, component, or circuit to perform a required function under stated conditions for a stated period of time. This may be expressed as a probability. ('Time' may be considered as distance, cycles, or other appropriate units.) *Reliability characteristics* are those quantities used to express reliability in numerical terms. Electronic items are usually approached from the viewpoint of *failure, because of their extremely high reliability.

reluctance *Syn.* magnetic resistance. Symbol: R; unit: henry^{-1}. The ratio of the magnetomotive force, F_m, to the total magnetic flux, Φ,

$$R = F_m/\Phi$$

reluctivity The reciprocal of magnetic *permeability.

remanence *Syns.* retentivity; residual magnetism. *See* magnetic hysteresis.

remote control A system of electrical or electronic control that has the controlling equipment located at a distance from the controlled equipment.

remote job entry *See* batch processing.

repeater A device that receives signals in one circuit and automatically delivers corresponding signals to one or more other circuits. A repeater is most often used with telephonic or telegraph circuits, usually amplifying the signal, and in pulse telegraphy performs pulse regeneration on the transmitted pulses. Repeaters can operate either on signals in one direction only or on two-way signals; telephone repeaters operate on four-wire circuits or two-wire circuits. A *terminal repeater* is a repeater that is used at the end of a trunk feeder or transmission line.

repeating coil An audiofrequency transformer that is used to couple two sections of telephone line.

replay head *See* magnetic recording.

reproduction of sound The reproduction of sound information from a source of audiofrequency electrical signals. A complete sound reproduction system contains the original source of audio information, preamplifier and control circuits, audiofrequency power amplifier(s), and loudspeaker(s). The source of sound may be a compact disc, gramophone record, magnetic tape, a broadcast transmission, or a sound-on-film recording. The last system is not normally found in domestic sound reproduction systems.

Monophonic sound reproduction uses only a single audiofrequency channel. One or more loudspeakers may be used in parallel at the output. *Stereophonic sound reproduction* uses two channels to carry the audio information and at least two loudspeakers. Stereophony utilizes the ability of the human brain to detect the direction of sounds and to some extent to discriminate between wanted and unwanted sounds by assessing the time taken by the sound to reach each ear and to detect the difference in volume and phase caused by the screening effect of the head and ears.

Stereophonic signals require at least two microphones. A coincident pair of matched directional microphones may be used at the same location or a pair of separated microphones may be used. In the latter case the outputs are divided in the correct proportion between the two channels by means of a suitable potential divider – a panoramic potentiometer. In order to provide a signal that is compatible with monophonic sound reproduction the stereophonic signals are combined into sum and difference signals. If the two channels are A and B, the sum $(A + B)$ and the difference $(A - B)$ of the signals are used. A monophonic system produces an output using only the $(A + B)$ signal. The stereo system combines them to produce two signals that correspond to the original A and B information. In *stereophonic broadcast transmission* the sum signal is used to modulate the main carrier and the difference signal is used to modulate a subcarrier, separated from the main carrier frequency.

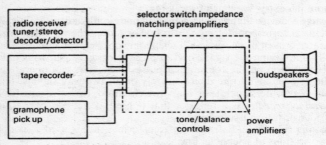

Stereophonic sound reproduction system

A sound-reproduction system that is composed of high-quality expensive parts and reproduces the original audio information faithfully and with very low noise levels is referred to as a *high-fidelity* (hi-fi) system. In such a system several different inputs may be available, which share the use of the power amplifiers and loudspeakers. Domestic systems usually have facilities for reproducing gramophone records, tape recordings, and broadcast radio transmissions. Suitable impedance matching circuits are required for each input. The system may be modular, i.e. each unit is separately boxed and interconnections made between units by means of wires and plugs; alternatively all the units may be combined in a single housing, an arrangement sometimes referred to as a *music centre*. A block diagram of a complete stereophonic system is shown.

reset To restore an electrical or electronic device or apparatus to its original state following operation of the equipment. *See also* clear.

residual charge The portion of the charge stored in a capacitor that is retained when the capacitor is discharged rapidly and may be withdrawn from it subsequently. It results from viscous movement of the dielectric under charge causing some of the charge to penetrate the dielectric and hence become relatively remote from the plates. Only the charge near to the plates is removed by rapid discharge.

residual current Current that flows for a short time in the external circuit of an active electronic device after the power supply to the device has been switched off. The residual current results from the finite velocity of the charge carriers passing through the device.

residual induction *See* magnetic hysteresis.

residual magnetism *Syn. for* remanence. *See* magnetic hysteresis.

residual resistance The inherent resistance of a conductor that is independent of temperature variations. It is usually ascribed to irregularities in the molecular structure of the material.

resist An energy-sensitive material used in *lithography. Resists are applied to the substrate material as a thin film and selectively exposed to an energy beam (light, electrons, etc.), which causes chemical changes in portions of the resist. The exposed film is then developed to selectively remove either the exposed portions (positive resist) or the unexposed portions (negative resist). *See also* photoresist; electron-beam lithography; X-ray lithography; ion-beam lithography.

resistance (1) Symbol: *R;* unit: ohm. The tendency of a material to resist the passage of an electric current and to convert electrical energy into heat energy. It is the ratio of the applied potential difference across a conductor to the current flowing through it (*see* Ohm's law). If the current is an alternating current the resistance is the real part – the *resistive component* – of the electrical *impedance, *Z:*

$$Z = R + \mathrm{i}X$$

where i is equal to $\sqrt{-1}$ and X is the *reactance.

(2) *See* resistor.

resistance brazing Electrical brazing that utilizes the heating effect of a current to provide the heat source.

resistance-capacitance coupling (RC coupling) *See* coupling.

resistance coupling *See* direct coupling.

resistance drop *See* voltage drop.

resistance furnace A furnace that utilizes the heating effect of a current in order to provide the heat source.

resistance gauge A gauge that is used to measure high fluid pressures by measuring the change in electrical *resistance of a sample of manganin or mercury when the sample is subjected to the pressure. The gauge must be calibrated against known pressures before use.

resistance lamp An electric lamp that is used to limit the current flowing in a circuit.

resistance strain gauge *See* strain gauge.

resistance thermometer An electrical thermometer that utilizes the change in electrical *resistance with temperature (*see* temperature coefficient of resistance) of a wire to measure the temperature of its surroundings.

It consists of a small coil of wire (usually platinum but other metals may be used at low temperatures) wound on a mica former and enclosed in a sheath of silica or porcelain (*see* diagram). The change in resistance is determined by placing the coil in one arm of a *Wheatstone bridge. Compensating leads are usually added to the other arm of the bridge to compensate for temperature variations in the leads since the coil is usually remote from the measuring instru-

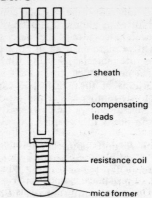

sheath

compensating
leads

resistance coil

mica former

Resistance thermometer

ment. Resistance thermometers can be used over a wide range of temperatures from −200°C to over 1200°C.

resistance welding Electrical welding in which the heating effect of a current is utilized to provide the heat source. An electric current is passed across the contact surface between the components to be welded and local melting of the metals occurs. The components are subjected to external pressure to maintain the contact between the surfaces. The current is usually applied across the surface in controlled short bursts.

resistance wire Wire constructed from a material, such as nichrome or constantan, that has a high resistivity and low temperature coefficient of resistance. It is used for accurate wire-wound resistors.

resistive component (of a complex *impedance) *See* resistance.

resistive coupling *Syn. for* resistance coupling. *See* coupling.

resistivity *Syn.* volume resistivity. Symbol: ρ; unit: ohm metres. An intrinsic property of a material equal to the resistance per metre of material with cross-sectional area of one square metre:

$$\rho = RA/L$$

where R is the resistance, A the cross-sectional area, and L the length. Resistivity depends only on the nature of the material whereas resistance depends not only on the material but on its length and cross-sectional area.

Resistivity is the reciprocal of *conductivity: the lower the resistivity of a material the better conductor it is. Materials can be classified as conductors, semiconductors, or insulators according to their resistivities (*see* table). In semiconductors, the higher the *doping

Material	Resistivity (ohm metres)
conductors	$10^{-8} - 10^{-6}$
semiconductors	$10^{-6} - 10^{-7}$
insulators	$10^{7} - 10^{23}$

level the lower the resistivity. *See also* mass resistivity; surface resistivity.

resistor *Syn.* resistance. An electronic device that posseses *resistance and is selected for use because of that property. There are several different types of resistor in common use; the type of resistor chosen depends on the particular application for which it is designed. The three main types of resistance element are *carbon, wire-wound,* and *film resistors. These may be produced as fixed-value resistors or adjustable resistors.

Carbon resistors consist of finely ground particles of carbon mixed with a ceramic material and encapsulated in an insulating material. The encapsulation has a set of coloured stripes or dots – the *colour code – denoting the value of the resistance (*see* Table 2, backmatter). Carbon resistors are compact, robust, and relatively cheap to manufacture and are widely used in electronic circuits in which the resistance value is not critical. The value of the resistance however is a function of the operating voltage and of temperature and close tolerances cannot be maintained over a wide range of load and ambient conditions. When used at power levels above one megohm the level of thermal *noise becomes high, precluding their use at high levels of power. The relatively short length of these resistors causes them to have a noticeable shunted capacitance and therefore at high (VHF and higher) operating frequencies the effective resistance is reduced as a result of the dielectric losses involved (*see* Boella effect).

Adjustable carbon composition resistors may be formed on an insulating base or moulded at a high temperature on to a moulded plastic base. The composition is formed with a linear rotating contact or it may be tapered to produce a nonlinear characteristic. The resistance change is continuous. These resistors, particularly the thinner ones, tend to be noisy and are subject to mechanical wear in frequent use.

Wire-wound resistors are formed from a wire of uniform cross section wound on a suitable former. The value of resistance may be determined very accurately and for uses where the resistance value is critical, wire-wound resistors are usually preferred. Wire-wound resistors however are unsuited to use above 50 kilohertz as they have marked inductive and capacitive effects even when specially wound

491

(*see* noninductive). At high frequencies the *skin effect causes an increase in the effective resistance.

Adjustable wire-wound resistors are produced in a wide range of values and types. Linear or circular types are made with single-turn or multiturn contacts. By tapering or producing other shapes of the former the contact rotation versus resistance curve may be altered to generate various functions: logarithmic, sine wave, or other characteristics are available. Wire-wound resistors almost invariably change resistance values in steps. This can sometimes provide an unwanted pulse in the output. The motion of the contact across the turns also tends to generate noise.

Resistive *film resistors are the most suitable type of resistor for high-frequency applications since the inductance, even in high-value spiral films, is very much less than with wire-wound resistors. Special high-frequency film resistors with very low values of inductance are available.

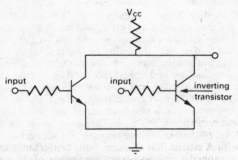

RTL two-input NOR gate

resistor-transistor logic (RTL) A family of integrated *logic circuits in which the input is through a resistor into the *base of an inverting transistor. The basic NOR gate is shown. The output is high (corresponding to a logical 1) only if both the inputs are low (corresponding to a logical 0). If either input is high the transistor conducts and saturates and the voltage at the output is low. RTL circuits tend to be slow low *fanout circuits that are susceptible to noise but the power dissipated is low compared to emitter-coupled, diode-transistor, and transistor-transistor logic circuits. RTL is of interest because it was the first integrated logic family to be developed. It is now little used.

resolving time *See* counter.

resonance A condition existing when an oscillatory circuit responds with maximum amplitude to a periodic driving force so that a relatively small amplitude of the driving force produces a large amplitude of oscillation. Resonance is achieved when the frequency of the driving force coincides with the natural undamped frequency of the oscillatory system. *See also* forced oscillations; resonant frequency; tuned circuit.

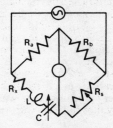

Resonance bridge

resonance bridge A four-arm bridge that has a tuned circuit in one of the arms (*see* diagram). The bridge will only balance at the resonant frequency of the circuit and at balance the frequency and resistance of the circuit are given by

$$\omega^2 LC = 1$$
$$R_x = R_s R_a / R_b$$

where ω is the angular frequency.

resonant cavity *See* cavity resonator.

resonant circuit A circuit that contains both *inductance and *capacitance so arranged that the circuit is capable of *resonance. The frequency at which resonance occurs – the *resonant frequency – depends on the value of the circuit elements and their arrangement.

A *series resonant circuit* contains the inductance and capacitance in series. Resonance occurs at the minimum combined impedance of the circuit and a very large current is produced at the resonant frequency; the circuit is said to be an *acceptor* at that frequency. A *parallel resonant circuit* (or *rejector*) contains the circuit elements arranged in parallel. Resonance occurs at or near the maximum combined impedance of the circuit (*see* resonant frequency). The overall current in the circuit is a minimum at the resonant frequency, a voltage maximum being produced; the circuit is said to reject that frequency. *See also* tuned circuit. *Compare* aperiodic circuit.

resonant frequency Symbols: ω_0, ν_0. The frequency at which *resonance occurs in a particular circuit or network. Resonance occurs in a circuit containing both capacitance and inductance when the imaginary component of the complex combined impedance of the circuit is zero, i.e. when the supply current and voltage are in *phase and the circuit has unit power factor.

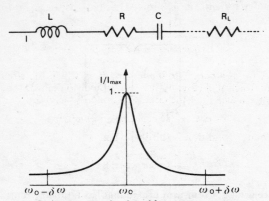

a Series resonant circuit and frequency response

In a *series resonant circuit,* which contains the capacitive and inductive elements in series (Fig. *a*), the combined impedance, *Z,* is given by

$$Z = R + i\omega L - i/\omega C$$

where R is the ohmic resistance, ω the angular frequency ($\omega = 2\pi \times$ frequency), L the inductance and C the capacitance, and i equals $\sqrt{-1}$. The resonance condition is fulfilled when

$$\omega_0 L = 1/\omega_0 C$$

i.e. when

$$\omega_0 = 1/\sqrt{(LC)}$$

In a series resonant circuit resonance occurs therefore when the combined impedance is purely resistive and is a minimum. The resistance can be low even for large values of L and C. In this case the current flowing through the circuit will be high, and although large voltages are developed across the individual elements these are out of phase with each other so that the total voltage developed across the circuit is relatively low; maximum current will then flow in a load resistor, R_L, in series with the circuit.

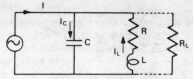

b Parallel resonant circuit

In the case of a *parallel resonant circuit,* in which the circuit elements are in parallel (Fig. *b*), it is convenient to consider the combined admittance, Y, of the circuit ($Y = Z^{-1}$), given by

$$Y = i\omega C + (R - i\omega L)/(R^2 + \omega^2 L^2)$$

The resonance condition is fulfilled when

$$R^2 + \omega_0^2 L^2 = L/C$$

i.e. when

$$\omega_0^2 = [1 - (R^2 C/L)][1/LC]$$

Since the term $R^2 C/L$ is usually very small this approximates to

$$\omega_0 = 1/\sqrt{(LC)}$$

which is the value of the series resonant frequency; the term however cannot always be neglected.

In a parallel resonant circuit therefore resonance occurs when the combined admittance is low. At resonance $Z = 1/Y$ is high and is termed the *parallel resistance* of the circuit. The overall current is low but the voltage developed across the circuit, and therefore across a load resistor, R_L, in parallel with it, is high. The individual currents developed in the inductance and capacitance at resonance can be very large but are out of phase with each other, resulting in the low combined current.

The above consideration leads to a unique solution for the resonant frequency. The resonant frequency may also be defined as that frequency at which the complex impedance passes through a minimum (in a series resonant circuit) or a maximum (parallel resonant circuit). In the series resonant case the solution is the same as above but in the case of parallel resonance a unique value is not necessarily found. The resonant frequency may have slightly different values that depend on the particular circuit parameter that is varied in order to achieve resonance. The differentials of Y with respect to C, L, or ω may therefore lead to slightly different resonance conditions.

resonant line A *transmission line that exhibits *resonance at the operating frequency.

restore *See* clear.

retention time *See* storage time.

retentivity *Syn. for* remanence. *See* magnetic hysteresis.

return current *Syn.* reflected current. *See* reflection coefficient.

return-current coefficient *See* reflection coefficient.

return interval *Syn. for* flyback. *See* sawtooth wave; time base.

return stroke *See* lightning stroke.

return trace *Syn. for* flyback. *See* time base.

reverse bias *Syn.* reverse voltage. *See* reverse direction.

reverse blocking triode thyristor *See* silicon-controlled rectifier.

reverse current *See* reverse direction.

reverse direction *Syn.* inverse direction. The direction of operation of an electrical or electronic device in which the device exhibits the larger resistance. A voltage applied in the reverse direction is a *reverse bias* and the current flowing is the *reverse current.*

A device such as a semiconductor *diode or thermionic diode exhibits an extremely high reverse resistance and can therefore be used as a *rectifier or *switch. Such a device exhibits a very small reverse saturation current below *breakdown due to the movement of a few charge carriers across the p-n junction or due to *reverse emission* of electrons from the anode when it becomes negative with respect to the cathode. The *reverse recovery time* of such a device, particularly a semiconductor diode, is the time interval between instantaneous switching from a forward bias to a reverse bias and the reverse current reaching the saturation value.

reverse emission *See* reverse direction.

reverse recovery time *See* reverse direction.

reverse saturation current *See* p-n junction.

reverse voltage *Syn. for* reverse bias. *See* reverse direction.

reversible transducer *See* transducer.

rewrite *See* refresh.

r.f. *Abbrev. for* radiofrequency.

r.f. heating *See* induction heating.

RHEED *Abbrev. for* reflection high-energy electron diffraction. *See* diffraction.

rheostat A variable *resistor that may be connected, in series, into a circuit and used to alter the current flowing through the circuit. A common arrangement is a linear or circular wire-wound resistor with a sliding contact whose position can be altered. Alternatively a number of small resistors may be used with a rotary switch that selects the appropriate value. The term is usually applied to physically large devices. Small rheostats are usually called *potentiometers.

rhumbatron *See* cavity resonator.

ribbon microphone A *microphone that consists of a very thin ribbon of aluminium alloy a few millimetres wide loosely fixed in a strong magnetic flux density parallel to the plane of the strip. A sound wave incident on the ribbon causes a pressure difference to be established between the front and back edges of the ribbon and it therefore experiences a force. The resultant motion causes a corresponding e.m.f. to be induced in the ribbon (*see* electromagnetic induction).

If the acoustic path difference across the ribbon is much smaller than a quarter wavelength, the pressure on the ribbon (and hence the e.m.f.) is proportional to the particle velocity and to the frequency. If the resonant frequency of the ribbon is made smaller than the frequency of the sound waves the frequency dependence becomes negligible and the induced e.m.f. is proportional to the particle velocity. The microphone has strong directional properties since sound waves that originate in the plane of the ribbon arrive at the front and back edges in phase and therefore no resultant force is produced.

The ribbon microphone may be used to measure sound intensities. An alternating current of the same frequency as the sound wave is passed through the ribbon and the phase and amplitude are varied until the force due to the current just balances that due to the sound wave. The sound intensity is then calculated from the current amplitude and the magnetic field strength.

Rice neutralization *See* neutralization.

Richardson's equation *Syns.* Dushman's equation; Richardson–Dushman equation. The fundamental equation of *thermionic emission that relates the number of electrons emitted from a surface to the temperature of the body:

$$j = AT^2 \exp(-b/T)$$

where j is the emitted current density, A and b are constants, and T is the thermodynamic temperature. A is a function of the surface state of the material; b equals Φ/k, where Φ is the *work function and k the *Boltzmann constant.

ridged waveguide *See* waveguide.

Rieke diagram A diagram that shows the performance data of a microwave oscillator, particularly a *magnetron or *klystron. It shows the variation of power output, anode voltage, efficiency, and frequency with changes in the ratio of the maximum and minimum voltages of the standing wave and phase angle in the load at fixed typical operating conditions (as of magnetic flux density, anode current, pulse duration, pulse repetition rate, and tuner setting for tunable tubes).

The Rieke diagram is plotted on polar coordinates: the radial coordinate represents the *reflection coefficient measured in the transmission line that connects the tube to the load; the angular coordinate represents the angular distance of the minimum standing-wave voltage from a designated reference plane on the output

497

terminal. Lines of constant power output, frequency, anode voltage, efficiency and output may be drawn on the diagram.

Righi effect *See* Leduc effect.

right-hand rule (for a dynamo) *See* Fleming's rules.

ringing (1) An unwanted low-frequency resonant tone that occurs in a radio receiver due to low-frequency oscillations produced in the receiver by the received radiofrequency waves.

(2) *See* pulse.

ring junction *Syns.* hybrid ring junction; rat-race. *See* hybrid junction.

ring winding *Syns.* toroidal winding; Gramme winding. A method of fabricating a coil, particularly for a *winding in an electrical machine, in which the coil is wound on an annular *core. One side of each turn of the coil is threaded through the ring to form a toroid.

ripple (1) An alternating-current component superimposed on a direct-current component resulting in variations in the instantaneous value of a unidirectional current or voltage. The term is particularly applied to the output of a *rectifier. The frequency of the a.c. component is the *ripple frequency;* for a full-wave rectifier it is twice the frequency of the input signal.

The magnitude of the ripple is given by the ratio of the root-mean-square value of the a.c. component to the mean value of the total and is usually expressed as a percentage. This is known as the *ripple factor.* Some form of *smoothing circuit is normally used in order to reduce the amount of ripple present on the output of a rectifier, generator, etc.

(2) *See* pulse.

ripple factor *See* ripple.

ripple filter *See* smoothing circuit.

ripple frequency *See* ripple.

rise time (of a pulse) *See* pulse.

rising sun magnetron *See* magnetron.

RMM *Abbrev. for* read-mostly memory. *See* read-only memory.

r.m.s. *Abbrev. for* root mean square.

Rocky point effect *U.S. syn. for* flash arc.

rod thermistor *See* thermistor.

ROM *Abbrev. for* read-only memory.

root-mean-square (r.m.s.) value *Syns.* effective value; virtual value. The square root of the mean value of the squares of the instantaneous values of a periodically varying quantity averaged over one complete cycle. In the case of a sinusoidally varying quantity the r.m.s. value is equal to the peak value divided by $\sqrt{2}$.

rope *See* confusion reflector.

rotary encoder A form of *photoconverter that is used as an *analog/digital converter. It usually consists of a transparent disc that carries an optical pattern in a suitably coded form and is mounted on a rotating shaft. The analog signal causes the shaft to rotate so that the angular position of the shaft corresponds to the analog amplitude at any instant. The coded pattern is used to modulate the intensity of a light beam that is subsequently detected by an array of *photocells. The output of the photocells is obtained in a coded digital form.

rotating-anode tube *See* X-ray tube.

rotational energy levels *See* energy levels.

rotator *See* waveguide.

rotor *See* motor.

R-S flip-flop *See* flip-flop.

R-S-T flip-flop *See* flip-flop.

RTL *Abbrev. for* resistor-transistor logic.

ruby laser *See* laser.

rumble Unwanted low-frequency noise heard on the loudspeaker of a gramophone. It is caused by mechanical vibrations of the record adding to the signal produced from the *pick-up cartridge.

Rutherford back scattering (RBS) The scattering through 180° of an ion when it impinges on an atom, the energy E of the scattered ion being given by

$$E = [(M - m)/(M + m)]^2 E_0$$

where E_0 is the initial energy of the ion and M and m are the masses of the atom and incident ion respectively. Similar expressions are found for angles other than 180°. If the incident ion travels through a material before back scattering takes place, additional energy losses occur both incoming and outgoing. Such losses can be determined for different materials and different depths. Detection of the back-scattered ion and measurement of its energy can therefore determine both the mass and the depth of the scattering atom.

Use of a monoenergetic beam of ions and a high energy resolution detector can thus yield a picture of not only the impurity elements in a semiconductor crystal but also the perfection of the crystal structure, since back scattering also occurs from lattice defects. A depth resolution of about 10 nanometres is possible.

S

SAINT process *Short for* self-aligned implantation for n$^+$ layer technology. A method of forming *field-effect transistors for digital logic circuits on gallium arsenide slices. The process was devised to allow annealing of the slice to take place before the ohmic source and drain contacts are formed and the Schottky barrier gate electrode is fabricated. (The annealing activates the donors.) The process is complicated and utilizes a combination of ion implantation, multilevel resist techniques, sputtering, and etching.

SAM *Abbrev. for* scanning Auger microprobe. *See* Auger electron spectroscopy.

sample intelligence *See* sampling.

sampling (1) A technique in which only some portions of an electrical signal are measured and are used to produce a set of discrete values that is representative of the information contained in the whole. In order that the output values represent the input signal without significant loss of information the rate of sampling of a periodic quantity must be at least twice the frequency of the signal.

   A *sampling circuit* is used to produce a set of discrete values representative of the instantaneous values of the input signal. The output may be in the form of a set of instantaneous values (*instantaneous sampling*) or in a coded form. The technique is widely used in *analog/digital converters, *digital voltmeters, *multiplex operation, *pulse modulation, etc.

   (2) A technique in which intermittent measurements of an electrical signal are made. The technique is used in *feedback control systems where the controlled variable is sampled intermittently and a correction applied if necessary as a result of the instantaneous value (the *sample intelligence*). The technique is also used in radio-navigation systems when information from the navigation signal is extracted only when the *sampling gate* is activated by a selector pulse and therefore interrogates the signal and produces a corresponding output *pulse or *waveform.

   (3) A system of quality control of mass-produced electronic components, circuits, devices, or other equipment. Random samples of the manufactured items are removed from the manufacturing point and tested exhaustively. The sampling process is usually carried out for each processing stage during manufacture.

sampling circuit *See* sampling.

sampling gate *See* sampling.

sat *Short for* saturated mode.

satellite An artificial body that is projected from earth to orbit either the earth or another body of the solar system. There are two main classes: *information satellites* and *communications satellites*.

Information satellites transmit signals containing many different types of information to earth. Typical uses include the provision of atmospheric and meteorological data, infrared, ultraviolet, gamma- and X-ray studies of celestial objects, surveys of the earth's shape, surface, and resources, and as navigational aids. Communications satellites receive radiofrequency signals from earth by means of highly directional aerials and return them to another earth location for purposes of long-distance telephony, TV broadcasting, etc.

saturable reactor *See* transductor.

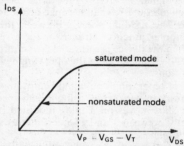

Saturated and nonsaturated modes of an FET

saturated mode The operation of a *field-effect transistor in the portion of its characteristic beyond pinch-off, i.e. $V_P \leqslant V_{DS}$, where V_P is the pinch-off voltage (*see* diagram) and V_{DS} the drain voltage. The *drain current is independent of the drain voltage in this region. *Nonsaturated mode,* sometimes known as *triode-region operation,* is operation of the device in the portion of its characteristic below pinch-off, i.e. $V_P \geqslant V_{DS}$.

saturation (1) A condition where the output current of an electronic device is substantially constant and independent of voltage. In the case of a device such as a field-effect transistor or thermionic valve, saturation is an inherent function of the device and produces the maximum current inherent to the device. In the case of a bipolar junction *transistor, saturation occurs because the output from the collector electrode is limited by the circuit elements of the external circuit and changing these alters the magnitude of the *saturation current drawn from the device.

(2) (magnetic) The maximum possible degree of magnetization of a material; it is independent of the strength of the magnetic flux density applied to the material. All the domains in the material at

501

saturation are assumed to be fully orientated with respect to the magnetic flux density (*see* ferromagnetism).

saturation current The current shown on the portion of the static characteristic of an electronic device where it is substantially constant and independent of voltage. Very little further increase of current with voltage occurs until *breakdown is reached. The value of the saturation current is a function of the device and the external circuit.

saturation resistance The resistance that occurs between the *collector and *emitter electrodes of a bipolar junction *transistor for a specified value of *base current when the device is saturated (*see* saturation) due to the external circuit.

saturation signal A signal, received by a *radar receiver, that has an amplitude larger than the *dynamic range of the receiver.

saturation voltage The residual voltage between the *collector and *emitter of a bipolar junction *transistor for a specified value of base current when the device is saturated (*see* saturation) due to the external circuit. It is smaller than the operating voltage under non-saturated conditions.

sawing A method of separating individual *chips on a semiconductor *wafer preparatory to packaging, using a high-speed precision circular saw. The slice moves on a table mounted below the blade, which is typically between 3 and 15 micrometres wide and coated with diamond. *Compare* scribing.

sawtooth oscillator A *relaxation oscillator that produces a *sawtooth waveform.

sawtooth pulse A *pulse that has a geometrical shape similar to one complete period of a *sawtooth waveform.

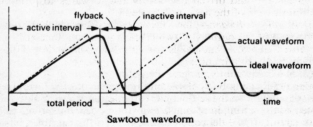

Sawtooth waveform

sawtooth waveform A periodic waveform whose amplitude varies approximately linearly between two values, the time taken in one direction, the *active interval,* being very much greater than the time taken in the other. The shorter period is termed the *flyback.* An ideal sawtooth waveform is linear with a sharp change of direction; in practice this is not achieved (*see* diagram) and the transition stage

502

departs from linearity, often with a short unwanted *inactive interval* before the next cycle. Sawtooth waveforms are commonly produced by suitably designed *relaxation oscillators and are frequently used to provide a *time base.

S band A band of microwave frequencies ranging from 1·55 to 5·20 gigahertz. *See* frequency band.

scaler A device, commonly used in counting circuits, that produces an output pulse whenever a designated number of input pulses have been received. The *scaling factor* is the number of input pulses per output pulse. Scaling factors of 10 or two are most commonly used: a *decade scaler* has a factor of 10 and a *binary scaler* has a factor of two. The latter usually consists of a *flip-flop or *trigger circuit. Scaler circuits are most commonly used with radiation *counters and in digital *computers.

scaling factor *See* scaler.

scaling tube A *gas-discharge tube that contains several anodes (usually ten) and that is used in a *scaler circuit. An electrical signal input to the tube causes each successive anode to be activated in turn so that the glow discharge moves in position. The tube may be arranged so that a spot of light moves around the circumference or the glow may take up the shape of the digits 0–9 in turn.

scalloping *See* travelling-wave tube.

scan converter *See* storage tube.

scanner (1) *Syn.* rectilinear scanner. A device that visualizes the distribution of a radioactive compound in a particular system, usually the human body. The scanner consists of a *scintillation crystal, *photomultiplier tube, and amplifying circuits. The crystal is collimated, so that only the radiation from a small area is received at any given instant, and driven backwards and forwards to produce a rectilinear scan of the area investigated.

Double-headed scanners have been developed in which the output is the sum of the outputs of two crystals that simultaneously scan the area of interest. Anomalies arising from a source deep within a human body are minimized by this technique.

(2) *See* flying-spot scanner.

scanning (1) Causing one complete horizontal or vertical traverse of the spot of light on the screen of a *cathode-ray tube (CRT), i.e. one sweep of the screen, in response to a voltage generated by a *time-base circuit. If the deflection extends beyond the usable physical dimensions of the phosphor on the screen, *overscanning* occurs; if it is less than the usable size *underscanning* results. In the former case information can be lost from the ends of the scan; in the latter the image does not fill the screen.

A particular type of scanning sometimes used with a CRT is *spiral scanning*. In this case the electron beam is made to execute a spiral trace on the screen.

(2) Exploring a particular area or volume in a methodical manner in order to produce a variable electrical signal whose instantaneous values are a function of the information contained in the small area examined at each instant. The information scanned may then be reproduced by a suitable receiver. The technique is most often used in *television, *radar, and *facsimile transmission.

The most common type of scanning used in television and facsimile is *rectilinear scanning*, in which the target area is rectangular and scanning proceeds in a set of narrow parallel strips (*see also* television).

High-velocity scanning is a type of *electron scanning*, i.e. scanning of the target with a beam of electrons, in which the energy of the electrons in the beam is sufficient to produce a *secondary-emission ratio at the target that is greater than unity. If the energy of the electrons is less than the minimum velocity required to produce a secondary-emission ratio of unity, the scanning is termed *low-velocity scanning*.

Coarse scanning is frequently used in order to produce a 'rough' picture of the target before carrying out a more detailed investigation. In this case the size of the scanning spot is comparable to the image detail and is the diameter of the electron beam, light beam, or beam of radiowaves used to carry out the scan. Coarse scanning is most often used in radar systems.

Radar systems may also employ *circular scanning*, in which one complete scan is a horizontal rotation of the radar beam through 360°, or *conical scanning*, when the major lobe of the transmitted *radiation pattern generates a cone.

Most forms of scanning assume that the scanning speed is uniform throughout each scan. In some applications however it is useful to employ a variable speed of the scanning spot if one section of the scanned area is to be investigated in more (or less) detail than the rest. Some form of *distribution control* is used to achieve a variable speed distribution during the trace interval by controlling the output of the time-base circuit used.

scanning Auger microprobe (SAM) *See* Auger electron spectroscopy.

scanning electron microscope (SEM) *See* electron microscope.

scanning transmission electron microscope (STEM) *See* electron microscope.

scanning yoke *Syn.* deflecting yoke. *See* cathode-ray tube.

scattering loss Energy loss from a beam of electromagnetic radiation because of deflection of the radiation by scattering. The individual particles or photons in the beam can interact with the nuclei or electrons present in the medium through which they are propagated, with photons of another field of radiation, or irregularities in a reflective surface.

schematic *Colloq.* A circuit diagram.

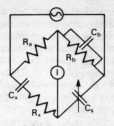

Schering bridge

Schering bridge A four-arm bridge for the measurement of capacitance (*see* diagram). At balance,

$$C_x = C_s R_b / R_a$$
$$R_x = R_a C_b / C_s$$

SCH laser *Short for* separate confinement heterostructure laser. *See* injection laser.

Schmitt trigger A *bistable circuit in which the output voltage level is binary, is determined by the magnitude of the input signal, and is independent of the input signal waveform. The output level changes to the high level when the input signal exceeds a predetermined value. It falls to the low level when the input signal magnitude drops below a predetermined value. The circuit inevitably exhibits hysteresis; the amount of hysteresis is determined by the components in the circuit and can be altered so that the desired switching values can be selected.

The Schmitt trigger can be used in binary logic circuits in order to maintain the integrity of the logical one and zero levels. It can also be used with a variety of analog waveforms as a level detector, i.e. it acts as a trigger for other circuits or devices when the magnitude of the input waveform exceeds or falls below the predetermined levels. The device may also be used to generate a rectangular pulsetrain from a variety of input waveforms.

Schockley diode *Syn. for* four-layer diode. *See* pnpn device.

Schockley equation *Syn.* ideal diode equation. *See* p-n junction.

Schottky barrier *See* Schottky effect; Schottky diode.

Schottky clamp A *Schottky diode that is used to prevent the voltage at a particular point in a circuit from exceeding a predetermined value. The forward bias characteristic of a Schottky diode is such that in the conducting state the diode has an essentially constant small voltage drop across it. Carrier storage is negligible, leading to very fast switching between the 'on' and 'off' states of the diode. The most common application of Schottky clamps is in integrated *logic circuits in which the bipolar transistors forming the logic gates are operated in *saturation during part of the switching cycle. A Schottky diode connected across the base and collector prevents the collector-base voltage swinging too far in the forward direction and hence controls the depth of saturation of the transistor; the speed of operation of the gate is thus optimized.

Schottky diode *Syn.* hot-carrier diode. A rectifying diode formed from a junction between a metal and a semiconductor – a Schottky barrier (*see* Schottky effect). The term is reserved for Schottky barriers in which the energy gap and doping level of the semiconductor are such that minority carriers in the semiconductor do not contribute significantly to the current flowing in the diode.

If a forward bias is applied across the junction, majority carriers (electrons in an n-type semiconductor) that have energies greater than the Schottky barrier height – hot carriers – can cross the barrier and current flows (*thermionic emission). Carriers may also cross the barrier by *tunnelling if it is sufficiently narrow (*field emission). A mixture of these mechanisms often occurs and is known as *thermionic-field emission*, but thermionic emission is the dominant process in Schottky diodes. As the forward bias increases more hot carriers are present and the current increases rapidly. With sufficiently large voltages all the free carriers in the semiconductor can cross the barrier; the current can then only increase with increasing voltage as a result of acceleration of free majority carriers. The current-voltage characteristic is then linear in this region. Under reverse bias conditions the current falls to a small reverse saturation current. Since the minority carriers have a negligible contribution to the current in either the forward or reverse directions *carrier storage at the junction is negligible and the diode has very fast switching speeds.

Schottky effect A reduction in the effective *work function of a solid when an external accelerating electric field, *E*, is applied in vacuo to the surface. In the case of a metal, image charges (*see* electric images) contribute to the effect, which is therefore sometimes termed *image-force lowering*. The external field lowers the potential energy of electrons outside the solid with a consequent distortion of the potential barrier at the surface. Electrons just inside the surface are liberated by surmounting the barrier (*compare* tunnel effect). A slight increase in electron emission from a thermionic cathode results from this effect.

A similar effect is observed when the metal surface is in contact with a semiconductor. Such a metal-semiconductor junction is termed a *Schottky barrier;* the energy levels associated with the junction are shown in the diagram. The magnitude of the lowering of the work function depends critically upon the surface state of the semiconductor but is usually less than the vacuum case described above. Schottky barriers are used to form *Schottky diodes, in Schottky

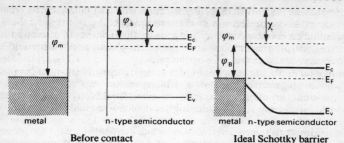

Energy diagrams before and after forming ideal Schottky barrier

TTL and Schottky I²L logic circuits, and to form the gate electrode of one type of junction *field-effect transistor.

Schottky-gate field-effect transistor *See* MESFET.

Schottky I²L (STL) *See* I²L.

Schottky noise *Syn.* shot noise; flicker noise. Variations in the current output of an electronic device caused by the random manner in which electrons or holes are emitted from an electrode, such as the *collector of a *transistor or the *source of a *field-effect transistor.

Schottky photodiode *See* photodiode.

Schottky TTL *See* transistor-transistor logic.

Schrödinger's equation The fundamental equation of wave mechanics (*see* quantum mechanics). It describes the behaviour of a particle moving in a field of force and takes account of the *de Broglie waves associated with the moving particle. Schrödinger's equation in three dimensions is given by

$$(h^2/8\pi^2 m)\nabla^2\psi + (E - V)\psi = 0$$

where h is the *Planck constant, m the mass of the particle, E the total energy of the particle, and V the potential energy. ∇^2 is the *Laplace operator and ψ the wave function of the particle. The wave function is a complex mathematical function of the spatial coordinates x, y, and z of the particle. *See also* energy bands; direct-gap semiconductor.

scintillation (1) A flash of light produced in certain materials (*scintillators*) when exposed to ionizing radiation. The frequency of the light emitted in the crystal is a function of the energy of the incident radiation. Each incident ionizing event produces one flash.

(2) A rapid fluctuation of the image of the target of a *radar system about its mean position on the radar display.

(3) A small random fluctuation of the received signal in a radio transmission system about the mean signal value. This effect is anal-

ogous to the twinkling of light from a star and is caused by small variations in the density of the atmosphere.

scintillation counter A radiation *counter that consists of a scintillator (*see* scintillation), *photomultiplier, *amplifier, and *scaler and is used to measure the activity of a radioactive source. The scintillator crystal emits flashes of light of a characteristic frequency when exposed to gamma rays from a radioactive source. Each scintillation produces an output pulse from the photomultiplier. The count rate produced from the photomultiplier is measured and the activity of the source calculated.

The scintillation counter is energy-dependent since the frequency of the emitted light in dependent on the energy of the incident radiation. The energy of the electrons emitted from the *photocathode is a function of the light frequency and hence pulses produced by radiation of energy other than that being considered may be excluded using a suitable *discriminator circuit. The *signal-to-noise ratio is therefore improved relative to other radiation counters, such as the *Geiger counter, since background radiation and scattered radiation may be excluded, and one radioactive nuclide may be counted in the presence of another.

The energy dependence of a scintillation counter may also be used to study the energy distribution of radioactive nuclides, thus using it as a *scintillation spectrometer*.

scintillation crystal *Syn. for* scintillator. *See* scintillation.

scintillation spectrometer *See* scintillation counter.

scintillator *Syn.* scintillation crystal. *See* scintillation.

scotophor A material, such as an alkali halide, that can be used on the screen of a *cathode-ray tube instead of the usual *phosphor when daylight viewing and long persistence are required. The material darkens under electron bombardment to produce a black-on-white picture that can be erased by heating. *See* skiatron.

Scott connection *See* three-phase to two-phase transformer.

SCR *Abbrev. for* silicon-controlled rectifier.

scrambler A circuit or device that is used in communication systems to produce an unintelligible version of the signal to be transmitted, in a predetermined manner. The received signal is rendered intelligible by an unscrambling circuit used at the receiver, in sympathy with the scrambler.

screen (1) The surface of a cathode-ray tube, suitably coated, on which the visible pattern is produced.

(2) *Syn.* shield. Material that is suitably arranged so as to prevent or reduce the penetration of an electric or magnetic field into a particular region. *See also* electric screening; magnetic screening.

screened pair *See* shielded pair.

screen grid *See* thermionic valve; tetrode.

screening reactor *See* line choking coil.

screening test *See* life test.

scribing Scoring a wafer of *semiconductor with a precision diamond tool in order to separate individual *chips, each containing a circuit or component, preparatory to packaging. *Laser scribing* uses an *injection laser to score the wafer rather than a mechanical tool. *Compare* sawing.

scribing channel A gap left between the areas on a *semiconductor wafer during the manufacture of several circuits or components on the wafer, in order to allow for *scribing into individual *chips.

SCS *Abbrev. for* silicon-controlled switch. *See* pnpn device.

scumming A phenomenon that can occur during *photolithography. Unexposed photoresist slowly dissolves in the developer and is then deposited in spaces left where developed photoresist has been removed. Such 'scum' must be removed by a suitable descumming method. One method is dry etching using an oxygen plasma (ashing).

search coil *Syn. for* exploring coil. *See* flip-coil.

SECAM Acronym from *SEquential Couleur À Memoire*. A *line-sequential colour-television system developed and adopted in France. *Compare* PAL.

second Symbol: s. The *SI unit of time defined as the duration of 9 192 631 770 periods of the radiation corresponding to the transition between two hyperfine levels of the ground state of the caesium-133 atom. It was formerly defined as 1/86 400 of the mean solar day.

secondary cell *Syns.* accumulator; storage cell. *See* cell.

secondary electron An electron emitted from a material as a result of *secondary emission.

secondary emission The emission of electrons from the surface of a material, usually a metal, as the result of bombardment by high-velocity electrons or positive ions. The probability of secondary emission occurring when an ion approaches the surface is given by the *Massey formula.*

The total energy of the incident primary electrons is often sufficient to liberate several secondary electrons per incident particle. The *secondary emission ratio*, δ, is the number of secondary electrons emitted per incident particle. Secondary emission is mainly used in the *electron multiplier and in *storage tubes.

secondary-emission ratio *See* secondary emission.

secondary failure *See* failure.

secondary-ion mass spectroscopy (SIMS) A method of analysing the surface of a material to detect impurities. An ion beam is used to sputter material in the form of secondary ions from the surface of a semiconductor. The secondary ions are electrostatically accelerated and then analysed using a mass spectrometer. The ion beam may be kept small and scanned across the sample and the term *ion* *microprobe* is sometimes used. Over 90% of the secondary ions are emitted from the two top atomic layers. A depth profile can be obtained by sputtering continuously in the vertical direction but the accuracy decreases with increasing depth. This is a destructive analysis technique.

secondary radiator *Syn. for* passive aerial. *See* directive aerial.

secondary service area *See* service area.

secondary standard (1) A copy of a primary standard that has a known difference from the primary standard.

(2) A quantity that is accurately known in terms of a primary standard and that may be used as a unit.

(3) A measuring instrument that has been accurately calibrated in terms of a primary standard and that can be used for accurate measurements or calibration of other equipment.

Compare primary standard.

secondary voltage (1) The voltage developed across the secondary (output) winding of a *transformer.

(2) The voltage developed by a secondary *cell.

secondary winding *See* transformer.

secondary X-rays *See* X-rays.

second-channel frequency *See* image frequency.

Seebeck effect *See* thermoelectric effects.

seed crystal A small single crystal that can act as a focus for crystallization, as from a supersaturated solution or a supercooled liquid. It is used to produce large single crystals of a *semiconductor during the manufacture of solid-state electronic components, circuits, or devices.

selective fading *See* fading.

selective interference Interference to radiofrequency signals that occurs only within a particular narrow band of signal frequencies.

selectivity The ability of a *radio receiver to discriminate against radiowaves with a *carrier frequency different from that to which the receiver is tuned (*see* tuned circuit). The selectivity is usually expressed on a graph that shows the power ratio E/E_0 expressed in decibels against frequency, at specified values of the *modulation factor or index and modulation frequency; E_0 is the output power at the resonant frequency, v_0, and E the output power at frequency v for the same value of input power as that producing E_0.

The graph indicates the factor by which the input power must be increased, as the carrier frequency is varied from the resonant frequency to which the receiver is tuned, in order to maintain a constant output. It is usual to render any *automatic gain control inoperative when selectivity curves are being obtained.

selenium Symbol: Se. A *semiconductor element, atomic number 34. In the form of its grey allotrope selenium is markedly light-sensitive and is extensively used in photoconductive *photocells.

selenium rectifier A *Schottky diode that consists of a selenium-iron junction and is used as a *rectifier. It is usual to construct a stack of such junctions in series.

self-aligned gate *See* recessed gate FET.

self-bias A required magnitude of bias developed from the main power supply at a point in a circuit by means of a *dropping resistor rather than supplied by a separate battery.

self-capacitance The inherent distributed capacitance associated with an inductance coil or resistor. To a first approximation the self-capacitance of the coil or resistor may be represented as a single capacitance connected in parallel with it.

self-excited Denoting an *oscillator that produces a build-up of output oscillations to a steady value upon application of power to the circuit. No separate input at the required output frequency is needed in order to produce the oscillations.

self-inductance *See* inductance; electromagnetic induction.

self-locking relay *See* relay.

self-sealing capacitor *Syn.* Mansbridge capacitor. A type of capacitor that is relatively little affected by puncture of the dielectric arising from an excessive voltage surge. The heat generated as a result of dielectric puncture causes local oxidation of the metal electrode adjacent to the puncture and the insulation is consequently restored with very little loss of capacitance.

self-sustaining oscillations *See* oscillation.

SEM *Abbrev. for* scanning electron microscope. *See* electron microscope.

semiconductor A material having a *resistivity in the range between conductors and insulators and having a negative temperature coefficient of resistance. The conductivity increases not only with temperature but is also affected very considerably by the presence of *impurities in the crystal lattice. Semiconductors are used in a wide variety of solid-state devices including *transistors, *integrated circuits, *diodes, *photodiodes, and *light-emitting diodes.

An *intrinsic semiconductor* is a perfect-crystal semiconductor in which the energy gap, E_g, between the conduction band and the valence band (*see* energy bands) is comparable to thermal energies. The simplified energy bands of an intrinsic semiconductor are shown in Fig. *a*. At a given temperature T_1 some electrons will be thermally excited into the conduction band. The number, n, of these electrons is given by

$$n = \int_{E_c}^{E_2} g_c(E)\,dE\,/\,[(\exp(E-E_F)/kT) + 1]$$

where $g_c(E)$ is the number of energy levels in the conduction band (the density of states), E_c and E_2 are the lower and upper limits of the band, respectively, and are termed the *band-edge energies*, E_F is

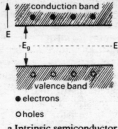

a Intrinsic semiconductor

the *Fermi level of the material, k is the Boltzmann constant, and T the thermodynamic temperature.

Each electron that is thermally excited into the conduction band leaves behind a vacant energy level in the valence band and in a pure crystal the number of vacancies equals the number of electrons in the conduction band. The number of electron-vacancy pairs per second produced in the semiconductor is the *generation rate.* Bombardment of the material by electromagnetic radiation can also produce electron-vacancy pairs in excess of the thermal equilibrium number (*see* photoconductivity).

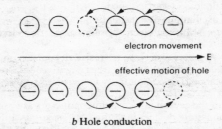

b Hole conduction

The action of an applied electric field causes conduction both in the conduction band and in the valence band. The conduction electrons are accelerated by the applied field and carry negative charge through the semiconductor. The electrons in the valence band move to occupy each vacancy and the net effect is that a vacancy moves through the material as if it were a positive charge (Fig. *b*). The vacancies are known as *holes and are treated as positive charge *carriers.

In practice absolutely pure crystals do not exist. The presence of defects in the crystal lattice of a semiconductor, such as vacancies in the lattice or interstitial atoms, distort the energy bands near the defect position and cause extra energy levels to be formed in the forbidden band. If these occur near the valence band electrons without sufficient energy to cross the forbidden band may have sufficient energy to enter the extra energy levels formed. These levels act as traps for electrons and conduction can take place predominantly in the valence band, as empty levels will be left as holes. Conduction due to the presence of defects is known as *defect conduction. (A similar type of conduction occurs in *insulators.) If the extra levels occur near the conduction band, electrons in the conduction band can be trapped and the overall conductivity is decreased. If the extra levels occur deep in the valence band they can act as *hole traps* and conduction takes place predominantly in the conduction band.

The presence of *impurities in a semiconductor affects the conductivity very significantly. An *extrinisic semiconductor* is one whose properties depend on the presence of impurities and on the type and concentration of impurity.

Donor impurities are atoms that, when present in the crystal lattice, have more valence electrons than are required to complete the bonds with neighbouring atoms. The presence of these atoms affects the distribution of energy states in the immediate vicinity and extra levels are formed in the forbidden band, close to the conduction band (Fig. *c*). These *donor levels* can give up their electrons to the conduction band very easily. As the energy difference is small, at room temperatures the donor atoms are usually ionized. The positive charge due to the vacant level is bound to the atom site. The number of conduction electrons is therefore greater than the number of mobile holes and the semiconductor is said to be *n-type*.

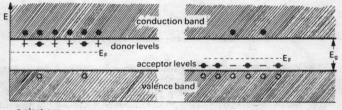

● electrons
○ holes

c Energy bands in an n-type (left) and p-type (right) semiconductor

Acceptor impurities are atoms that have fewer valence electrons than are required to complete the bonds with neighbouring atoms and they therefore accept electrons from any available source to complete the bonds. These extra electrons are almost as tightly bound to the atom as the valence electrons and the presence of acceptor impurities results in energy levels just above the valence band (Fig. *c*). The electrons in the valence band only need a small increment of energy to occupy the *acceptor levels* and provide the source of electrons for the acceptor atom. Mobile holes are then left in the valence band but the electrons are bound to the acceptor atom, which is ionized by the electron capture; mobile holes therefore predominate and the semiconductor is known as *p-type*.

The presence of ionized acceptor levels in a crystal can result in the production of an *exciton* when excess holes are injected into the material: the negative ion attracts a positive hole, which becomes bound to the lattice site; this results effectively in a bound electron-hole pair.

The carriers that predominate in a particular semiconductor are called the *majority carriers (e.g. electrons in n-type) and the others are *minority carriers (e.g. holes in n-type). The effect of the impurity is to move the Fermi level near to the conduction band in n-type semiconductors and near to the valence band in p-type as the distribution of available energy levels has changed. If sufficient impurity levels are present so that the Fermi level falls in the conduction or valence band, the semiconductor becomes *degenerate and exhibits metallic properties.

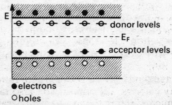

d Energy bands in a compensated semiconductor

The conductivity of an extrinsic semiconductor will depend on the type and amount of impurities present and this may be controlled by adding impurities of a particular sort to achieve the desired type of conductivity. This process is known as *doping and the amount of impurity is the *doping level. This is usually carried out by *diffusion or *ion implantation of the impurity into the crystal. *Doping compensation compensates for the effects of one type of impurity by diffusing the opposite type of impurity into the material. It is possible to produce a semiconductor with properties similar to an intrinsic semiconductor using doping compensation (Fig. *d*). This material is sometimes known as a *compensated intrinsic semiconductor*.

At thermal equilibrium a dynamic equilibrium exists in a semiconductor. Mobile charges move around the crystal in a random manner due to scattering by the nuclei in the crystal lattice. The crystal retains overall charge neutrality and the number of charge carriers remains essentially constant. A continuous process of re-generation and recombination occurs however as thermally-excited electrons enter the conduction band and other electrons fall back into the valence band and combine with holes (*see also* recombination processes).

*Fermi-Dirac statistics shows that the number of charge carriers in an extrinsic semiconductor is related by the product

$$n_e n_p = n_i^2$$

where n_e is the number of electrons, n_p the number of holes, and n_i the number of electrons (or holes) in the intrinsic material. The number of majority carriers is approximately equal to the number of impurity atoms since ionization of impurity atoms is achieved at lower energies than thermal generation of electron-hole pairs. The lower energy levels in the conduction band of n-type material will therefore be occupied by electrons from donor impurities and fewer thermally generated pairs are possible at a given temperature because of the effectively larger energy required to produce them. In p-type material a similar effect is caused by the upper levels of the valence band being 'occupied' by mobile holes due to acceptor impurities. If extra charge carriers are generated by, for example, light energy, these will have a limited lifetime in the material.

The lifetime of excess carriers is defined in terms of minority carriers. In the bulk of a homogeneous semiconductor the *bulk lifetime* is the average time interval between generation and recombination of minority carriers. Recombination also takes place near the surface of a semiconductor; the mechanism is rather different from the direct recombination occurring in the bulk of the material. It is defined by the *surface recombination velocity,* which is the ratio of the component, normal to the surface, of the electron (hole) current density to the excess electron (hole) volume charge density close to the surface. *Injection of excess minority carriers by any means causes the conductivity of the material to change until thermal equilibrium has been re-established. This effect is termed *conductivity modulation.*

If an electric field is applied to the semiconductor, charge carriers move under the influence of the field but still undergo scattering processes. The result of the field is to impose a drift in one direction on to the random motion of the carriers (Fig. *e*). The *drift mobility* of a charge carrier is the average drift velocity of excess minority carriers per unit electric field. The average distance travelled by a minority carrier during its lifetime is the *diffusion length.* The mobility of electrons has been shown to be about three times that for holes and

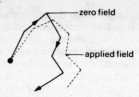

e Movement of a carrier in a semiconductor

depends on the semiconductor material. Gallium arsenide, for example, has a higher carrier mobility than silicon.

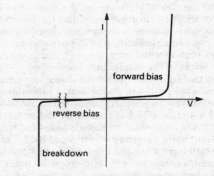

f I–V characteristic of a p-n junction

Junctions between semiconductors or between a metal and a semiconductor form a fundamental part of modern electronic components and circuits. A *p-n junction is formed between a p-type and an n-type semiconductor. A typical current-voltage characteristic is shown in Fig. f. If the junction is formed between two dissimilar semiconductors it is a heterojunction. An *abrupt junction and Schottky barrier (see Schottky diode) are specialized junctions.

Types of semiconductor material commonly used are elements falling into group 4 of the periodic table, such as silicon or germanium. The donor and acceptor impurities are group 5 and group 3 elements, respectively, differing in valency by only one electron. Certain compounds such as gallium arsenide, which has a total of eight valence electrons, also make excellent semiconductors. These materials are classified as 3–5 or 2–6 depending on their position in the periodic table. Suitable impurities in these cases would be from groups 2, 4, and 6 or from groups 3 and 5, respectively.

See also energy bands; direct-gap semiconductor.

semiconductor counter A radiation counter that uses a *photodiode as the radiation detection element together with a suitable counting circuit.

semiconductor device Any electronic circuit or device that depends for its operation on the flow of charge *carriers within a *semiconductor.

semiconductor diode A *diode constructed from semiconducting material.

semiconductor laser *Syn. for* injection laser.

semiconductor memory *See* solid-state memory.

semitransparent cathode A *photocathode that exhibits *photoemission from both surfaces in response to electromagnetic radiation incident on one of the surfaces. This type of photocathode has the advantage that electron emission occurs on the opposite side to the incident radiation and it is often used in television *camera tubes.

sense amplifier *See* random-access memory.

sensing element *See* transducer.

sensitivity (1) In general, the change produced in the output of a physical device per unit change in the input.

(2) The magnitude of the change in the indicated value or deflection of a measuring instrument produced by a specified change in the measured quantity. It is usually quoted as the magnitude of the measured quantity required to produce full-scale deflection.

(3) The ability of a *radio receiver to respond to weak input signals. It is the minimum signal input to the receiver that produces a particular output value under stated conditions, particularly a designated signal-to-noise ratio.

sensor *See* transducer.

separate confinement heterostructure laser (SCH laser) *See* injection laser.

sequential-access memory *See* serial memory.

sequential control Operation of a *computer in which a sequence of instructions is produced and input to the computer during the solution of a problem.

sequential scanning *See* television.

serial memory *Syn.* sequential-access memory. A *memory used in a digital *computer in which information is retrieved by sequential access, i.e. in order to retrieve information stored as the nth record the preceding $(n-1)$ records must first be scanned. The *major cycle* of such a memory is the time interval during which a particular memory location is not accessible for *read operation. The major cycle determines the maximum access time (*see* memory) of the memory. The access time for a given memory location may be less than the major cycle; the point during the cycle at which a read pulse is received determines the practical access time. The cycle time may be reduced by using a serial parallel serial arrangement (*see*

517

CCD memory) of the storage locations. Devices such as CCD memory and *magnetic bubble memory are often organized in this manner. *Compare* random-access memory.

serial transfer Transfer of information along a single path in a computer or data processing system in which the characters move one after the other along the path.

serial transmission A communication system in which the characters of a word are transmitted sequentially along a single line. *Compare* parallel transmission.

series Circuit elements are said to be *in series* if they are connected so that one current flows in turn through each of them (*see* diagram).

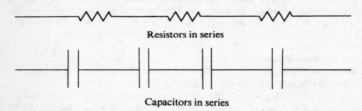

Resistors in series

Capacitors in series

For *n* resistors in series the total resistance, *R*, is given by:

$$R = r_1 + r_2 + \ldots r_n$$

where $r_1, r_2, \ldots r_n$ are the values of the individual resistors. The resistors thus behave collectively as one large resistance.

For *n* capacitors in series the total capacitance, *C*, is given by:

$$1/C = 1/c_1 + 1/c_2 + \ldots 1/c_n$$

where $c_1, c_2, \ldots c_n$ are the individual capacitances.

Machines, transformers, and cells are said to be in series when terminals of opposite polarity are connected together to form a chain. Several cells connected together in series add the values of the emfs and can therefore supply a larger voltage. *Compare* parallel.

series feedback *Syn. for* current feedback. *See* feedback.

series-gated ECL *See* emitter-coupled logic.

series network *See* network.

series-parallel connection (1) An arrangement whereby electronic devices or circuits may be connected either in *series or in *parallel.

(2) A method of connecting the elements of a circuit or network containing resistors, inductors, and capacitors so that some are in *series and some in *parallel with each other.

series resonant circuit *Syn.* acceptor. *See* resonant circuit; resonant frequency.

series stabilization *See* stabilization.

series supply A method of applying bias to an electrode of an active device so that the bias is applied across the same impedance in which the signal current flows (Fig. *a*).

a Series supply   b Parallel supply

Parallel supply is a method of applying the bias so that the bias is supplied across an impedance in *parallel with that in which the signal current flows (Fig. *b*).

series transformer *Syn. for* current transformer. *See* transformer.

service area The region covered by the useful range of a radiofrequency broadcast transmitter, such as a radio or television transmitter. The region is often represented pictorially on a *service-area diagram.*

The *primary service area* is the area within which satisfactory reception is possible, day or night, as a result of reception of the *ground wave. The strength of the ground wave within this area is large compared to that of interference and *indirect waves.

The *secondary service area* is the area within which satisfactory reception is possible using indirect waves. The strength of the ground wave within this area is attenuated to substantially less than the indirect waves.

The *mush area* is an area within which substantial *fading or *distortion of the received signal occurs. The unsatisfactory reception in this area results from interference either between waves from two or more synchronized transmitters or between direct and indirect waves from a single transmitter.

set-up scale instrument *See* suppressed-zero instrument.

shading (1) The generation of a nonuniform background level in the image produced by a television *camera tube that was not present in the original scene.

(2) Compensation for the spurious signals generated by a television *camera tube during the flyback interval.

519

shadow effect An effect observed when electromagnetic waves are transmitted, as during broadcast transmissions, due to the topography of the region between the transmitter and receiver. A loss of signal strength at the receiver is usually observed compared to the expected signal strength when transmission is over a uniformly flat region.

shadow mask *See* colour picture tube.

shaped-beam tube A *cathode-ray tube that is used to display characters or numerals. Suitable electric and magnetic fields are applied to the tube in order to produce a beam whose cross section is that of the desired character. All parts of the character are therefore shown on the screen simultaneously.

sheath (of electrons) *See* magnetron.

sheet resistance The resistance of a unit square of a thin-film material, such as a metal or thin layer of semiconductor, defined as $R_s = \rho/t$ where R_s is the sheet resistance, ρ the *resistivity, and t the thickness. A film of length L, width W, and thickness t therefore has a total resistance equal to $R_s(L/W)$. R_s has dimensions of resistance but is commonly given the unit 'ohms per square'.

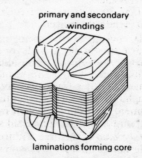

primary and secondary windings

laminations forming core

Single-phase shell-type transformer

shell-type transformer A *transformer in which most of the windings are enclosed by the *core (*see* diagram). The core is made from laminations and usually the windings are assembled and then the laminated core built up around them. *Compare* core-type transformer.

SHF *Abbrev. for* superhigh frequency. *See* frequency band.

shield *See* screen.

shielded pair *Syn.* screened pair. A *transmission line that consists of two wires enclosed in a metal sheath.

shift register A *digital circuit that can store a set of information in the form of pulses and displace it either to the left or right upon application of a *shift pulse*. If the information consists of the digits of a numerical expression, a shift of position to the left (or right) is equivalent to multiplying (or dividing) by a power of the base. Shift registers are extensively used in *computers and data processing systems as storage or delay elements. *See also* delay line.

shock excitation *Syn.* impulse excitation. The production of *free oscillations in a system, such as a *resonant circuit, by suddenly introducing electric energy into the system; the energy may be in the form of a relatively short-lived impulse voltage or the shock may be the switching-on of a source of constant direct voltage.

short *Colloq.* A short circuit.

short circuit An accidental or deliberate electrical connection of relatively very low resistance between two points in a circuit.

short-circuit impedance *See* network.

short-time duty *See* duty.

short-wave Denoting a radiowave that has a wavelength in the range 10 to 100 metres, i.e. in the high-frequency band. *See* frequency band.

short-wave converter A *frequency changer in a radio receiver that is tuned to receive a broadcast transmission in the short-wave frequency band and that converts it to a frequency band within the dynamic range of a standard receiver.

shot noise *See* Schottky noise.

shunt (1) In general, *syn. for* parallel.

(2) *Syn.* instrument shunt. A resistor, usually of a relatively low value, that is connected in parallel with a measuring instrument, such as a galvanometer. Only a fraction of the current in the main circuit passes through the instrument so that the shunt increases the range of the instrument and also protects it from possible damage caused by current *surges. *See also* universal shunt.

shunt feedback *Syn. for* voltage feedback. *See* feedback.

shunt network *Syn. for* parallel network. *See* network.

shunt stabilization *See* stabilization.

sideband *See* carrier wave.

side circuit *See* phantom circuit.

side frequency *See* carrier wave.

siemens Symbol: S. The *SI unit of electrical *conductance, *susceptance, and *admittance. An element possesses a conductance of one siemens if it has electrical *resistance of one ohm. The siemens has replaced the *mho.

Siemen's electrodynamometer An *electrodynamometer that may be calibrated as an ammeter, voltmeter, or wattmeter. The signal to be measured produces an electromagnetic torque on the movable coil that in turn is balanced against the torque of a spiral spring connected to it by adjusting a calibrated torsion head attached to the spring.

At the balance position the deflection of the movable coil is zero, and the value of the measured parameter is given by the setting of the torsion head.

signal A variable electrical parameter, such as current or voltage, that is used to convey information through an electronic circuit or system.

signal electrode *Syn.* back plate. *See* vidicon; iconoscope.

signal generator Any electronic circuit or device that produces a variable and controllable electrical parameter. The term is most commonly applied to a device that supplies a specified voltage of known variable amplitude, frequency, and waveform. A generator that produces *pulse waveforms is normally referred to as a *pulse generator, the term signal generator being reserved for a continuous-wave generator, particularly of a sinusoidal wave.

signal level The magnitude of a signal at a point in a transmission system, with reference to an arbitrarily chosen reference signal.

signal-to-noise ratio At any point in an electronic circuit, device, or transmission system, the ratio of one parameter of a desired signal to the same or a corresponding parameter of the *noise. In broadcast communication the signal-to-noise ratio is often quoted in *decibels and the noise parameter taken as its *root-mean-square value.

signal winding *See* transductor.

silent discharge An electrical discharge that occurs at high voltages and is inaudible to the human ear. Such a discharge involves a relatively high dissipation of energy and takes place most readily from a sharply pointed conductor.

silent zone *Syn.* skip zone. The portion of the skip area (*see* skip distance) surrounding a particular transmitter that falls outside the range for ground-wave propagation of the transmitter. In practice there is a weak residual signal within the silent zone that results from scattering, localized reflection, or some abnormal means of propagation.

silica Symbol: SiO_2. An extremely abundant compound occurring in several different natural forms, the best known of which are probably quartz and common sand (in which the silica is discoloured by ferric oxides). Silica is important as a source of silicon for the manufacture of electronic components, devices, and integrated circuits, as a grown oxide for the *passivation of such equipment, and as natural *quartz, which has marked piezoelectric properties.

silica gel Deliquescent crystals consisting mainly of silica (SiO_2) that are used as a drying agent, particularly during dispatch and delivery of electronic and electrical equipment.

silicon Symbol: Si. A *semiconductor element, atomic number 14. It is very abundant in nature in the form of silicon dioxide (*silica) and is the most widely used semiconductor in solid-state electronics. It is cheap and extremely versatile and rapidly replaced germanium except for a very few specialized applications.

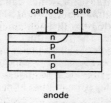

a Silicon-controlled rectifier

silicon-controlled rectifier (SCR) *Syn.* reverse blocking triode thyristor. A *pnpn device in which the forward anode-cathode current is controlled by means of a signal applied to a third electrode, called the *gate*. The device is constructed as a four-layer chip of semiconductor material forming three *p-n junctions, with *ohmic contacts to three of the layers (Fig. *a*).

The operation of the device is most easily understood by representing it as a combination of two bipolar junction *transistors with two common electrodes (Fig. *b*). The gate electrode forms the collector of the p-n-p transistor and the base of the n-p-n transistor. A voltage is applied across the device between the anode and the cathode. A reverse bias, i.e. positive bias applied to the n-type cathode, results in a very small reverse *blocking current* through the device and the device is effectively 'off'.

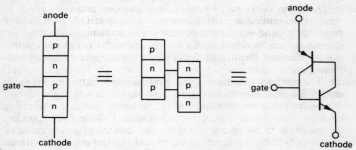

b Construction of an SCR

With forward bias, if no signal is applied to the gate electrode ($V_{GC} = 0$), the n-p-n transistor will not conduct since no current is supplied to its base. The p-n-p transistor will not conduct either since the base current, which must flow to allow transistor action, is supplied via the n-p-n transistor (which is 'off'). This is the *forward*

523

blocking state of the device. If a positive current is supplied to the gate electrode the n-p-n transistor turns on and draws current from the base of the p-n-p transistor, which can also conduct. When the transistors are both turned on, current flows through the device from anode to cathode provided that the anode is positive with respect to the cathode. The magnitude of the gate current determines the *breakover point*. This is the anode voltage at which the device switches from the blocking state to the conducting state (Fig. *c*). The two transistors provide a positive current feedback loop for each other and current continues even after the cessation of applied signal to the gate.

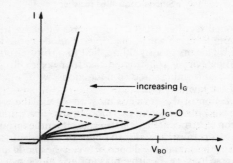

c Effect of increasing gate current on breakover point

The current may only be cut off by reducing the anode voltage to near zero, when the emitter-base junction of the p-n-p transistor becomes reverse biased and current ceases to flow. Alternatively it can be cut off by reducing the current through the device to a low value, when the gains of each transistor become so low that insufficient current is supplied to the bases to allow conduction to continue. The minimum current, I_h, for continuation of conduction is the *holding current*. The device cannot immediately return to the foward blocking state after turn off. Carrier storage occurs at the junctions and the *turn-off time* is the delay required for these excess carriers to be removed. If the anode voltage is reduced below zero a pulse of reverse current is obtained as the excess carriers are removed; the effective turn-off time is thus reduced, compared to that obtained by reducing the anode current. The SCR is the solid-state equivalent of the *thyratron valve and was originally called a thyristor. The current-voltage characteristic (Fig. *d*) is similar to that of the thyratron.

The most important applications of SCRs are in a.c. control systems and solid-state *relays. If an a.c. signal is applied to the anode

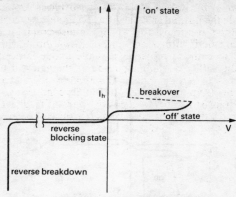

d Current-voltage characteristic of an SCR

the device may be switched on at any desired portion of the positive half-cycle using a *trigger pulse or by illuminating the gate region causing electron-hole pairs to be formed. The device will automatically switch off at the end of the half-cycle when the anode voltage drops below the turn-off level. This type of operation is known as *phase control*. If the gate turn-on current is supplied at the beginning of the positive half-cycle, a single SCR acts as a *half-wave rectifier. *Full-wave rectification can be achieved by using a bridge rectifier circuit in conjunction with an SCR or by using two SCRs in antiparallel connection (Fig. *e*).

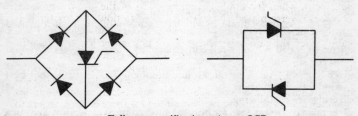

e Full-wave rectification using an SCR

A version of the SCR for switching a.c. is known as the *triac*. It is almost equivalent to two antiparallel SCRs made in a single chip but is able to operate with a single gate connection; its operation relies on a complex pattern of current paths within the device. The triac acts as a bidirectional switch: a typical current-voltage characteristic is shown in Fig. *f* together with the junction diagram of a triac. Since it is effectively an antiparallel arrangement of two SCRs

525

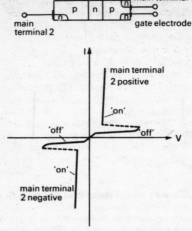

f Construction and I–V characteristic of a triac

neither direction is the forward direction, and both blocking states are termed 'off'.

silicon-controlled switch (SCS) A semiconductor-controlled switch fabricated in a silicon chip. *See* pnpn device.

silicon diode A semiconductor *diode fabricated in silicon.

silicon-gate technology *See* MOS integrated circuit.

silicon-on-sapphire C/MOS *See* MOS logic circuit.

silicon rectifier (1) A *silicon diode used as a *rectifier.

(2) A *Schottky diode, in which the semiconductor is silicon, used as a *rectifier.

silicon solar cell A *solar cell constructed as alternate layers of p- and n-type silicon.

silver Symbol: Ag. A metal, atomic number 47. The metal that best conducts electricity.

silvering *See* metallizing.

silver mica capacitor *Syn. for* bonded silvered mica capacitor. *See* mica capacitor.

simple magnet *See* permanent magnet.

simplex operation Operation of a communications channel in one direction only. *Compare* duplex operation.

SIMS *Abbrev. for* secondary-ion mass spectroscopy.

simulator A device, such as an analog *computer, that simulates the behaviour of an actual physical system and can therefore be used to solve complex problems associated with the operation of the sys-

526

tem. A simulator is usually fabricated from components that are easier, cheaper, or more convenient to manufacture than the system itself.

sine galvanometer *See* tangent galvanometer.

sine potentiometer *See* potentiometer.

sine wave *See* sinusoidal; equivalent sine wave.

singing Unwanted self-sustained oscillations occurring in a *telecommunication system. The term is particularly applied to a telephone line that contains one or more *repeaters. A musical note often results, hence the term 'singing'.

singing arc An *arc that emits a musical note as a result of impressed oscillations that cause a variation in the heating effect. The oscillations are set up when the arc is shunted with an inductance and capacitance in series, as is often the case in a practical circuit. *See also* speaking arc.

singing point *See* feedback.

single crystal A crystal in which the corresponding atomic planes are effectively parallel. This is demonstrated by the single spot diffraction pattern produced by such a crystal using a collimated beam of electromagnetic radiation. The crystal may either be a *perfect crystal or a *mosaic crystal.

single-current system A telegraph system that uses a unidirectional electric current for the transmission of the signal. *Compare* double-current system.

single drift device *See* IMPATT diode.

single-ended Denoting a method of supplying an input signal or obtaining an output signal from a circuit in which one side of the input or output is connected to earth. A *single-ended amplifier* is one in which both the input and output are single-ended. A *double-ended* input (or output) is one in which neither side of the input (or output) is connected to earth and a differential signal is applied (or obtained).

An example of a circuit that may be used with either single- or double-ended input or output is the simple *long-tailed pair (*see* diagram). When used as an *inverter, input B to the base of transistor T_2 is earthed to give single-ended input; the output is taken from output P across a load resistor, R_L, connected to earth.

The circuit may also be used as a *differential amplifier. The difference signal (double-ended) is applied across A and B and the output produced between P and Q. Single-ended output may also be obtained using either P or Q; with Q the output voltage developed is approximately half the value using P, assuming the components are all matched. Differential output may also be obtained using single-ended input.

single-ended amplifier *See* single-ended.

single-gun storage tube *See* storage tube.

527

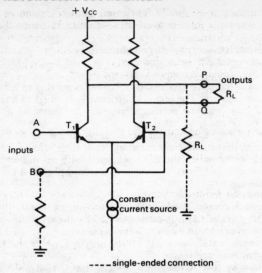

---- single-ended connection

Single-ended and double-ended long-tailed pair

single heterostructure laser *See* injection laser.

single-loop feedback *Syn. for* multistage feedback. *See* feedback.

single-phase Denoting an electrical system or apparatus that has only one alternating voltage. *Compare* polyphase system.

single-pole *See* number of poles.

single-pole switch A switch that operates in only one circuit. *Compare* double-pole switch.

single-shot multivibrator *Syn. for* one-shot multivibrator. *See* monostable.

single-shot trigger A *trigger circuit in which each triggering pulse initiates only one complete cycle of events in a driven circuit, the cycle ending in a stable condition.

single-sideband transmission (SST) The transmission of only one of the two sidebands produced by *amplitude modulation of a *carrier wave. The carrier wave is usually suppressed at the transmitter (in addition to the suppressed sideband) and it is therefore necessary to reintroduce the carrier artificially at the receiver by means of a locally generated oscillation. The frequency of the original carrier must be reproduced in the local oscillation as nearly as possible but this requirement is not as stringent as in *double-sideband transmission.

The main advantages of SST compared to transmission of the carrier and both sidebands are the reduction in transmitter power required for the transmission and the reduced *bandwidth required for the transmission of signals within a designated *frequency band.

single-tuned circuit A *resonant circuit that may be represented as a single *capacitance and a single *inductance with the associated values of resistance.

sink (1) *Short for* heat sink.
(2) *See* source.

sinusoidal Denoting a periodic quantity that has a waveform graphically identical in shape to a sine function; waveforms represented by the functions $\sin x$ and $\cos x$ would both be described as sinusoidal or as *sine waves*.

SI units The internationally agreed system of *units intended for all scientific and technical purposes. The system is based on the *MKS system and replaces the *CGS and Imperial systems of units. There are three types of units in the SI system: *base units, derived units,* and supplementary units. All are *absolute units.

The base units are an arbitrarily defined set of dimensionally independent physical quantities. In any purely mechanical system of units only three base units – of mass, length, and time – are required. In a consistent electric and magnetic system four base units are needed. In the SI system there are seven base units: the *metre, *kilogram, *second, *ampere, *kelvin, candela (the standard of luminous intensity), and mole (the standard of amount of substance).

The derived units are formed by combining, by multiplication and/or division, two or more base units. Thus the *coulomb, which is the derived unit of charge, is formed from a combination of one ampere times one second. The base units, the named derived units relevant to electronics, and the supplementary units are given in Tables 6–8, backmatter. There are at present two supplementary units: the radian and steradian, which are the units of plane and solid angle, respectively.

A set of 14 prefixes, including micro- and kilo-, are used with the SI units to form multiples and submultiples of the units. The prefixes are listed in Table 9, backmatter.

When considering electric and magnetic quantities a fourth term is required in addition to the fundamental units of mass, length, and time, for their complete definition. In the MKS system the fourth quantity is the permeability of free space, μ_0, which is defined as $4\pi \times 10^{-7}$ henry per metre. In the SI system it is the ampere that is the fundamental unit and μ_0 then has the value $4\pi \times 10^{-7}$ henry per metre.

skew A time delay between any two electrical signals.

skiatron A type of *cathode-ray tube that has a screen coated with a *scotophor, such as potassium chloride, and therefore produces a black-on-white picture.

skin depth *See* skin effect.

skin effect A nonuniform distribution of current over the cross section of a conductor when carrying alternating current, with the greater current density located at the surface (or 'skin') of the conductor. The skin effect is caused by electromagnetic induction in the wire and increases in magnitude with increasing frequency. At sufficiently high frequencies the current is almost entirely confined to the surface of the conductor and results in a greater *I^2R loss than when the current is uniformly distributed. The *skin depth* of the conductor, d, is that distance from the surface that the current has decreased to $1/e$ of its surface value. It is given by

$$d = (2/\mu\omega\sigma)^{\frac{1}{2}}$$

for a metal, where μ is the magnetic permeability, ω the angular frequency, and σ the conductivity.

The *effective resistance of the conductor is therefore greater than the d.c. or ohmic resistance, when carrying alternating current, and for high-frequency applications the *high-frequency resistance* of a conductor can be substantially greater than the nominal d.c. value. The *surface resistance* of a metal, $R_{surface}$, is given by

$$R_{surface} = 1/\sigma d$$

A metallic conductor of thickness less than one skin depth will have a radiofrequency current distributed fairly uniformly throughout. The high-frequency resistance will be at a minimum in a conductor of a few skin depths cross-section, with most of the current flow within one skin depth of the surface.

The skin effect may be minimized for high-frequency uses by employing hollow conductors or stranded conductors, such as *Litzendraht wire.

skinny zero *See* fat zero.

skip distance The least distance, for a specified operating frequency, at which radiowaves are received in a given direction from the transmitter by reflection from the ionosphere. Reflection from the sporadic E-layer is customarily ignored. The area surrounding a particular transmitter swept out by a complete rotation of a radius vector equal in length to the skip distance is known as the *skip area*. This area is not necessarily circular as the skip distance is not always the same in all directions.

skip zone *See* silent zone.

sky wave *Syn. for* ionospheric wave. *See* ionosphere.

slave circuit An electronic circuit that requires a *trigger or a *clock pulse from an external source (that may be common to several different slaves) in order to perform its cycle of operation. Examples include a clocked *flip-flop and a *slave sweep*, i.e. a triggered time base.

slew rate The rate at which the output from an electronic circuit or device can be driven from one limit to the other over the *dynamic range.

slice *Syn.* wafer. A large single crystal of semiconductor material that is used as the *substrate during the manufacture of a number of *chips. Very large single crystals are grown and then sliced into wafers before processing. The number of viable chips that can be produced from a single slice, typically up to 10 cm in diameter in the case of silicon, depends on the size and complexity of the circuits or components on the chips.

slicer *See* limiter.

slicing *See* compact disc system.

slide wire A wire of uniform resistance provided with a sliding contact that can make a connection at any desired point along the length. Slide wires are used to provide a variable resistance, as a potentiometer, or to provide a desired resistance ratio (*see* Wheatstone bridge). An overall length of one metre is commonly chosen.

slope resistance *Syns.* electrode a.c. resistance; electrode differential resistance. The ratio of an incremental voltage change applied to a specified electrode in an electronic device to the corresponding current increment of that electrode, the voltages applied to all other electrodes being maintained constant at known magnitudes.

For example, the slope resistance, r_c, of the *collector of a bipolar junction *transistor is given by

$$r_c = \partial V_c / \partial I_c \qquad V_b, V_e = \text{constant}$$

where V_c is the collector voltage, I_c the collector current, and V_b and V_e the base and emitter voltages, respectively.

slot matrix tube *See* colour picture tube.

slotted waveguide *See* waveguide.

slow-break switch A manually operated switch whose speed of operation depends upon the speed at which the operating handle or lever is moved. *Compare* quick-break switch.

slow-operating relay *Syn.* time-delay relay. *See* relay.

slow-wave structure *See* travelling-wave tube.

slow-wave tube *See* travelling-wave tube.

slug tuning *Syn.* permeability tuning. *See* tuned circuit.

small-scale integration (SSI) *See* integrated circuit.

small-signal parameters *See* network; transistor parameters.

smearer *See* pulse.

smoothing choke *See* choke.

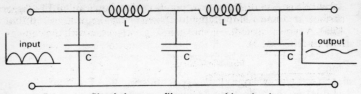

Simple low-pass filter as smoothing circuit

smoothing circuit *Syns.* ripple filter; rectifier filter. A circuit that is designed to reduce the amount of *ripple present in an essentially unidirectional current or voltage. A typical smoothing circuit consists of a low-pass *filter (*see* diagram) but a single inductance may be used.

snapback diode *Syn.* snap-off diode. *See* charge-storage diode.

snow An unwanted pattern appearing on the screen of a *television or *radar receiver that resembles falling snow. It may appear when the received signal is absent or transmitted at a lower power than is usual and is caused by noise associated with the circuits and equipment forming the receiver. In the case of a *colour picture tube, the snow is coloured.

soft ferromagnetic material *See* ferromagnetism.

soft tube *Syn.* soft valve. *See* electron tube.

software The programs associated with a *computer. *Compare* hardware.

soft X-ray appearance potential spectroscopy (SXAPS) *Syn.* particle-induced X-ray emission. A method of analysing the material in the surface of a semiconductor by detecting the soft X-rays emitted by atoms in response to stimulation by an incident beam of low-energy electrons. The energy of the beam is gradually increased from zero to approximately 1 keV. Elements present in the sample will begin to emit X-rays only when the electron beam energy is sufficiently high, and the value of the threshold energy (the *appearance potential*) for X-ray emission is a characteristic of the particular element. As the electron beam energy is increased, peaks in the emitted X-ray spectrum are observed, characteristic of the elements present in the surface of the semiconductor. The process is nondestructive; because of the low energies involved the spectra are comparatively simple and elements below approximately 10 nanometres of the surface are not involved. Unfortunately some elements are not detected by this method, but it is particularly sensitive to other elements, especially the rare earths.

soft X-rays *See* X-rays.

solar cell A device that utilizes the *photovoltaic effect in order to convert radiation from the sun directly into electrical energy. The solar cell is essentially a *p-n junction that has a relatively large surface area and relatively high efficiency (of the order of 10% to 15%). A typical structure is shown in Fig. *a* together with the equivalent circuit (Fig. *b*).

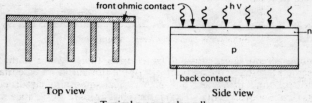

a Typical n-on-p solar cell

Solar cells are the most important long-duration power supply for satellites and space vehicles. They have been fabricated from a range of different semiconductors including silicon, gallium arsenide, selenium-cadmium sulphide, and thin-film cadmium sulphide. Gallium arsenide provides the optimum inherent efficiency at lower temperatures (from 0°C). At higher temperatures (above 100°C) a 50–50 gallium arsenide/gallium phosphide material produces the optimum.

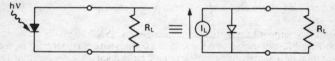

b Equivalent circuit of an idealized solar cell

The output of a solar cell is given by the short-circuit current, I_{sc}, produced for a given incident flux of radiation. This is very strongly dependent on the wavelength, λ, of the radiation due to several factors. Assuming that the incident flux is of photons of energy, $h\nu$, greater than the band-gap energy, E_g, the number of charge carriers generated in the material depends on the photon density within the material, which is given by

$$\phi = \phi_0 \exp(-\alpha x)$$

where ϕ_0 is the incident flux, α the absorption coefficient, and x the distance below the surface. ϕ_0 and α are both a function of wavelength, since the energy of an individual photon depends on frequency. Under steady-state conditions the effect of the radiation is that of a current source in parallel with the cell. The output current is proportional to the number of the carriers that cross the junction.

Some of the generated carriers are lost by *recombination as they diffuse to the junction. The number crossing the junction depends on the depth, d, of the junction below the surface and on the diffusion lengths, L_p and L_n, of minority carriers on each side of the p-n junction. Assuming that

$$\alpha L_n, \alpha L_p \ll 1$$

and

$$d/L_p, d/L_n \gg 1$$

the output response may be simplified to

$$dI_{sc}(\lambda)/d\lambda \simeq \alpha\lambda(L_n + L_p)\exp(-\alpha d)$$

A typical spectral response characteristic is shown in Fig. c for a p-on-n silicon junction with d equal to 2.0 micrometres.

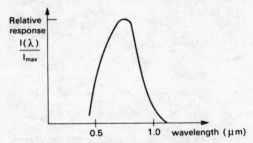

c Spectral response of a typical silicon cell

The absorption coefficient, α, is large for shorter wavelengths so the short-wavelength response may be improved by forming the junction relatively close to the surface. The long-wavelength response may be improved by forming a comparatively deep junction. The depth is usually chosen to given maximum response near blue light for optimum response.

In a practical solar cell the output current may be reduced by recombination occurring within the depletion region and by the series resistance of the cell. Both these factors are dependent partially on the impurity concentration; the former also depends on other defects within the lattice and the latter on the device geometry. Cells with a low series resistance will produce a higher output.

The presence of a large number of defects in the lattice reduces the output. The *radiation effect*, whereby the solar cell output is reduced with time, is due to the production of lattice defects by bombardment with the high-energy particulate radiation (cosmic rays) in outer space. This effect is the major limiting factor affecting the useful life of the cell.

solenoid A coil of wire that has a long axial length relative to its diameter. The coil is usually tubular in form and is used to produce a known magnetic flux density along its axis.

At a point on the axis inside the solenoid, ignoring any end effects, the magnitude of the magnetic flux density, B, is given by

$$B = \tfrac{1}{2}\mu_0 nI(\cos\theta_1 + \cos\theta_2)$$

where μ_0 is the magnetic constant, n the number of turns per unit length, I the current flowing through the solenoid, and θ_1 and θ_2 the semiangles subtended at the point by the ends.

A solenoid is often used to demonstrate *electromagnetic induction and a bar or rod of iron that is free to move along the axis of the coil is usually provided for this purpose.

solid conductor A conductor that is composed of a single wire or uniform thin metal rather than being stranded or otherwise divided.

solid-state camera A *television camera in which the light-sensitive target area consists of an array of *charge-coupled devices (CCDs). Exposure to light energy results in the generation of electron-hole pairs in the semiconductor substrate (*see* photoconductivity); the number of electron-hole pairs generated is a function of the light intensity. The majority carriers migrate into the bulk material and the minority carriers accumulate in the potential wells at the electrodes of the CCDs. The accumulated charge is transferred into other nonphotosensitive CCD storage sites during the retrace intervals of the television display system and is then transferred to the output device while further signals are being generated.

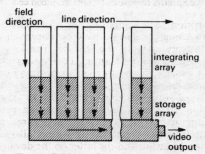

a Frame/field transfer solid-state camera

Two basic methods of obtaining the video signal are used. A *frame/field transfer device* is shown in Fig. *a*. A set of charge-transfer devices – the *integrating array* – is arranged in the vertical (field) direction and exposed to the optical image. During the vertical retrace interval of the television display system the charge pattern that has accumulated is clocked at high speed into the storage area. During the normal horizontal (line) blanking periods the pattern of charges in the storage area is moved downwards by one line into the bottom horizontal register and clocked horizontally to the output to form the video signal.

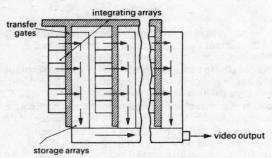

b Interline transfer device

An *interline transfer device* is shown in Fig. *b*. In this type of solid-state camera the storage arrays are arranged alternately with the photosensitive arrays and are connected to them by a transfer gate. During the vertical retrace period the information is transferred horizontally from the photosensitive electrodes to the storage electrodes by means of the transfer gates and then read out line by line in a similar manner to the frame/field device.

Both types of device may be used for interlaced scanning (*see* television). Interlacing is achieved in the frame/field device by adjusting the potentials applied to the gate electrodes of the integrating array so that information accumulating during the second field period is physically offset by half of a storage cell length compared to that accumulated during the first field period (Fig. *c*). The entire array is completely emptied of information during each field sweep and the integrating period is equal to the field period.

Interlacing in the interline transfer device is achieved by adjusting the potentials of the transfer gates so that the information in alternate sites (Fig. *b*) is transferred during each vertical retrace interval. The integrating time in each site is therefore for the full frame interval.

The frame/field transfer device requires only half the number of integrating sites required by the interline transfer device but the location of the storage area is such that additional area of semiconductor chip is required. The interline device has a more economical

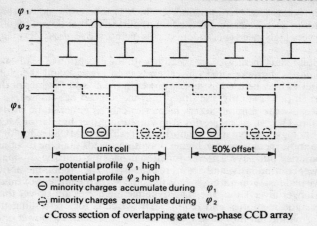

φ_1

φ_2

φ_s

|← unit cell →| |← 50% offset →|

——— potential profile φ_1 high
------- potential profile φ_2 high
⊖ minority charges accumulate during φ_1
⊜ minority charges accumulate during φ_2

c Cross section of overlapping gate two-phase CCD array

arrangement of electrodes but can be less sensitive than the frame/field device since the transfer gates and storage sites must be shielded from the optical image and twice as many storage sites are required. Integration for the full frame period can result in a slower response to fast moving objects; this can be overcome by reading out both sets of integration sites during each field period and pairing the outputs in a suitable manner. Extra sensitivity can be achieved using buried-channel CCDs for applications where a low light intensity is used.

solid-state device An electronic component or device that is composed chiefly or exclusively of solid materials, usually semiconducting, and that depends for its operation on the movement of charge carriers within it. A solid-state device has no moving parts.

solid-state maser *See* maser.

solid-state memory *Syn.* semiconductor memory. A *memory formed as an integrated circuit in a chip of semiconductor. Solid-state memories are widely used as internal memory in computers; they are cheap, robust, compact, and operate on relatively low voltages. The memory capacity that can be stored on a single chip is increasing by a factor of four approximately every three years.

There are several different types of solid-state memory. One of the most important types is the read-write *random-access memory (RAM) that is used as the main working memory of computer systems and stores data and programs, which are accessed by the central processing unit of the system. Solid-state *read-only memories are used for the permanent and semipermanent storage of informa-

537

tion. Solid-state *serial memory, which is inherently slower than random-access memory, is used for applications that do not require the very high speeds of random-access memory. *See also* CCD memory.

Solid-state memories are volatile, with the exception of read-only memories, but since the power levels used are low and therefore batteries may be used to supply power in the event of failure of the normal power supply, the stored information can be retained.

Read-write random-access memories can be either static or dynamic. The *static memories are realized in either bipolar or MOS technology The dynamic memories are realized in MOS technology. The static memories have higher operating speeds than the dynamic memories but have a lower functional packing density, dissipate power continuously, and are more expensive. The bipolar memories have had faster operating speeds than the MOS memories but this is no longer always the case. The dynamic memories are slower than the static memories and require extra circuits for *refreshing the information but they have a larger density and dissipate power only when operating; the stand-by power dissipation is very low.

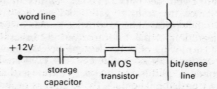

word line

+12V

storage
capacitor

MOS
transistor

bit/sense
line

a Equivalent circuit of single memory cell

The general purpose dynamic RAM suitable for most applications, except when the highest speeds of operation are required, is the *MOS random-access memory* (MOSRAM). The basic very simple memory cell consists of an MOS capacitor, in which the information is stored as electronic charges, and an MOS transistor, which is used as a switch in order to connect the appropriate capacitor to the sense amplifier. The interconnections between memory cells form the rows and columns of a rectangular matrix so that each memory cell has a unique address (*see also* random-access memory). The equivalent circuit of the basic memory cell is shown in Fig *a*. One plate of the storage capacitor is connected to the source of the MOS transistor; the gate and drain electrodes are connected to the rectangular matrix.

A cross section through a typical memory cell is shown in Fig. *b*. The MOS capacitor consists of a layer of highly doped polysilicon that forms the upper plate (POLY 1). The lower plate is formed

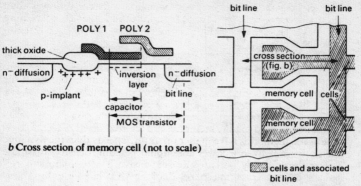

b Cross section of memory cell (not to scale)

c Schematic top section showing memory cell packing

when a positive potential of + 12 volts is applied to the polysilicon: an inversion layer is produced at the surface of the substrate and acts as the second plate. The MOS transistor is formed with a second polysilicon layer (POLY 2), which is the gate electrode. Overlapping the two layers of polysilicon allows the inversion layer of the capacitor to act as the source of the transistor. The bit line is formed from an n+ diffusion into the substrate and also acts as the drain of the transistor. The thick oxide layer is produced using the *coplanar process and extra p-type ions are implanted below it, in order to provide isolation of the memory cells and prevent *spurious MOST formation.

The memory cells are packed as closely as possible into the chip by suitable design of the arrangement of the cells and associated bit lines and the patterns of the various layers that make up the array. A small section is shown schematically in Fig. *c;* Fig. *d* shows a plan of eight memory cells with parts of the various layers emphasized.

A particular storage location is selected by applying a high voltage level to the word line, which is connected to the gate electrodes of all the transistors in the column. This causes these switch transistors to be 'on'. Each memory cell is connected to a different bit line, which in turn is connected to a sense amplifier. The stored charge produces an appropriate output level on the chosen bit line by charge sharing. Built-in positive feedback is provided by a suitable arrangement of switch transistors (Fig. *e*) so that the condition of the selected storage capacitor is automatically regenerated following a read operation, i.e. logical 1 or logical 0. Leakage of charge occurs during the periods when the storage location is inactive and

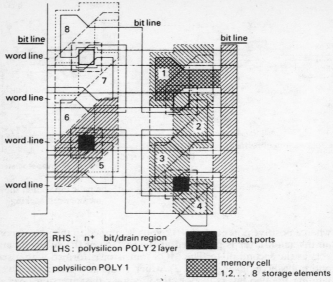

d Schematic top section of 8 bit section of MOSRAM memory

the data must therefore be regenerated periodically. This is achieved by periodically performing a read-write cycle automatically on every storage location. Information is input to the memory by switch transistors that produce the correct voltage levels for charge to be stored (or not stored) at a particular site on the memory.

A memory with a storage capacity of 16 kilobits can be produced on a chip that measures approximately 6mm by 3mm. The chip contains various logic circuits in addition to the memory cells and sense amplifiers: address decoders select the desired location; automatic refresh circuits regenerate the data; clocking circuits control the various functions and operate the switch transistors in order to perform read or write operations.

Static memories are used as a buffer memory (*see* computer) in large computer systems that require extremely high speeds of operation. They usually consist of an array of *flip-flops connected to the address lines. A typical memory cell is shown in Fig. *f* for both bipolar and MOS circuits. In the latter case extra MOS transistors may be used as the load resistors. A high logic level (logical 1) at points A causes the transistors T_2 to be 'on' and the voltage at points B therefore drops. This in turn causes transistors T_1 to be 'off',

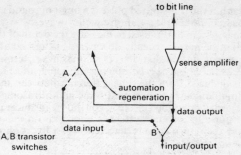

e Read-write circuit

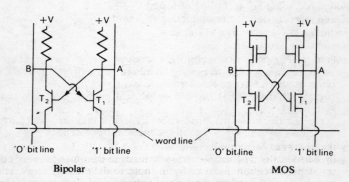

Bipolar MOS

f Basic static memory cell

which maintains the high voltage level at points A and the circuit is latched in that state. A low logic level at points A causes the reverse situation and transistors T_1 are 'on'. Power is continuously dissipated since one or other of the transistors is always conducting. The data is read using sense amplifiers connected to the bit lines.

solid-state physics The branch of physics that studies the structure and properties of solids and any associated phenomena. Properties and associated phenomena that are dependent on the structure of the solid include electrical conductivity, semiconduction, superconductivity, photoconductivity, the photoelectric effect, and field emission.

solid-state relay *See* relay.

sonar Acronym from *so*und *na*vigation *r*anging. A method of detecting and locating underwater objects that operates on a similar principle to *radar but the transmitted pulse is a burst of sound energy, usually ultrasonic, rather than radiofrequency waves.

A sonar system that is used to measure the depth of the sea bed by projecting the pulse vertically downwards is termed an *echo-sounder*. Any sonar system that detects submarines is termed *asdic*. One type of asdic employs *echo-ranging*, in which the location of the target is deduced from the time difference between the two echoes from simultaneously transmitted sonic and ultrasonic pulses.

The specially designed electro-acoustic *transducer used to transmit the pulse in a sonar system is known as an *underwater sound projector;* the return echo is detected with a hydrophone.

sonde *See* radiosonde.

sound carrier *See* television.

sound-on-film *See* recording of sound.

sound recording *See* recording of sound.

sound reproduction *See* reproduction of sound.

soundtrack *See* recording of sound.

sound wave *See* acoustic wave.

source (1) The point in a vector field at which lines of flux originate; an example is a positive electric charge in an electrostatic field. A point at which lines of flux terminate is a *sink*.

(2) Any device that produces electrical energy, e.g. a current source.

(3) The electrode in a *field-effect transistor that supplies charge *carriers (holes or electrons) to the interelectrode space.

source follower *See* emitter follower.

source impedance The *impedance presented to the input terminals of an electronic circuit or device by any source of electrical energy. The source impedance of an ideal voltage source is zero (i.e. $dV_s/dI_s = 0$) whereas that of an ideal current source is infinity (i.e. $dV_s/dI_s = \infty$).

space charge In any device, a charge density that is significantly different from zero in any given region. The region containing the space charge is a *space-charge region*. In a *semiconductor it is the region containing the depletion layers associated with the junction between two dissimilar conductivity types. In a *thermionic valve the space-charge region surrounds the cathode and contains electrons not immediately attracted to the anode. These two examples of space-charge regions can exist in the devices in equilibrium under conditions of zero applied bias; they constitute potential barriers that must be overcome, when bias is applied, before the device can conduct.

Space charge also causes divergence of a beam of electrons and *debunching in *velocity-modulated tubes. A radial field is often used to counteract the space-charge divergence of an electron beam and results in a cylindrical beam. This type of flow of the electrons is *Harris flow*.

space-charge density The net electric charge per unit volume in a space-charge region (*see* space charge).

space-charge limited region *See* thermionic valve.

space-charge region *See* space charge.

space diversity *See* diversity system.

space wave A radiowave that travels between a transmitting and a receiving aerial situated above the ground and that includes the *direct wave and the *ground-reflected wave. It is the component of the *ground wave that does not travel along the surface of the earth. If the two aerials are placed at a sufficient height above the ground the *surface wave is negligible and only the space wave needs to be considered.

spark A visible disruptive discharge of electricity between two points of high potential difference, preceded by ionization of the path. A sharp crackling noise occurs because of the rapid heating of the air through which the spark passes. The distance travelled is determined by the shape of the electrodes and the potential difference between them.

Under specified conditions the distance between the electrodes is termed the *spark gap*. Specially designed electrodes are used to produce a spark over a given spark gap under particular conditions, as for ignition purposes in an internal-combustion engine. The insulation is self-restoring when the potential across the spark gap falls below that required to produce the spark. When the electrodes are in the form of needle points the spark gap is termed a *needle gap*. A *sphere gap* is one that has spherical electrodes. A spark is of much shorter duration than an *arc.

spark channel The path followed by a *spark between two electrodes. The channel becomes established by ionization of the gas (usually air) between the electrodes and may persist for some time. It is not necessarily the shortest possible path.

spark coil *See* ignition coil.

spark counter A radiation detector that is used to detect and count alpha particles. The counter consists of two electrodes: a wire or mesh anode and a metal plate cathode in close proximity. A high potential difference, just less than that required to cause a *spark, is applied across the electrodes. When an alpha particle approaches the anode the electric field between the electrodes rises sufficiently to cause a spark, which is heard as a sharp crackling noise. At the moment of discharge the voltage across a load resistor, connected in series with the anode, drops significantly.

The alpha particles may be detected by ear because of the noise associated with the sparks. They may be counted by using a suitable counting circuit that detects the voltage drops across the anode load resistor or by photography.

Alpha counters are completely unaffected by beta particles and X- or gamma rays and are therefore very useful for the detection of leakage of alpha-emitters (such as radium) from sealed containers that normally completely absorb the alpha particles.

spark gap *See* spark.

sparkover *See* flashover.

spark photography Any form of photography that uses a *spark discharge as the source of illumination.

speaker *Short for* loudspeaker.

speaking arc A means of reproducing sounds that uses an *arc discharge to amplify the original sound. The small induced current from a microphone is transformer-coupled to a circuit containing an arc-discharge device. The resulting variations of the current in the arc cause variations in the amount of heat evolved by the arc and hence variations of the pressure in the surrounding air. The resulting pressure variations are heard as the original sounds considerably intensified.

specific conductance *Obsolete syn. for* conductivity.

specific contact resistance *See* contact.

specific resistance *Obsolete syn. for* resistivity.

spectral characteristic A graph that shows the sensitivity or relative output as a function of frequency for any frequency-dependent device, circuit, or other equipment.

spectrum analyser *See* multichannel analyser.

speech coil *Syn.* voice coil. *See* loudspeaker.

speech recognition device A device that produces from the human voice coded signals suitable for a given *computer.

speed of light *See* Table 5, backmatter.

sphere gap *See* spark.

spike *See* pulse.

spill The loss of information from an element of a storage device, such as a charge *storage tube, due to unwanted redistribution of charge within the device.

spin The intrinsic angular momentum possessed by all elementary particles, including the electron and proton. An atomic electron also possesses angular momentum as a result of its orbital motion. Spin is a quantized quantity, the *spin quantum number* (or *spin*), s, of particles having either a half integral or integral value. Both electron and proton have spin $\frac{1}{2}$.

The magnetic fields produced by an atomic electron's spin and orbital motion interact in such a way that the electron can exist in either of two closely spaced energy levels. The spin in one level is

$+\frac{1}{2}$ and in the other it is $-\frac{1}{2}$. Because of its spin an electron has an intrinsic *magnetic moment.

spiral scanning *See* scanning.

split-anode magnetron *See* magnetron.

split-electrode technique *See* CCD filter.

spontaneous transition *See* maser.

spot (1) An area, such as on the screen of a cathode-ray tube, that is immediately affected by an electron beam impinging on or near to it.

(2) A local imperfection on the surface of an electrode.

spot speed (1) The product of the number of spots in the scanning line and the number of scanning lines per second in a television picture tube.

(2) The number of spots scanned or recorded per second in *facsimile transmission.

sprat Acronym from *small portable radar torch*. A hand-held radar system that is based on the *Gunn diode. It is light (about 2·5 kg) and has a range of about 600 metres.

spreading resistance The component of the series resistance of a semiconductor device that is due to the bulk material located some way from the junctions and metal contacts.

spurious MOST An unwanted MOS *field-effect transistor formed in an *MOS integrated circuit by the interconnections between the parts of the circuit. If the voltage on an interconnection crossing a region of thick (or field) oxide between diffusions exceeds the field threshold voltage, an unwanted conducting *channel will form below the oxide.

spurious response An unwanted output from an electronic circuit, device, or *transducer in the absence of an input signal or as a result of the presence of an input signal other than the required signal.

sputter etching *See* etching.

sputtering The use of a high-voltage low-pressure glow discharge to deposit material in the form of a film. *See* cathode sputtering.

square-law detector A *detector that produces an output voltage that is proportional to the square of the input voltage. Relatively large variations in output voltage result from minor variations in the input voltage and the sensitivity of this type of detector is therefore relatively high.

square-loop material A ferromagnetic material that exhibits a substantially rectangular *magnetic hysteresis curve. The ratio of the remanence to the saturation flux density of such a material is the *squareness ratio*.

square wave A *pulse train that consists of rectangular pulses the mark-space ratio of which is unity (*see* diagram).

Square wave

square-wave response In general, the response of any electronic circuit or device to a square-wave input signal, given as the wave shape and peak-to-peak amplitude of the output signal.

In particular, it is the peak-to-peak amplitude obtained from a television *camera tube in response to a test pattern of alternate black and white bars.

squegging oscillator A type of oscillator in which the main oscillations output from it alter the electrical conditions of the oscillator circuit so that the output amplitude periodically builds up to a peak value and then falls to zero. The *blocking oscillator is a special type of squegging oscillator and is frequently used as a *pulse generator for use, for example, in *radar systems.

squelch circuit A circuit used in a radio receiver that suppresses the audiofrequency output unless an input signal of a predetermined character is received.

SQUID (or **squid**) Acronym from *s*uperconducting *qu*antum *i*nterference *d*evice. An extremely sensitive magnetic-field detector that consists of two Josephson junctions (*see* Josephson effect). The presence of an external magnetic field is detected by measuring the voltage change across the junctions.

SSB *Abbrev. for* single sideband.

SSI *Abbrev. for* small-scale integration. *See* integrated circuit.

SST *Abbrev. for* single-sideband transmission.

stabilivolt A gas tube, such as a neon lamp, that produces an essentially constant voltage drop across its terminals for a relatively wide range of currents. Stabilivolts have been used as voltage regulators for a wide range of uses but have now been replaced for many applications by *catching diodes.

stabilization The provision of negative *feedback in a circuit, such as an *amplifier, that contains inductance and capacitance and introduces substantial *gain between the input and the output, in order to provide overdamping (*see* damped) in the circuit and thus prevent any oscillations occurring.

Shunt stabilization is a type of stabilization in which the amplifier and the feedback circuit are connected in parallel. *Series stabilization* is provided by a feedback circuit connected in series with the amplifier.

stable circuit A circuit that does not produce any unwanted oscillations over the entire dynamic operating range.

stable oscillations Oscillations that tend to decrease in amplitude with respect to time.

stage *Short for* amplifier stage.

stage efficiency *See* amplifier stage.

staggered aerial *Syn. for* endfire array. *See* aerial array.

stagger tuned amplifier *See* tuned amplifier.

standard cell An electrolytic *cell that is used as a voltage reference standard. *See* Clark cell; Weston standard cell.

standardization (1) The process of relating a physical magnitude or the indicated value of a measuring instrument, such as a voltmeter, to the primary standard unit of that quantity.

(2) A nationally or internationally agreed system for the standardization of electronic or electrical components or devices. If electronic devices, etc., are produced haphazardly by different manufacturers the ease of use of such components is severely limited. The convenience of the user is greatly increased when there are several sources from which devices may be chosen and when the devices are interchangeable. *Interchangeability* is greatly increased if the devices are produced only in certain physical sizes and in a range of predetermined values that are easily identifiable, usually by a *colour code.

These values are the *preferred values* and are chosen by certain technical bodies although other values for special applications are not precluded. Broadcast transmissions and other means of international telecommunications are also greatly facilitated by international agreement on the frequency bands to be used and on harmonization of equipment between the sender and receiver.

The main international bodies that determine standards for telecommunications are the *International Telecommunication Union* (ITU) with its two associated international committees: the *International Telegraph and Telephone Consultative Committee* (CCITT) and the *International Radio Consultative Committee* (CCIR). The main international standardizing bodies for electronic components and parts are the various technical committees of the *International Electrotechnical Commission* (IEC), with some broad areas defined by the *International Standards Organization* (ISO).

In the United Kingdom the body that determines national standards is the *British Standards Institute* (BSI), acting on advice from its various technical committees. The *Institute of Electrical Engineers* (IEE) is concerned with standardization for electronic equipment.

In the United States there are various bodies and committees that determine the standards on behalf of the *USA Standards Institute* (USASI). These include the Electronic Industries Association (EIA), the Institute of Electrical and Electronics Engineers, Inc. (IEEE), the Joint Electron Device Engineering Council (JEDEC), the National Bureau of Standards (NBS), National Aeronautical and Space Administration (NASA), the Federal Communications Committee (FCC), and the National Television System Committee (NTSC).

standing wave *Syn.* stationary wave. A wave that remains stationary, i.e. the displacement at any given point is always the same and a given displacement, such as that of a node, is not propagated along the wave. Standing waves result from the superimposition of two or more waves of the same period and usually occur when a wave is reflected totally or partially from a given barrier. *Compare* travelling wave.

standing-wave accelerator *See* linear accelerator.

starter electrode An auxiliary electrode that is used in a glow-discharge tube (*see* gas-discharge tube) in order to initiate the glow discharge; it is also used in an arc-discharge tube, such as a *mercury-arc rectifier, to initiate the arc discharge. A sufficiently high potential difference – the *starter voltage* – is applied between the starter electrode and the cathode and once the glow discharge is established the necessary voltage is applied to the anode. The maintaining voltage is considerably lower than the voltage required to initiate the discharge and this method of operation protects the main circuit from excessively high voltages. The starter electrode may also be used to maintain the discharge during a period when the main anode is 'off'.

The conduction path between the cathode and the starter electrode is termed the *starter gap;* that between cathode and anode is termed the *main gap.*

starter gap *See* starter electrode.

starter voltage *See* starter electrode.

starting current The magnitude of the current in an *oscillator at which self-sustaining oscillations are initiated for specified load conditions.

start-oscillation current *See* travelling-wave tube.

start-stop apparatus A device used in *telegraphy in which each set of coded signal elements corresponding to a transmitted character is preceded by a start signal that activates the receiving apparatus in preparation for the character and is followed by a stop signal that brings the receiving apparatus to rest. Start-stop apparatus is used in *telex systems. *See also* teleprinter.

stat- A prefix to a unit, indicating its use in the obsolete CGS electrostatic system of units.

1 *statampere* = $3 \cdot 336 \times 10^{-10}$ ampere
1 *statvolt* = $2 \cdot 998 \times 10^2$ volts
1 *statohm* = $8 \cdot 988 \times 10^{11}$ ohms
Compare ab-.

548

static (1) Unwanted random noises produced at the loudspeaker of a *radio receiver. It is usually heard as an unpleasant crackling sound and is caused by atmospheric conditions, usually by the presence of static electricity in the air through which the radiowave is propagated.

(2) *See* dynamic.

static characteristic *See* characteristic.

static memory A *solid-state memory that does not require refresh operations in order to retain the stored information.

stationary orbit *See* communications satellite.

stationary state The state of an atom or other system that in the quantum theory or quantum mechanics is described by a given set of quantum numbers. Each of the various energy states that may be assumed by an atom is a stationary state of that atom.

stationary wave *See* standing wave.

statistical weight *See* degeneracy.

stator *See* motor.

steady state A state reached by a system under steady operating conditions after any transient effects resulting from a change in the operating conditions have died away. A steady state occurs, for example, with *forced oscillations.

steel-tank rectifier *See* mercury-arc rectifier.

steepness (of a wavefront) *See* impulse voltage.

steerable aerial *Syn.* musa (acronym from *multiple unit steerable aerial*). A *directive aerial that consists of several fixed units in which the direction of maximum sensitivity (i.e. the direction of the major lobe of the *radiation pattern) can be altered by adjusting the phase relationship between the units from which it is formed.

step-down transformer *See* transformer.

step function *See* unit step function.

stepped leader stroke *See* lightning stroke.

stepping relay A *relay with a contact arm that rotates through 360° in two or more discrete steps.

step-recovery diode *See* charge-storage diode.

step stress life test A *life test consisting of several stress levels applied sequentially for periods of equal duration to one sample. During each period a stated stress level is applied and increased from one step to the next.

step-up transformer *See* transformer.

stereophonic radio receiver *See* radio receiver.

stereophonic sound reproduction *See* reproduction of sound.

sticking potential *See* cathode-ray tube.

stimulated transition *See* maser.

STL *Abbrev. for* Schottky I^{2}L. *See* I^{2}L.

stochastic process A random process.

Stokes' law *See* luminescence.

stopper (1) *See* parasitic oscillations. (2) *See* channel stopper.

stopping potential *See* photoelectric effect.

storage battery A *battery that is formed from secondary *cells.

storage capacity *Syn. for* memory capacity. *See* memory.

storage cathode-ray tube *See* storage tube.

storage cell *Syn. for* secondary cell. *See* cell.

storage device Any device that holds information by a physical or chemical method. The term is particularly applied to a *memory in a computer, which holds information usually in binary form.

storage effect *See* carrier storage.

storage element (1) *See* storage tube. (2) *See* memory.

storage location *See* memory location.

storage time (1) *Syn.* retention time. In any device that stores information, such as a *storage device or a *storage tube, the maximum time that the information may be stored without significant loss of information.

(2) *See* carrier storage.

storage tube An electron tube that is used to store information for a determined and controllable time and from which the information may be extracted as required. Various principles are used to operate storage tubes and there are many different types of tube. The most common types of tube in general use are *charge-storage tubes* in which the information is stored as a pattern of electrostatic charges. The information may be extracted as a visual display, as in the storage cathode-ray tube, or as an electronic signal, as in the electrostatic charge-storage tube.

Charge-storage tubes contain a *target plate* on which the information is stored. The information to be stored is used to modulate the intensity of an electron beam and the beam is made to scan the target. *Secondary emission of electrons from the target plate occurs and an electrostatic charge pattern is left on the target. Each small area of the target that retains information and is distinguishable from neighbouring small areas forms a *storage element*. The number of secondary electrons emitted from each storage element is a function of the energy of the electron beam, the tube design, and the intensity of the beam. If the unmodulated beam produces a *secondary-emission ratio greater than unity a positive charge image is produced; if it is less than unity a negative charge image results. The information-modulated electron beam is called the *writing beam;* the rate at which information can be written to successive elements is the *writing speed.*

Storage cathode-ray tubes produce a visual display of controllable duration. The tube has two *electron guns – the *writing gun* and the *flooding gun;* it also has a phospher viewing screen and two fine mesh metal screens. One of the metal screens, the *storage screen,* is

550

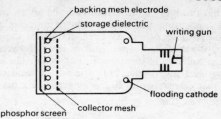

a Storage cathode-ray tube

coated with a thin dielectric material to form the target and the other serves as an electron collector (Fig. *a*). A positive charge image is produced on the storage screen by scanning with a high-resolution intensity-modulated writing beam from the writing gun. It remains until it decays or is erased. Information is extracted by *flooding* the storage screen with an electron beam from the flooding gun. Each storage element effectively forms an elemental electron gun with each mesh hole forming a control element of one of the guns. The value of positive charge deposited at each aperture determines the amount of current (from the flooding beam) that can pass through to the phosphor viewing screen. At the viewing screen a light output is produced that is a function of the original information. The stored charges are erased by flooding the storage surface with low-velocity electrons, thereby depositing negative charge on each element, until the pattern is erased and the surface prepared for storing a new image.

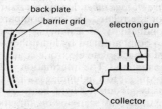

b Single-gun barrier-grid tube

Electrostatic charge-storage tubes do not produce a visual image. The target of the *barrier-grid storage tube* consists of a mesh barrier grid placed in front of a backplate. A thin layer of dielectric material may sometimes be placed between them. The target acts essentially as an array of elemental capacitors that are charged or discharged by an electron beam from the electron gun. Information is stored by scanning the target with a high-resolution intensity-modulated writ-

551

ing beam that produces a pattern of charged areas. Information is extracted by scanning the target with an unmodulated electron beam – the *reading beam*. As each elemental capacitor is approached by the reading beam electrons leave the area and are attracted to the collector to form the output signal. The same electron gun may be used to produce both the writing and reading beam (*single-gun tube*), as shown in Fig. *b*, or separate guns may be used (*two-gun tube*).

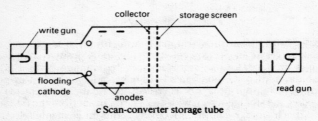

c Scan-converter storage tube

Another type of electrostatic storage tube is the *scan converter*. This tube consists of two electron guns, which are mounted coaxially at opposite ends of the tube, a collector mesh, and a storage mesh screen coated with dielectric (Fig. *c*). Each electron beam is focused and caused to scan by its own deflection system. One beam writes and the other reads; both operations may be carried out simultaneously. A charge pattern may be read out many times (up to several thousand) before being erased by flooding with low-velocity electrons from either of the guns.

Photoconductive storage tubes depend for their operation on *photoconductivity or the related electron-bombardment conductivity, in which the conductivity of a material is temporarily increased when exposed to bombardment by light photons or electrons. The target consists of a back-plate electrode coated with a thin layer of photoconductive material. The information is deposited on the target either by exposing it to a light image or by scanning with a high-resolution intensity-modulated light beam or electron beam. The information is extracted by scanning the target with an unmodulated reading beam; the numbers of electrons, i.e. the signal intensity, reaching the back-plate electrode depends on the conductivity of each small element. The output is produced as a current in a load resistor in series with the back-plate electrode. The *graphecon* is a photoconductive tube that uses a high-velocity electron beam as the writing beam. It is used in *radar or as a translating mechanism.

The *decay time* of any storage tube is the time during which the stored information decays to a specified fraction (usually 1/e) of the original value; it limits the useful time for which information may be stored. In most cases the charge pattern may be regenerated by exposing the target to a diffuse beam of electrons – the *holding beam*.

Spurious pulses in the output are sometimes generated. *Cloud pulses* are caused by space-charge effects when the electron beams are switched on or off; *blemish* pulses are due to imperfections of the target surface. Storage tubes may also suffer from *spill.

There has also been development of storage tubes that depend for their operation on photoemission, luminescence, and photochromism, but these are relatively uncommon.

store *See* memory.

strain gauge An instrument that measures strain at the surface of a solid body by means of changes in the electrical properties of associated circuits. There are several types of strain gauge.

A *resistance strain gauge* consists of a fine wire attached to the surface. The length of the wire is altered by the strain, causing an associated variation in the resistance.

An *electromagnetic strain gauge* consists of a small soft-iron armature attached to the surface and free to move inside a fixed inductive coil. Movement of the surface alters the position of the armature causing an associated variation in the inductance of the coil.

A *variable capacitance gauge* consists of a parallel plate capacitor that has one fixed plate and the other attached to the surface. The strain causes the separation of the plates, and hence the capacitance, to be varied.

Piezoelectric or *magnetostriction strain gauges* have a suitable piezoelectric or magnetostrictive crystal attached to the surface. The strain is measured by the associated piezoelectric or magnetostrictive changes that it induces.

strapping A method of suppressing unwanted modes of oscillation of a multicavity *magnetron by connecting together resonant segments of the same polarity.

stray capacitance Any capacitance in an electronic circuit or device that is due to interconnections, electrodes, or the proximity of elements in the circuit and is additional to the intentional capacitance of the circuit or device. Stray capacitance is usually unwanted but can sometimes be utilized as part of the tuning of a *tuned circuit.

striking *See* thyratron.

striking potential The potential difference necessary to initiate an *arc discharge between two electrodes.

string electrometer An electrometer that consists of a fine metallized quartz fibre stretched between two parallel conducting plates. The plates are oppositely charged by the potential difference to be measured and the deflection of the quartz fibre observed.

string galvanometer *See* Einthoven galvanometer.

strip core A magnetic *core that is fabricated from a continuous strip of the magnetic material. This form of manufacture is necessitated by a rolling treatment that is applied to the material in order to improve its magnetic properties but that causes grain orientation within the material and hence results in directional magnetic properties.

stroke pulse A *pulse used in a digital *computer or *radar system for timing or indicating purposes. A stroke pulse may either be used as a *gate, by allowing a signal to pass only during the pulse duration (*see also* clock), or it may be superimposed on a signal and act as a marker. The latter application is most often used with radar systems.

strong electrolyte *See* electrolyte.

strong inversion *See* MOS capacitor.

Strowger exchange An automatic telephone exchange in which the necessary connections between telephone lines and exchanges are effected by a series of Strowger switches in order to establish a telephone channel.

The Strowger switch (or uniselector) is a switch that makes a connection between a telephone line and any one of 100 outlet positions. The outlets are arranged in a number of circular groups one above the other. The coded signals from the dial of a telephone set control the movement of a contact causing it to move in vertical steps between the groups, and rotate in a horizontal plane to select the chosen outlet.

stub A device used for *impedance matching of a *transmission line to a *load in microwave and ultrahigh-frequency applications. It most often consists of a short variable-length section of transmission line of similar properties to the line to be matched. The position and length of the stub is adjusted to give optimum energy transfer.

subassembly Two or more parts that form a portion of an assembly but that may be individually replaceable. An example is a terminal board with mounted parts.

subcarrier A *carrier wave that is used to modulate a second different carrier.

subcarrier modulation *See* telegraphy.

subharmonic *See* harmonic.

subroutine *See* program.

subscriber station *See* telephony.

subsonic frequency A frequency of value less than the audiofrequency range.

substandard Denoting a measuring instrument that is used as a standard instrument but is not quite so accurate as the *primary standard. It may be used to check or calibrate a device but in turn needs itself to be checked against the primary standard.

substation A complete assemblage of plant, equipment, and the necessary buildings at a place where electrical energy is received and where it may be either converted from alternating current to direct current, stepped up or stepped down by means of transformers, or used for control purposes. The substation usually receives power from one or more *power stations.

substrate A single body of material on or in which circuit elements or *integrated circuits are fabricated. The substrate may be passive, as with a *printed circuit board, or active, as with a bulk *semiconductor.

subsystem An interconnected set of related circuits (which may be integrated circuits) that form a logical subdivision of a piece of electrical equipment or an operational system and may be manufactured as an assembly or subassembly.

Suhl effect *See* Hall effect.

summation instrument A single instrument that receives signals, such as current, power, or energy, from a number of separate circuits and measures the aggregate.

superconductivity A phenomenon that occurs in certain metals and a large number of compounds and alloys when cooled to a temperature close to the absolute zero of thermodynamic temperature. At temperatures below a critical *transition temperature, T_c,* the electrical resistance of the material becomes vanishingly small and the material behaves as a perfect conductor. Currents induced in superconducting material have persisted for several years without significant decay.

The material also exhibits perfect *diamagnetism in weak magnetic flux densities, i.e. the flux inside the material is zero. If the material is in the form of a hollow cylinder, the magnetic flux density contained in the hollow region remains constant and trapped in the state existing at the transition temperature, while the flux density within the material becomes zero. These magnetic effects are termed *Meissner effects.* If the value of applied magnetic flux density rises to a value greater than a critical value, B_c, the superconductivity is destroyed. The value of B_c – the *transition flux density* – is a function of the temperature of the material and its nature. A superconducting current in the material itself can produce an associated magnetic flux density greater than the critical value; there is therefore an upper limit to the current density that may be sustained by the material in the superconducting state. Certain alloys have relatively high transition temperatures and high critical field values and are used in superconducting magnets. Niobium-tin (Nb_3Sn) for example can produce a magnetic field of about 12 tesla at 4·2 kelvin, i.e. at the boiling point of liquid helium. Other transition metal compounds such as Nb_3Ge have transition temperatures at around the boiling point of liquid hydrogen (20 K). Such compounds fall into the Al5 series, i.e. they have a crystallographic structure similar to beta-tungsten.

Superconductivity with transition temperatures of 90 K or more has recently been demonstrated using complex metallic oxides that contain rare earth elements or transition metals and have the general composition

$$RBa_2Cu_3O_{9-y}$$

where R is a rare earth ion or transition metal. Scandium, lanthanum, neodymium, ytterbium, and several other elements have all been successfully used to obtain high-temperature superconducting samples.

The transition temperature of these samples is not clear cut. The resistivity of the sample decreases significantly at temperatures above the zero-resistance temperature, and a resistivity curve is obtained. The value of T_c also depends sensitively on the barium concentration and the heat treatment of the sample during its production.

One of the most successful theories of superconductivity is the *BCS theory* in which electron pairs – *Cooper pairs* – can form in the presence of other electrons. The pairing results from interactions between the electrons and the quantized vibrations of the crystal lattice – phonons – and produces a highly ordered state with no dissipation of energy in the electron movement. The BCS theory, however, proposes a singlet pair state. At higher temperatures this is not sufficient to fully explain the phenomenon and various ideas have been proposed to explain the postulated attractive interactions between electrons and the configuration of the electron pairs. An understanding of the theoretical basis for superconductivity will be determined by careful analysis of data from experiments carried out on single crystal samples, and should help the search for materials with even higher critical temperatures. There have been predictions of superconductivity at 240 K in the foreseeable future. Such a phenomenon would have a wide variety of applications in both the electrical and the electronics fields.

The *Josephson effect occurs when an extremely thin layer of insulating material is introduced into a superconductor. A current, below a certain critical value, can flow across the insulator in the absence of an applied voltage.

superconductor A material that exhibits *superconductivity.

superheterodyne reception The most widely used type of radio reception in which the incoming signal is fed into a *mixer and mixed with a locally generated signal from a *local oscillator*. The output consists of a signal of *carrier frequency equal to the difference between the locally generated signal and the carrier frequencies but containing all the original modulation. This signal, the *intermediate* *frequency* (or IF) signal, is amplified and detected in an *intermediate-frequency amplifier* and passed on to the audiofrequency amplifier.

The advantages of superheterodyne reception are directly attributable to the use of the intermediate frequency. Stable high-gain amplification is very much easier at intermediate frequencies than at *radiofrequencies. The use of the intermediate frequency allows much greater selectivity and hence easier elimination of any unwanted signals, since it is a difference signal and the percentage difference between the signal and other signals is much greater than at the original radiofrequency.

The overall sensitivity and selectivity may also be increased by using a *preselector*. This is a tuned radiofrequency preamplifier that amplifies the incoming signal before it is mixed with the intermediate frequency. If the mixer and local oscillator are combined in one circuit, the combination is an *autodyne mixer*. Stable operation is more easily obtained when a separate oscillator circuit is used.

Double superheterodyne reception employs two intermediate frequencies in order to improve the overall performance. The output of the first intermediate-frequency amplifier is mixed with a second locally generated carrier frequency before further amplification and detection of the audiofrequency signals. This method of superheterodyne reception requires two local oscillators. It is useful in areas where the received signal is particularly weak.

Superheterodyne reception may be employed with *suppressed-carrier transmission. The receiver in this case contains an extra local oscillator and mixer that reproduces the original radiofrequency *carrier wave and mixes it with the received signal before preselection and mixing with the intermediate frequency.

Compare beat reception.

superhigh frequency (SHF) *See* frequency band.

Supermalloy *Tradename* An alloy of iron, nickel, and molybdenum that is similar to *permalloy but that has a higher magnetic permeability.

Supermendur *Tradename* A *square-loop material used as a magnetic *core.

super-regenerative reception A method of reception used for ultrahigh-frequency radiowaves in which the *detector is a *squegging oscillator. The frequency at which the oscillations are quenched – the *quench frequency* – is a function of the frequency of the received radiowaves. Very large values of amplification can be obtained using this method of reception as a result of the positive *feedback employed in the detector. Compared however to *superheterodyne reception the *selectivity is relatively poor.

supersensitive relay An electromechanical *relay that operates with currents of less than about 250 microamps.

supersonics *See* ultrasonics.

supervisory control A remote-control system that uses relatively few transmission channels.

suppressed-carrier transmission A method of transmission of radio-waves in which the carrier component of the modulated wave is not transmitted. One or both of the sidebands only are transmitted. Suppressed-carrier transmission is used in *single-sideband transmission and *double-sideband transmission. It requires a local oscillator at the receiver that regenerates the carrier frequency and mixes it with the received signal in order to detect the modulating wave. This method of detection is termed *synchronous detection.*

suppressed-zero instrument *Syn.* set-up scale instrument. A measuring or recording instrument in which the zero position falls outside the *dynamic range of the instrument; the moving part is not deflected until a predetermined value of the measured signal is reached.

suppressor *See* interference.

suppressor grid *See* thermionic valve; pentode.

surface acoustic wave *See* acoustic wave device.

surface-barrier transistor A *transistor in which the usual *p-n junctions are replaced by Schottky barriers (*see* Schottky effect). *Carrier storage is zero for the Schottky barriers and these transistors are therefore used for high-frequency operation. They may be used for microwave-frequency switching applications.

surface charge density *See* charge density.

surface-charge transistor *See* charge-coupled device.

surface conductivity The reciprocal of *surface resistivity.

surface leakage A leakage current that results from the flow of charge at the surface of a material rather than in the bulk material.

surface noise *See* frying.

surface passivation *See* passivation.

surface potential *See* MOS capacitor.

surface recombination velocity *See* semiconductor.

surface resistance *See* skin effect.

surface resistivity The resistance between two opposite sides of a unit square of the surface of a material. The measured value can vary greatly depending on the method of measurement.

surface wave A *radiowave that travels along the surface separating the transmitting and receiving aerials. The surface wave is affected by the properties of the ground along which it travels. *See also* ground wave. *Compare* space wave.

surge An abnormal transient electrical disturbance in a conductor. Surges are produced from many sources, such as a lightning stroke, sudden faults in electrical equipment or transmission lines, or switching operations.

surge absorber *See* surge modifier.

surge current *See* fault current.

surge-current indicator *Syn.* magnetic link. A device that determines the approximate magnitude of a surge current. It consists of a specimen of ferromagnetic material with high remanence (*see* magnetic hysteresis), such as steel, placed in close proximity to the conductor concerned. A surge in the conductor, caused for example by a *lightning stroke, produces associated changes in the magnetic state of the specimen. The high remanence of the material allows it to be examined for a considerable elapsed time after the surge has occurred.

surge diverter A device that is connected between a conductor and earth and diverts the major part of any excessively large voltage surge. A *lightning arrester* is a particular example of a surge diverter. The most common arrangement used consists of one or more *spark gaps connected in series with a material, such as silicon carbide, that exhibits a decrease in electrical resistance with increasing voltage. This material assists the spark gaps to return quickly to their normal condition following the passage of a surge.

surge generator *See* impulse generator.

surge impedance The impedance presented by a transmission line to an electrical *surge that is propagated along the line as a travelling wave. The surge impedance, Z_s, is the ratio of the voltage to the current in the surge and is given approximately as

$$Z_s = \sqrt{(LC)}$$

L and C are the inductance and capacitance, respectively, of unit length of the line. For an ideal lossless line the above expression is exact.

surge modifier A device connected to a power transmission system or telecommunication system that alters the waveshape of any surges that might occur in order to protect other equipment, circuits, or devices associated with the system. For example, a surge modifier that reduces the steepness of a surge wavefront is frequently used to reduce electrical stress in transformer windings.

A *surge absorber* is a surge modifier that also contains some resistance and therefore dissipates a substantial fraction of the energy in a surge.

susceptance Symbol: B; unit: siemens. The imaginary part of the *admittance, Y, which is given by

$$Y = G + iB$$

where G is the *conductance. For a circuit containing both resistance, R, and reactance, X, the susceptance is given by

$$B = -X/(R^2 + X^2)$$

559

susceptibility (1) *Syn.* magnetic susceptibility. Symbol: χ_m. A dimensionless quantity given by

$$\chi_m = \mu_r - 1$$

where μ_r is the relative *permeability of a material. Magnetic susceptibility describes the response of a material to a magnetic field, being the ratio of *magnetization to *magnetic field strength:

$$\chi_m = M/H$$

χ is a tensor when M is not parallel to H, otherwise it is a simple number. For crystalline material χ may depend on the direction of the field with respect to the crystal axes because of anisotropic effects. It has a wide range of values: diamagnetic materials have a negative value; paramagnetic materials (*see* paramagnetism) have a small positive value; ferromagnetic materials can have a very large variable value (up to about one), which is dependent on the magnetic field strength (*see* magnetic hysteresis).

(2) *Syn.* electric susceptibility. Symbol: χ_e. A dimensionless quantity given by

$$\chi_e = \varepsilon_r - 1$$

where ε_r is the relative *permittivity. Electric susceptibility measures the ease of polarization of a dielectric and is given by the ratio

$$\chi_e = P/\varepsilon_0 E$$

where P is the *dielectric polarization, E the electric field strength, and ε_0 the permittivity of free space.

sweep *See* time base.

sweep frequency *See* time base.

sweep generator *Syn. for* time-base generator. *See* time base.

sweep voltage The voltage output from an internal or external *time base that when applied to the appropriate deflector plates or coils of a *cathode-ray tube causes a horizontal or vertical deflection of the electron beam.

swing The limits of the values of a varying electrical parameter, such as amplitude or frequency.

swinging choke *See* choke.

swiss cheese A technique of manufacturing circuits that uses a thin board containing a lattice of holes into which circuit elements, suitably encapsulated to form pellet shapes, are inserted and attached. The interconnection pattern is formed by metallization on to the board.

switch (1) A device that opens or closes a circuit.

(2) A device that causes the operating conditions of a circuit to change between discrete specified levels.

(3) A device that selects from two or more components, parts, or circuits the desired element for a particular mode of operation.

In general, a switch may consist of a mechanical device, such as a *circuit-breaker, or a solid-state device, such as a *transistor, *Schottky diode, or *field-effect transistor. *See also* anticapacitance switch; commutation switch; glow switch; flip-flop; muting switch; single-pole switch; time switch; transmit-receive switch; vacuum switch.

switching tube An arc-discharge tube, such as an *ignitor, that contains at least two anodes and is used as a switch. The current in the discharge flows to that anode held at the required voltage to sustain the arc; if the voltage is switched to a different anode the discharge path also moves. Switching tubes are used as transmit-receive switches and as scaling tubes.

SXAPS *Abbrev. for* soft X-ray appearance potential spectroscopy.

syllable articulation score *See* intelligibility.

symmetrical quadripole *See* quadripole.

symmetrical transducer *See* transducer.

sync *Abbrev. for* synchronous or synchronizing signal or synchronism.

synchrocyclotron *Syn.* frequency-modulated cyclotron. A form of *cyclotron that takes account of the relativistic mass change of the accelerated particles. Above a particular energy (about 15 MeV/proton) the relativistic mass change causes the orbital time of the particles to fall out of synchronism with the accelerating alternating voltage. This is overcome by frequency modulation of the alternating voltage so as to restore the synchronism between the circulation time and the accelerating supply. *See also* isochronous cyclotron.

synchronism The relationship between two periodically varying quantities when they are in *phase.

synchronizing pulses *See* television.

synchronometer A device that counts the number of cycles of a periodically varying quantity that occur during a predetermined time interval.

synchronous *Syn.* clocked. Denoting any circuit or device that is operated by means of clock pulses. *See also* clock.

synchronous alternating-current generator *Syns.* alternator; synchronous generator. An alternating-current generator that consists of one or more coils that are made to rotate in the magnetic fields produced by several electromagnets excited by a direct-current source. The frequency, f, of the alternating currents and emfs induced in the coils is equal to the product of the speed, n_s, at which the coils rotate and the number of pairs of magnetic poles, p. This type of generator can operate and produce electrical power independently of any other source of alternating current (*compare* induction generator) and is the type of generator most commonly employed in *power stations.

synchronous clock A mains-operated electric clock in which the speed of the driving motor is a function of the mains frequency and therefore the time-keeping is controlled by the mains supply.

synchronous communications satellite (syncom) *See* communications satellite.

synchronous computer A *computer in which the timing of all operations is controlled by a *clock. Such a computer often employs *fixed-cycle operation* in which a fixed time is assigned in advance to each operation performed.

synchronous detection *Syn.* coherent detection. *See* suppressed-carrier transmission.

synchronous gate *Syns.* clocked gate; time gate. A *gate the output of which is synchronized to the input signal. The synchronizing signals may be derived from an independent clock so that the gate is operative during predetermined intervals of time; alternatively the input signal itself may be used as the trigger so that the gate operates only when an input signal is present.

synchronous generator *See* synchronous alternating-current generator.

synchronous logic A logic system that operates with *synchronous timing*, i.e. the timing of all the switching operations is controlled by *clock pulses. A logic system is said to be *asynchronous* when all the switching operations are triggered by a free-running signal so that successive stages or instructions are triggered by the completion of operation of the preceding stage.

Synchronous logic in general is slower and the timing more critical than asynchronous logic, but usually fewer and simpler circuits are required.

synchronous orbit *See* communications satellite.

synchronous timing *See* synchronous logic.

synchronous vibrator *See* vibrator.

synchronous voltage *See* travelling-wave tube.

synchrotron A cyclic particle *accelerator that is used to accelerate a beam of electrons – *electron synchrotron* – or protons – *proton synchrotron* – to very high energies. The particles travel along a circular evacuated tube of fixed radius: an applied magnetic flux density causes them to travel in the circular orbit. They are accelerated by a radiofrequency electric field applied across a gap in a metallic cavity inside the evacuated chamber. Acceleration also occurs as a result of electromagnetic induction within the acceleration chamber between the electric vector associated with the magnetic flux density and the beam current. In order to maintain the correct relationship between the magnetic flux density, the energy of the particles, and the radius of the chamber, the magnetic flux density is made to increase with time and the frequency of the applied r.f. field modulated. This prevents loss of synchronism due to the relativistic mass increase.

The particles are usually injected at high energies from a *linear accelerator in order to minimize the range of modulation of the r.f. field. An alternative method is to operate the machine without the r.f. field at lower energies; acceleration is then the result of the electromagnetic induction only.

sync separator *See* television.

T

TAB *Abbrev. for* tape automated bonding.

tandem A method of connecting two *quadripole networks so that the two output terminals of one network are connected to the input terminals of the other. *See also* cascade.

tandem exchange *See* telephony.

tandem generator *See* Van de Graaff accelerator.

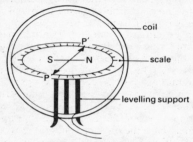

Tangent galvanometer

tangent galvanometer A moving-magnet *galvanometer that has a small magnet in the centre of a large fixed coil. The magnet carries a long light nonmagnetic pointer, PP', usually aluminium, that moves over the scale (*see* diagram). The instrument must be set up correctly with the plane of the coil vertical and along the earth's magnetic meridian. The current, I, is then proportional to the tangent of the angle of deflection, α, i.e.

$$I = k \tan\alpha$$

where k is a constant known as the *reduction factor*.

The *sine galvanometer* is similar to the tangent galvanometer except that the coil and scale are rotated together in order to restore the needle to zero. The current in the coil is then proportional to the sine of the angle of rotation of the coil.

Tangent galvanometers are little used now: they can be used to verify Ohm's law, of which they are independent, and with a known current to determine the earth's magnetic field.

tantalum Symbol: Ta. A metal, atomic number 73, that has an extremely high resistance to corrosion and is used for applications where this property is desirable.

tantalum capacitor *See* electrolytic capacitor.

tantalum rectifier *See* electrolytic rectifier.

tap changer *Syn.* ratio adjuster. A device that alters the ratio of a *transformer by selecting the desired *tapping. One that is designed to be used only when the supply voltage is switched off is an *off-circuit tap changer*. If it is designed to operate when the supply voltage is switched on it is an *on-load tap changer,* although the transformer need not necessarily be on load for operation of the tap changer.

tape *Short for* magnetic tape; paper tape.

tape automated bonding (TAB) A method used during the packaging of *integrated circuits, usually when a large number of interconnections (over 100) is required between the chip and the *leadframe. The leadframe is formed from plated copper on a strip of plastic (the tape), and extends to reach the *bonding pads of the chip. Metal bumps are formed at the points to be connected to the bonding pads, and the pads themselves have metallic caps formed on them by *sputtering. The chip is positioned on the tape and all the bonds formed simultaneously by thermal compression. The process is automated by producing the tape as a long strip of successive leadframes. The bonded assemblies are then encapsulated in moulded plastic and separated to produce the finished packages. The packages may have the form of *pin grid arrays, *leadless chip carriers, or *dual in-line packages. *Compare* wire bonding.

tape punch *See* paper tape.

tape reader *See* paper tape.

tape recording *See* magnetic recording.

tapping A conductor, usually a wire, that makes an electrical connection with a point between the ends of a winding or coil. The number of turns included in the active portion of the coil can then be selected. More than one tapping may be made to a particular winding, such as that of a *transformer.

target (1) *See* storage tube.
(2) *See* camera tube.
(3) *Syn.* anticathode. *See* X-ray tube.
(4) An object detected by a *radar or *sonar system.

target plate *See* storage tube.

target voltage The potential difference between the cathode and the signal electrode of a low-electron-velocity *camera tube. The minimum value that is required to produce a discernible video output is the target *cut-off voltage.*

Tchebyshev filter *See* filter.

tearing An effect observed in the picture on the screen of a television receiver in which the picture appears to break up. It is most commonly caused by faults in the synchronizing circuits.

In the case of a *colour picture tube tearing can be observed as *colour breakup:* the picture appears to break up into its colour primaries. Colour break-up can also result from rapid changes in viewing, such as blinking.

teaser transformer *See* three-phase to two-phase transformer.

TED *Abbrev. for* transferred-electron diode. *See* Gunn diode.

Teflon *Tradename* Polytetrafluoroethylene (PTFE). An insulator that has an extemely high resistivity and is very resistant to moisture and temperature.

TEGFET *See* high electron mobility transistor.

Telecom Gold *See* teletext.

telecommunications The study and practice of the transfer of information by any electromagnetic means, such as wire or radiowaves.

telecommunication system The complete assembly of apparatus and circuits required to effect a desired transfer of information. Systems include television, radio, and telephony.

telegraphy Communication by means of a *telecommunication system that transmits documentary matter, such as written or printed matter or fixed images, and reproduces it at a distance. The matter is transmitted as a suitable signal code, such as international *Morse code, either by means of wire or by radio (*radio telegraphy*). Pictures are transmitted using *facsimile transmission. A telegraph network is a complete system of stations, installations, and communication channels that provides a telegraph service.

Telegraphy may be effected automatically or manually. A synchronous system is one in which the sending and receiving instruments operate at substantially the same frequency and are maintained with a desired phase relationship. Transmission may be in the form of a direct current applied to the line by the sending apparatus (*direct-current telegraphy*) or a modulated carrier wave (*carrier telegraphy*). Amplitude, frequency, or pulse-code modulation of the carrier can be used. *Subcarrier modulation* is a method that may be employed in radio telegraphy when a low-frequency carrier wave (the subcarrier) is frequency-modulated by the telegraph signal and then this modulated wave is used to modulate a second *radiofrequency carrier wave.

telemeter *See* telemetry.

telemetry Measurement at a distance. Data is transmitted over a particular telecommunication channel from the measuring point to the recording apparatus. A measuring instrument that measures a quantity and transmits the measured data as an electrical signal to a distant recording point is known as a *telemeter.* Space exploration and physiological monitoring in hospitals both require the use of telemetry.

telephone line *See* telephony.

telephone set *See* telephony.

telephone station *See* telephony.

telephony Communication by means of a *telecommunication system that is designed to transmit speech or sometimes other sounds. A complete telephone system contains all the circuits, switching apparatus, and other equipment necessary to establish a communication channel between any two users connected to the main system. Communication between two points takes place along suitable cables (*telephone lines*) except where this is inappropriate; a particular access point may then be connected to the main system by means of a radio link (*radio telephone*), as in ship-to-shore telephony or in Cellnet.

A *telephone set* is an assembly of apparatus that includes a suitable *handset* containing the transmitter and receiver, and usually a switch hook and the immediately associated wiring. A telephone set connected to a telephone system is a *telephone station;* one that has access to the public telephone network is a *subscriber station.* It is mainly large organizations that have a *private exchange* to interconnect telephone stations. Usually such an exchange is also connected to the public telephone system and is known as a *private branch exchange:* it is either a *PABX* (private automatic branch exchange) or *PMBX* (private manual branch exchange).

The public telephone system is organized in such a way as to facilitate the establishment of telephone channels. It consists of a large number of *local exchanges* that are switching stations where subscribers' lines terminate and where facilities exist for interconnection of local lines or for connection to the *trunk* (long-distance) network. Local exchanges are treated in a similar way to subscribers and are connected together through higher-order switching stations, termed *tandem exchanges.* These in turn are interconnected by means of toll centres, each serving a city, then group centres and zone centres so that a large region, such as Great Britain, is served economically. Usually connection is made automatically by suitable switching networks (*see also* Strowger exchange; Crossbar exchange). These networks are actuated by signals from the caller's telephone station; facilities also exist for manual connection when difficulty is experienced by the user. In the trunk network a system of repeaters is used that include one or more amplifiers to prevent loss of signal strength.

Communication between points in a telephone system may be made by means of a modulated carrier wave (*carrier telephony*). If the carrier wave is suppressed when no audiofrequency signals are present the system is *quiescent-carrier telephony.* *Voice-frequency telephony is also used. In the public telephone system voice-frequency telephony is employed for local connections between two lines in the same local exchange and carrier telephony for the trunk system. *Multiplex operation allows the same telephone line to be used by several channels simultaneously.

The public telephone system is also used for electronic data transmission, such as *electronic mail, *telex, and *facsimile transmission. The data is transmitted in digital form using *pulse code modulation. The telephone system is currently being converted to operate on a totally digital system. *Analog/digital converters are required to convert the analog signals from telephones into digital form for transmission. At present both analog and digital methods are in use.

teleprinter A form of start-stop typewriter that comprises a *keyboard transmitter,* which converts keyboard information into electrical signals, and a *printing receiver,* which reverses the process. Teleprinters are used in *telex systems and in some computing systems.

teletext *Syn.* videotex. An information service in which information can be displayed as pages of text on the screen of a commercial television receiver. The information may be transmitted as part of the commercial television broadcast signal or as coded telephone signals. It is in the form of pulse-code modulated signals that use two of the unused lines in an ordinary television video signal transmitted during the normal vertical retrace period. Special decoding circuits are required for the extraction of the teletext signals from the normal television signals or from the telephone line and for decoding them.

A typical teletext page consists of 24 rows with up to 40 characters in a row. A limited amount of colour information can be used and a flashing facility is also provided. The television systems transmit the coded lines during each field blanking period.

Teletext decoders contain facilities enabling the user to select the page required and to store and display the information. They also have facilities for inserting news flashes into the normal television picture, or to insert subtitles for people with impaired hearing. Some allow the full text to be superimposed on the normal picture or can store the required information for later display. It is also possible to display the current television programme in much reduced format as a small insert in the teletext display.

The two compatible television systems in current use in the UK are *CEEFAX,* used by the British Broadcasting Corporation, and *Oracle,* used by the Independent Broadcasting Authority. The systems used by British Telecom in the UK are *Prestel* (previously known as *viewdata*) and *Telecom Gold.* The teletext information in a telephone system is transmitted as coded telephone signals, on demand from the user, and is displayed on a television screen. Television systems are basically sequential transmitting systems and are limited by the time taken for the required information to become available to the user. A telephone system however is capable of providing much more information to the user, since any information in the central store may be demanded by the user, and it is limited only by the size of the central store. It also differs from television systems in that it is an interactive two-way system in which data may be transmitted to the central system by the user employing a suitable interface with the system, such as a microcomputer.

television A telecommunication system in which both visual and aural information is transmitted for reproduction at a receiver. The basic elements of the system are:

*television cameras and *microphones that convert the original visual and aural information into electrical signals, i.e. into *video signals* and *audio signals* respectively;

amplifiers and control and transmission circuits that transmit the information along a suitable communication channel: broadcast television uses a modulated radiofrequency *carrier wave;

a *television receiver that detects the signals and produces an image on the screen of a specially designed *cathode-ray tube and a simultaneous sound output from a *loudspeaker.

The information on the target in the television *camera tube is extracted by *scanning and the spot on the screen of the receiver tube is scanned in synchronism with it to produce the final image. A process of rectilinear scanning is used in which the electron beam traverses the target area in both the horizontal and vertical directions. The horizontal direction is termed the *line* and the vertical direction the *field.* *Sawtooth waveforms are used to produce the deflections of the beam and in both the camera and receiver the flyback period is blanked out.

The *synchronizing pulses* synchronize the camera and the receiver and are transmitted during the flyback. The line synchronizing pulses are transmitted during the line flyback period and the field synchronizing pulses during the field flyback. A combination of overscanning (*see* scanning) and blanking is often used in order to allow a sufficiently long interval for synchronization without loss of picture information. During the blanked interval the level of the signal is held at a reference value, termed the *blanking level,* that represents the blackest elements of the picture except for the synchronizing pulses. This allows easy recognition of the synchronizing pulses at the receiver. The interval immediately preceding the synchronizing pulse is termed the *front porch* and the interval immediately succeeding the sync pulse is the *back porch.* The synchronizing pulses are extracted from the video signal by means of a *sync separator* in the receiver.

For the maximum information to be obtained from the target area of the receiver the number of horizontal scans is made larger than the number of vertical scans so that as much of the target area is covered as possible. The number of lines traversed per second is the *line frequency;* the number of vertical scans per second is the *field frequency.* The scanning process is most important since imperfections in scanning or synchronization between transmitter and receiver can result in geometric distortion of the picture or in other faults, such as *tearing or *pairing. A method of scanning that produces the entire picture in a single field (or *raster*) is termed *sequential scanning.*

Most broadcast television systems use a system of *interlaced scanning.* In this system the lines of successive rasters are not superimposed on each other but are interlaced; two rasters constitute a complete picture or *frame.* The number of complete pictures per second is the *frame frequency,* which is half the number of rasters per second, i.e. half the *field frequency.* The field frequency needs to be relatively slow to allow as many horizontal lines as possible but sufficiently fast to eliminate *flicker. Various compromises are used. European television systems use a 50 hertz field frequency (25 Hz frame frequency) system with 625 lines per frame. American television uses a 60 Hz field frequency and 525 lines per frame.

Definition in television is a measure of the resolution of the system, which in turn depends on the number of lines per frame. High-definition systems have more lines. Some *closed-circuit television systems use as many as 2000 lines per frame. The relationship between the total number of scanning lines per field and the corresponding bandwidth of the video signal is given by the *Kell factor.*

Positive or *negative transmission* may be employed for transmitting the video signals, positive transmission being most often used. In positive transmission an increase in amplitude or frequency above the reference *black level* of the received signal is proportional to the light intensity. The peak value of the video signal that corresponds to the lightest area of the picture is the *white peak.* In negative transmission the carrier wave value decreases below the black level in proportion to the light intensity.

The basic television system transmits images in black and white only (*monochrome television*). *Colour television is now widely used and the broadcast signal is received on special colour receivers. Modern monochrome receivers use the brightness information transmitted as part of the colour signal – the luminance signal – but the image produced is black and white (*see also* colour television).

The transmitted signal contains both video and audio information. The *picture carrier* is the *carrier wave modulated by the video information. The audio signal modulates a second carrier wave, termed the *sound carrier.* The sound-carrier frequency differs from that of the picture carrier and is chosen so that both signals fall within a designated radiofrequency band but do not overlap each other.

television camera The device used in a *television system to convert the optical images from a lens into electrical signals. The optical image formed by the lens system of the camera falls on to a photosensitive target. This is scanned, usually by a low-velocity electron beam and the resulting output is modulated with video information obtained from the target area. The output signal can be considered as an essentially constant d.c. signal with a superimposed a.c. signal. The amplitude variations of the latter correspond to the brightness, i.e. the *luminance,* of the target. The d.c. component is the value of the video signal that corresponds to the average luminance of the picture with respect to a fixed reference level.

The camera consists of three major parts housed in one container: optical lens system, *camera tube, and preamplifier. The resulting output is further amplified and transmitted in the broadcasting network. Some cameras are self-contained, with the amplifier and transmitter in the same container. These cameras are usually employed in a closed-circuit system, or for special applications, such as an outside broadcast. Several types of camera tube have been developed; the major differences between the various types of tubes are in the composition of the photosensitive material used and the means of extracting the electrical information produced.

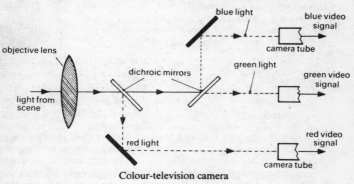

Colour-television camera

The camera used in *colour television consists of three camera tubes each of which receives information that has been selectively filtered to provide it with light from a different portion of the visible spectrum. Light from the optical lens system is directed at an arrangement of *dichroic mirrors each of which reflects one colour band and allows other frequencies to pass through. The original multicoloured signal is split into red, green, and blue components, and the video output from the three camera tubes represents the red, green, and blue components of the image (*see* diagram). The scanning systems in the three tubes are driven simultaneously by a master oscillator to ensure that the output of each tube corresponds to the same image point. The three outputs are then combined to provide brightness (luminance) and colour (chrominance) information (*see also* colour television).

television receiver A device that receives a television broadcast signal and effects the sound and vision reproduction of the original scene. The complete television receiver contains detecting and amplifying circuits that extract the video and audio signals from the received signal; circuits that extract the synchronizing signals from the received signal and control the appearance of the image; a *picture tube that reproduces the picture, and audio circuits that reproduce the transmitted sound. A desired broadcast channel may be selected using a *tuner. A colour television receiver also contains extra decoding circuits that extract the chrominance (colour) information from the received signal.

Various control circuits are provided in order to compensate for minor variations in the received signal level. *Automatic brightness control* is used to maintain automatically a constant value for the average luminance (brightness) level of the image. *Automatic contrast control* maintains a constant value between the black peak and the white peak of the image despite input variations. Various manual controls are also usually available:

horizontal hold and *vertical hold* are controls that adjust the starting points of the line sweep and field scan, respectively, relative to the television screen;

brightness and *contrast controls* adjust the average brightness and contrast of the image. They adjust the average luminance level and contrast range between the black and white peaks, respectively;

colour saturation control and *tint control* are provided in colour television receivers. The colour saturation alters the total intensity of all three electrons beams and makes the final image more coloured or less coloured. The tint control alters the relative intensities of the three electron beams to adjust the colour balance of the picture.

See also colour picture tube; colour television; television.

telewriter A device used in *telegraphy that converts manually controlled movements of a pen over a plane surface into two currents whose instantaneous values are a function of the position of the pen. These are used to cause automatically corresponding movements of a similar pen at the receiver.

telex A *telegraphy system that enables users to communicate directly and temporarily among themselves using *teleprinters and the public telegraph system.

Telstar *See* communications satellite.

temperature coefficient of resistance The incremental change in the resistance of any material as a result of a change in *thermodynamic temperature. In general conductors exhibit a positive coefficient of resistance; semiconductors and insulators have a negative coefficient of resistance.

In a conductor the distribution of electronic energy levels in the material is such that conduction levels are always available (*see* energy bands). An increase in the thermodynamic temperature causes an increase in the vibration of nuclei in the crystal lattice. This leads to an increase in the amount of scattering of conduction electrons as they drift through the material and causes the resistance to increase.

In a semiconductor and an insulator the existence of a forbidden gap between the valence and conduction bands has the effect that as the temperature increases more charge carriers become available for conduction by crossing the forbidden gap. The resistance therefore decreases. The increased numbers of carriers more than offsets the effect of scattering by lattice nuclei.

For a given material at thermodynamic temperature T the resistance R_T is given by

$$R_T = R_0 + \alpha T + \beta T^2$$

where R_0 is the resistance at absolute zero ($T = 0$) and α and β are constants characteristic of the material. In general β is negligible and the temperature coefficient of resistance is given by α.

temperature saturation *See* thermionic valve.

temporary memory *See* memory.

TEM wave *See* mode.

tera- Symbol: T. A prefix to a unit, denoting a multiple of 10^{12} of that unit: one terahertz is 10^{12} hertz.

terminal (1) A device that provides *input/output facilities to a computer, often from a remote location. It may be used *interactively and usually contains a keyboard and/or *visual display unit. An *intelligent terminal* contains some local storage and processing ability and can perform simple tasks independently of the main computer.

(2) Any of the points at which interconnecting leads may be attached to an electronic circuit or device and at which signals may be input or output.

terminal impedance The complex *impedance at a pair of terminals of a *transmission line or other device under normal operating conditions but under no-load conditions.

terminal repeater *See* repeater.

termination A load impedance connected across the output of a transmission line or transducer that completes the circuit while ensuring *impedance matching and preventing unwanted reflections.

tertiary winding An additional secondary winding on a *transformer. It can be used to supply a load when a different voltage is required from that of the main secondary or when a load must be kept electrically insulated from that of the normal secondary. It may also be used to interconnect supply systems that operate at different voltages.

tesla Symbol: T. The *SI unit of *magnetic flux density. One tesla is defined as one weber of magnetic flux per square metre of circuit area.

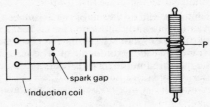

Tesla coil

Tesla coil An induction coil used to generate very high frequency high-voltage oscillatory currents (*see* diagram). The induction coil, I, discharges across a spark gap and feeds the primary, P, of a transformer through two large capacitors. The primary of the transformer has very few turns compared to the secondary, which is wound on a very large former. Tesla coils are used to test for leakage in a vacuum system. A glow discharge will be induced by a Tesla coil in any gas entering the system.

test pattern A chart used in television broadcasting that is transmitted at certain times when no programme is being transmitted. The pattern on the chart can be used for general testing purposes.

tetrode Any electronic device that has four electrodes. The term is most commonly applied to a *thermionic valve that contains a second grid. This auxiliary grid is usually a *screen grid* designed to decrease the anode-grid capacitance and hence to increase the resistance to high-frequency currents. It may also be used either to decrease the anode-cathode resistance or to modulate the main electron stream by injecting an alternating voltage.

TE wave *Syn.* H wave. *See* mode.

T flip-flop *See* flip-flop.

theoretical cut-off frequency *See* cut-off frequency.

thermal battery *See* thermocouple.

thermal breakdown *Syn.* thermal runaway. *Breakdown of a reverse-biased *p-n junction caused by the generation of excess free charge carriers due to the cumulative interaction between increasing junction temperature and increasing power dissipation. As the temperature is increased the effect is to reduce the voltage at which *avalanche breakdown occurs.

thermal imaging *Syn.* thermography. Producing an image of an object by means of the infrared radiation emitted by it. A *camera tube with a suitable lens system may be used to produce the image. Thermal imaging does not require any external source of illumination and is used to produce images in the dark, for example at night. It is also used for diagnostic purposes to discover any areas of the body that have an unusual temperature distribution.

thermal instrument *Syn.* electrothermal instrument. A measuring instrument that utilizes the heating effect of a current in a conductor. The conductor may be a *bimetallic strip, a hot wire (*see* thermoammeter), or a *thermocouple.

thermal noise *Syn.* Johnson noise; Brownian-motion noise. *See* noise.

thermal resistance In general, the ratio of the difference in temperature between two specified points to the heat flow between these points, under thermal equilibrium conditions. In particular, in a semiconductor device, it is the ratio of the temperature difference between a region in the device and the ambient temperature to the power dissipation in the device. The thermal resistance of the device depends on the material used and the geometry of the device and affects the ease of cooling the device.

thermal runaway *See* thermal breakdown.

thermionic cathode *Syn.* hot cathode. A *cathode in which *thermionic emission provides the source of electrons. A *directly heated cathode* is one that acts as its own source of heat: the cathode is in the form of a filament and the heater current is passed through it, superimposed on the normal cathode potential (usually earth potential). An *indirectly heated cathode* is one that has a separate heater. The heater is usually a coil of wire formed around the cathode and to which a current is applied.

The electron emission from the cathode may be increased by coating it with a thin layer of a suitable material. A *coated cathode* usually consists of a cylinder of platinum coated with an oxide of barium, strontium, or calcium. It operates at a lower temperature than the clean metal. An electron tube containing a coated cathode is known as a *dull emitter. Compare* photocathode.

thermionic emission Electron emission from the surface of a solid that is a result of the temperature of the material. An electron can escape from the surface with zero kinetic energy if it has thermal energy just equal to the *work function of the material (*compare* photo-

575

emission). The numbers of electrons emitted increases sharply with temperature (*see* Richardson's equation). *See also* Schottky effect.

thermionic-field emission *See* Schottky diode.

thermionic valve A multielectrode evacuated *electron tube that contains a *thermionic cathode as the source of electrons. Thermionic valves containing three or more electrodes are capable of voltage amplification: the current that flows through the valve between two electrodes, usually the anode and the cathode, is modulated by a voltage applied to one or more of the other electrodes. Thermionic valves have rectifying characteristics, i.e. current will flow in one direction only (the forward direction) when positive potential is applied to the anode.

The simplest type of thermionic valve is the *diode,* which has been most often used in rectifying circuits. Electrons are released from the cathode by thermionic emission. Under zero-bias conditions electrons released by the cathode form a *space charge region in the vacuum surrounding the cathode and exist in dynamic equilibrium with the electrons being emitted. If a positive potential is applied to the anode, electrons are attracted across the tube to the anode and a current flows. The maximum available current, the *saturation current, is given by

$$I_{sat} = AT^2 \exp(-B/T)$$

where A and B are constants and T is the thermodynamic temperature of the cathode. The current does not rise rapidly to the saturation value as the anode voltage is increased but is limited by the mutual repulsion of electrons in the interelectrode region. This is the *space-charge limited region* of the characteristic and the current obeys *Child's law* approximately where Child's law is given by

$$I = KV_a^{3/2}$$

where V_a is the anode voltage and K is a constant determined by the device geometry. Increasing the temperature of the cathode has very little effect on the current in this region of the characteristic curve and *temperature saturation* is said to occur. The motion of the electrons may be affected by the magnetic field associated with the current flowing in the heater and the electrons will be deflected from a linear path. This effect is the *magnetron effect* and it contributes to the delay in reaching the saturation current.

Under conditions of reverse bias (*see* reverse direction) no current flows in the valve until the field across the valve is sufficient to cause *field emission from the anode or *arc formation; *breakdown of the device will then occur. The characteristics of a simple diode (Fig. *a*) can be compared with the characteristics of a simple *p-n junction diode (Fig. *b*), which is the solid-state analogue of the device.

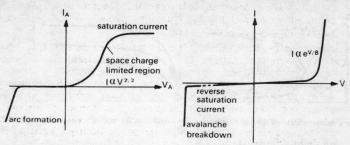

a Characteristic of a valve diode　　　b Characteristic of p-n junction diode

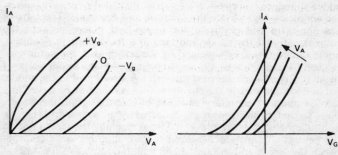

c Anode characteristics of a triode　　　d Transfer characteristics of a triode

The diode characteristic can be modified by interposing extra electrodes, called *grids since they are usually in the form of a wire mesh, between the anode and the cathode of a valve. The simplest such valve is the *triode* with only one extra electrode, a *control grid*. Application of a voltage to the grid affects the electric field at the cathode and hence the current flowing in the valve. A family of characteristics is generated for different values of grid voltage, similar in shape to the diode characteristic. The anode current at a given value of anode voltage is a function of grid voltage and amplification may be achieved by feeding a varying voltage to the grid; comparatively small changes of grid voltage cause large changes in the anode current. In normal operation the grid is held at a negative potential and therefore no current flows in the grid since no electrons are collected by it. Anode and transfer characteristics of a triode are shown in Figs. *c* and *d*. Triodes have been extensively used in *amplifying and *oscillatory circuits.

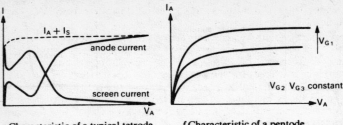

e Characteristic of a typical tetrode   f Characteristic of a pentode

A disadvantage of the triode is the large grid-anode capacitance, which allows a.c. transmission, and extra electrodes have been added to reduce this effect. Such valves are called *screen grid valves,* the simplest of which are the *tetrode and *pentode. The tetrode has one extra grid electrode, the *screen grid,* placed between the control grid and the anode and held at a fixed positive potential. Some electrons will be collected by the screen grid, the number of electrons being a function of anode voltage. At high anode voltages the majority of electrons pass through the screen grid to the anode. An undesirable kink in the characteristics is therefore observed in a tetrode due to *secondary emission of electrons from the anode, these secondary electrons being collected by the screen grid (Fig. *e*). Secondary electrons are prevented from reaching the screen grid in the pentode by introducing another grid, the *suppressor grid,* between the screen and the anode and maintaining it at a fixed negative potential, usually cathode potential. This eliminates the kink of the characteristic of the tetrode. The pentode characteristics (Fig. *f*) are similar to those observed in *field-effect transistors, which are the solid-state analogues.

Thermionic valves with even more electrodes, such as the *hexode, *heptode, and *octode, have been designed to produce particular characteristics. Multipurpose valves, such as the diode-triode, have an arrangement of electrodes so that the functions of several simpler valves are combined in a single envelope. Thermionic valves containing gas at low pressure are also used, the most important of these being the *thyratron.

In everyday applications, such as amplification, thermionic valves have been almost completely replaced by their solid-state equivalents. In applications requiring high voltages and currents valves are still used but these are usually special-purpose valves as with the cathode-ray tube, magnetron, and klystron. For most applications solid-state devices, such as the p-n junction diode, bipolar junction *transistor, and field-effect transistor, frequently in the form of *integrated circuits, have the advantages of small physical size, cheapness, robustness, and safety as the power required is very much less than for valves.

thermistor A resistor, usually fabricated from semiconductor material, that has a large nonlinear negative *temperature coefficient of resistance. An electrolyte, such as a viscous solution of waterglass, is sometimes used as the temperature-sensitive element. The thermistor is usually shaped as a rod, bead, or disc and named accordingly. Applications include compensation for temperature variations in other components, use as a nonlinear circuit element, and for temperature and power measurements. A *thermistor bridge* is an arrangement of thermistors for measuring power. *Compare* ballast resistor; barretter.

thermoammeter An *ammeter that measures a current by means of the heating effect produced by the current. The two main types of thermoammeter are the *thermocouple ammeter* and the *hot-wire ammeter*. Both these instruments may be used for either direct-current or alternating-current measurements since the heating effect is proportional to the square of the current. They must be calibrated empirically against known values of current.

The thermocouple ammeter consists of a *thermocouple placed in contact with a metallic strip or wire and connected to a sensitive galvanometer. The current to be measured flows through the metallic strip and the consequent rise in temperature is detected by the thermocouple.

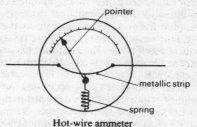

Hot-wire ammeter

The hot-wire ammeter measures current in terms of the thermal expansion of a metallic strip or wire. The current to be measured is passed through a metallic strip or wire that is clamped rigidly at its ends. The heating effect of the current causes the wire to elongate. The elongation is magnified mechanically and causes a pointer to move over a scale (*see* diagram). The *Cardew voltmeter* is a hot-wire ammeter that has an extremely large resistance in series, allowing it to be used as a voltmeter.

thermocouple Two dissimilar metals joined at each end to form an electrical circuit. If the two junctions are maintained at different

579

temperatures an electromotive force (e.m.f.) is developed between them as a result of the Seebeck effect (*see* thermoelectric effects). The e.m.f. is not affected by the presence of other metal junctions provided that they are all maintained at the same temperature. The thermocouple may therefore be connected to a suitable measuring instrument and used as a thermometer. It is convenient in use since it functions over a wide temperature range, it can be used to measure temperature at a very small area, and may be used remotely from the indicating instrument. The range of temperatures measured depends on the materials used.

In practice one junction is held at 0°C and then the e.m.f. E is given by

$$E = \alpha T^2 + \beta T$$

T is the temperature of the other junction – the 'hot' junction – and α and β are constants dependent on the metals used. At a temperature T_N where

$$T_N = -\beta/2\alpha,$$

E is a maximum. T_N is termed the *neutral temperature* and the use of the thermocouple is normally restricted to temperatures in the range $0 - T_N$°C. Copper/constantan or iron/constantan thermocouples can be used up to 500°C. Temperatures up to about 1500°C may be measured using an platinum/platinum-rhodium alloy thermocouple and even higher temperatures may be measured with an iridium/iridium-rhodium alloy thermocouple.

Thermocouple instruments are measuring instruments that use a thermocouple to measure electric current, voltage, or power by means of the heating effect in a metallic strip or wire (*see* thermoammeter). A *vacuum thermocouple* is a thermoammeter that is designed to measure very small currents. The metallic current-carrying strip and the thermocouple are sealed into an evacuated container in order to minimize errors in the measurement due to conduction of heat by the air surrounding the conductor.

The sensitivity of a thermocouple instrument is increased by connecting several couples together in series to form a *thermopile*. An easily detectable output is produced by heat radiation impinging on the hot junctions. The *thermal battery* is a thermopile that is used to generate an e.m.f. when heat is supplied to the hot junctions.

thermocouple ammeter *See* thermoammeter.

thermocouple instrument *See* thermocouple.

thermodynamic temperature *Syn.* absolute temperature. Symbol: T. Temperature that is measured as a function of the energy possessed by matter and as such is a physical quantity that can be expressed in units, termed *kelvin. In the thermodynamic temperature scale changes of temperature are independent of the working substance used in a thermometer. The zero of the scale is *absolute zero. The triple point of water is defined as 273·16 kelvin.

thermoelectric effects Phenomena that occur as a result of temperature differences in an electrical circuit.

The *Seebeck effect* is the development of an electromotive force between two junctions formed by joining two dissimilar metals if the two junctions are at different temperatures. The circuit constitutes a *thermocouple. In general, the e.m.f. E is given by

$$E = a + b\theta + c\theta^2$$

where a, b, and c are constants and θ is the temperature difference between the junctions. If the colder junction is maintained at 0°C then

$$E = \alpha T^2 + \beta T$$

where α and β are constants dependent on the metals used and T is the temperature of the hot junction. At temperatures below the neutral temperature (*see* thermocouple) if α is small (as is usually the case) then E is directly proportional to the temperature of the hot junction.

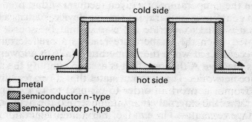

Peltier element

The *Peltier effect* is the converse of the Seebeck effect. If a direct current is passed round a circuit formed from two dissimilar metals or from a metal and a semiconductor, one junction gives off heat and is cooled and the other absorbs heat and becomes warm. The effect is reversible, i.e. if the current is reversed the cool junction becomes warm and the hot junction cools. Larger temperature differences are produced with metal-semiconductor junctions than with metal-metal junctions. A metal-n-type junction produces a temperature difference in the opposite sense to that of a metal-p-type junction for the same direction of current flow. A number of such junctions can be used to form a *Peltier element* (*see* diagram), which may be used as a heating or cooling element.

In the *Kelvin effect* a temperature difference between different regions of a single metal causes an e.m.f. to be developed between them. Also a current that flows along a wire in which a temperature gradient exists causes heat to flow from one region of the wire to another. The direction of heat flow is a function of the particular metal used.

thermoelectric module *Syn. for* Peltier element. *See* thermoelectric effects.

thermoelectric series An ordering of the metal elements so that if a *thermocouple is formed from two of them, the direction of current flow in the hotter junction is from the metal appearing earlier in the series to the other.

thermoelectron An electron emitted from the surface of a solid as a result of *thermionic emission.

thermogalvanometer *See* galvanometer.

thermography *See* thermal imaging.

thermojunction A junction of a *thermocouple.

thermoluminescence *See* luminescence.

thermomagnetic effect *See* magnetocaloric effect.

thermopile *See* thermocouple.

thermostat An automatic temperature-control switch that is used in conjunction with heating systems, such as an immersion heater, to maintain the temperature of a given medium within predetermined limits. It contains a temperature-sensing device, such as a bimetallic strip, that is used to operate a *relay so that the source of heat is interrupted when the temperature reaches a predetermined value and is reconnected when the temperature falls to a lower value.

Thévenin's theorem A theorem that is used to simplify the analysis of resistance networks. The theorem states that if two terminals (A, B) emerge from a network in order to connect to an external circuit, then as far as the external circuit is concerned the network behaves as a voltage generator. The e.m.f. of the voltage generator is equal to $V_{A,B}$, where $V_{A,B}$ is measured under open-circuit conditions, and it has an internal resistance given by

$$R_{A,B} = V_{A,B}/I_{A,B}$$

where $I_{A,B}$ is the short-circuit current.

Norton's theorem is an equivalent theorem to Thévenin's theorem and states that the network, under similar circumstances, can also be represented by a current generator shunted by an internal conductance. Proofs of both these theorems depend upon Ohm's law. Both theorems can also be applied to alternating-current linear networks but the resistance and conductance must be replaced by the complex impedance $Z_{A,B}$ or admittance $Y_{A,B}$, respectively.

thick-film circuit A circuit that is manufactured by thick-film techniques, such as silk screen printing, and usually contains only *passive components and interconnections. A thick film, up to about 20 micrometres thick and composed of a suitable glaze or cement, such as a ceramic/metal alloy, is deposited on a glass or ceramic substrate; the desired pattern for interconnections and passive components is then produced on it.

Thick-film circuits may be used to form hybrid *integrated circuits: silicon *chips containing *active components or devices are wire-bonded to the film circuit to form the completed circuit, which is then packaged.

thin-film circuit A circuit that is manufactured by thin-film techniques, such as *cathode sputtering or *vacuum evaporation, and usually contains only *passive components and interconnections. A thin film only a few micrometres thick is deposited on a suitable glass or ceramic substrate and formed into the desired pattern for interconnections and components. Thin-film techniques have been used to form some *active components, such as the *thin-film transistor.

thin-film memory A magnetic *memory that is fabricated using thin-film techniques, such as *cathode sputtering or *vacuum evaporation. The magnetic material forming the elements of the memory is deposited on to the glass or ceramic substrate in the presence of a strong magnetic flux density that is orientated parallel to the surface of the substrate so that the magnetic dipoles of the deposited material are all orientated with the same sense.

thin-film transistor (TFT) An insulated-gate *field-effect transistor (IGFET) that is fabricated using thin-film techniques on an insulating substrate rather than on a semiconductor *chip.

The insulating substrate reduces the bulk capacitance of the device and hence the operating speed can be increased. The technique was originally used for the fabrication of discrete cadmium sulphide transistors and the films could be deposited on the substrate in the order semiconductor, insulator, metal or vice versa. The technique is now used mainly for the construction of silicon-on-sapphire C/MOS circuits (*see* MOS logic circuit).

Thomson bridge *See* Kelvin double bridge.

Thomson effect *Syn. for* Kelvin effect. *See* thermoelectric effects.

thoriated tungsten filament A tungsten filament that contains a small amount of thorium and is used as a thermionic cathode. The amount of thermionic emission of electrons at any given temperature is greater for the mixture than for pure tungsten.

thou *Colloq.* One thousandth of an inch.

three-level maser *See* maser.

three-phase CCD *See* charge-coupled device.

three-phase system *See* polyphase system.

three-phase to two-phase transformer A *transformer that is designed to operate with a three-phase supply in the primary circuit and produce a two-phase supply at the secondary circuit, or vice versa.

Methods of winding that are used to achieve this include the *Leblanc connection* and *Scott connection*. In the former case the pri-

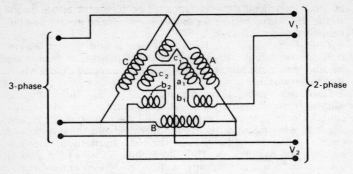

3-phase

2-phase

V₁

V₂

C

A

c₁

c₂

a₁

b₂

b₁

B

a Leblanc connection

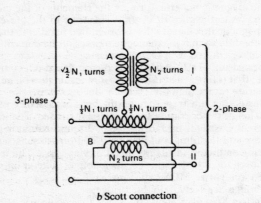

3-phase

2-phase

A

$\sqrt{\frac{1}{2}}N_1$ turns

N_2 turns

I

$\frac{1}{2}N_1$ turns

$\frac{1}{2}N_1$ turns

B

N_2 turns

II

b Scott connection

mary windings A, B, and C are wound on three separate cores (Fig. *a*). In the latter case two transformers, A and B, with similar secondary windings are interconnected (Fig. *b*). A is termed the *teaser transformer* and B the main transformer in order to distinguish them.

three-phase transformer A *transformer that has three independent sets of windings, each usually with the same turns ratio, and is suitable for use with a three-phase (*see* polyphase) mains input.

threshold current (1) *See* gas-discharge tube. (2) *See* injection laser.

threshold frequency The frequency at which a particular phenomenon, such as the *photoelectric effect or *photoconductivity, just occurs or the frequency below which an electronic device, such as a high-pass *filter, does not operate. In the latter case the term *cut-off frequency* is also used.

threshold signal (1) *See* minimum discernible signal. (2) *See* automatic control.

threshold voltage The voltage at which an electronic device first functions in a specific manner. In particular it is the voltage at which the conducting channel of an insulated-gate *field-effect transistor is just formed. *See also* MOS capacitor.

throat microphone *See* microphone.

through path *See* feedback control loop.

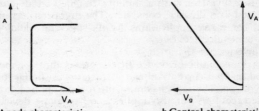

a Anode characteristic   b Control characteristic

thyratron *Syn.* gas-filled relay. A *gas-filled tube with three electrodes in which the voltage applied to one of the electrodes – the grid (or control) electrode – is used to initiate the discharge but does not limit it. The anode characteristic is shown in Fig *a*. A positive potential, greater than the ionization potential of the gas, is applied to the anode and a large negative potential applied to the grid. This neutralizes the effect of the anode potential at the thermionic cathode of the tube and prevents current flowing in the tube. If the grid voltage is made less negative the field at the cathode is increased until the discharge starts. This is termed *striking* the tube. Positive ions in the gas, usually mercury vapour, are attracted to the grid and effectively form a shield around it so that it no longer affects the main discharge even if it is made very negative. At the moment of striking, the anode potential falls approximately to the ionization potential of the gas (about 10–15 volts) until the saturation current is achieved. In order to stop the discharge the anode potential must be reduced below this value: this is most commonly achieved by means of a resonant circuit in the external anode circuit.

The value of the grid voltage at which the tube will strike is a function of the applied anode voltage and the device geometry. The *control characteristic* is the curve showing anode voltages and the corresponding critical grid voltages for striking; a typical curve is shown in Fig. *b*. The *control ratio* is the ratio $|V_A/V_g|$ of the straight line portion of the control characteristic.

The thyratron has been used as a *relay and as a *counter for radioactive particles. A small positive potential, derived from a trigger circuit or radiation detector and applied to the grid, causes the tube to strike; extinction of the discharge is obtained with a suitable anode circuit. The speed of operation of the device is limited by the recombination time of the ionized gas in the tube. This is known as the *dead time of the tube. The *hydrogen thyratron* is a hydrogen-filled thyratron that is specially designed for high-speed applications. It produces a large peak current at relatively high anode voltages and is frequently used in *radar systems. The thyratron has rectifying properties and can be used as a *rectifier. The *thyratron inverter* is a thyratron with a suitably designed anode circuit that can be used as a d.c./a.c. converter. The thyratron has now been replaced for many applications by its solid-state analogue, the *silicon-controlled rectifier.

thyratron inverter *See* thyratron.

thyristor A semiconductor device that contains three or more junctions and that has current-voltage characteristics similar to those of the *thyratron. The term has most often been applied to *silicon-controlled rectifiers but also applies to other *pnpn devices.

tight-coupled inductor A mutual inductance that has very efficient coupling between the two self-inductances so that the square of the mutual inductance is nearly equal to the product of the individual self-inductances.

time base A voltage that is a predetermined function of time and that is used to deflect the electron beam of a *cathode-ray tube so that the luminous spot traverses the screen in a desired manner. One complete traverse of the screen, usually in a horizontal direction, is termed a *sweep* (or time base).

The most common type of time base is one that produces a linear sweep: a sawtooth waveform is used to effect this. The circuit that produces the required voltage is a *time-base generator;* it may be free-running, in which a periodic sawtooth waveform is produced, or it may be clocked, when one sweep is produced on application of a trigger pulse to the circuit. The period during which the spot returns to the starting point is the *flyback* and in many applications, such as in television receivers, the flyback is suppressed, i.e. no luminous spot is observed on the screen during the return interval. The *sweep frequency* is the repetition rate of the sweeps across the screen.

A *Miller sweep generator* is a time-base generator that contains a *Miller integrator in the circuit in order to improve the linearity of the sweep. In some applications it is necessary to produce a time base in which the electron beam moves at a faster speed during part of the sweep, termed an *expanded sweep.* An *expanded-sweep generator* is one in which the incremental output voltage with respect to time is greater during a portion of the sweep in order to produce such an expanded sweep. A *delayed sweep* is a sweep produced by a synchronous time-base generator in which a predetermined delay time is introduced between application of the trigger pulse and commencement of the sweep on the screen.

Time bases are used to control the spot on the screen in many applications. *Cathode-ray oscilloscopes usually contain a circuit that generates a free-running sawtooth waveform with an adjustable sweep frequency. *Radar systems use a synchronized time base that is controlled by the transmitter so that each sweep is synchronized with the transmitted pulses and the return echo appears at a distance along the trace determined by the distance of the target from the transmitter. *Television systems employ time bases in the camera tubes and in the receivers to scan the lines and frames. The time base in the receiver is usually controlled by synchronizing pulses in order to retain the correct relationship to the transmitter; alternatively a *flywheel time base* is sometimes used in which the frame frequency is controlled by the electrical inertia of the circuit. This eliminates the need for frame-synchronizing pulses.

time-base generator *Syn.* sweep generator. *See* time base.

time constant The time required for a unidirectional electrical quantity, such as voltage or current, to decrease to $1/e$ (approximately 0·368) of its initial value or to increase to $(1 - 1/e)$ (approximately 0·632) of its final value in response to a change in the electrical conditions in an electronic circuit or device. Thus at any instant following initiation of the change in electrical equilibrium, such as switching a d.c. supply voltage on or off, the instantaneous change of a quantity, V, is given by

$$dV/dt = (V_f - V)/\tau$$

where V_f is the final value of the quantity and τ the time constant. In the case of a decreasing quantity V_f is usually zero and the expression becomes

$$dV/dt = -V/\tau$$

The time constant is a measure of the speed of operation of any circuit or device; circuits that contain capacitance or inductance can have very long time constants (of about a few seconds). For example, the time constant of a circuit containing a resistance of R ohms in series with a capacitance of C farads is given by

$$\tau = RC \text{ seconds}$$

One that contains a resistance, R, in series with an inductance of L henries has a time constant given by

$$\tau = L/R \text{ seconds}$$

587

time delay *See* time lag.

time-delay relay *Syn. for* slow-operating relay. *See* relay.

time-delay switch A switch that automatically causes a predetermined series of operations to be carried out with a specified time interval between each operation.

time discriminator A circuit in which the magnitude of the output is a function of the time interval between two input pulses and the sense of the output is determined by the order in which the inputs are received.

time-division multiplexing A form of *multiplex operation that samples and transmits each of the input signals sequentially so that each input signal is transmitted over the common channel during a set of predetermined time intervals. Each set of communicating terminal equipment is connected intermittently to the common channel, usually at regular intervals, by means of an automatic distribution system, such as a *commutation switch.

In order to maintain the correct connections between terminal equipment, the transmitter and receiver at each end of the common channel must be synchronized. *Marker pulses* are transmitted at regular intervals in order to maintain the synchronism. The switching speed between terminals must be sufficiently great so that each signal is sampled at least at the minimum sampling frequency. *Pulse modulation is the form of transmitted-signal modulation commonly employed. A *varioplex* system is a time-division multiplex system that controls the time interval allotted to each user according to the total number of users. *Compare* frequency-division multiplexing.

time gate *See* synchronous gate.

time lag *Syn.* time delay. The time that elapses between operation of a circuit-breaker, relay, or similar apparatus and the response of the current in the main circuit. A time lag may be deliberately introduced into a particular system. A definite time lag is a predetermined interval, sometimes adjustable, that is independent of the magnitude of the electrical quantity causing the operation. An inverse time lag is a delay time that is an inverse function of that quantity. *See also* overcurrent release.

time sharing A form of *time-division multiplexing by which a number of users may communicate directly with a *computer by means of a number of individual *terminals. The speed of operation of the machine is such that each user receives the impression of being the sole user of it. *See also* interactive. *Compare* batch processing.

time switch A device, such as a switch, circuit-breaker, or relay, that incorporates a clock mechanism and automatically operates at predetermined times.

time to flashover *See* impulse voltage.

time to puncture *See* impulse voltage.

time to trip *Syn.* opening time. The natural *delay, apart from any intentional delay, between the instant at which a predetermined signal is applied to a switch, circuit-breaker, or other similar device and the instant at which the device operates.

tin Symbol: Sn. A metal, atomic number 50, that has good resistance to corrosion and alloys readily with other metals, such as copper. It is extensively used to make permanent ohmic contacts in electronic circuits, devices, or other equipment.

tint control *See* television receiver.

T-junction *Syns.* hybrid T-junction; magic-T junction. *See* hybrid junction.

TM wave. *Syn.* E wave. *See* mode.

tolerance The maximum permissible error or variation permitted in the electrical properties or physical dimensions of any component or device.

tone control A device that alters the relative frequency response of an audiofrequency amplifier used in the reception or production of sound, in order to produce a subjectively more pleasing sound at the output.

topside ionosphere *See* ionosphere.

toroidal winding *See* ring winding.

torque The moment exerted by a force acting on a body and tending to cause rotation about an axis. It is given by the product of the perpendicular distance from the axis to the point of application of the force, and the component of the force in the plane perpendicular to the axis.

total capacitance The capacitance between one conductor in a system and all the other conductors in the system when electrically connected together.

total electron binding energy *See* binding energy.

total emission The peak value of the current that may be obtained by *thermionic emission from a cathode under normal heating conditions. The anode of the valve containing the cathode, together with all other electrodes, must be raised to sufficiently high potentials so that saturation is ensured.

touch control A manually operated switch that is activated by touch and that presents no moving parts to the operator. Two metallic areas are bridged by the finger or hand and cause a relay to operate.

Townsend avalanche *See* avalanche.

Townsend discharge *See* gas-discharge tube.

trace The image traced out by the luminous spot on the screen of a cathode-ray tube.

trace interval *Syn. for* active interval. *See* sawtooth waveform.

track The portion of a moving storage medium, such as magnetic tape or disk (*see* moving magnetic surface memory), that is accessible to a given reading device in a computer.

track-following servo system *See* compact disc system.

tracking (1) Maintaining a prescribed relationship between an electrical parameter of one electronic circuit or device and the same or a different parameter of a second device, suitably arranged, when both are subjected to the same stimulus. In particular it is the maintaining of a predetermined frequency relation in *ganged circuits, especially a constant difference between the *resonant frequencies of two ganged *tuned circuits.

(2) Maintaining a radar or radio beam set on a target while its position is determined.

(3) The formation of unwanted conducting paths on the surface of a dielectric or insulator when it is subjected to a high electric field. Such paths often result from carbonization on the surface.

(4) Maintaining the stylus of a *pick-up accurately within the grooves of a gramophone record.

(5) Maintaining the spot of light from the laser accurately on the track of a *compact disc.

tracking error signal *See* compact disc system

trailing edge (of a pulse) *See* pulse.

transadmittance *See* mutual conductance.

transconductance *See* mutual conductance.

transducer *Syn.* sensor. Any device that converts a nonelectrical parameter, e.g. sound, pressure, or light, into electrical signals or vice versa. The variations in the electrical signal parameter are a function of the input parameter. Transducers are used in a wide range of measuring instruments and have a variety of uses in the electroacoustic field. Gramophone pick-ups, microphones, and loudspeakers are all *electroacoustic transducers.* The term is also applied to a device in which both the input and output are electrical signals. Such a device is known as an *electric transducer.*

The physical quantity measured by the transducer is the *measurand,* the portion of the transducer in which the output originates is the *transduction element,* and the nature of the operation is the *transduction principle.* The device in the transducer that responds directly to the measurand is the *sensing element* and the upper and lower limits of the measurand value for which the transducer provides a useful output is the *dynamic range.

Several basic transduction elements can be used in transducers for different measurands. They include capacitive, electromagnetic, electromechanical, inductive, photoconductive, photovoltaic, and piezoelectric elements. Most transducers require external electrical excitation for their operation; exceptions are self-excited transducers, such as piezoelectric crystals, photovoltaic, and electromagnetic types.

Most transducers provide linear (analog) output, i.e. the output is a continuous function of the measurand, but some provide digital output in the form of discrete values. Most transducers are *linear transducers,* i.e. they are designed to provide an output that is a linear function of the measurand since this allows easier data handling. If the measurand varies over a stated frequency range the output of the transducer varies with frequency. The frequency response of the transducer is the change with frequency of the output to measurand amplitude ratio. The portion of the response curve over which attenuation of the measurand is significant is the *rejection band of the transducer.

Like many networks transducers may be considered as *quadripole devices but one pair of terminals is not necessarily electrical. A *symmetrical transducer* is one in which the input terminals and output terminals may each be simultaneously reversed without affecting the operation of the device; otherwise it is *dissymmetrical.* A transducer that operates in one direction only is a *unilateral transducer;* otherwise it is *bilateral.* If the energy loss in a bilateral transducer is the same for both directions of operation, it is a *reversible transducer.*

An *active transducer* is one that introduces gain, i.e. it derives energy from a source that is independent of the input signal energy. The transducer gain is defined as the ratio of the energy delivered to a suitable predetermined load to the available power at the input. In the case of a *passive transducer,* in which no gain is introduced, the loss is defined as the ratio of the available power at the input to the power delivered to the load under specified operating conditions.

transduction element *See* transducer.

transduction principle *See* transducer.

transductor *Syn.* saturable reactor. Acronym from *trans*fer in*ductor.* A device that consists of a magnetic core carrying several windings. The state of magnetic flux density in the core is controlled by a fixed alternating current in one of the windings. This current is sufficiently large to cause saturation of the core. Small variations in the current of one of the other windings – the *signal winding* – then cause large variations in the power in another circuit coupled by another winding – the *power winding.* The device thus operates by *magnetic modulation* and is used in control circuits, such as lighting circuits, and, particularly, in aircraft.

Variations of the current in the signal winding are caused by the *control circuit* and must be slow relative to the frequency of the alternating current in the supply circuit; frequencies up to about 2000 hertz are possible and therefore control of signals in the lower audiofrequency range is available. The controlled signal may be output directly from the power winding.

transfer To transmit or copy information from one device in a computer to another.

transfer characteristic (1) *See* characteristic.

(2) The relation between the degree of illumination of a television camera tube and the corresponding output current under specified conditions.

transfer constant *Short for* image transfer constant.

transfer current (1) The current applied to the *starter electrode of a glow-discharge tube (*see* gas-discharge tube) in order to initiate a glow discharge across the main gap.

(2) The current at one electrode of a *gas-filled tube that causes *gas breakdown to occur at a different electrode.

transfer gate *See* charge-coupled device.

transfer inefficiency *See* charge-coupled device.

transfer layer *See* multilevel resist.

transfer length *See* contact.

transfer parameter *Syn.* mutual parameter. The tangent at any point on a given transfer *characteristic of any network, amplifying circuit, transducer, or device. At the point the transfer parameter is the incremental change in the electrical output quantity to the incremental change in the input quantity. The most common transfer parameter used is the *mutual conductance, the transfer characteristic being the output current, I_{out}, versus the input voltage, V_{in}. The *transistance* is the parameter derived from the characteristic showing the active output voltage, V_{out}, versus input current, I_{in}; other transfer parameters, such as the transfer impedance, are derived by plotting the appropriate transfer characteristics.

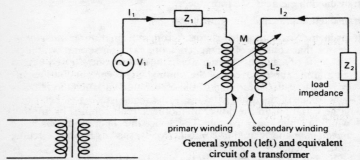

General symbol (left) and equivalent
circuit of a transformer

transformer An apparatus that has no moving parts and that transforms electrical energy at one alternating voltage into electrical energy at another usually different alternating voltage without change of frequency. It depends for its action upon mutual induction (*see* electromagnetic induction) and consists essentially of two electric circuits coupled together magnetically. The usual construction is of two coils (or windings) with a magnetic core suitably arranged between them. One of these circuits, called the *primary*, receives energy from an a.c. supply at one voltage; the other circuit, called the *secondary*, delivers energy to the load, usually at a different voltage.

The general symbol for a transformer and the circuit diagram for a typical transformer feeding a load impedance, Z_2, is shown in the diagram. In the case of an *ideal transformer* there is complete *coupling between the primary and secondary windings and therefore

$$M^2 = L_1 L_2$$

where M is the mutual inductance and L_1 and L_2 are the self-inductances of the two coils. The self-inductances are related to the squares of the numbers of turns, n_1, n_2, of the two coils by the equation

$$L_1 L_2 = (n_1/n_2)^2 = n^2$$

where n is the *turns ratio of the given transformer.

In the ideal case energy dissipation in the core can be ignored and it can be shown that

$$Z_p = V_1/I_1 = Z_1 + Z_2/n^2$$

where Z_p is the impedance of the primary circuit and Z_2/n^2 represents the effect of the secondary circuit on the primary and is known as the *reflected impedance*. The current in the secondary circuit is I_2 and the impedance is Z_s. They are given by

$$I_2 = -nV_1/Z_s$$
$$Z_s = Z_2 + Z_1 n^2$$

From these equations it can be seen that

$$Z_s = n^2 Z_p \qquad \text{(impedance transformation)}$$
$$V_2 = -nV_1 \qquad \text{(voltage transformation)}$$
$$I_2 = I_1/n \qquad \text{(current transformation)}$$

In practice complete coupling is not achieved and the mutual inductance is given by

$$M = k(L_1 L_2)^{1/2}$$

and

$$V_2 \simeq -knV_1$$

where k is the coupling coefficient.

Suitable design of the device can optimize the coupling and minimize the energy dissipation: values of k almost equal to unity may be achieved. Energy dissipation is kept to a minimum by winding the coils around laminated cores; the magnetic circuit of the cores is completed by forming them as a yoke. The two coils are sometimes interwound but insulation can be a problem where high voltages are used and they are then wound side by side. Uniformity of the magnetic flux density within the coils can be maintained by using an extra core (a *limb*) that surrounds the coils.

The property of voltage transformation is used in the *voltage transformer*. This can be used as an *instrument transformer in order to measure voltages. The primary winding is connected in parallel with the main circuit and the secondary winding connected to a suitable measuring instrument. The voltage transformer may also be used as a *power transformer* in order to supply electrical power at a predetermined voltage to a circuit. The primary winding is usually connected to the electrical mains supply. The secondary winding consists either of several windings or of one winding with several *tappings in order to supply different magnitudes of voltage or to supply more than one circuit. The transformer is described as *step-up* or *step-down* according to whether the secondary voltage is respectively greater or less than the primary voltage.

The *current transformer* utilizes the current transformation property and is most often used as an *instrument transformer. The primary winding is connected in series with the main circuit and the secondary to the appropriate measuring instrument. Current transformers are also used to operate protective *relays in order to prevent the current in the main circuit rising above a predetermined value.

See also autotransformer; core-type transformer; shell-type transformer; variocoupler.

transformer ratio *See* turns ratio.

transient A phenomenon, such as damped oscillations or a voltage or current surge, that occurs in an electrical system following a sudden change in the dynamic conditions of the system and that is usually relatively short-lived. A transient may be caused by the application of an impulse voltage or current to the system or by the application or removal of a driving force. The nature of the transient is a function of the system itself but the magnitude depends on the magnitude of the impulse or the driving force.

The *transient response* of an electronic device, such as an amplifier, is the change in output that occurs as a result of a specific sudden change in the input.

transient response *See* transient.

transistance Acronym from *trans*fer re*sistance*. *See* transfer parameter.

transistor A multielectrode *semiconductor device in which the current flowing between two specified electrodes is modulated by the voltage or current applied to one or more specified electrodes. The semiconductor material is usually silicon. Transistors have now replaced *thermionic valves as the general-purpose active electronic device except for some very specialized uses.

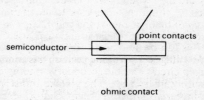

a Point-contact transistor

Some confusion has arisen in the terminology of transistors due to the historical development of the device. The first transistors were invented in 1948. These were *point-contact transistors,* which are now obsolete. The point-contact transistor consisted of a small crystal of semiconductor (usually germanium) with two rectifying point contacts attached in close proximity to each other and a single large-area *ohmic contact at some distance from the point contacts (Fig. *a*). On the application of small voltages, current flowed in the device and could be modulated by signals fed onto the ohmic contact. The *junction transistor* was developed in 1949 and its action was fully described by Schockley in 1950. *Field-effect transistors were developed more recently and have a different principle of operation to the point-contact or junction transistors.

Modern transistors fall into two main classes: bipolar devices, which depend on the flow of both minority and majority *carriers through the device, and unipolar transistors (*see* field-effect transistor), in which the current is carried by majority carriers only.

The basic bipolar junction transistor (usually simply called a *transistor*) consists of two *p-n junctions that are in close proximity with either the n or p regions common to both junctions, thus forming either a *p-n-p* or *n-p-n transistor,* respectively (Fig. *b*). The electrodes are called the emitter, base, and collector. The *energy bands of an n-p-n transistor under zero bias conditions are shown in Fig *c.* The central p-region of the transistor is called the *base* region. If a voltage is applied across the transistor, one junction becomes forward biased (*see* forward direction) and the other junction reverse biased (*see* reverse direction). Current flows across the forward-biased junction: electrons from the n-type region cross into the base and *holes from the base cross into the n-region. This junction is

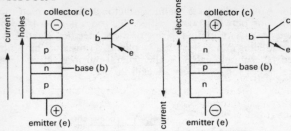

b p-n-p (left) and n-p-n bipolar junction transistors

called the emitter-base junction, the n-region being the *emitter*. The electrons entering the base diffuse across it towards the reverse-biased junction. Once they enter the *depletion layers associated with this junction they are swept across into the other n-region, which is called the *collector,* and enter the lower energy bands associated with this electrode. The hole concentration in the base falls due to holes leaving the base and entering the emitter across the forward-biased emitter-base junction and to holes recombining with injected electrons from the emitter as they diffuse across the base. This effect tends to reduce the forward voltage across the junction until the current ceases.

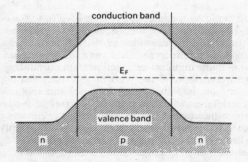

c Energy bands in n-p-n transistor under zero bias

If the base region is connected to a suitable point in the circuit so that the emitter-base junction remains forward biased, electrons flow out of the base region to maintain the hole concentration and current continues to flow across the device, from collector to emitter (i.e. in the opposite direction to the flow of electrons). The total current flowing is related by:

596

$$I_e = I_b + I_c$$

where I_e is emitter current, I_b base current, and I_c collector current.

Voltage amplification is possible using the emitter as the input terminal. This method of operation is called *common-base connection; the characteristics are shown in Fig. *d*. The term 'transistor' is an acronym from *trans*fer re*sistor* and derives from this original application of the device since the transfer characteristic indicates V_{out} against I_{in} and the associated *transfer parameter is the transfer resistance (transistance).

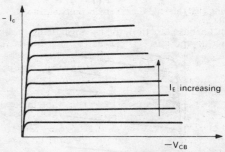

d Common-base output characteristics of a junction transistor

For efficient amplification, the collector current must be as nearly equal to the emitter current as possible; this may be achieved by reducing the base current to as low a value as is practicable. The device then has a low input impedance and high output impedance. Several factors contribute to the base current, the most important being holes crossing into the emitter and holes recombining with electrons in the base. Increasing the *injection efficiency of the emitter reduces the hole current as most of the current across the junction is due to electrons from the emitter. This is achieved by making the concentration of charge carriers in the emitter much larger than in the base, i.e. a high *doping level is used in the emitter. Compared to the base the emitter is therefore an n^+ semiconductor. The recombination current in the base is reduced by making the base region narrower than the diffusion length of the minority electrons so that most of them reach the collector before recombination occurs. The concentration of carriers through the base may also be altered by varying the doping level through the base region (*see* drift transistor). The base resistance is relatively high due to the comparatively low doping levels required in the base region. The *heterojunction bipolar transistor (HJBT) uses heterojunctions to allow relatively high doping levels in the base region and hence reduced values of the base resistance.

The common-base forward-current transfer ratio, α, is given by $\alpha = I_c/I_e$, and is a function of a particular device. Values of near unity are possible although α varies with frequency. The base current, I_b, is given by

$$I_b = (1 - \alpha)I_e$$

For any given transistor the ratio

$$I_e:I_b:I_c = 1:(1 - \alpha):\alpha$$

is constant.

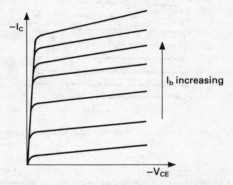

e Common-emitter output characteristics of a junction transistor

Current amplification is possible by applying the input signal to the base rather than to the emitter. Provided the emitter-base junction is always forward biased, carriers travel through the device as described above, but a variation in the base current causes a corresponding change in the collector current to maintain the ratio

$$\alpha:(1 - \alpha)$$

This is known as *common-emitter connection and is the most usual method of operating the device. The characteristics are shown in Fig. *e*. The *beta-current gain factor is defined as

$$\beta = I_c/I_b = \alpha/(1 - \alpha)$$

Since α can approach unity, β can be very large. Large collector currents are therefore generated by small base currents. This leads to saturation of the transistor since the current that can flow out of the collector is limited by the components in the external circuit.

When the collector current is saturated, the collector-emitter voltage drops to a small value; this is the *saturation voltage of the device. It is dependent on the value of base current and the external circuit components. If the base current falls the transistor ceases to be saturated when

$$\beta I_b < V/R_L$$

where R_L is the resistance of a load resistor connected between the collector and the power supply V; the collector voltage then rises to a value determined by R_L. The transistor may therefore be used as a switch by driving it into saturation using the base current as the driver. It acts as an *inverter when used in this way and has many switching applications in *digital circuits (*see* transistor-transistor logic; diode-transistor logic).

One of the most important differences between holes and electrons is the difference between their mobilities, electrons being about three times as mobile as holes. This makes devices depending mainly on electron flow much faster in operation than those depending on the flow of holes, and capable of being used at higher frequencies. This accounts for some of the small differences between n-p-n and p-n-p transistors. However in principle they are the same except that for p-n-p transistors holes replace electrons in the above description.

Bipolar junction transistors rapidly replaced the original point-contact transistors. The different types of junction transistor are classified by the manufacturing method used, many slightly different types of transistor having been produced. The two most commonly used methods of forming junctions are *diffusion and *ion implantation. *Alloyed junctions and *grown junctions are now becoming obsolete. *Epitaxy is often used during manufacture, the junctions being formed in an epitaxial layer (*see also* planar process).

Transistors have now replaced *thermionic valves as the general-purpose active electronic device, except for some very specialized uses. They are small, robust, require small supply voltages, are relatively easy to manufacture, cheap, and ideally suited for most applications: n-p-n transistors are most commonly used and the semiconductor material is usually silicon. For very high frequency applications gallium arsenide (GaAs) devices are often used since the carrier mobility is greater than in silicon. Diffusion techniques are however inappropriate for GaAs leading to difficulty in manufacture.

transistor parameters A transistor is a nonlinear device whose behaviour is difficult to represent exactly by a set of mathematical equations. When designing transistor circuits the behaviour of the transistor is represented approximately by *equivalent circuits that act as models of the device. The particular equivalent circuit used will be the one that is most appropriate for the type of circuit being designed, i.e. for use with large signals, small signals, as switches, etc.

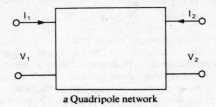

a Quadripole network

The transistor is represented by an equivalent circuit with two input terminals and two output terminals (Fig. *a*). This is a *quadripole network. Over a small portion of its operating characteristic the device is assumed to behave linearly. This is particularly true in small-signal operation but is only an approximation for large-signal operation. The input and output voltages and currents are related by two simultaneous equations of the general matrix form

$$[A] = [p][B]$$

where A and B represent current or voltage and p the particular transistor parameter used. The most common *matrix parameters* are the *h, y, z,* and *g* parameters (*see also* network).

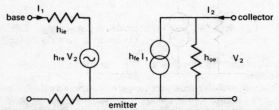

b Hybrid equivalent circuit with common-emitter connection

The *h* or *hybrid parameters* are the most commonly used for bipolar junction *transistors; they are so called because the dimensions are mixed and a true matrix representation is not possible. A typical equivalent circuit is shown in Fig. *b* with the transistor used in the *common-emitter connection. The transistor is assumed to consist of a voltage source with series impedance at the input and a current source with shunt impedance at the output. The equations relating the input and output are

$$V_1 = h_{11}I_1 + h_{12}V_2$$
$$I_2 = h_{21}I_1 + h_{22}V_2$$

The values of the constants h_{11}, etc., are characteristic of the transistor used and may be measured: if the output terminals are short circuited ($V_2 = 0$) then

$$h_{11} = V_1/I_1$$
$$h_{21} = I_2/I_1$$

h_{11} is an impedance and h_{21} is dimensionless. If the input terminals are open circuited ($I_1 = 0$) then

$$h_{12} = V_1/V_2$$
$$h_{22} = I_2/V_2$$

h_{12} is dimensionless and h_{22} is an admittance. A standard nomenclature has been adopted:

$h_i = h_{11}$ = input impedance with output short circuited
$h_r = h_{12}$ = reverse voltage feedback ratio
$h_f = h_{21}$ = forward current transfer ratio
$h_o = h_{22}$ = output admittance with input open circuited

The method of connecting the transistor in the circuit is indicated by a second subscript. These are b for *common-base, e for common-emitter, and c for *common-collector connections; for example, h_{fe} is the common-emitter forward-current transfer ratio.

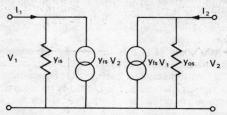

c y-equivalent circuit with common-source connection

The *y parameters* are frequently used for circuits containing *field-effect transistors. The equivalent circuit is a current source with shunt impedance at both input and output (Fig. c). The simultaneous equations corresponding to the circuit are

$$[I] = [y][V]$$

where y has dimensions of admittance.

Other parameters used are z *parameters,* which are impedance parameters and are the inverse of y parameters (voltage sources with series impedances); g *parameters* are the inverse of h parameters.

601

If the circuit is for use with small signals, the *small-signal parameters* are written h_{fe}, etc.; if for use with large signals, capitals are used for the subscripts: h_{FE}, etc.

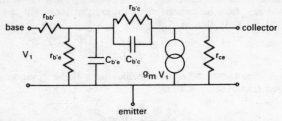

d Hybrid-π equivalent circuit with common-emitter connection

Other equivalent circuits are made up of components relating to the actual physical nature of the device rather than the more abstract linear networks described above. The most common circuit which is useful for high-frequency circuits is the *hybrid-π equivalent circuit* (Fig. *d*). The components are labelled with subscripts which are unprimed and primed. The letters *b*, *c*, and *e* represent the transistor regions (base, etc.). The unprimed subscripts represent the external portions of the region (connections and some semiconductor material) while the primed subscripts represent the intrinsic or internal portion of the electrode that is close to the junction and actually involved in the transistor action. Thus $r_{bb'}$ represents the base *spreading resistance and $C_{b'e}$ represents the intrinsic base to emitter terminal capacitance.

transistor-transistor logic (TTL) A family of high-speed integrated *logic circuits in which the input is through a multiemitter *transistor; usually the output stage is *push-pull. *Diode-transistor logic operates in a similar manner but the input is through a number of *diodes. A typical circuit (a three input NAND circuit) is shown in Fig. *a*. If the input levels are all high, the emitter-base junctions of the input transistor T_1 will all be reverse biased: current through the base will flow across the forward-biased collector junction to the phase-splitting transistor T_2, which will switch on. Current flows through T_2 to T_4 and turns on T_4. T_3 will remain off as current is shunted away from the base and the output voltage will be low. If any one or more of the inputs is low, the emitter-base junction of T_1 will be forward biased and current flows out of the base through the emitter of T_1. The current is therefore diverted away from T_2, and T_2 and T_4 will be turned off. The current through R_2 flows into the base of T_3 and T_3 is switched on and the output voltage is high. The output voltage will change rapidly when the input conditions change since the transistors drive the level in both directions.

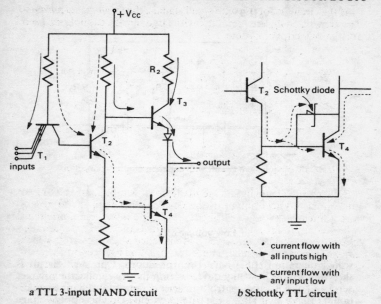

a TTL 3-input NAND circuit　　**b Schottky TTL circuit**

The biggest limitation in speed is caused by the *delay time due to hole storage in the saturated output transistor T_4. The speed may be improved by adding a *Schottky diode with low forward diode voltage across the base-collector junction of T_4. This circuit is called *Schottky TTL* and part of the circuit is shown in Fig. *b*. This diode prevents T_4 saturating. Hole storage therefore does not occur in the collector-base junction and since it does not occur in Schottky diodes, T_4 will be turned off very rapidly when the base current is cut off.

TTL is one of the most widely used types of integrated logic circuit for high-speed applications and, together with *emitter-coupled logic (ECL), tends to be regarded as a standard against which all other logic circuits are judged. TTL is also characterized by medium power dissipation and *fan-out and good immunity to *noise. TTL circuits may have high operating speeds but at the expense of power dissipation, since the higher the speed the greater the power consumed. They are also medium-scale integration (MSI) circuits. They are therefore not suitable for applications where low power dissipation and large functional packing density is required. Simplified low-voltage versions of TTL that are more suitable for large-

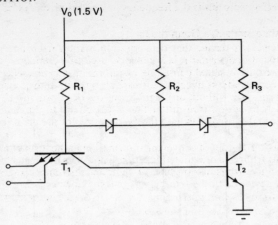

c Low-voltage TTL

scale integration (LSI) have been produced. One such circuit is shown in Fig. *c*. Schottky diodes clamp the base-collector voltages across T_1 and T_2 and control the level of saturation. R_2 is an additional resistor that allows the gate to operate from a supply voltage less than or equal to 1.5 volts. *MOS logic circuits are usually the circuits of choice for such applications but they operate at much lower speeds. C/MOS circuits, in particular, have very low power dissipation and C/MOS versions of TTL gates have been produced. *I^2L circuits are now being used for applications where a greater speed than that of MOS is required.

transition A sudden change in the energy state of an atom between two of its *energy levels. A transition may be caused by absorption of energy, as in the photoelectric effect, or may occur when an atom in an excited state reverts to a state of lower energy. In the latter case it is accompanied by the emission of a photon of energy. Both these effects are utilized in the *laser.

transition flux density *See* superconductivity.

transition temperature *See* superconductivity.

transit time The time required for a charge *carrier to travel directly from one designated point or region in an electronic device to another, under specified operating conditions. The transit time depends on the operating conditions, the device geometry, and in semiconductor devices upon the *drift mobility of the carriers within the semiconductor; in the absence of *carrier storage the transit time effectively limits the frequency at which a given device may be operated.

transit time mode *See* Gunn diode.

translator (1) A device that converts information from one form to another. In particular it is a network, used in a computer, that accepts an input signal consisting of information expressed in a certain code and produces a corresponding output expressed in a different code without any significant loss of information.

(2) A television transmitter that is peripheral to the main transmitter and is used to extend the *service area of the main transmitter.

transmission The act of conveying information in the form of electrical signals from one designated location to another by means of a wire, waveguide, transmission line, or radio channel and using any circuits, devices, or other equipment that may be necessary.

transmission electron microscope *See* electron microscope.

transmission gain *See* transmission loss.

transmission level The ratio of the electrical power at any point in a *telecommunication system to the power at some arbitrary reference point, usually the sending point in a two-wire system.

transmission line (1) *Syn.* power line. An electric line, such as an *open wire, that is used to convey electrical power from a power station or substation to other stations or substations.

(2) An electric *cable or *waveguide that conveys electrical signals from one point to another in a *telecommunication system and that forms a continuous path between the two points.

(3) *Syn.* feeder. The one or more conductors that connect an aerial to a transmitter or receiver and that are substantially nonradiative.

Any of the above transmission lines is described as smooth or uniform if its electrical parameters are distributed uniformly along its length. A *balanced line* is a transmission line that has conductors of the same type, equal values of resistance per unit length, and equal impedances from each conductor to earth and to other electrical circuits. The transmission characteristics of a particular transmission line are usually frequency-dependent and *loading* is often used to improve the transmission characteristic throughout a given frequency band. Loading is the addition of inductance to a line: in *coil loading* inductance coils (*loading coils*) are placed in series with the line conductors at regular intervals; in *continuous loading* a continuous layer of magnetic material is wrapped along each conductor.

The presence of a discontinuity in a transmission line can cause some signal energy to be reflected back along the line. The presence of such *line reflections* can lead to the production of standing waves in the transmission line. A transmission line that has no line reflections and therefore no standing waves is a *nonresonant line.*

The most efficient transfer of signal power from a transmission line to the *load occurs when the load is matched to the transmission line, i.e. the load impedance, Z_L, is equal to the *characteristic impedance,* Z_0, of the transmission line. Z_0 is given by the ratio V/I at every point along a transmission line when no standing waves are present in the line. V and I are the complex voltage and current at each point along the line.

transmission loss The reduction in power between any two points 1 and 2 in a telecommunication system given by the ratio P_2/P_1, where P_1 and P_2 are the powers at points 1 and 2 respectively and 2 is further away from the source of power than 1. For a given transmission path at a designated frequency the transmission loss is given by the ratio of the available power at the input of a receiver to the power transmitted from the output of the transmitter.

In any system where *gain is introduced to the signal, the *transmission gain* is given by the ratio of the powers P_2/P_1. This represents the increase in power between the points 1 and 2 where 2 is again further from the source of power than 1.

transmission mode *See* mode.

transmission primary Any of the set of three primary colours used to provide the chrominance signals in a *colour-television system. The colours chosen are red, green, and blue.

transmit-receive switch (TR switch) A switch used in a *radar system that has a common aerial for both transmission and reception. The TR switch automatically decouples the receiver from the aerial during the transmitting period. It is commonly used in conjunction with an *anti-transmit-receive switch* (ATR switch) that automatically decouples the transmitter from the aerial during the receiving period; extra protection is thus provided for the transmitter from return echos received by the aerial.

A common type of transmit-receive switch is a combination of a *gas-discharge tube and a *cavity resonator. The cavity resonator is used to connect the aerial to the transmitter (or receiver). When a discharge is established in the gas-discharge tube, i.e. when the tube is fired, the electrical conditions of the cavity resonator become such that resonance is not possible and the transmitter (or receiver) is therefore disconnected. A *keep-alive circuit* is frequently incorporated in the switch to maintain the discharge in the tube during periods when it is not operative.

transmitted-carrier transmission A telecommunication system that uses *amplitude modulation of a carrier wave in which the carrier is transmitted. *Compare* suppressed-carrier transmission.

transmitter The device, circuit, or apparatus used in a *telecommunication system to transmit an electrical signal to the receiving part of the system.

transmitting aerial *See* aerial.

transponder A combined transmitter and receiver system that automatically transmits a signal when a predetermined *trigger is received by it. The trigger, which is often in the form of a *pulse, is known as the *interrogating signal*. The minimum amplitude of the trigger signal that initiates a response from the transponder is the *trigger level*.

transposed transmission line A transmission line in which the conductors of which it is composed are interchanged in position at regular intervals of distance, in order to reduce the electric and magnetic coupling between the line and other lines.

transverse wave A *wave in which the displacement at each point along the wave is in a direction perpendicular to the direction of propagation. An electromagnetic wave is an example of a transverse wave. *See also* mode.

trapezium distortion *See* distortion.

trapping recombination *See* recombination processes.

travelling wave An electromagnetic wave that is propagated along and is guided by a *transmission line. In the case of a hypothetical lossless line of uniform cross section and infinite length, a sinusoidal a.c. supply at one end of the line (the sending end) causes electrical energy to be transmitted along the line with instantaneous values of current and voltage at any given point varying sinusoidally. If the line is situated in a medium of relative permittivity ε_r and relative permeability μ_r, the sine waves are propagated along the line with a velocity v given by

$$v = c/\sqrt{(\varepsilon_r \mu_r)}$$

where c is the velocity of light. In the case of a lossless line of finite length terminated with a load impedance equal to the characteristic impedance of the line the above equation holds. In practice dissipation in the line causes both a reduction in the velocity and *attenuation. Nonsinusoidal travelling waves, such as an *impulse or *surge voltage, may also be propagated.

Impedance discontinuities at any point on the line result in partial reflection of the initial wave, which then consists of two waves: the *reflected wave*, which travels in a backward direction towards the sending end, and the transmitted wave, which travels forward towards the receiving (output) end of the line.

travelling-wave accelerator *See* linear accelerator.

travelling-wave maser *See* maser.

travelling-wave tube (TWT) *Syn.* slow-wave tube. A *microwave tube in which an electron beam interacts continuously with a radiofrequency (r.f.) electromagnetic field in order to produce amplification or, in some types of tube, oscillation at microwave frequencies.

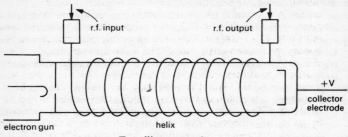

Travelling-wave tube

The basic tube is an O-type (linear-beam) tube in which an axial electron beam produced from the electron gun and attracted by a collector electrode is surrounded by a suitable structure capable of propagating an r.f. electromagnetic wave. A helix structure is shown in the diagram. The propagating circuit is proportioned so that the phase velocity of the r.f. wave is small (typically about one tenth to three tenths of the velocity of light); the circuit is therefore termed a *slow-wave structure*. The longitudinal component of an r.f. wave propagated along the slow-wave structure interacts with the electron beam to produce *velocity modulation of those electrons that are approximately in synchronism with the axial phase velocity. The velocity-modulated beam in turn interacts with the slow-wave structure to produce additional waves in it. Suitable design of the structure causes the induced waves in successive portions of the structure to be in phase in the forward direction (along the direction of flow of the electron beam) and out of phase in the backward direction. The process of mutual interaction occurs along the length of the tube with the net result being a transfer of energy from the electron beam to the radiofrequency wave. This results in amplification of the wave, termed the *forward wave*.

The forward travelling-wave tube can, with suitable design, have bandwidths of greater than an octave. For optimum transfer of energy approximate synchronism is required between the electron velocity and the phase velocity of the wave and the tube will operate correctly only for a limited range of applied voltage, termed the *synchronous voltage*. The *Millman travelling-wave tube* uses as the slow-wave structure a series of equally spaced cavity resonators that are electrically connected and separated by field-free drift regions. The helix structure however has the advantage of a phase velocity almost independent of frequency over a wide range and hence a substantially constant operating voltage.

The *backward-wave oscillator* is a form of O-type travelling-wave tube in which the optimum transfer of power produces radiofrequency waves in the backward direction, i.e. the group velocity and phase velocity differ by 180°. A beam current of sufficient magnitude interacts with the slow-wave structure to produce radiofrequency oscillations that are delivered as microwave power at the electron-gun end of the structure. The minimum beam current required to produce oscillations is the *start-oscillation current.* At currents below this value the tube may be used as an amplifier by supplying a radiofrequency-wave input at the collector end of the structure. The interaction efficiency of this type of tube is increased by using an electron beam with a hollow cross section. This is achieved by magnetically confining the electron flow from the cathode.

Focusing of the electron beam in the above types of tube is usually achieved using a periodically reversing magnetic field produced by a permanent-magnet structure; axial magnetic fields produced by an electromagnet may also be used. Variations in the output of the tube may sometimes result from *scalloping,* when axial variations of the focusing field result in corresponding variations in the diameter of the electron beam.

The *carcinotron* is a crossed-field (*see* microwave tube) backward-wave oscillator in which the electric and magnetic fields on the slow-wave structure are perpendicular to each other and cause the electron beam to bend and travel approximately parallel to the slow-wave structure. The carcinotron is usually a circular device and offers very high efficiency typical of a crossed-field device as compared to the equivalent linear-beam device.

treble compensation *See* radio receiver.

treble response *See* radio receiver.

triac *Syn.* bidirectional triode thyristor. *See* silicon-controlled rectifier.

triboelectricity *See* frictional electricity.

triboluminescence *See* luminescence.

trickle charge A small continuous charge applied to a secondary battery in order to maintain it in a fully charged condition during storage. The charging current is maintained at a value that just compensates for internal dissipation due to local action within the battery.

trigger Any stimulus that initiates operation of an electronic circuit or device. Also the act of initiating operation in the circuit or device. In general the response to a trigger continues after the cessation of the stimulus. A circuit, such as a flip-flop or multivibrator, that is used to trigger other circuits is known as a *trigger circuit.* The output signal is often in the form of a *trigger pulse,* such as a *clock pulse.

trigger level *See* transponder.

trilevel resist *See* multilevel resist.

trimmer *Syn.* trimming capacitor. A relatively small variable capacitor used in *parallel with a large fixed capacitor in order to adjust the total capacitance of the combination over a limited range of values.

trimming capacitor *See* trimmer.

Trinitron *Tradename. See* colour picture tube.

triode Any electronic device with three electrodes, such as a bipolar junction *transistor, *field-effect transistor, or *thyratron. The term is particularly applied to a *thermionic valve that has three electrodes.

triode-hexode A multiple (or multielectrode) *thermionic valve that contains both a triode and a hexode within the same envelope. The triode-hexode is most often used as a frequency changer: it acts as a combined oscillator (triode portion) and *mixer (hexode portion).

triode-region operation (of a field-effect transistor) *See* saturated mode.

trip coil *See* tripping device.

triple detection *Syn. for* double superheterodyne detection. *See* superheterodyne reception.

triplen harmonic A *harmonic that has a frequency of $3nf$ when n is an integer and f the frequency of the fundamental. A frequency multiplier in which the output is the first triplen harmonic ($n = 1$) of the input frequency is a *tripler.*

tripler *See* frequency multiplier; triplen harmonic.

tripping device A device that normally constrains a *circuit-breaker in the 'on' position until actuated, when the circuit-breaker is allowed to break the circuit. Manual operation is common in many types of tripping device. Other types are operated electromagnetically. A typical example is the *trip coil,* which consists of a coil that controls a movable plunger or armature; the plunger controls the action of the circuit-breaker.

TR switch *Abbrev. for* transmit-receive switch.

truncated test *See* life test.

trunk *See* trunk feeder.

trunk feeder *Syns.* interconnecting feeder; interconnector; trunk; trunk main. A *transmission line that is used to interconnect two electric *power stations or two electric power distribution networks.

trunk main *See* trunk feeder.

trunk telephony *See* telephony.

truth table A table used in formal logic that lists the truth or falsity of the outcome when a logical operator, such as 'and' or 'or', is applied to combinations of logical statements. The truth table has been

610

adapted to describe the operation of *logic circuits by listing the outputs of a binary logic gate, such as a *flip-flop, for all possible combinations of inputs. The 'true' state corresponds to the voltage level representing a logical 1 and 'false' to logical 0.

T-section *See* quadripole; filter.

TTL *Abbrev. for* transistor-transistor logic.

tube *Syn.* valve. *US, short for* electron tube.

tube heating time *See* heating time.

tunable magnetron *See* magnetron.

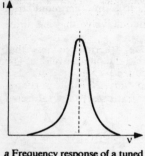
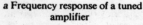
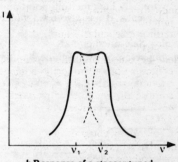

| *a* Frequency response of a tuned amplifier | *b* Response of a stagger tuned amplifier |

tuned amplifier An amplifier that is designed to operate only with a very narrow range of input frequencies. A typical frequency response curve is shown in Fig. *a*. The response may be modified to give an essentially flat response to the frequencies in the pass band by using stagger tuning. A *stagger tuned amplifier* consists of two or more *amplifier stages each of which is tuned to slightly different frequencies, ν_1 and ν_2. The overall response is essentially flat within the pass band (Fig. *b*), the centre of the peak occurring at the frequency corresponding to

$$|\nu_1 - \nu_2|/2$$

tuned circuit A *resonant circuit that contains some form of *tuning* so that the natural *resonant frequency of the circuit may be varied. The resonance condition for *forced oscillations can thus be changed. A tuned circuit is used to select the frequency of resonance of a variety of frequency-selective devices, such as audiofrequency amplifiers. Tuning may be carried out by adjusting the value of the capacitance (*capacitive tuning*), or the inductance (*inductive tuning*), or both. Inductive tuning is commonly achieved by altering the position of a suitably shaped piece of soft ferromagnetic material (a slug) relative to a coil in the circuit (*slug tuning*).

tuner (1) A device, such as a variable capacitor or inductor, that is used to alter the resonant frequency of a *tuned circuit.

(2) The first stage of a *radio or *television receiver that is used to select a particular broadcast channel. It contains either a *tuned circuit in which the resonant frequency may be altered to accept the desired channel frequency or a set of *resonant circuits each of which is tuned to an individual channel and may be selected (*see* turret tuner).

tungsten Symbol: W. A heavy metal, atomic number 74, that has an extremely high melting point and is extensively used to form lamp filaments. It has also been widely used to form *thermionic cathodes.

tuning *See* tuned circuit.

tuning screw *See* waveguide.

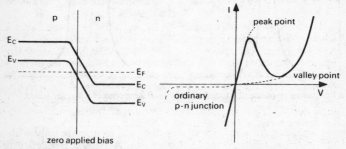

a Energy bands of tunnel diode   *b* Characteristic of tunnel diode

tunnel diode *Syn.* Esaki diode. A p-n junction diode that has extremely high doping levels on each side of the junction, i.e. it is doped into *degeneracy. The energy bands are shown in Fig. *a* for zero applied bias. At such high doping levels tunnelling of electrons (*see* tunnel effect) across the junction can occur in the forward direction (positive applied voltage on the p-region). As the positive bias is increased the height of the potential barrier at the junction is decreased and the width increased; the diode therefore exhibits a negative resistance portion of the characteristic as the tunnel effect contributes progressively less towards the conductance. A minimum current is reached at the *valley point* when the tunnel effect ceases and for voltages above this point the diode behaves as a normal p-n junction diode. Tunnelling also occurs in the *reverse direction in a similar manner to that in the *Zener diode but the effective Zener breakdown voltage can be considered to occur at a small positive value of voltage, termed the *peak point*. A typical tunnel diode characteristic is shown in Fig. *b*.

The *backward diode* is similar to the tunnel diode but the doping levels are slightly lower so that the semiconductor regions are not quite degenerate (Fig. *c*). The tunnel effect occurs readily for small values of reverse bias but the negative-resistance portion of the characteristic disappears. The current flowing in the reverse direction is larger than in the forward direction (Fig. *d*).

c **Energy bands of backward diode**   d **Characteristic of backward diode**

The backward diode can be used for rectification of small signals, when the conventional forward-direction positive bias on the p-region becomes the reverse direction for this device. *Carrier storage at the junction does not occur and the backward diode therefore has a high speed of response and may be used at microwave frequencies. Very little variation of the current-voltage characteristics occurs with temperature or incident radiation. Unfortunately the high doping levels required for both tunnel diodes and backward diodes make reliable manufacture difficult.

tunnel effect The crossing of a potential barrier by a particle that does not have sufficient kinetic energy to surmount it. The effect is explained by wave mechanics. Each particle has an associated wave function that describes the probability of finding the particle at a particular point in space. As a particle approaches the potential barrier the wave function is considered to extend inside the region of the potential barrier. Provided that the barrier is not infinitely thick a small but finite probability exists that the particle will appear on the other side of the barrier. As the thickness of the barrier decreases the probability of tunnelling through it increases.

The tunnel effect is the basis of *field emission and the *tunnel diode but the probability of it occurring in macroscopic systems is extremely small.

turn-off time *See* silicon-controlled rectifier.

turns ratio *Syn.* transformer ratio. Symbol: *n*. The ratio of the number of turns, n_2, active in the secondary circuit of a transformer to the number of turns, n_1, in the primary winding. In the case of a simple transformer with no *tappings n_2 is the number of turns in the secondary winding. In the case of a current transformer the ratio of the currents is the inverse of the turns ratio, i.e.

$$I_{out}/I_{in} = 1/n$$

See also transformer.

turret tuner A tuning device used in a *television or *radio receiver. It contains a set of resonant circuits each tuned to the frequency of one of the separate broadcast channels. One or more manually operated switches, termed *band switches,* allow the particular circuit corresponding to the desired channel to be selected by the user.

tweeter A physically small loudspeaker that reproduces sounds of relatively high frequency, for example, frequencies from five kilohertz upwards. In a high-fidelity sound-reproduction system a tweeter and a *woofer are used together.

twin cable *See* paired cable.

twin-T network *Syn.* parallel-T network. *See* quadripole.

twisted pair *See* pair.

two-gun storage tube *See* storage tube.

two-phase CCD *See* charge-coupled device.

two-phase system *Syn.* quarter-phase system. *See* polyphase system.

two-phase to three-phase transformer *See* three-phase to two-phase transformer.

two-port network *See* quadripole.

two-way communication A *telecommunication system in which signals may be transmitted between two points in both directions. One direction is usually arbitrarily designated 'go' and the other 'return'.

two-wire circuit A circuit that consists of two conductors insulated from each other and that provides simultaneously a two-way communication channel in the same frequency band between two points in a *telecommunication system. The circuit may be a *phantom circuit in which case it is formed from two groups of conductors. A circuit that operates in the same manner but that is not limited to only two conductors (or groups of conductors) is termed a *two-wire type circuit. Compare* four-wire circuit.

TWT *Abbrev. for* travelling-wave tube.

U

UHF *Abbrev. for* ultrahigh frequency. *See* frequency band.

ultrahigh frequency (UHF) *See* frequency band.

ultrasonic communication Underwater communication at ultrasonic frequencies using suitably modified *sonar.

ultrasonic delay line An acoustic *delay line that uses acoustic waves of ultrasonic frequencies as the delay element.

ultrasonic light valve A device that can be used to transmit video information. It consists of a piezoelectric quartz crystal immersed in a transparent liquid. The crystal is excited by ultrasonic-frequency alternating current and the resulting mechanical vibrations set up compressive waves in the liquid. If the crystal is fed a modulated video signal, corresponding changes in the compression result. The system acts as a liquid diffraction grating to a light beam that is shone through it. The video information may be recorded on photographic film or detected with a suitable *photodetector.

ultrasonics *Syn.* supersonics. The study and application of sound frequencies that lie above the limits of audibility of the human ear, i.e. frequencies above about 20 kilohertz.

Ultrasonic waves may be generated electronically by *magnetostriction generators* or *piezoelectric generators*. In the former case a high-frequency electric oscillation is applied to a material that exhibits *magnetostriction so as to produce ultrasonic waves. In the latter the *piezoelectric effect transforms the electric oscillations into ultrasonic waves.

Ultrasonic waves may be detected using the *hot-wire microphone. The only electrical receiver that is sensitive to ultrasonic waves is the *piezoelectric quartz crystal. When excited by an ultrasonic wave that is propagated in a suitable direction and corresponds in frequency to the natural frequency of the receiving crystal, mechanical vibrations are set up in the crystal and produce varying electric potentials. Since two crystals of the same natural frequency are very difficult to obtain, the receiving crystal must be electrically tuned to the frequency of the received wave. A type of detector due to Pierce uses the same piezoelectric crystal as both transmitter and receiver.

Ultrasonic waves have many applications. In electronics the main applications include degassing of melts during the manufacture of electronic components and circuits, cleaning of electronic components, *echo sounding and underwater communication, and picture transmission using an *ultrasonic light valve. Ultrasonics is used in medicine for the location of a brain tumour or a foetus.

ultraviolet photoelectron spectroscopy (UPS) *See* photoelectron spectroscopy.

ultraviolet radiation Electromagnetic radiation lying between light and X-rays on the electromagnetic spectrum. Radiation of frequency close to that of light is *near ultraviolet* radiation; that at the high-frequency end of the range is *far ultraviolet*.

unabsorbed field strength The field strength of a radiowave that would exist at the receiving point in the absence of any *absorption between the transmitting and receiving aerials.

unclocked flip-flop *See* flip-flop.

underbunching *See* velocity modulation.

undercoupling *See* coupling.

undercurrent release A switch, circuit-breaker, or other tripping device that operates when the current in a circuit falls below a predetermined value. A current that causes the release to operate is termed an *undercurrent*. *Compare* overcurrent release.

underdamping *See* damped.

underscanning *See* scanning.

undershoot An initial transient response to a change in an input signal that precedes the desired response and is in the opposite sense. *Compare* overshoot.

undervoltage release A switch, circuit-breaker, or other tripping device that operates when the voltage in a circuit falls below a predetermined value. A voltage that causes the release to operate is an *undervoltage*. An undervoltage release is sometimes used in conjunction with a *circuit-breaker in order to prevent the circuit-breaker closing when the supply voltage is too low. This application is known as *undervoltage no-close release*.

underwater sound projector A piezoelectric loudspeaker that is used to produce sound in water. *See* sonar.

undistorted wave A periodically varying quantity that consists of sinusoidal components of which none is present at one point that is not present at all points and in which both the attenuation and velocity of propagation are the same for all the sinusoidal components.

uniconductor waveguide *See* waveguide.

unidirectional current *See* current.

unidirectional transducer *Syn. for* unilateral transducer. *See* transducer.

unifilar suspension A form of construction of an instrument in which the movable part is suspended by a single thread, wire, or strip. The restoring force may or may not be produced mainly by torsion in the suspension. *Compare* bifilar suspension.

uniform cable *See* cable.

uniform line A transmission line with electrical properties that are substantially identical along its length.

uniform waveguide A waveguide that has constant electrical and physical characteristics along its axial length.

unijunction transistor *Syn.* double-base diode. A bipolar *transistor that has three terminals and one junction. A schematic diagram is shown in Fig. *a* and the general symbol in Fig. *b*.

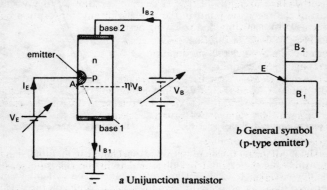

a Unijunction transistor

b General symbol
(p-type emitter)

The device consists of a bar of lightly doped (high-resistivity) *semiconductor, usually n-type, with an opposite-polarity region of highly doped (low-resistivity) material located near the centre of the bar. Ohmic contacts are formed to each end of the bar (base 1 and base 2) and to the central region (emitter). Originally the emitter region was alloyed into the bulk material but planar diffused or planar epitaxial structures are now produced.

Under normal operating conditions base 1 is earthed and a positive bias, V_B, applied to base 2. A point A on the least positive side of the emitter junction is considered. The voltage on the n side of the junction at point A is given by ηV_B, where η is the *intrinsic stand-off ratio*:

$$\eta = R_{B1}/R_{BB}$$

R_{B1} is the resistance between point A and base 1 and R_{BB} the resistance between base 1 and base 2.

If the voltage, V_E, applied to the emitter is less than ηV_B the junction is *reverse biased and only a small reverse saturation current flows. If V_E is increased above ηV_B the junction becomes *forward biased at point A and holes are injected into the bar. The electric field within the bar causes the holes to move towards base 1 and increase the conductivity in the region between point A and base 1. Point A becomes less positive and therefore more of the junction becomes forward biased; this causes the emitter current, I_E, to increase rapidly. As I_E increases the increased conductivity causes the emitter voltage to drop and the device exhibits a negative-resistance portion of the current-voltage characteristic (Fig. *c*).

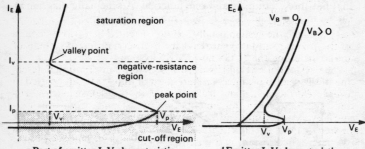

c Part of emitter I–V characteristic   *d* Emitter I–V characteristics

The emitter voltage at which the device starts to conduct, V_p, is known as the *peak point*. At voltage V_v, known as the *valley point*, the device ceases to exhibit negative resistance. The switching time between the peak point and the valley point depends on the device geometry and the biasing voltage, V_B, applied to base 2. It has been found to be proportional to the distance between point A and base 1.

If base 2 is open circuit, the I–V curve is essentially that of a simple *p-n junction (Fig. *d*). As V_B is increased the peak point, V_p, and the current, I_v, of the valley point also increase. The characteristics show little temperature dependence.

The most common application of the unijunction transistor is in *relaxation oscillator circuits.

unilateral network *See* network.

unilateral transducer *Syn.* unidirectional transducer. *See* transducer.

unintelligible crosstalk *See* crosstalk.

uninterrupted duty *See* duty.

unipolar transistor A transistor in which current flows as a result of the movement of majority carriers only. *See* field-effect transistor.

unipotential cathode *Syn. for* indirectly heated cathode. *See* thermionic cathode.

unit A precisely specified quantity, such as the ampere or second, in terms of which the magnitude of other physical quantities of the same kind can be stated:

physical quantity = numerical value × unit
current = *n* amperes
charge = *n* amperes per second = *n* coulombs

In a purely mechanical system the units can be defined in terms of certain units of the three fundamental quantities of mass, length, and time: other units are derived by multiplying and/or dividing sets of these fundamental units. In any system concerned with electric and magnetic properties a fourth fundamental quantity is necessary to make the system totally self-consistent. The various choices of this quantity and its value have led to several different systems of units, including the *CGS and *MKS systems.

The system of *SI units was formulated, by international agreement, in order to produce a coherent set of units that can be used in every field of science and technology without inconsistencies arising. There are seven fundamental or base SI units from which almost every other unit can be derived.

unit-step function An electrical signal that is zero for all times before a predetermined instant and unity for times following that instant.

unitunnel diode *Syn. for* backward diode. *See* tunnel diode.

universal motor An electric motor that may be used with a direct-current power supply or an alternating-current supply. It incorporates a commutator in order to achieve this operation. Small motors of this type are commonly used in domestic appliances, such as vacuum cleaners or portable drills.

universal shunt A galvanometer *shunt that is tapped so that it can pass designated fractions (0·1, 0·01, etc.) of the main current and can thus be used with galvanometers of widely varying internal resistance. An example is the *Ayrton shunt,* which is a relatively large resistance with suitable tappings.

univibrator *See* monostable.

unpaired electron *See* Pauli exclusion principle.

unstable oscillation *See* oscillation.

upper sideband *See* carrier wave.

UPS *Abbrev. for* ultraviolet photoelectron spectroscopy. *See* photoelectron spectroscopy.

USASI *Abbrev. for* USA Standards Institute. *See* standardization.

USA Standards Institute (USASI) *See* standardization.

useful life *See* failure rate.

V

vacancy A site in a crystal lattice that is not occupied by an atomic nucleus. A vacancy is distinct from a *hole.

vacuum evaporation A technique used during the manufacture of electronic circuits and components in order to produce a coating or thin film of one solid material on the clean surface of another, in particular a metal on a semiconductor. The solid to be deposited is heated in a vacuum in the presence of the cool substrate material. Atoms evaporated from the material (either in solid or liquid form) suffer few collisions from the residual low-pressure gas and travel directly to the substrate. They condense on the substrate surface and form a thin film. *See also* thin-film circuit.

vacuum phototube *See* phototube.

vacuum switch A *switch in which the contacts are contained in a substantially evacuated container in order to minimize *spark formation.

vacuum thermocouple *See* thermocouple.

vacuum tube An *electron tube evacuated to a sufficiently low pressure that its electrical characteristics are independent of any residual gas. *Compare* gas-filled tube.

valence band The band of energies in a solid in which the electrons cannot move freely but are bound to individual atoms. In a metal electrons can be readily excited into the *conduction band. In a *semiconductor at ambient temperature very few electrons are excited into the conduction band. Mobile *holes are formed in the valence band. *See* energy bands.

valence electrons Electrons that occupy the outermost (lowest energy) energy levels of an atom and are involved in chemical and physical changes.

valency The number of hydrogen (or equivalent) atoms with which an atom will combine or that it will displace. The two most important types of valency are electrovalency and covalency.

An *electrovalent bond* is formed when one or more electrons are transferred from one atom to another to complete the outermost shells of both atoms and form two charged *ions, one positively charged and the other negatively charged. The ions are then held together by electrostatic attraction.

A *covalent bond* is formed when the outermost electrons (usually three or four) of two atoms are shared between the two atoms and orbit both nuclei. No ions are formed in this type of bond.

valley *See* pulse.

valley point (1) *See* unijunction transistor. (2) *See* tunnel diode.

valve *Syn.* electron tube. An *active device in which two or more electrodes are enclosed in an envelope, usually of glass, one of the electrodes acting as a primary source of electrons. The electrons are most often provided by thermionic emission (in a *thermionic valve) and the device may be either evacuated (vacuum tube) or gas-filled (gas-filled tube). The name derives from the rectifying properties (*see* rectifier) of the devices, i.e. current flows in one direction only. The word *valve* is gradually becoming obsolete and is being replaced by *electron tube*. Valves have been replaced by semiconductor devices for everyday applications.

valve characteristics *See* characteristics.

valve diode *See* diode; thermionic valve.

valve rectifier A *rectifier in which a *valve, usually a *diode, acts as the rectifying element.

valve voltmeter A type of *voltmeter that has now generally been superseded by the *digital voltmeter. It consists of an *amplifier of extremely high *input impedance and containing one or more *thermionic valves, with a measuring instrument in the output circuit. Direct voltage or alternating voltage may be measured with this instrument.

Van de Graaff accelerator An *accelerator in which the high-voltage terminal of a *Van de Graaff generator acts as a charged-particle source. Ions derived from the high-voltage terminal are injected into an evacuated accelerating tube. An appropriate voltage gradient is maintained along the tube. Van de Graaff accelerators can produce ion currents of several hundred microamps and electron currents of one milliamp at voltages of about five megavolts.

The *tandem generator* is a multistage version of the Van de Graaff accelerator that accelerates particles through twice the applied voltage. The negative ions are accelerated by an applied voltage, *V*, as in the Van de Graaff accelerator. At the end of the accelerating tube they pass through a stripping foil that removes electrons and therefore forms positive ions. These are further accelerated from +*V* to earth potential in a second tube. Protons with energies of about 20 MeV and heavy ions with energies greater than 100 MeV have been produced in this device.

Van de Graaff generator An electrostatic *voltage generator that can produce potentials of millions of volts. An electric charge from an external source is applied to a continuous insulated belt at points A (*see* diagram). The belt travels vertically up into a large hollow metallic sphere and the charge is collected at points B. The charge resides on the exterior of the sphere and the possible voltage generated is limited only by leakage.

Van Vleck paramagnetism *See* paramagnetism.

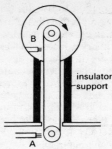

Van de Graaff generator

vapour phase epitaxy The most common method of *epitaxy. The material to be deposited on the substrate is heated to gaseous form in a deposition furnace that also contains the substrate material. The substrate is held at a temperature just below the solidification point. As the gas molecules reach the substrate they are deposited on the surface, replicating the substrate crystal structure. The conditions in the deposition furnace can be adjusted with respect to temperature and pressure to allow particular desired combinations of substrate and deposited material to be produced. *See also* MOCVD.

vapour plating A technique for producing a thin film of one solid material on the clean surface of another solid. A compound of the material to be deposited is vaporized and thermally decomposed in the presence of the substrate material. Atoms from the vapour condense on the surface to be plated and form a thin film on it.

var The unit of reactive *power of an alternating current. It is identical to the *watt, being the product of volts times amperes.

varactor Acronym from *var*iable re*actor*. A semiconductor *diode or *Schottky diode operated with reverse bias so that it behaves as a voltage-dependent capacitor. The *depletion layer at the junction acts as the dielectric and the n- and p-regions form the plates (*see* diagram). A diode intended for use as a varactor generally has a particular impurity profile designed to give an unusually large variation of junction capacitance and to minimize the series resistance.

In a diode with an *abrupt junction the voltage dependence is given by

$$C \propto V^{-1/2}$$

If the diode has a graded junction then

$$C \propto V^{-1/3}$$

where C is the capacitance and V the reverse voltage.

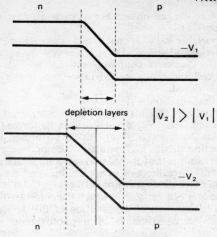

depletion layers

$$|v_2| > |v_1|$$

Energy bands in a varactor

Varactors may be used at low power for *varactor tuning. High-power varactors are used as nearly lossless frequency multipliers. Gallium arsenide may be used to form the varactor when required for very high frequency applications. Otherwise silicon is used.

varactor tuning Capacitive tuning (*see* tuned circuit) employed in *receivers, such as television receivers, in which the variable capacitance element is provided by a *varactor.

variable capacitance gauge *See* strain gauge.

variable impedances *Impedances, such as capacitors, that are adjustable so as to present variable values of the impedance. The method of adjustment is usually provided by a movable contact or by physical adjustment of the size of the device.

variable inductance gauge *Syn. for* electromagnetic strain gauge. *See* strain gauge.

variable mu valve A *thermionic valve in which the *amplification factor, μ, is a function of the grid voltage. The grid is usually constructed from nonuniformly spaced wires in order to achieve this effect.

variable reluctance pick-up *See* pick-up.

variable resistance gauge *Syn. for* resistance strain gauge. *See* strain gauge.

variable resistance pick-up *See* pick-up.

variation *See* deviation.

variocoupler A *transformer used in radio circuits that is constructed in such a way that the mutual inductance between the windings can be varied while the self-inductance of each winding remains substantially constant.

variometer A variable inductor that usually consists of two coils connected in *series and arranged so that one coil can rotate relative to the other. The degree of coupling between the coils is varied by moving one coil and hence the self-inductance of the series combination is also varied.

varioplex *See* time-division multiplexing.

varistor Acronym from *vari*able res*istor*. A *resistor that has a markedly nonohmic characteristic. A varistor may be formed from a *p-n junction diode. A symmetrical varistor can be formed by connecting two p-n junction diodes in *parallel, with opposite polarity. This arrangement exhibits the forward current-voltage characteristic of a p-n junction in either direction of applied bias and may be used as a voltage limiter (*see* diode forward voltage).

A *photovaristor* is a varistor in which the current-voltage relation depends on the amount of illumination of the device. A suitable material for the construction of a photovaristor is cadmium sulphide.

varying duty *See* duty.

V band A band of microwave frequencies ranging from 46·0 to 56·0 gigahertz. *See* frequency band.

V-beam radar *See* radar.

VDU *Abbrev. for* visual display unit.

vector field *See* field.

vector scanning *See* electron-beam lithography.

velocity microphone A *microphone in which the electrical output is a function of the particle velocity of the detected sound waves. *See* ribbon microphone.

velocity-modulated tube *See* velocity modulation.

velocity modulation A process that introduces a radiofrequency (r.f.) component into an electron stream and thereby modulates the velocities of the electrons in the beam. Individual electrons will be either accelerated or retarded by the radiofrequency signal depending on the relative phase of the r.f. component at the point of interaction with the electrons. The velocity modulation therefore causes *bunching* of the electron beam as it travels down the electron tube since the faster electrons catch up with preceding slower ones (and conversely). The bunching may be represented graphically on an *Applegate diagram.*

The amount of bunching of an electron stream varies with the distance travelled from the point of first interaction with the r.f. field. *Ideal bunching* is the production of small sharply defined bunches of electrons with no electrons in the regions between bunches. In practice however this is not achieved. *Optimum bunching* occurs at a particular distance down the tube for any given tube and is the condition when the minimum size of bunches containing the maximum possible numbers of electrons is achieved. *Underbunching* is the condition of less than optimum bunching. If the electron beam is allowed to travel beyond the point of optimum bunching faster electrons begin to leave the slower ones behind. This condition is termed *overbunching*. If the electron-beam direction is reversed by a reflector during the transit interval the resulting bunching is known as *reflex bunching*.

Velocity-modulated tubes are used as microwave oscillators and amplifiers. They include the *klystron and the *travelling-wave tube. The r.f. field can be made to interact with the electron stream in one sharply defined region, such as a *cavity resonator; the region is known as the *buncher*. The electron beam is then allowed to travel through a field-free drift space. This method is employed in the klystron. An alternative method is used in travelling-wave tubes: the r.f. field is propagated along the length of the electron tube and interaction between the electron beam and the field is a continuous two-way process.

velocity of light *See* Table 5, backmatter.

vertical blanking *See* blanking.

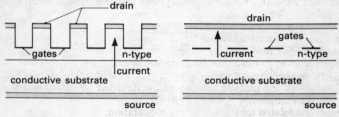

a Crossed section of vertical FET   *b* Cross section of permeable base transistor

vertical FET (VFET) An FET (*field-effect transistor) in which the *source is formed on the back of a semiconductor slice and current flows vertically through the device rather than horizontally as in the more usual structure. The cross section of a VFET is shown in Fig. *a*. A variation of the VFET is the *permeable base transistor* in which an epitaxial layer is formed over the *gate electrodes before the *drain is formed (Fig. *b*).

vertical hold *See* television receiver.

vertical recording A method of recording a gramophone record or magnetic disk in which the groove modulations are orthogonal to both the surface of the recording medium and the motion of the cutter.

very high frequency (VHF) *See* frequency band.

very large scale integration (VLSI) *See* integrated circuit.

very low frequency (VLF) *See* frequency band.

vestigial sideband *See* carrier wave.

vestigial-sideband transmission A form of *suppressed-carrier transmission in which one sideband and the corresponding vestigial sideband (*see* carrier wave) are transmitted.

VFET *Short for* vertical FET (field-effect transistor).

V-groove technique An etching technique used to produce a very precise edge in a silicon crystal. The technique is used to produce mesa layers or during the manufacture of *V/MOS circuits.

A silicon crystal with (100) orientation is etched from the surface by a suitable etch, such as potassium hydroxide, that will not etch perpendicularly to the (lll) plane. The etching process therefore follows the direction of the (lll) crystal planes and stops at any point where the (lll) planes intersect. This results in a very precise V-groove in the material. The depth of the groove depends on the size of the original opening in the surface oxide layer and etching to a precise predetermined depth is therefore possible with this technique.

VHF *Abbrev. for* very high frequency. *See* frequency band.

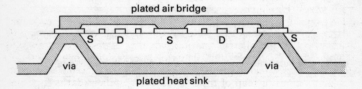

plated air bridge

S   D   S   D   S

via   via

plated heat sink

Vias formed through the back of a slice

via A plated hole used to provide low-inductance connections on an *integrated circuit. A via may be formed through the dielectric layer on the top of a slice to interconnect two metallization patterns. A via may also be formed through the back of a slice containing FETs where a low-inductance earth is required; this is especially useful where the FET is not near the edge of the chip or where a power FET is concerned, and metal plating is used on the back of the slice to provide a heat sink and earth plane simultaneously (*see* diagram).

vibrational energy levels *See* energy levels.

vibration galvanometer *See* galvanometer.

vibrator *Syn.* chopper. A device that produces an alternating current by periodically interrupting or reversing a continuous steady current from a direct-current source. The vibrator consists of an electromagnetic *relay with a vibrating armature that alternately makes and breaks one or more pairs of contacts.

The most common application is in a power-supply unit in which a high-voltage direct current must be produced from a low-voltage d.c. source, such as a battery. The vibrator produces a low-voltage periodically varying current that is transformed (*see* transformer) into a high-voltage a.c. supply and then rectified to produce a high-voltage d.c. supply. A *rectifier circuit may be used to produce the direct-current output or the vibrator itself can be used for this purpose. In the latter case the vibrator (termed a *synchronous vibrator*) is fitted with an extra pair of contacts that are used to reverse the connections to the secondary winding of the transformer in synchronism with the reversal of the current in the primary winding so that the output from the transformer secondary circuit is a direct current.

video *Colloq.* (1) A complete videotape recorder system that is used with domestic television receivers. *See* videotape.

(2) A prerecorded videotape.

video amplifier *See* video frequency.

video camera A *television camera that is designed for use with *videotape rather than direct broadcasting. The term most commonly applies to cameras used to prerecord outside broadcasts or for small portable domestic systems.

video frequency The frequency of any component of the output signal from a *television camera. Video frequencies are within the range 10 hertz to two megahertz. An amplifier that is designed to operate with video-frequency signals is termed a *video amplifier*.

video mapping A technique used in the display of a *radar system in which a chart or map of the area covered by the radar is superimposed electronically on the radar display.

Videophone *Tradename* A telephone receiver that can receive and display visible images simultaneously with the telephone signals.

video recorder *Syn. for* videotape recorder. *See* videotape.

video signal *Syn.* picture signal. *See* camera tube; television.

videotape A form of magnetic tape that is suitable for use with a *television camera. Simultaneous recording of the video signal from the camera and the audio signal from the microphone system is carried out on separate tracks on the tape. The signals may later be output directly to the modulating circuits of the transmission system.

A form of *videotape recorder* is available for use with domestic television receivers and can be used either to record a received broadcast on videotape or to replay a prerecorded videotape directly into the television receiver.

Many television programmes are recorded on to videotape prior to transmission rather than being broadcast live.

videotex *See* teletext.

vidicon A low-electron-velocity photoconductive *camera tube that is widely used in closed-circuit television and as an outside broadcast camera since it is smaller, simpler, and cheaper than the image orthicon.

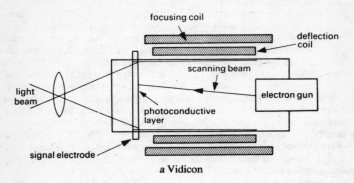

a Vidicon

The photosensitive target area of the vidicon consists of a transparent conducting film placed on the inner surface of the thin glass faceplate (Fig. *a*). A thin layer of photoconductive material (*see* photoconductivity) is deposited on the conducting layer and a fine mesh grid placed in proximity to the photoconductive layer. The conducting layer acts as the *signal electrode* from which the output is obtained and is held at a positive potential. The photoconductor can be considered as an array of leaky capacitors with one plate electrically connected to the signal electrode and the other floating except when subjected to an electron beam (Fig. *b*). The surface of the photoconductive layer is charged to cathode potential by a beam from the electron gun and each elemental capacitor therefore becomes charged.

The optical image is produced on the target electrode, and the value of each effective leakage resistor is determined by the intensity of illumination of the corresponding element on the target area (*see* photoconductivity). During the frame period, each elemental capacitor discharges by an amount that depends on the value of the corresponding leakage resistor; a positive potential pattern thus appears on the electron-gun side of the target, corresponding to the optical image.

The target is scanned with a low-velocity electron beam and each elemental capacitor is again charged to cathode potential by electrons from the electron beam. As the target is scanned current flows in the signal electrode; the magnitude of the current developed is a

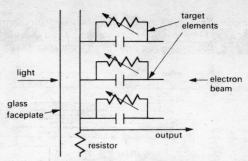

b Photosensitive target area of vidicon

function of the charge on the target area and hence of the illumination from the optical-lens system.

In the *plumbicon,* a modern development of the vidicon tube, the target elements may be considered as semiconductor current sources controlled by light energy. When not illuminated, the target elements are essentially reverse-biased Schottky diodes with a very low reverse saturation current. The principle advantages of these tubes compared to the vidicon are the low dark current and good sensitivity and light-transfer characteristics.

viewdata *See* teletext.

virtual cathode The surface, located in a space-charge region (*see* space charge) between the electrodes of a *thermionic valve, at which the electric potential is a mathematical minimum and the potential gradient is zero. It can be considered to behave as if it were the source of electrons.

virtual value *See* root-mean-square value.

visibility factor The ratio of the *minimum discernible signal (mds) of a *television or *radar receiver that can be detected by ideal instruments to the mds that can be detected by a human operator.

visual display unit (VDU) A display device used with a *computer that displays information in the form of characters and line drawings on the screen of a *cathode-ray tube. A VDU is most often associated with a keyboard and/or *light pen, which allow the information in the display to be altered or new data to be input, and a copying device for a permanent record of the display. *See also* terminal.

VLF *Abbrev. for* very low frequency. *See* frequency band.

VLSI *Abbrev. for* very large scale integration. *See* integrated circuit.

V/MOS MOS circuits or transistors that are fabricated using the *V-groove technique. Regions of the required conductivity type are formed in (100) orientation silicon crystals using a combination of planar diffusion and epitaxy. V-groove etching is then performed,

629

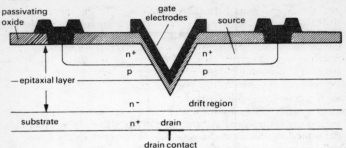

a V/MOS power transistor

and the gate electrode formed in the groove, in order to produce vertical MOS transistors of the same structure as *D/MOS. This technique is used to produce MOS devices of very precise channel length; as with D/MOS the channel length is determined by the diffusion rather than by *photolithography. V/MOS devices also contain an n^- drift region (*see* D/MOS) in order to prevent *punch-through. The length of the drift region determines the breakdown voltage of the device, as with D/MOS devices, and the structure shown in Fig. *a* is used to form discrete high-power MOS transistors. Operating voltages of about 300 volts can be achieved using a drift length of about 25 micrometres although voltages of about 100 V are more usual. The 'on' impedance can be kept low (of the order of a few ohms) using several long grooves on each transistor chip.

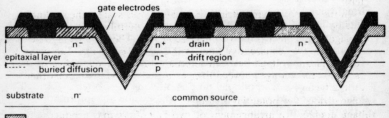

passivating oxide

metal

b Common-source V/MOS

The technique is also used to produce an array of transistors for applications, such as *read-only memories, that are used with a common source (Fig. *b*). In this case the device is physically inverted and the groove must be taken right through to the substrate. The drift region can therefore only be about two micrometres in length in order to accommodate the gate electrodes. An inverted V/MOS circuit is only suitable for low-power high-speed applications up to microwave frequencies. It has however an increased functional packing density compared to D/MOS.

voice coil *Syn. for* speech coil. *See* loudspeaker.

voice frequency *See* audiofrequency.

voice-frequency telephony A form of *telephony in which the electric signals are transmitted as *audiofrequency signals of substantially the same frequencies as the voice frequencies producing them.

volatile memory *See* memory.

volt Symbol: V. The *SI unit of *electric potential, *potential difference, and *electromotive force. It is defined as the potential difference between two points on a conductor when the current flowing is one ampere and the power dissipated between the points is one watt. In practice volts are measured by comparison with a *Weston standard cell using a *potentiometer.

This unit was formerly called the *absolute volt* and replaced the *international volt* (V_{int}) as the standard unit of voltage: one V_{int} equals 1·000 34 V.

voltage Symbol: V; unit: volt. The potential difference between two points in a circuit or device. *See also* active current (or voltage); reactive voltage.

voltage amplifier *See* amplifier.

voltage between lines *Syns.* line voltage; voltage between phases. The voltage between the two lines of a single-phase electrical power system or between any two lines of a symmetrical three-phase system.

In the case of a symmetrical six-phase power system the six lines can be considered as arranged around the periphery of a regular hexagon in correct order of phase sequence. Then the voltage between any two consecutive lines is the *hexagon voltage*. The *delta voltage* is the voltage between alternate lines and the *diametrical voltage* is the voltage between lines arranged opposite to each other on the hexagon.

voltage divider *See* potential divider.

voltage doubler An arrangement of two *rectifiers that produces an output voltage amplitude twice that of a single rectifier. In a typical circuit using *diode rectifiers (*see* diagram) each rectifier separately rectifies alternate half cycles of the input alternating voltage and the two outputs are then summed.

voltage drop The voltage between any two specified points of an electrical conductor, such as the terminals of a circuit element or component, due to the flow of current between them.

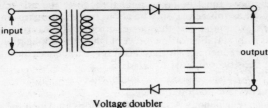

Voltage doubler

In the case of direct current the voltage drop is equal to the product of the current in amperes and the resistance in ohms. In the case of alternating current the product of the current in amperes and the resistance in ohms is the *resistance drop,* which is in *phase with the current. The product of the current and the reactance gives the *reactance drop,* which is in *quadrature with the current.

voltage feedback *Syn.* shunt feedback. *See* feedback.

voltage gain *See* gain.

voltage generator Any source of electrical power that is used to supply voltage to a circuit or device. *See* constant-voltage source.

voltage jump An abrupt unwanted change or discontinuity in the operating voltage of any electronic circuit or device, particularly a glow-discharge tube (*see* gas-discharge tube).

voltage level The ratio of the voltage at a point in a transmission system to the voltage of an arbitrary reference point. The arbitrary reference point is specified by the CCITT (*see* standardization) as a point of *zero level. The voltage level is expressed in dBm0, i.e. decibels measured with reference to the zero relative level. Absolute measurements of received signal magnitudes or noise levels at various receivers may be compared by converting them to the zero relative values in order to assess the performance of the receiving circuits.

voltage multiplier An arrangement of *rectifiers that produces an output voltage amplitude that is an integral multiple of that of a single rectifier, i.e. of the peak value of applied alternating voltage. *See also* voltage doubler.

voltage reference diode *See* diode forward voltage.

voltage regulator *See* regulator.

voltage-regulator diode *See* Zener diode.

voltage relay *See* relay.

voltage selector *Syn. for* clipper. *See* limiter.

voltage sensing *See* charge-coupled device.

voltage stabilizer A circuit or device that produces an output voltage that is substantially constant and independent of variations either in the input voltage or in the load current, i.e. it acts as a *constant-

632

voltage source. Such a device is most often used as a voltage *regulator.

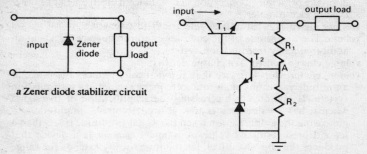

a Zener diode stabilizer circuit

b Series stabilizer circuit

An early form of voltage stabilizer was based on the *gas-discharge tube. Such circuits have now largely been superseded by circuits based on solid-state devices, such as the *Zener diode (Fig. *a*) or the bipolar junction transistor (Fig. *b*). In the series stabilization circuit shown in Fig. *b* the load impedance is in series with the circuit.

voltage transformer *Syn.* potential transformer. *See* transformer; instrument transformer.

voltaic cell *See* cell.

voltameter *Obsolete syn. for* coulombmeter.

volt-ampere Symbol: VA. The *SI unit of apparent *power, defined as the product of the *root-mean-square values of voltage and current in an alternating-current circuit. *See* power; active volt-amperes; reactive volt-amperes.

volt box A *potential divider that consists of a series of resistors by means of which a set of discrete fractions of the applied voltage are made available. It can be used for the measurement of a voltage falling outside the dynamic range of a particular voltmeter.

volt efficiency *See* cell.

voltmeter A device that measures voltage. Voltmeters in common use include d.c. instruments (such as permanent-magnet *moving-coil instruments), *digital voltmeters, and *cathode-ray oscilloscopes.

In order to provide the minimum disturbance in the circuit containing the voltage to be measured, voltmeters are required to pass very little current and therefore require a very high input impedance. Digital voltmeters, cathode-ray oscilloscopes, and the now little used *valve voltmeter comply with this requirement. A large series resistance however is required in the case of the moving-coil voltmeter and the *electrostatic voltmeter* in order to increase their input impedances. The electrostatic voltmeter is a voltmeter based upon the principle of operation of a *quadrant electrometer or other type of electrometer.

volt per metre The *SI unit of *electric field strength.

volume The magnitude of the complex *audiofrequency signals in an audiofrequency transmission system.

volume charge density *See* charge density.

volume compressor A device that automatically reduces the range of amplitude variations of an *audiofrequency signal in a transmission system. It operates by decreasing the amplification of the signal when it has a value greater than a predetermined amplitude and increasing the amplification when the signal amplitude is less than a second predetermined value. A *volume expander* is a device that produces the opposite effect, i.e. it automatically extends the range of amplitude variations of the transmitted audiofrequency signal.

A suitably designed expander used at one point of a transmission system can be made to compensate for the effect of a compressor in another part of the system and thus restore the original audiofrequency signal. A compressor and expander used together in this manner are termed a *compandor.*

In the *recording of sound a compressor may be used to reduce the volume range of the signals recorded on a gramophone record or on a film track. The sound-reproducing apparatus will then include an expander to compensate.

A telecommunication system, such as a radio-telephone system, often employs a compandor to improve the *signal-to-noise ratio of the system; the compressor is used at the transmitter and the expander at the receiver. The relative increase in the smaller transmitted signals reduces the effect of noise on these signals.

volume control *See* automatic gain control; gain control.

volume expander *See* volume compressor.

volume lifetime *Syn. for* bulk lifetime. *See* semiconductor.

volume limiter A *limiter that operates on an *audiofrequency signal.

volume resistivity *See* resistivity.

volumetric radar *See* radar.

W

wafer *Syn.* slice. A large single crystal of *semiconductor material that is used as the substrate during the manufacture of a number of *chips. Very large single crystals are grown and then sliced into wafers before processing. The number of viable chips that can be produced from a single wafer, typically up to 10 cm in diameter in the case of silicon, depends on the size and complexity of the circuits or components on the chips.

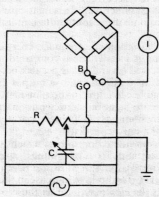

Wagner earth connection

Wagner earth connection A method of connection used with an alternating-current *bridge circuit that minimizes the *admittance to earth of the bridge. The earth connection is formed by means of the centre tapping of a three-terminal adjustable resistor, R, in *parallel with the a.c. supply (*see* diagram). The input of the a.c. supply is usually transformer-coupled. If frequencies greater than the audio-frequency range are involved an adjustable capacitor, C, is also used as shown. The indicating instrument, I, is connected across the bridge with a two-position switch. The bridge is balanced with the switch in position B and then rebalanced, using R and C, in position G. The original balance is then rechecked.

walkie-talkie *Colloq.* A light portable radio set designed to be carried and capable of allowing two-way communication by the user while in motion.

walk-out A phenomenon observed in semiconductor devices that are subjected to repeated *avalanche breakdown. It is a progressive increase in the value of the avalanche-breakdown voltage of the device and results from the injection of *hot electrons into the surface passivating oxide layer. The change in the breakdown characteristics are caused by the change in the surface field due to the injected electrons.

wall effect Any effect that is due to the inside wall of a container and contributes to the behaviour of the system contained by it.

Gas-filled tubes, for example, suffer from wall effects. In gas-filled radiation detector tubes some of the energy of the primary ionizing event may be lost to the wall of the tube rather than in the gas, resulting in a loss of *ionization. In an *ionization chamber an extra contribution to the current may result from electrons that are liberated from the chamber wall. In a *gas-discharge tube the wall of the tube can promote the recombination of ions in the gas to give the neutral species.

warble The periodic variation of a frequency between two limits. The frequency variation is usually small compared to the nominal frequency and is repeated several times per second. A *warble tone generator* consists of an oscillator whose output frequency is varied using a small variable capacitor in the tuned circuit. A warble tone is often used in a reverberation chamber in order to produce a uniform sound field containing no standing waves.

warm-up The period after a particular electronic device, circuit, or apparatus has been switched on during which it reaches a state of thermal equilibrium with its surroundings. A circuit may not be fully operational until after the warm-up period, particularly if it contains a *thermionic cathode. Alternatively the electrical characteristics of its components may tend to drift to their steady value during this period. The warm-up includes the *heating time of any individual device contained in the apparatus under consideration.

watt Symbol: W. The *SI unit of *power. It is defined as the power resulting when one joule of energy is dissipated in one second. In an electric circuit one watt is given by the product of one ampere and one volt.

watt-hour One thousandth of a *kilowatt-hour.

watt-hour meter *Syns.* integrating wattmeter; recording wattmeter. An *integrator that measures and records electrical energy in watt-hours or more usually in *kilowatt-hours.

wattless component (1) (of current) *See* reactive current.

(2) (of volt-amperes) *See* reactive volt-amperes.

(3) (of voltage) *See* reactive voltage.

wattmeter An instrument that measures electrical power and is calibrated in watts, multiples of a watt, or submultiples of a watt. The most commonly used type of wattmeter is the electrodynamic wattmeter (*see* electrodynamometer). In circuits that have substantially constant currents and voltages an induction wattmeter (*see* induction instrument) may be used. For standardization and calibration purposes an *electrostatic wattmeter* is used. This consists of a *quadrant electrometer that is arranged in a suitable circuit containing noninductive resistors so that it measures power directly. The National Physical Laboratories, London, use this type to calibrate other wattmeters.

wave A periodic disturbance, either continuous or transient, that is propagated through a medium or through space and in which the displacement from a mean value is a function of time or position or both. Sound waves, water waves, and mechanical waves involve small displacements of particles in the medium; these displacements return to zero after the disturbance has passed. With electromagnetic waves (*see* electromagnetic radiation) it is changes in the intensities of the associated magnetic and electric fields that represent the disturbance and a medium is not required for propagation of the wave.

The instantaneous values of the periodically varying quantity plotted against time gives a graphical representation of the wave that is known as the *waveform*. If the waveform is sinusoidal in shape it is usually described as undistorted; a nonsinusoidal waveform is distorted. The *wavefront* is the imaginary surface over which the displacements are all of the same *phase. The *amplitude* of the wave is the peak value of the displacements relative to the equilibrium state or to some arbitrary reference level, usually zero. If a wave suffers *attenuation the amplitude of successive periods is continuously reduced (*see also* propagation coefficient).

The *wavelength* is the distance between two displacements of the same phase along the direction of propagation. If v is the velocity of the wave and λ its wavelength, then the frequency of vibration, ν, is given by

$$\nu = v/\lambda$$

The frequency is the reciprocal of the *period, T, of the wave. The frequency (or wavelength) of electromagnetic radiation is commonly used to describe particular regions of the *electromagnetic spectrum, such as the visible or radio regions (*see also* frequency band).

An alternating current propagated through a long chain *network or *filter behaves as if it were a wave. Elementary particles, such as electrons, have associated wavelike characteristics. *See also* Doppler effect.

wave analyser A device, such as a spectrum analyser (*see* multichannel analyser) or *frequency analyser*, that resolves a complex waveform into its fundamental and harmonic components.

The *heterodyne analyser* is a form of wave analyser in which the wave under investigation is used for *amplitude modulation of an internally generated *carrier wave. The modulated signal is input to a highly selective tuned amplifier. The wave under analysis produces sidebands of the carrier wave corresponding to each of the harmonic components. The carrier frequency is adjusted until one of the sidebands corresponds to the frequency of the amplifier. The frequencies and relative amplitudes of all the components of the wave may thus be obtained by adjusting the frequency of the carrier wave over a suitable range.

wave duct *See* waveguide.

waveform *Syn.* waveshape. *See* wave.

wavefront (1) *Syn.* wave surface. *See* wave. (2) *See* impulse voltage.

waveguide A *transmission line that consists of a suitably shaped hollow conductor, which may be filled with a dielectric material, and that is used to guide ultrahigh-frequency electromagnetic waves propagated along its length. The transmitted wave is reflected by the internal walls of the waveguide and the resulting distribution within the guide of the lines of electric and magnetic flux associated with the wave is the transmission *mode. At any instant the phase and amplitude of the wave are given by the appropriate *propagation coefficient.

For any given transmission mode there is a lower limit to the frequency that may be propagated through the waveguide. This *cut-off frequency* occurs when the complex propagation coefficient becomes real. The electromagnetic wave is then attenuated exponentially and soon becomes substantially zero. (A short length of waveguide used below cut-off may sometimes be employed as a known *attenuator.) The waveguide therefore acts effectively as a high-pass *filter. In an ideal (lossless) waveguide, above the cut-off frequency, the propagation coefficient is purely imaginary and no attenuation occurs. In practice however some attenuation of the wave always occurs due to energy dissipation in the guide.

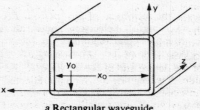

a **Rectangular waveguide**

The most common shapes of waveguide are rectangular (Fig. *a*) and cylindrical; the most common dielectric is air. A cylindrical waveguide is sometimes known as a *wave duct;* one that contains a solid rod of dielectric is a *uniconductor waveguide.* If a wide range of frequencies is to be transmitted a *ridged waveguide* (Fig. *b*) may be used. The presence of the ridges extends the possible range of frequencies that may be propagated in a particular transmission mode but the attenuation is greater than in the equivalent rectangular waveguide.

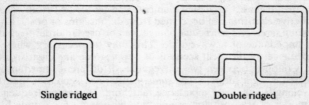

Single ridged   Double ridged

b Cross section of ridged waveguides

Electromagnetic waves may be excited in a waveguide by the electric and magnetic fields associated with electromagnetic waves present in another device, such as a *cavity resonator or *microwave tube. A suitable *mount* is used to connect the waveguide to the source of radiation in order to achieve the optimum transfer of energy and to excite the desired transmission mode. Energy may also be extracted from a waveguide in a similar manner.

Energy may also be transferred using either a probe to which a voltage is applied or a coil that carries an electric current. The alignment of the exciting probe or coil depends on the desired transmission mode. The probe must be aligned along the direction of lines of electric flux in order to excite the electric field; the coil is aligned so as to excite the desired magnetic field. Energy may be similarly extracted from the waveguide. Probes are also used to examine the distribution of the fields within the waveguide.

A *slotted waveguide* is one that is provided with nonradiative slots at intervals along its length that allow the insertion of a probe. The presence of the slots affects the distribution of the fields inside the guide, as can junctions (flanges) between sections of a waveguide. Junctions and slots in a guide contribute towards the distributed capacitance of the device and also contribute to the energy dissipation in the guide. A waveguide in which no reflected waves occur at any of the transverse sections is known as a *matched waveguide.*

Bends in a waveguide are usually made smoothly in order to prevent unwanted reflections. An *E-bend* is a smooth bend in a waveguide in which the direction of polarization is parallel to the axis (i.e. a horizontally polarized TM wave). An *H-bend* is a smooth bend in a guide in which the direction of polarization is perpendicular to the axis (i.e. a vertically polarized TE wave). If the direction of polarization has to be changed a special joint, known as a *rotator*, is used. In a rectangular guide twisting of the guide structure can be used to rotate the plane of polarization.

It is possible to produce local concentrations of electric or magnetic energy within a waveguide by inserting suitably shaped metallic or dielectric pieces. These act essentially as lumped inductances or capacitances. Capacitive elements can be formed from screws; inductive elements can be formed from diaphragms or posts. Either a *tuning screw* or a *waveguide plunger* can be used in order to change the impedance of a waveguide. This may be necessary when the impedances of different sections of a waveguide are slightly different. Coupling between two waveguides of different impedances is effected using a suitable *waveguide transformer,* such as a *quarter-wavelength line, for *impedance matching. Unwanted frequencies or modes of transmission are eliminated using a waveguide filter. An *iris* consists of a shaped inductive diaphragm and capacitive screw inserted into a guide to act as a reactance, susceptance, or waveguide filter.

Waveguides are widely used for the transmission of electromagnetic waves. Electromagnetic energy may also be stored by using a section of waveguide short-circuited at each end to form a *cavity resonator. A section of coaxial transmission line may also be used. If the length of the section of waveguide or transmission line is h, resonance occurs when

$$h = n(\lambda_g/2)$$

where n is an integer and

$$\lambda_g = \lambda[\varepsilon - (\lambda/\lambda_c)^2]^{1/2}$$

λ is the wavelength in free space, λ_c the wavelength of the waveguide cut-off and ε the relative dielectric constant of the waveguide dielectric. A cavity resonator formed in this way is termed either a *waveguide resonator* or a *coaxial resonator*.

waveguide plunger *See* waveguide.

waveguide resonator *See* waveguide.

waveguide transformer *See* waveguide.

wave heating A method of heating a material by allowing it to absorb energy from a *travelling wave of electromagnetic radiation.

wavelength *See* wave.

wavelength constant *Syn. for* phase-change coefficient. *See* propagation coefficient.

wavelength dispersive spectroscopy (WDS) *See* electron microprobe.

wave mechanics *See* quantum mechanics.

640

wavemeter An apparatus that measures the frequency or wavelength of a *radiowave. It consists essentially of a capacitively tuned circuit together with a current-detecting instrument. The variable capacitor is calibrated directly to read frequency or wavelength. A current maximum is obtained when the resonant frequency of the tuned circuit corresponds to the frequency of the received radiowave.

waveshape *Syn. for* waveform. *See* wave.

wave surface *Syn. for* wavefront. *See* wave.

wavetail *See* impulse voltage.

wavetrain A succession of *waves, particularly a small group of waves of limited duration (*see* pulse).

wavetrap A tuned circuit, usually a rejector (*see* resonant circuit) that is included in a *radio receiver in order to reduce interference from unwanted radiowaves of the particular frequency to which the circuit is tuned.

W band A band of microwave frequencies ranging from 56·0 to 100 gigahertz. *See* frequency band.

WDS *Abbrev. for* wavelength dispersive spectroscopy. *See* electron microprobe.

weak electrolyte *See* electrolyte.

wear-out failure *See* failure.

wear-out failure period *See* failure rate.

weber Symbol: Wb. The *SI unit of *magnetic flux. One weber is the magnetic flux that, linking a circuit of one turn, produces an electromotive force of one volt when it is reduced to zero at a uniform rate in one second.

Wehnelt cathode *Syn. for* coated cathode. *See* thermionic cathode.

Weiss constant *Syn.* paramagnetic Curie temperature. *See* paramagnetism.

welding The process of joining together two pieces of the same metal using either no extra metal or extra metal of the same material. Welding is performed electrically using either *resistance welding, in which no extra metal is required, or *arc welding, in which a filler rod of the same metal is used.

Wenner winding A method of forming a wire-wound resistor in order to reduce the *reactance. The resistor is wound on a former and alternate turns of wire are looped along the former. *Compare* bifilar winding.

Wertheim effect *Syn. for* Wiedemann effect. *See* magnetostriction.

Weston standard cell *Syn.* cadmium cell. A cell that has a substantially constant terminal voltage and is used as a reference standard for electromotive force. This cell is constructed in an H-shaped glass vessel. The positive electrode is mercury and the negative electrode is a cadmium and mercury amalgam with a saturated cadmium sulphate solution as the electrolyte.

The e.m.f. developed by the cell at 20°C is 1·018 58 volts. The cell has a very low temperature coefficient of e.m.f. *Compare* Clark cell.

wet cell *See* cell.

wet etching *See* etching.

wet flashover voltage *See* flashover voltage.

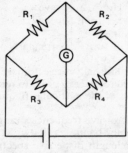

Wheatstone bridge

Wheatstone bridge A four-arm *bridge used for measuring resistance. Each arm contains a resistance (*see* diagram), with the unknown and reference resistances, R_1 and R_2, being connected at a point. At balance, when a null response is obtained from the indicating instrument,

$$R_1/R_2 = R_3/R_4$$

The two arms R_3 and R_4 are known as the ratio arms and may take the form of a wire of uniform resistance, which is tapped by a sliding contact. R_3 and R_4 are proportional to the lengths of wire, l_1 and l_2, on each side of the contact, so that

$$R_1/R_2 = l_1/l_2$$

whistle *See* heterodyne interference.

white compression Compression (*see* volume compressor) applied to the components of a *television video signal that correspond to the light areas of the picture.

white noise *See* noise.

white peak *Syn.* picture white. *See* television.

white recording *See* recording of sound.

wideband *Syn.* broadband. (1) A frequency band that extends over a relatively large range.

(2) Denoting an electronic device or circuit, such as an *amplifier, that operates satisfactorily over a large range of input signal frequencies.

Wiedemann effect *Syn.* Wertheim effect. *See* magnetostriction.

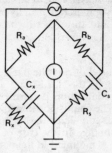

Wien bridge

Wien bridge A four-arm bridge that is used for the measurement either of capacitance or frequency. A typical network is shown in the diagram. At balance, when a null response is obtained from the indicating instrument,

$$C_x/C_s = (R_b/R_a) - (R_s/R_x)$$
$$C_sC_x = 1/\omega^2 R_s R_x$$

where ω is the angular frequency. For the measurement of a frequency f, it is convenient to make $C_s = C_x$, $R_s = R_x$, and $R_b = 2R_a$ so that

$$f = (2\pi CR)^{-1}$$

Wien effect An increase in the conductivity of an electrolyte when subjected to a high voltage gradient (of the order of two megavolts per metre). At sufficiently high values of voltage gradient the ions in the solution move at such a rate that they pass very quickly out of their *ionic atmosphere and are not therefore subjected to the usual retarding effect.

Wilson effect Electrical polarization of an insulator when it is moved through a region containing a magnetic flux. The motion induces a potential difference across the material that results in its polarization since the insulating properties inhibit the creation of an electric current.

Wimshurst machine An early electrostatic *generator.

Winchester disk drive *See* moving magnetic surface memory.

winding A complete group of insulated conductors in an electrical machine, transformer, or other equipment that is designed either to produce a magnetic field or to be acted upon by a magnetic field. It may consist of a number of separate suitably shaped conductors,

643

electrically connected, or a single conductor shaped to form a number of loops or turns. *See* ring winding.

window (1) The thin sheet of material, usually mica, that covers the end of a *Geiger counter and through which the radiation enters the instrument.

(2) *See* radio window.

(3) *See* confusion reflector.

wire bonding A method used during the packaging of *integrated circuits to connect the chip to the *leadframe. The leadframe is formed by stamping the interconnection pattern required from a strip of thin copper sheeting. The chip is positioned in a small depression in the centre and the leads connected to the *bonding pads of the chip using individual wire bonds. In the case of packages with ceramic casings, the leadframe is mounted in the package before the chip is bonded in place. For plastic packages, the chip is bonded to the leadframe and then encapsulated in moulded plastic. The finished package may be in the form of a *pin grid array, *leadless chip carrier, or *dual in-line package. *Compare* tape-automated bonding.

wire broadcasting *Television or *radio broadcasting by means of *transmission lines, such as conducting wire, rather than by transmission of electromagnetic waves from an aerial. Wire broadcasting transmits the signals as audio- or video-frequency signals or as modulated carrier waves.

wire gauge A set of standard diameters of wires. Gauges in common use are, in the UK, *British Standard* (NBS) *wire gauge* and, in the US, *American* (B&S) *wire gauge.*

wireless *See* radio receiver.

wireless broadcasting *Obsolete syn. for* radio broadcasting. *See* broadcasting.

wireless telegraphy *Obsolete syn. for* radio telegraphy. *See* telegraphy.

wire-wound resistor *See* resistor.

wiring diagram A diagram of a piece of electronic equipment showing the interconnections between assemblies and subassemblies. Such a diagram is particularly useful for the maintenance or repair of such equipment.

wobbulator A *signal generator in which an automatic periodic variation is applied to the output frequency so that it varies over a predetermined range of values. It can be used to investigate the frequency response of an electronic circuit or device or as a test instrument for a *tuned circuit.

Wollaston wire Extremely fine platinum wire that is used for *electroscope wires, microfuses, and *hot-wire instruments. It is produced by coating platinum wire with a sheath of silver, drawing them together into a relatively fine uniform diameter wire, and then dissolving the silver with a suitable acid. Diameters down to about one micrometre may be produced by this method.

woofer A relatively large *loudspeaker that is used to reproduce sounds of relatively low frequency in a high-fidelity sound reproduction system. *Compare* tweeter.

word A string of *bits that stores a unit of information in a *computer. The length of the word depends on the particular machine. Typical word lengths contain 32, 36, 48, or 64 bits.

word-addressable *See* address.

word line *See* random-access memory.

work function Symbol: Φ; unit: electronvolt. The difference in energy between the Fermi level (*see* energy bands) of a solid and the energy of free space outside the solid, i.e. the vacuum level. It is the minimum energy required to liberate an electron from the solid at absolute zero temperature.

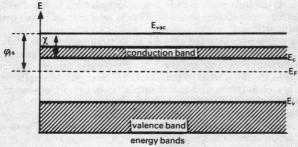

energy bands

Work function and electron affinity of a semiconductor

In a metal the *image potential that an electron just outside the surface would experience also contributes to the value of the work function. In a semiconductor the work function is greater than the *electron affinity, χ (*see* diagram).

wound core *See* core.

wow Low-frequency (up to about 10 hertz) audible periodic variations in the pitch of the sound output in a high-fidelity sound-reproduction system. Wow is an unwanted form of *frequency modulation and is usually caused by a nonuniform rate of reproduction of the original sound, such as by nonuniform rotation of a gramophone turntable. *See also* flutter.

write To enter information into a storage element of a *memory or *storage tube.

write time *See* memory.

writing beam *See* storage tube.

writing gun *See* storage tube.

writing speed *See* storage tube.

Wulff plot *See* edge profile.

X

X-axis The horizontal axis on the screen of a *cathode-ray tube.

X band A band of microwave frequencies ranging from 5·2 to 10·9 gigahertz. *See* frequency band.

xerography A photographic copying process that uses electrical effects to form the image. An image of the document to be copied is reproduced on a uniformly charged plate, usually coated with selenium: different areas of the plate are discharged to different extents by an intensity-modulated beam of ultraviolet radiation to leave a charge pattern corresponding to the brightness information of the original document.

An oppositely charged powder is then applied to the plate and this accumulates on the various discharged regions. The areas of greatest accumulation are the least discharged regions and correspond to the dark areas of the original. The powder, consisting of a mixture of graphite and a thermoplastic resin, is transferred to a charged paper and fixed by heat treatment.

X-guide A *transmission line that is used for the propagation of surface waves and consists of a length of dielectric material with an X-shaped cross section.

XPS *Abbrev. for* X-ray photoelectron spectroscopy. *See* photoelectron spectroscopy.

X-ray crystallography *See* X-rays.

X-ray fluorescence (XRF) *See* electron microprobe.

X-ray lithography A method of *lithography that uses X-rays rather than light beams. The basic process uses soft X-rays to expose appropriate resists. Because of the low wavelength (0.2–1.0 nanometre), diffraction effects are virtually nonexistent, and back scattering or reflection from the substrate material is almost nonexistent. Masks can be placed proximal to the substrate rather than in contact with it, and most dust is transparent to X-rays and therefore does not affect the process. X-ray lithography offers the advantages of excellent resolution combined with a large depth of field, vertical-walled patterns, and simplicity of the system – no complex lenses, mirrors, or electron-beam lenses are required. Disadvantages have arisen in the development of suitable X-ray sources, masks, and alignment procedures.

The optimum X-ray source is the *synchrotron, which produces highly collimated intense beams of X-rays; these machines, however, are scarce and expensive. Masks must be opaque to X-rays and therefore must be made from high atomic number materials such as gold. The resulting masks are very fragile and require careful handling. X-ray resists also present problems. With the exception of synchrotron-produced X-rays, most other sources of X-rays do not produce a collimated intense beam and few suitable resist materials exist with sufficient sensitivity, particularly positive resists. *Multilevel resist techniques must be used, where the pattern is produced in the very thin top layer of resist and transferred to the lower layers by dry *etching.

Despite the disadvantages, X-ray systems are being developed and reliable low-cost systems are potentially likely in the near future.

X-ray photoelectron spectroscopy (XPS) *See* photoelectron spectroscopy.

X-rays Electromagnetic radiation of frequencies between that of ultraviolet radiation and gamma rays, and sometimes higher. They are produced when matter is bombarded with sufficiently energetic electrons and were first observed by Roentgen in 1895. X-rays can be produced as a result of electron transitions from higher to lower energy levels within an atom. X-rays resulting from transitions have a frequency that is characteristic of the material and are therefore termed *characteristic X-rays*. X-rays are also produced during the rapid decelerations of electrons as they approach the nucleus. These X-rays, a form of *bremsstrahlung radiation, have a relatively wide frequency range and are known as *continuous X-rays*. X-rays of relatively low energy are termed *soft X-rays;* those at the high-energy portion of the frequency spectrum are *hard X-rays*.

X-rays can be reflected, refracted, and polarized and also exhibit interference and diffraction. They interact with matter to produce relatively high energy electrons: the mechanism is the same as in the *photoelectric effect. These electrons are usually of sufficiently high energy to ionize a gas or to produce *secondary X-rays* from the matter. The ionization of a gas can be utilized, in the *ionization chamber, to measure the intensity of an X-ray beam.

X-rays are widely used in *radiography* to investigate materials that are opaque to visible light but relatively transparent to X-rays. Radiography is used to detect flaws in structures and for diagnostic examinations. Sufficiently high intensities of X-rays of higher energies damage human tissue and X-rays are used in *radiotherapy* for the therapeutic destruction of diseased tissue. X-rays are also used in *X-ray crystallography* to investigate and determine crystal structures by using the crystal as a three-dimensional *diffraction grating. *X-ray lithography systems are being developed.

X-ray topography *See* diffraction.

X-ray tube An *electron tube that produces X-rays. Early X-ray tubes were gas-filled tubes but only electron beams of relatively low energy could be produced because of disruptive discharges occurring

647

through the gas. Modern X-ray tubes are invariably hard tubes (*see* electron tube) developed from the Coolidge type of tube; they are more stable than the earlier gas-filled sort.

A beam of electrons, produced by *thermionic emission from the cathode, is accelerated and focused on to a *target electrode*. The target electrode is usually the anode of the tube. The X-ray spectrum produced from the target contains frequencies that are characteristic of the target material and in addition a continuous range of frequencies of which the short-wave (high-energy) limit is determined by the energy of the electrons and hence by the accelerating voltage across the tube. The current in a vacuum tube can be controlled by varying the temperature of the cathode and hence the numbers of electrons emitted from it.

In order to produce high-energy X-rays the voltage across the tube must be very large: values of greater than one megavolt have been used. These extremely high voltages are usually supplied by a step-up transformer: if the secondary winding is connected across the tube it can act as its own rectifier. If this mode of operation is not suitable, for example when a continuous steady output is required, separate rectifying and smoothing circuits are necessary.

When operated at such energies cooling of the target is important. The extremely small size of the focal spot causes the generation of localized high temperature. The target is therefore constructed of a metal, such as tungsten, that has an extremely high melting point. The rest of the target electrode is usually made from a metal, such as copper, that is a good conductor of heat. A *rotating-anode tube* contains an anode that rotates about a centre that is not the focal spot of the electron beam; the same part of the target is thus not always bombarded. A rotating hollow cylinder with water cooling can also be used.

XRF *Abbrev. for* X-ray fluorescence. *See* electron microprobe.

x-y plotter A graphical instrument that produces a chart showing the relationship between two varying signals. One of the signals causes the pen to move in the direction of the x-axis and the other independently causes it to move in the direction of the y-axis. The corresponding instantaneous values of the two signals are plotted on the graph.

Y

Yagi aerial A sharply directional *aerial array from which most aerials used for *television and *radioastronomy have been developed. The active part of the aerial consists of one or two *dipole aerials together with a parallel reflector *aerial and a set of parallel directors. The directors are relatively closely spaced, being from 0·15 to 0·25 of a wavelength apart. When the aerial is used for transmission, the directors absorb energy from the back lobe of the dipole *radiation pattern and rereflect it in the forward direction; the major lobe is thus reinforced at the expense of the back lobe. When used for reception the inverse process occurs causing the signal to be focused on the dipole.

Y-axis The vertical axis on the screen of a *cathode-ray tube.

YIG *Abbrev. for* yttrium-iron-garnet. A *ferrite that is widely used for microwave applications. The magnetic properties are altered by the amount of trace elements in the material. The most common trace elements are calcium, vanadium, and bismuth. Calcium-germanium YIGs may be easily grown on a substrate. This material is used as a thin magnetic film on a nonmagnetic substrate, such as $*G^3$, for the production of solid-state magnetic circuits, particularly *magnetic bubble memories.

yoke A piece of ferromagnetic material that is used to connect permanently two or more magnetic *cores and thus complete a magnetic circuit without surrounding it by a winding of any kind.

y parameters *See* transistor parameters.

Z

Zener breakdown A type of *breakdown observed in a reverse-biased
*p-n junction that has very high *doping concentration on both
sides of the junction. At sufficiently high doping levels, the *tunnel
effect is the dominant mechanism under reverse bias. The built-in
field (*see* p-n junction) is high and the depletion layer narrow as a
result of the high level of doping. The application of a small reverse
voltage (of up to about six volts) is sufficient to cause electrons to
tunnel directly from the valence band to the conduction band (*see*
energy bands). At the Zener breakdown voltage a sharp increase in
the reverse current is obtained. Above the breakdown value the
voltage across the diode remains substantially constant. This process is reversible since the dielectric properties of the material remain unchanged. Unlike *avalanche breakdown no multiplication
of charge carriers occurs.

Avalanche breakdown is the dominant breakdown mechanism in
most semiconductor devices except those with high doping concentrations. Zener breakdown is utilized in *Zener diodes.

Zener diode *Syn.* voltage-regulator diode. A *p-n junction *diode that
has sufficiently high doping concentrations on each side of the junction for *Zener breakdown to occur. The diode therefore has a well-
defined reverse *breakdown voltage (of the order of a few volts
only) and can be used as a voltage regulator. Unlike *tunnel diodes,
the doping level is not high enough for the semiconductor to become
*degenerate and the diode behaves like a normal p-n junction in the
forward direction.

The term is also applied to less highly doped p-n junction diodes
that have higher breakdown voltages (up to 200 volts) and undergo
*avalanche breakdown. Most so-called Zener diodes are in fact
avalanche breakdown diodes. True Zener diodes have a low value of
reverse breakdown voltage.

zero error *See* index error.

zero level An arbitrary reference level used in telecommunication systems to compare the relative intensities of transmitted signals or of
*noise.

zero potential *See* earth potential.

zinc Symbol: Zn. A metal, atomic number 30, that is used as an electrode in some electrolytic *cells.

zirconium Symbol: Zr. A metal, atomic number 40, that can be used as
a *getter in hard vacuum electron tubes.

z-modulation (of a cathode-ray tube) *See* intensity modulation.

zone levelling *See* zone refining.

zone purification *See* zone refining.

zone refining A method of redistributing impurities within a solid material, such as a semiconductor, by melting parts of the material and causing the molten zones to move along the sample. Impurities travel along the sample in a direction determined by their effect on the freezing point of the material. If the impurity depresses the freezing point it will travel in the same direction as the molten zone; if it raises the freezing point it will travel in the opposite direction. *Zone levelling* is the use of zone refining in order to distribute impurities evenly throughout the bulk of the material. *Zone purification* is the application of zone refining in order to reduce the concentration of an impurity in a material.

In zone refining the sample of the material is moved slowly past a heater so that the molten zone effectively moves along the length of the bar. Melting is achieved by induction heating, electron bombardment, or the heating effect of a current through a resistance coil. Zone purification carried out using a large number of heaters can reduce the impurity concentration to about one part in 10^{10}.

z parameters *See* transistor parameters.

Table 1 Graphical Symbols—mainly solid-state components also some electron tubes; IEC approved symbols are indicated O, popular usage is indicated □

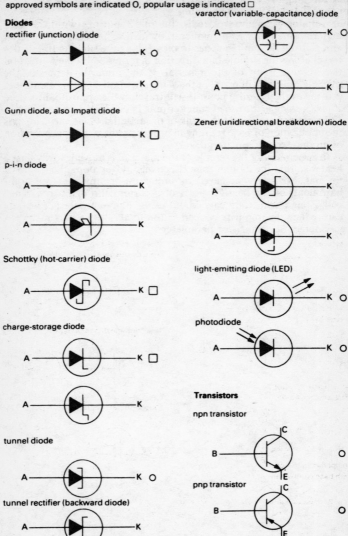

Diodes

rectifier (junction) diode

Gunn diode, also Impatt diode

p-i-n diode

Schottky (hot-carrier) diode

charge-storage diode

tunnel diode

tunnel rectifier (backward diode)

varactor (variable-capacitance) diode

Zener (unidirectional breakdown) diode

light-emitting diode (LED)

photodiode

Transistors

npn transistor

pnp transistor

multiple-emitter npn transistor

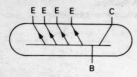

npn Darlington transistor

npn Schottky transistor

unijunction transistor (UJT) with
p-type base

unijunction transistor (UJT) with
n-type base

npn photo transistor,
no base connection

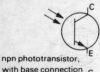

npn phototransistor,
with base connection

Field-effect transistor (FETs)

n-channel   p-channel
junction-gate (JFET)

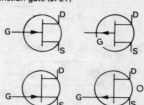

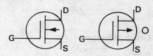

three-terminal depletion-type
insulated-gate (IGFET)

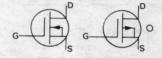

three-terminal depletion-type
IGFET, substrate tied to source

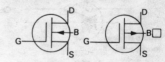

four-terminal depletion-type IGFET

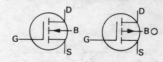

four-terminal enhancement-type IGFET

653

neon lamp, a.c. type

fluorescent lamp, two-terminal

Crystals
piezoelectric crystal

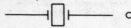

Circuit protectors
fuse

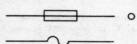

circuit-breaker

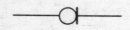

Audio devices
loudspeaker

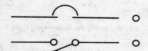

microphone

Amplifiers
single-ended amplifier

differential amplifier (or comparator)

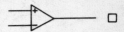

Norton (current) amplifier

Measuring instruments

type indicated by letter in circle
A ammeter
G galvanometer
V voltmeter

Filters
general

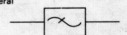

bandpass filter

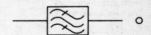

band-reject filter

low-pass filter

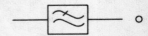

654

high-pass filter

frequency changer

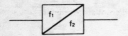

matched phase shifter

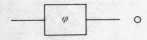

attenuator

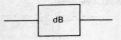

Aerials
receiving

transmitting

earth (ground)

chassis or frame connection

Interconnections
arrow indicates direction of signal flow;
other characteristics may be indicated

conducting path

pulsed signal

a.c. signal

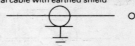

Cables
two-conductor cable with earthed shield

coaxial cable with earthed shield

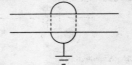

twisted pair

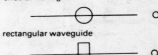

waveguides
circular waveguide

rectangular waveguide

flexible waveguide

twisted waveguide

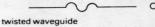

Thyristors

four-layer (pnpn, Schottky) diode

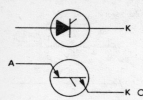

silicon-controlled rectifier (SCR)

silicon-controlled switch (SCS)

triac (gated bidirectional switch)

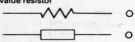

Circuit Components
Impedences

Resistors
fixed-value resistor

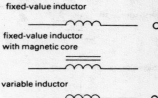

variable resistor

voltage-sensitive resistor (varistor)

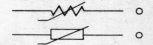

Capacitors
fixed-value capacitors

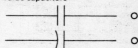

electrolytic capacitor

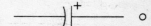

variable capacitor

Inductors
fixed-value inductor

fixed-value inductor
with magnetic core

variable inductor

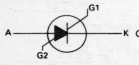

Transformers

transformer, air cored

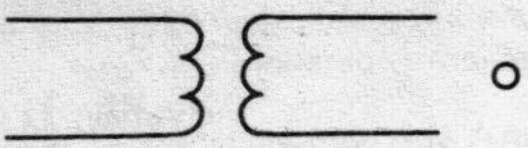

constant-current source

transformer with magnetic core

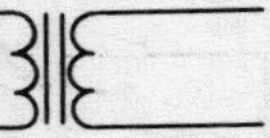

shielded transformer with magnetic core

a.c. oscillator source

Lamps

incandescent lamp

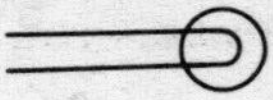

signal lamp

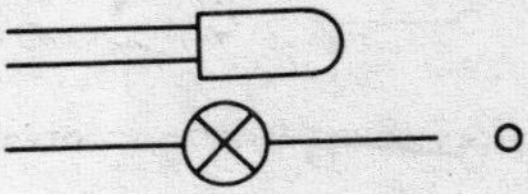

Batteries and sources

single-cell battery

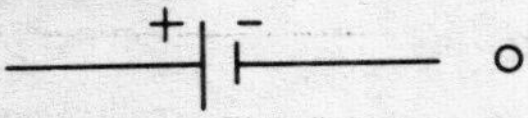

$+$ $-$

multiple-cell battery

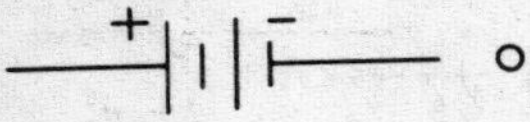

$+$ $-$

flashing signal lamp

constant-voltage source

$+$

$-$

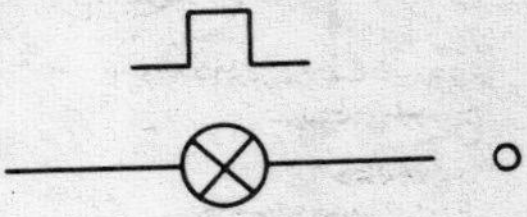

neon lamp, d.c. type

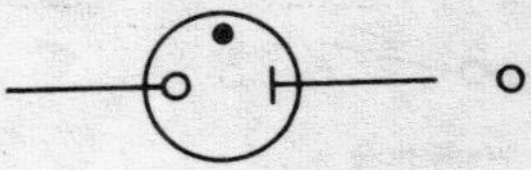

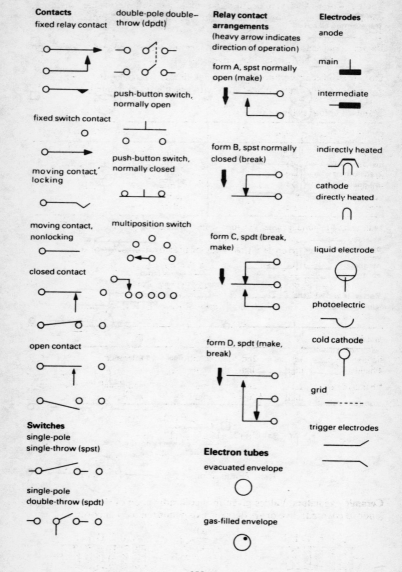

Contacts

fixed relay contact

double-pole double-throw (dpdt)

push-button switch, normally open

fixed switch contact

push-button switch, normally closed

moving contact, locking

moving contact, nonlocking

multiposition switch

closed contact

open contact

Switches

single-pole single-throw (spst)

single-pole double-throw (spdt)

Relay contact arrangements
(heavy arrow indicates direction of operation)

form A, spst normally open (make)

form B, spst normally closed (break)

form C, spdt (break, make)

form D, spdt (make, break)

Electron tubes

evacuated envelope

gas-filled envelope

Electrodes

anode

main

intermediate

indirectly heated

cathode directly heated

liquid electrode

photoelectric

cold cathode

grid

trigger electrodes

Table 2 Colour Codes: indications on components of their value, tolerance, voltage rating, etc., by means of circumferential coloured bands or coloured dots

Bands   axial lead components: to determine value, etc., start with band nearest end; band may cover endcap

Dots   axial lead components: start with dot nearest end
circular components: start with dot placed over one lead

Standard colour coding

Colour	Digit	Tolerance (%)		Colour	Digit	Tolerance (%)
black	0	± 20		blue	6	± 6
brown	1	± 1		violet	7	± 12·5
red	2	± 2		grey	8	± 30
orange	3	± 3		white	9	± 10
yellow	4	*		gold	–	± 5
green	5	± 5		silver	—	± 10

*guaranteed minimum value: variation from 0 to + 100 %

Resistors Values given in ohms; code of four or five bands or dots

ABCD

ABCD E

gap indicates L R direction

Band or dot	A	B	C	D	E
significance of 4 bands/dots	1st digit	2nd digit	additional zeros	tolerance	
5 bands/dots	1st digit	2nd digit	3rd digit	additional zeros	tolerance

red   green

example:

orange   green

R = 25 000 ohms ± 5 %

Ceramic capacitors Values given in picofarads; code of five or six bands (endcap covered), five or six dots, or five bands (endcap uncovered)

A BC DEF

C   D   E

B   F

A

A BCDEF

Table 2 Colour Codes (*continued*)

Band or dot	Significance of 5 band endcap covered	Significance of 6 band endcap covered	Significance of 5 uniform bands
A	temperature coefficient	↑ temperature coefficient ↓	1st digit
B	1st digit		2nd digit
C	2nd digit	1st digit	additional zeros
D	additional zeros	2nd digit	tolerance
E	tolerance	additional zeros	working voltage (digit × 100V)
F		tolerance	

Moulded mica capacitors Values given in picofarads; code of six identical dots or nine identical dots, three of which are on reverse side

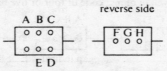

reverse side

Dot	A	B	C	D	E
Significance	white, silver, or black	1st digit	2nd digit	additional zeros	tolerance

Dot	F	G	H
Significance	d.c. working voltage	temperature range	white

Miniature foil capacitors Values coded in picofarad; multiply value by 10^{-6} to give capacitance in microfarad; code of five adjacent bands

Band	A	B	C	D	E
Significance	1st digit	2nd digit	additional zeros	tolerance	working voltage (digit × 100)

Table 2 Colour Codes *(continued)*

Tantalum bead capacitor Values in microfarads

Region	A	B	*C	*D
Significance	1st digit	2nd digit	additional zeros	working voltage

*see table below

Special colour coding for tantalum bead capacitors

Colour	Multiplier	Voltage	Colour	Multiplier	Voltage
black	× 1μF	10	blue	—	20
brown	× 10	—	grey	× 0·01	25
red	× 100	—	white	× 0·1	3
yellow	—	6·3	pink	—	35
green	—	16			

Table 3 Properties of Important Semiconductors

Semiconductor	Type	Energy gap at 300K (eV)	Drift mobility at 300K (cm² V⁻¹ s⁻¹)		Dielectric constant
			electrons	holes	
silicon (Si)	element	1·09	1500	600	11·8
germanium (Ge)		0·66	3900	1900	16
selenium (Se)		2·3	0·005	0·15	6·6
gallium arsenide (GaAs)	III–V compound	1·43	8600–11 000	3000	10·9
indium phosphide (InP)		1·29	4800–6800	150–200	14
cadmium sulphide (CdS)	II–VI compound	2·42	300	50	10

Table 4 Electric and Magnetic Quantities

Quantity	Symbol	SI unit	SI symbol
electric current	I	ampere	A
electric charge, quantity of electricity	Q	coulomb	C
charge density, volume density of charge	ρ	coulomb per cubic metre	C/m^3
surface charge density	σ	coulomb per square metre	C/m^2
electric field strength	E	volt per metre	V/m
electric potential	V, ϕ	volt	V
potential difference	U	volt	V
electromotive force	E	volt	V
electric displacement	D	coulomb per square metre	C/m^2
electric flux	ψ	coulomb	C
capacitance	C	farad	F
permittivity	ϵ	} farad per metre	F/m
permittivity of free space	ϵ_0		
relative permittivity	ϵ_r	—	—
electric susceptibility	X_e	—	—
dielectric polarization	P	coulomb per square metre	C/m^2
electric dipole moment	$p, (p_e)$	coulomb metre	C m
permanent dipole moment of a molecule	p, μ	coulomb metre	C m
induced dipole moment of a molecule	p, p_i	coulomb metre	C m
electric polarizability of a molecule	α	coulomb metre squared per volt	$C\ m^2/V$
electric current density	j, J	ampere per square metre	A/m^2
magnetic field strength	H	ampere per metre	A/m
magnetic potential difference	U_m	ampere	A
magnetomotive force	F, F_m	ampere	A
magnetic flux density	B	tesla	T
magnetic flux	Φ	weber	Wb
self inductance	L	henry	H
mutual inductance	M, L_{12}	henry	H
coupling coefficient	k	—	—
leakage coefficient	σ	—	—
permeability	μ	} henry per metre	H/m
permeability of free space	μ_0		
relative permeability	μ_r	—	—
magnetic susceptibility	X_m	—	—
magnetic moment	m	ampere metre squared	$A\ m^2$
magnetization	M	ampere per metre	A/m
magnetic polarization	B_i	tesla	T
velocity of light in a vacuum	c	metre per second	m/s
resistance	R	ohm	Ω
resistivity	ρ	ohm metre	Ω m
conductance	G	siemens	S
conductivity	$\kappa, \gamma, (\sigma)$	siemens per metre	S/m
reluctance	R, R_m	reciprocal henry	1/H
permeance	Λ	henry	H

Table 4 Electric and Magnetic Quantities *(continued)*

Quantity	Symbol	SI unit	SI symbol
phase displacement	ϕ	—	—
number of turns on winding	N	—	—
number of phases	m	—	—
number of pairs of poles	p	—	—
impedance (complex impedance)	Z	ohm	Ω
modulus of impedance	$|Z|$	ohm	Ω
reactance	X	ohm	Ω
quality factor	Q	—	—
admittance	Y	siemens	S
modulus of admittance	$|Y|$	siemens	S
susceptance	B	siemens	S
active power	P	watt	W
apparent power	$S.\ (P_s)$	watt	W ($=$ V A)
reactive power	$Q.\ (P_q)$	watt	W
Faraday constant	F	coulomb per mole	C/mol
wavelength	λ	metre	m
frequency	$\nu,\ f$	hertz	Hz
angular frequency	ω	hertz	Hz
period	T	second	s
relaxation time	τ	second	s
thermodynamic temperature	T	kelvin	K
energy	E	joule	J

Table 5 Fundamental Constants

Constant	Symbol	Value
velocity of light	c	$2 \cdot 997\,925 \times 10^8$ m s^{-1}
permeability of free space	μ_0	$4\pi \times 10^{-7} = 1 \cdot 256\,64 \times 10^{-6}$ H m^{-1}
permittivity of free space	$\epsilon_0 = \mu_0^{-1}c^{-2}$	$8 \cdot 854\,185\,3 \times 10^{-12}$ F m^{-1}
charge of electron or proton	e	$\mp 1 \cdot 602\,191\,7 \times 10^{-19}$ C
rest mass of electron	m_e	$9 \cdot 109\,56 \times 10^{-31}$ kg
rest mass of proton	m_p	$1 \cdot 672\,51 \times 10^{-27}$ kg
rest mass of neutron	m_n	$1 \cdot 674\,92 \times 10^{-27}$ kg
electron charge-to-mass ratio	e/m	$1 \cdot 758\,802 \times 10^{11}$ C kg^{-1}
electron radius	r_e	$2 \cdot 817\,939 \times 10^{-15}$ m
Planck constant	h	$6 \cdot 626\,196 \times 10^{-34}$ J s
Boltzmann constant	k	$1 \cdot 380\,622 \times 10^{-23}$ J K^{-1}
Faraday constant	F	$9 \cdot 648\,670 \times 10^4$ C mol^{-1}

Table 6 Base SI Units

Physical quantity	Name	Symbol
length	metre	m
mass	kilogram	kg
time	second	s
electric current	ampere	A
thermodynamic temperature	kelvin	K
amount of substance	mole	mol
luminous intensity	candela	cd

Table 7 Supplementary SI Units

Physical quantity	Name	Symbol
plane angle	radian	rad
solid angle	steradian	sr

Table 8 Derived SI Units with Special Names

Physical quantity	Name	Symbol	Dimensions of unit derived	base
frequency	hertz	Hz		s^{-1}
energy	joule	J		$kg\ m^2\ s^{-2}$
force	newton	N	$J\ m^{-1}$	$kg\ m\ s^{-2}$
power	watt	W	$J\ s^{-1}$	$kg\ m^2\ s^{-3}$
pressure	pascal	Pa	$N\ m^{-2}$	$kg\ m^{-1}\ s^{-2}$
electric charge	coulomb	C		$A\ s$
electric potential difference	volt	V	$J\ C^{-1}$	$kg\ m^2\ s^{-3}\ A^{-1}$
electric resistance	ohm	Ω	$V\ A^{-1}$	$kg\ m^2\ s^{-3}\ A^{-2}$
electric conductance	siemens	S	Ω^{-1}	$s^3\ A^2\ kg^{-1}\ m^{-2}$
electric capacitance	farad	F	$C\ V^{-1}$	$s^4\ A^2\ kg^{-1}\ m^{-2}$
magnetic flux	weber	Wb	$V\ s$	$kg\ m^2\ s^{-2}\ A^{-1}$
inductance	henry	H	$Wb\ A^{-1}$	$kg\ m^2\ s^{-2}\ A^{-2}$
magnetic flux density	tesla	T	$Wb\ m^{-2}$	$kg\ s^{-2}\ A^{-1}$

Table 9 Prefixes used with SI Units

Factor	Name of prefix	Symbol	Factor	Name of prefix	Symbol
10^{-1}	deci-	d	10	deca-	da
10^{-2}	centi-	c	10^2	hecto-	h
10^{-3}	milli-	m	10^3	kilo-	k
10^{-6}	micro-	μ	10^6	mega-	M
10^{-9}	nano-	n	10^9	giga-	G
10^{-12}	pico-	p	10^{12}	tera-	T
10^{-15}	femto-	f	10^{15}	peta-	P
10^{-18}	atto-	a	10^{18}	exa-	E

Table 10 Electromagnetic Spectrum

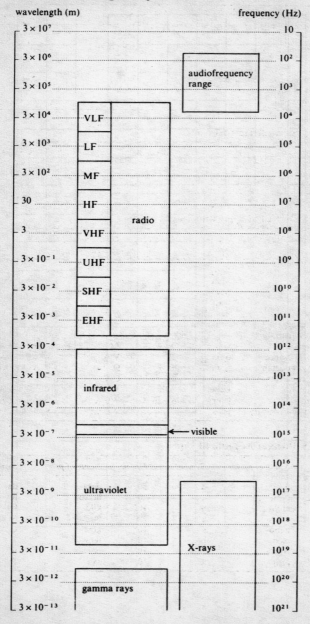

Table 11 Periodic Table of the Elements – giving group, atomic number, and chemical symbol

1A	2A	3B	4B	5B	6B	7B	8	8	8	1B	2B	3A	4A	5A	6A	7A	0
1 H																	2 He
3 Li	4 Be											5 B	6 C	7 N	8 O	9 F	10 Ne
11 Na	12 Mg											13 Al	14 Si	15 P	16 S	17 Cl	18 Ar
19 K	20 Ca	21 Sc	22 Ti	23 V	24 Cr	25 Mn	26 Fe	27 Co	28 Ni	29 Cu	30 Zn	31 Ga	32 Ge	33 As	34 Se	35 Br	36 Kr
37 Rb	38 Sr	39 Y	40 Zr	41 Nb	42 Mo	43 Tc	44 Ru	45 Rh	46 Pd	47 Ag	48 Cd	49 In	50 Sn	51 Sb	52 Te	53 I	54 Xe
55 Cs	56 Ba	57* La	72 Hf	73 Ta	74 W	75 Re	76 Os	77 Ir	78 Pt	79 Au	80 Hg	81 Tl	82 Pb	83 Bi	84 Po	85 At	86 Rn
87 Fr	88 Ra	89† Ac															

← transition elements →

*lanthanides

57 La	58 Ce	59 Pr	60 Nd	61 Pm	62 Sm	63 Eu	64 Gd	65 Tb	66 Dy	67 Ho	68 Er	69 Tm	70 Yb	71 Lu

†actinides

89 Ac	90 Th	91 Pa	92 U	93 Np	94 Pu	95 Am	96 Cm	97 Bk	98 Cf	99 Es	100 Fm	101 Md	102 No	103 Lr

Table 12 Electron Arrangement in Atomic Shells of neutral atoms in their lowest energy state

Atomic number	Element	K 1s	L 2s	L 2p	M 3s	M 3p	M 3d	N 4s	N 4p	N 4d	N 4f	O 5s	O 5p	O 5d	O 5f	O 5g	P 6s	P 6p	P 6d	P 6f	Q 7s	Q 7p
1	H	1																				
2	He	2																				
3	Li	2	1																			
4	Be	2	2																			
5	B	2	2	1																		
6	C	2	2	2																		
7	N	2	2	3																		
8	O	2	2	4																		
9	F	2	2	5																		
10	Ne	2	2	6																		
11	Na	2	2	6	1																	
12	Mg	2	2	6	2																	
13	Al	2	2	6	2	1																
14	Si	2	2	6	2	2																
15	P	2	2	6	2	3																
16	S	2	2	6	2	4																
17	Cl	2	2	6	2	5																
18	Ar	2	2	6	2	6																
19	K	2	2	6	2	6		1														
20	Ca	2	2	6	2	6		2														
21	Sc	2	2	6	2	6	1	2														
22	Ti	2	2	6	2	6	2	2														
23	V	2	2	6	2	6	3	2														
24	Cr	2	2	6	2	6	5	1														
25	Mn	2	2	6	2	6	5	2														
26	Fe	2	2	6	2	6	6	2														
27	Co	2	2	6	2	6	7	2														
28	Ni	2	2	6	2	6	8	2														
29	Cu	2	2	6	2	6	10	1														
30	Zn	2	2	6	2	6	10	2														
31	Ga	2	2	6	2	6	10	2	1													
32	Ge	2	2	6	2	6	10	2	2													
33	As	2	2	6	2	6	10	2	3													
34	Se	2	2	6	2	6	10	2	4													
35	Br	2	2	6	2	6	10	2	5													
36	Kr	2	2	6	2	6	10	2	6													
37	Rb	2	2	6	2	6	10	2	6			1										
38	Sr	2	2	6	2	6	10	2	6			2										
39	Y	2	2	6	2	6	10	2	6	1		2										
40	Zr	2	2	6	2	6	10	2	6	2		2										
41	Nb	2	2	6	2	6	10	2	6	4		1										
42	Mo	2	2	6	2	6	10	2	6	5		1										
43	Tc	2	2	6	2	6	10	2	6	6		1										
44	Ru	2	2	6	2	6	10	2	6	7		1										
45	Rh	2	2	6	2	6	10	2	6	8		1										
46	Pd	2	2	6	2	6	10	2	6	10												
47	Ag	2	2	6	2	6	10	2	6	10		1										
48	Cd	2	2	6	2	6	10	2	6	10		2										
49	In	2	2	6	2	6	10	2	6	10		2	1									
50	Sn	2	2	6	2	6	10	2	6	10		2	2									
51	Sb	2	2	6	2	6	10	2	6	10		2	3									
52	Te	2	2	6	2	6	10	2	6	10		2	4									
53	I	2	2	6	2	6	10	2	6	10		2	5									
54	Xe	2	2	6	2	6	10	2	6	10		2	6									
55	Cs	2	2	6	2	6	10	2	6	10		2	6				1					
56	Ba	2	2	6	2	6	10	2	6	10		2	6				2					
57	La	2	2	6	2	6	10	2	6	10		2	6	1			2					
58	Ce	2	2	6	2	6	10	2	6	10	2	2	6				2					
59	Pr	2	2	6	2	6	10	2	6	10	3	2	6				2					
60	Nd	2	2	6	2	6	10	2	6	10	4	2	6				2					
61	Pm	2	2	6	2	6	10	2	6	10	5	2	6				2					
62	Sm	2	2	6	2	6	10	2	6	10	6	2	6				2					
63	Eu	2	2	6	2	6	10	2	6	10	7	2	6				2					
64	Gd	2	2	6	2	6	10	2	6	10	7	2	6	1			2					
65	Tb	2	2	6	2	6	10	2	6	10	9	2	6				2					
66	Dy	2	2	6	2	6	10	2	6	10	10	2	6				2					
67	Ho	2	2	6	2	6	10	2	6	10	11	2	6				2					
68	Er	2	2	6	2	6	10	2	6	10	12	2	6				2					
69	Tm	2	2	6	2	6	10	2	6	10	13	2	6				2					
70	Yb	2	2	6	2	6	10	2	6	10	14	2	6				2					
71	Lu	2	2	6	2	6	10	2	6	10	14	2	6	1			2					
72	Hf	2	2	6	2	6	10	2	6	10	14	2	6	2			2					
73	Ta	2	2	6	2	6	10	2	6	10	14	2	6	3			2					
74	W	2	2	6	2	6	10	2	6	10	14	2	6	4			2					
75	Re	2	2	6	2	6	10	2	6	10	14	2	6	5			2					
76	Os	2	2	6	2	6	10	2	6	10	14	2	6	6			2					
77	Ir	2	2	6	2	6	10	2	6	10	14	2	6	9								
78	Pt	2	2	6	2	6	10	2	6	10	14	2	6	9			1					
79	Au	2	2	6	2	6	10	2	6	10	14	2	6	10			1					
80	Hg	2	2	6	2	6	10	2	6	10	14	2	6	10			2					
81	Tl	2	2	6	2	6	10	2	6	10	14	2	6	10			2	1				
82	Pb	2	2	6	2	6	10	2	6	10	14	2	6	10			2	2				
83	Bi	2	2	6	2	6	10	2	6	10	14	2	6	10			2	3				
84	Po	2	2	6	2	6	10	2	6	10	14	2	6	10			2	4				
85	At	2	2	6	2	6	10	2	6	10	14	2	6	10			2	5				
86	Rn	2	2	6	2	6	10	2	6	10	14	2	6	10			2	6				
87	Fr	2	2	6	2	6	10	2	6	10	14	2	6	10			2	6			1	
88	Ra	2	2	6	2	6	10	2	6	10	14	2	6	10			2	6			2	
89	Ac	2	2	6	2	6	10	2	6	10	14	2	6	10			2	6	1		2	
90	Th	2	2	6	2	6	10	2	6	10	14	2	6	10			2	6	2		2	
91	Pa	2	2	6	2	6	10	2	6	10	14	2	6	10	2		2	6	1		2	
92	U	2	2	6	2	6	10	2	6	10	14	2	6	10	3		2	6	1		2	

Table 13 Electromotive Series – giving reduction potentials relative to the standard hydrogen electrode

Element	Reaction	Potential volts	Element	Reaction	Potential volts
Li	$Li^+ + e^- \rightleftharpoons Li$	-3.045	Ga	$Ga^{3+} + 3e^- \rightleftharpoons Ga$	-0.56
Rb	$Rb^+ + e^- \rightleftharpoons Rb$	-2.925	S	$S + 2e^- \rightleftharpoons S^{2-}$	-0.508
K	$K^+ + e^- \rightleftharpoons K$	-2.924	Fe	$Fe^{2+} + 2e^- \rightleftharpoons Fe$	-0.409
Cs	$Cs^+ + e^- \rightleftharpoons Cs$	-2.923	Cd	$Cd^{2+} + 2e^- \rightleftharpoons Cd$	-0.4026
Ba	$Ba^{2+} + 2e^- \rightleftharpoons Ba$	-2.90	In	$In^{3+} + 3e^- \rightleftharpoons In$	-0.338
Sr	$Sr^{2+} + 2e^- \rightleftharpoons Sr$	-2.89	Tl	$Tl^+ + e^- \rightleftharpoons Tl$	-0.3363
Ca	$Ca^{2+} + 2e^- \rightleftharpoons Ca$	-2.76	Co	$Co^{2+} + 2e^- \rightleftharpoons Co$	-0.28
Na	$Na^+ + e^- \rightleftharpoons Na$	-2.7109	Ni	$Ni^{2+} + 2e^- \rightleftharpoons Ni$	-0.23
Mg	$Mg^{2+} + 2e^- \rightleftharpoons Mg$	-2.375	Sn	$Sn^{2+} + 2e^- \rightleftharpoons Sn$	-0.1364
Y	$Y^{3+} + 3e^- \rightleftharpoons Y$	-2.37	Pb	$Pb^{2+} + 2e^- \rightleftharpoons Pb$	-0.1263
La	$La^{3+} + 3e^- \rightleftharpoons La$	-2.37	Fe	$Fe^{3+} + 3e^- \rightleftharpoons Fe$	-0.036
Ce	$Ce^{3+} + 3e^- \rightleftharpoons Ce$	-2.335	D_2	$D^+ + e^- \rightleftharpoons \frac{1}{2}D_2$	-0.0034
Nd	$Nd^{3+} + 3e^- \rightleftharpoons Nd$	-2.246	H_2	$H^+ + e^- \rightleftharpoons \frac{1}{2}H_2$	0.000
H	$\frac{1}{2}H_2 + e^- \rightleftharpoons H^-$	-2.23	Re	$Re^{3+} + 3e^- \rightleftharpoons Re$	$+0.3$
Sc	$Sc^{3+} + 3e^- \rightleftharpoons Sc$	-2.08	Cu	$Cu^{2+} + 2e^- \rightleftharpoons Cu$	$+0.340$
Th	$Th^{4+} + 4e^- \rightleftharpoons Th$	-1.90		$Cu^+ + e^- \rightleftharpoons Cu$	$+0.522$
Np	$Np^{3+} + 3e^- \rightleftharpoons Np$	-1.9	I_2	$I_2 + 2e^- \rightleftharpoons 2I^-$	$+0.535$
U	$U^{3+} + 3e^- \rightleftharpoons U$	-1.80	Hg	$Hg_2^{2+} + 2e^- \rightleftharpoons 2Hg$	$+0.7961$
Al	$Al^{3+} + 3e^- \rightleftharpoons Al$	-1.706	Ag	$Ag^+ + e^- \rightleftharpoons Ag$	$+0.7996$
Be	$Be^{2+} + 2e^- \rightleftharpoons Be$	-1.70	Pd	$Pd^{2+} + 2e^- \rightleftharpoons Pd$	$+0.83$
Ti	$Ti^{2+} + 2e^- \rightleftharpoons Ti$	-1.63	Br_2	$Br_2(l) + 2e^- \rightleftharpoons 2Br^-$	$+1.065$
V	$V^{2+} + 2e^- \rightleftharpoons V$	-1.2	Pt	$Pt^{2+} + 2e^- \rightleftharpoons Pt$	$+1.2$
Mn	$Mn^{2+} + 2e^- \rightleftharpoons Mn$	-1.029	Cl_2	$Cl_2 + 2e^- \rightleftharpoons 2Cl^-$	$+1.358$
Te	$Te + 2e^- \rightleftharpoons Te^{2-}$	-0.92	Au	$Au^{3+} + 3e^- \rightleftharpoons Au$	$+1.42$
Se	$Se + 2e^- \rightleftharpoons Se^{2-}$	-0.78		$Au^+ + e^- \rightleftharpoons Au$	$+1.68$
Zn	$Zn^{2+} + 2e^- \rightleftharpoons Zn$	-0.7628	F_2	$\frac{1}{2}F_2 + e^- \rightleftharpoons F^-$	$+2.85$
Cr	$Cr^{3+} + 3e^- \rightleftharpoons Cr$	-0.74			

Table 14 Major Discoveries and Inventions in Electricity and Electronics

1745	capacitor: Leyden jar	Musschenbroek
1747 48	positive and negative electricity postulated	Franklin
1767	inverse square law postulated	Priestley
	demonstrated 1785	Coulomb
1800	electric battery: voltaic pile	Volta
1808	atomic theory	Dalton
1820	electromagnetism	Oersted
1820 21	Ampère's laws	Ampère
1821	thermoelectricity	Seebeck
1823	electromagnet: first made	Sturgeon
	improved 1831	Henry
1827	Ohm's law	Ohm
1831	electromagnetic induction	Faraday
1831	transformer	Faraday
1832	self-induction	Henry
1833 34	analytical engine conceived	Babbage
1834	electrolysis laws	Faraday
1845	Kirchhoff's laws	Kirchhoff
1847	magnetostriction	Joule
1855	gas-discharge tube: cold cathode	Gaugain
	low pressure 1856	Geissler
1860	microphone: diaphragm	J. P. Reis
	carbon-granule 1878	D. Hughes
1864	Maxwell's equations	Maxwell
1876	telephone	Bell
1876 79	cathode-rays studied	Crookes
1877	gramophone	Edison
1879	Hall effect	E. Hall
1880	piezoelectricity	P. & J. Curie
1887 88	electromagnetic waves (radio) first produced	Hertz
1887	aerial	Hertz
1887	photoelectric effect: observed	Hertz
	explained 1905	Einstein
1896	radio telegraphy: short distance	Marconi
	transatlantic 1901	Marconi
1897	electron discovered, proposed as constituent of all matter	J. J. Thomson
1897	cathode-ray oscilloscope	F. Braun
1898	magnetic recording: wire	V. Poulsen
	tape 1927	U.S. patent
1900	quantum theory	Planck
1901 02	ionosphere postulated	Kennelly; Heaviside
	demonstrated 1924	Appleton
1904	thermionic valve: diode	J. A. Fleming
	triode 1906	de Forest
	tetrode 1926	H. J. Round
	pentode 1928	Tellegen & Holst
1906	radio broadcasting: first successful	Fessenden
	commercial 1919 20	U.K., U.S.
1911	atomic nucleus	Rutherford
1913	Bohr atom	Bohr

Table 14 Major Discoveries and Inventions in Electricity and Electronics
(continued)

1923	television: electronic system (iconoscope)	Zworykin
	mechanical system 1926	Baird
	electronic system 1927	Farnsworth
	colour demonstrated 1928	Baird
1924/25	radar: first experiments	Appleton:
		G. Briet; et al
	developed 1935–45	Watson-Watt
1924/25	wave mechanics	de Broglie:
		Schroedinger: Born
		& Heisenberg: et al
1926	Yagi aerial	H. Yagi
1929–35	television broadcasting: mechanical scanning	BBC
	electronic system 1936	BBC
1933	frequency modulation	E. H. Armstrong
1935	transistor, unipolar (FET): conceived	O. Heil
	produced 1958	S. Teszner et al
1936	waveguide	various
1939	klystron	R. & S. Varians:
		W. C. Hahn
1939	magnetron	J. Randall &
		H. Boot
1939/40	computer: using valves	
	transistors 1956	
	integrated circuits 1964	
1948	transistor, bipolar: point contact	Bardeen,
	junction	Brattain &
		Schockley
1948	single crystal growth: germanium	G. Teal &
		J. Little
	silicon 1952	G. Teal &
		E. Buehler
1952	integrated circuits: conceived	G. Dummer et al
	produced 1960s	
1953	maser	C. Townes et al
1956	diffusion process	C. Fuller & H. Reis
1959	planar process	J. Hoerni
1960–64	logic circuits	
1962	communications satellite: Telstar	U.S.

Table 15 The Greek Alphabet

Letters		Name
A	α	alpha
B	β	beta
Γ	γ	gamma
Δ	δ	delta
E	ε	epsilon
Z	ζ	zeta
H	η	eta
Θ	θ	theta
I	ι	iota
K	κ	kappa
Λ	λ	lambda
M	μ	mu
N	ν	nu
Ξ	ξ	xi
O	o	omikron
Π	π	pi
P	ρ	rho
Σ ς	σ ς	sigma
T	τ	tau
Y	υ	upsilon
Φ	φ	phi
X	χ	chi
Ψ	ψ	psi
Ω	ω	omega

FOR THE BEST IN PAPERBACKS, LOOK FOR THE 🐧

QED Richard Feynman
The Strange Theory of Light and Matter

Quantum thermodynamics – or QED for short – is the 'strange theory' – that explains how light and electrons interact. 'Physics Nobelist Feynman simply cannot help being original. In this quirky, fascinating book, he explains to laymen the quantum theory of light – a theory to which he made decisive contributions' – *New Yorker*

God and the New Physics Paul Davies

Can science, now come of age, offer a surer path to God than religion? This 'very interesting' (*New Scientist*) book suggests it can.

Does God Play Dice? Ian Stewart
The New Mathematics of Chaos

To cope with the truth of a chaotic world, pioneering mathematicians have developed chaos theory. *Does God Play Dice?* makes accessible the basic principles and many practical applications of one of the most extraordinary – and mindbending – breakthroughs in recent years. 'Engaging, accurate and accessible to the uninitiated' – *Nature*

The Blind Watchmaker Richard Dawkins

'An enchantingly witty and persuasive neo-Darwinist attack on the anti-evolutionists, pleasurably intelligible to the scientifically illiterate' – Hermione Lee in the *Observer* Books of the Year

The Making of the Atomic Bomb Richard Rhodes

'Rhodes handles his rich trove of material with the skill of a master novelist ... his portraits of the leading figures are three-dimensional and penetrating ... the sheer momentum of the narrative is breathtaking ... a book to read and to read again' – Walter C. Patterson in the *Guardian*

Asimov's New Guide to Science Isaac Asimov

A classic work brought up to date – far and away the best one-volume survey of all the physical and biological sciences.

I: The Philosophy and Psychology of Personal Identity   Jonathan Glover

From cases of split brains and multiple personalities to the importance of memory and recognition by others, the author of *Causing Death and Saving Lives* tackles the vexed questions of personal identity. 'Fascinating … the ideas which Glover pours forth in profusion deserve more detailed consideration' – Anthony Storr

Minds, Brains and Science   John Searle

Based on Professor Searle's acclaimed series of Reith Lectures, *Minds, Brains and Science* is 'punchy and engaging … a timely exposé of those woolly-minded computer-lovers who believe that computers can think, and indeed that the human mind is just a biological computer' – *The Times Literary Supplement*

Ethics   Inventing Right and Wrong   J. L. Mackie

Widely used as a text, Mackie's complete and clear treatise on moral theory deals with the status and content of ethics, sketches a practical moral system and examines the frontiers at which ethics touches psychology, theology, law and politics.

The Penguin History of Western Philosophy   D. W. Hamlyn

'Well-crafted and readable … neither laden with footnotes nor weighed down with technical language … a general guide to three millennia of philosophizing in the West' – *The Times Literary Supplement*

Science and Philosophy: Past and Present   Derek Gjertsen

Philosophy and science, once intimately connected, are today often seen as widely different disciplines. Ranging from Aristotle to Einstein, from quantum theory to renaissance magic, Confucius and parapsychology, this penetrating and original study shows such a view to be both naive and ill-informed.

The Problem of Knowledge   A. J. Ayer

How do you *know* that this is a book? How do you *know* that you know? In *The Problem of Knowledge* A. J. Ayer presented the sceptic's arguments as forcefully as possible, investigating the extent to which they can be met. 'Thorough … penetrating, vigorous … readable and manageable' – *Spectator*

FOR THE BEST IN PAPERBACKS, LOOK FOR THE

PENGUIN REFERENCE BOOKS

The New Penguin English Dictionary

Over 1,000 pages long and with over 68,000 definitions, this cheap, compact and totally up-to-date book is ideal for today's needs. It includes many technical and colloquial terms, guides to pronunciation and common abbreviations.

The Penguin Spelling Dictionary

What are the plurals of *octopus* and *rhinoceros*? What is the difference between *stationary* and *stationery*? And how about *annex* and *annexe*, *agape* and *Agape*? This comprehensive new book, the fullest spelling dictionary now available, provides the answers.

Roget's Thesaurus of English Words and Phrases   Betty Kirkpatrick (ed.)

This new edition of Roget's classic work, now brought up to date for the nineties, will increase anyone's command of the English language. Fully cross-referenced, it includes synonyms of every kind (formal or colloquial, idiomatic and figurative) for almost 900 headings. It is a must for writers and utterly fascinating for any English speaker.

The Penguin Dictionary of Quotations

A treasure-trove of over 12,000 new gems and old favourites, from Aesop and Matthew Arnold to Xenophon and Zola.

The Penguin Wordmaster Dictionary
Martin H. Manser and Nigel D. Turton

This dictionary puts the pleasure back into word-seeking. Every time you look at a page you get a bonus – a panel telling you everything about a particular word or expression. It is, therefore, a dictionary to be read as well as used for its concise and up-to-date definitions.

FOR THE BEST IN PAPERBACKS, LOOK FOR THE

PENGUIN REFERENCE BOOKS

The Penguin Guide to the Law

This acclaimed reference book is designed for everyday use and forms the most comprehensive handbook ever published on the law as it affects the individual.

The Penguin Medical Encyclopedia

Covers the body and mind in sickness and in health, including drugs, surgery, medical history, medical vocabulary and many other aspects. 'Highly commendable' – *Journal of the Institute of Health Education*

The Slang Thesaurus

Do you make the public bar sound like a gentleman's club? Do you need help in understanding *Minder*? The miraculous *Slang Thesaurus* will liven up your language in no time. You won't Adam and Eve it! A mine of funny, witty, acid and vulgar synonyms for the words you use every day.

The Penguin Dictionary of Troublesome Words   Bill Bryson

Why should you avoid discussing the *weather conditions*? Can a married woman be *celibate*? Why is it eccentric to talk about the *aroma* of a cowshed? A straightforward guide to the pitfalls and hotly disputed issues in standard written English.

The Penguin Spanish Dictionary   James R. Jump

Detailed, comprehensive and, above all, modern, *The Penguin Spanish Dictionary* offers a complete picture of the language of ordinary Spaniards – the words used at home and at work, in bars and discos, at cafés and in the street, including full and unsqueamish coverage of common slang and colloquialisms.

The New Penguin Dictionary of Geography

From *aa* and *ablation* to *zinc* and *zonal soils*, this succinct dictionary is unique in covering in one volume the main terms now in use in the diverse areas – physical and human geography, geology and climatology, ecology and economics – that make up geography today.